Modeling, Identification & Control of Robots

KOGAN PAGE SCIENCE PAPER EDITION

Modeling, Identification & Control of Robots

W Khalil & E Dombre

KOGAN PAGE SCIENCE

To our families who supported us during the writing of this book...

First published in 2002 by Hermes Penton Ltd
Paperback edition published in 2004 by Kogan Page Science, an imprint of Kogan Page Ltd

Kogan Page Science
120 Pentonville Road
London N1 9JN
United Kingdom
www.koganpagescience.com

Kogan Page US
22883 Quicksilver Drive
Sterling VA 20166-2012
USA

British Library Cataloguing in Publication Data

A CIP record for this book is available from the British Library.

ISBN 1 9039 9666 X

Transferred to digital printing 2006
www.biddles.co.uk

Contents

Introduction

The control and simulation of robots requires the development of different mathematical models. Several levels of modeling – geometric, kinematic and dynamic – are needed depending on the objectives, the constraints of the task and the desired performance.

Obtaining these models is not an easy task. The difficulty varies according to the complexity of the kinematics of the mechanical structure and its degrees of freedom. The mathematical tools presented in this book are based on a description of mechanisms allowing a unified approach whatever the type of structure: serial, tree structured, or containing closed loops.

Using these models in control and simulation requires efficient and easy-to-use algorithms to estimate the values of the geometric parameters and the dynamic parameters of the robot. Besides, the on-line implementation of a control law on a robot controller requires efficient models with a reduced number of operations. The techniques proposed in this book have been developed to meet these requirements.

This book is a revised and augmented edition of the French version "Modélisation, identification et commande des robots" published by Hermès in 1999, whose first edition "Modélisation et commande des robots" was published in 1988. We consider it to be the third edition as it contains substantial modification and updating. The content is the following:

- Chapter 1 gives an introduction to the terminology and general definitions for the concepts used in this book: kinematic chains, types of joints, configuration space, task space, redundancy, singular configurations, architectures of robot manipulators, robot characteristics;
- Chapter 2 sets out the basic mathematical tools used in robot modeling: homogeneous transformations, differential transformations, screws, twists and wrenches;
- Chapter 3 deals with the direct geometric modeling of simple open chain robots (also termed serial robots). The Khalil-Kleinfinger notation is used to describe the geometry of the mechanical structure. This notation, which is a variation of the Denavit-Hartenberg one, also handles the description of

complex chains with tree structures or closed loops (Chapters 7 and 10). The various methods of describing the orientation of a solid in space are covered at the end of the chapter;

- Chapter 4 treats the inverse geometric model. Three approaches are described: the Paul method, which can be used for most industrial robots, the Pieper method, which deals with six degree-of-freedom robots having three prismatic joints or a spherical joint, and the Raghavan-Roth method, which is suitable for six degree-of-freedom robots with general geometry;
- Chapter 5 addresses the direct kinematic model. After developing efficient methods for calculating the Jacobian matrix, we present several applications: analysis of the robot workspace, determination of the degrees of freedom of structure, velocity and force ellipsoids, twist-wrench duality;
- Chapter 6 covers inverse kinematics. The main topics are: inversion at regular configurations, inversion close to singularities, inversion for redundant robots, and minimal task description;
- Chapter 7 examines the geometric and kinematic models of complex chain robots with tree or closed chain structures. The problem of solving the constraint equations of closed loop robots is treated using both geometric constraint equations and kinematic constraint equations;
- Chapter 8 introduces the geometric and kinematic models of parallel robots. The main architectures and features of these structures are given;
- Chapters 9 and 10 deal with dynamic modeling: simple open chains are considered in Chapter 9, whereas complex kinematic chains are presented in Chapter 10. Lagrangian and Newton-Euler formulations, which are linear in the dynamic parameters, are presented. The determination of the minimum inertial parameters, also termed base inertial parameters, is carried out using a direct symbolic method and by a numerical method, which is based on a QR decomposition. The number of operations of the inverse dynamic model are minimized thanks to the use of the base parameters and customized symbolic programming techniques. The models obtained allow on-line implementation with today's personal computers. We also give different methods for the direct dynamic model computation, more especially a method avoiding the inversion of the inertia matrix;
- Chapters 11 and 12 are devoted to identification of the geometric and dynamic parameters respectively. In Chapter 11, we present various geometric calibration methods. Some of them need external sensors, the others being autonomous. The construction of the observation matrix and the solution of the calibration equation are detailed for all the methods. A short subsection introduces the active field of research into parallel robot calibration. In Chapter 12, which concerns the dynamic parameters, several identification

methods based on the dynamic model or energy model are introduced. All of them consist in solving a model that is linear in the dynamic parameters;

- Chapter 13 introduces the problem of trajectory generation. Beginning with point-to-point trajectories both in the joint space and in the task space, the chapter then examines the problem of adding intermediate points. At the end, the trajectory generation on a continuous path is briefly treated;
- Chapters 14 and 15 deal with motion control and force control. The motion control chapter specifically covers PID control, computed torque control, passive control and adaptive control while the force control chapter addresses passive control, impedance control, hybrid force-position control and hybrid external control.

At the end of the book the reader will find eleven appendices, which give either detailed computations of examples or introductions to relevant mathematical methods. An abundant bibliography of more than 400 references related to this fast-growing field of research is also included.

This book is intended for researchers, university lecturers, engineers and postgraduates in the fields of automatic control, robotics and mechanics. It provides the necessary tools to deal with the various problems that can be encountered in the design, the control synthesis and the exploitation of robot manipulators. It can also be recommended as a textbook for students. It constitutes a complete course of about 70 lecture hours on modeling, identification and control of robot manipulators for engineering schools or Master of Science classes. For an introduction course of about 25 hours, the content could be reduced to: geometric and kinematic models of serial structures, trajectory generation between two points, and PID control (Chapters 1, 2, 3, 4, 5 and 6; partially Chapters 13 and 14). For a course of about 50 lecture hours, one could treat further dynamic modeling, calibration of geometric parameters, identification of dynamic parameters, and trajectory generation as well as the methods of motion control (Chapters 9, 11, 12, 13 and 14).

Chapter 1

Terminology and general definitions

1.1. Introduction

A robot is an automatically controlled, reprogrammable, multipurpose mechanical system with several degrees of freedom, which may be either fixed in place or mobile. It has been widely used so far in various industrial automation applications. Since the last decade, other areas of application have emerged: medical, service (spatial, civil security, ...), transport, underwater, entertainment, ..., where the robot either works in an autonomous manner or in cooperation with an operator to carry out complex tasks in a more or less structured environment. We can distinguish three main classes of robots: *robot manipulators*, which imitate the human arm, *walking robots*, which imitate the locomotion of humans, animals or insects, and *mobile robots*, which look like cars.

The terms *adaptability* and *versatility* are often used to highlight the intrinsic flexibility of a robot. Adaptability means that the robot is capable of adjusting its motion to comply with environmental changes during the execution of tasks. Versatility means that the robot may carry out a variety of tasks – or the same task in different ways – without changing the mechanical structure or the control system.

A robot is composed of the following subsystems:

- *mechanism*: consists of an articulated mechanical structure actuated by electric, pneumatic or hydraulic actuators, which transmit their motion to the joints using suitable transmission systems;
- *perception capabilities*: help the robot to adapt to disturbances and unpredictable changes in its environment. They consist of the internal sensors that provide information about the state of the robot (joint positions and velocities), and the external sensors to obtain the information about the environment (contact detection, distance measurement, artificial vision);

- *controller*: realizes the desired task objectives. It generates the input signals for the actuators as a function of the user's instructions and the sensor outputs;
- *communication interface*: through this the user programs the tasks that the robot must carry out;
- *workcell and peripheral devices*: constitute the environment in which the robot works.

Robotics is thus a multidisciplinary science, which requires a background in mechanics, automatic control, electronics, signal processing, communications, computer engineering, etc.

The objective of this book is to present the techniques of the modeling, identification and control of robots. We restrict our study to rigid robot manipulators with a fixed base. Thus, neither flexible robots for which the deformation of the links cannot be neglected [Cannon 84], [Chedmail 90a], [Boyer 94], nor mobile robots will be addressed in this book.

In this chapter, we will present certain definitions that are necessary to classify the mechanical structures and the characteristics of robot manipulators.

1.2. Mechanical components of a robot

The mechanism of a robot manipulator consists of two distinct subsystems, one (or more) end-effectors and an articulated mechanical structure:

- by the term *end-effector*, we mean any device intended to manipulate objects (magnetic, electric or pneumatic grippers) or to transform them (tools, welding torches, paint guns, etc.). It constitutes the interface with which the robot interacts with its environment. An end-effector may be multipurpose, i.e. equipped with several devices each having different functions;
- the role of *the articulated mechanical structure* is to place the end-effector at a given location (position and orientation) with a desired velocity and acceleration. The mechanical structure is composed of a kinematic chain of articulated rigid links. One end of the chain is fixed and is called the *base*. The end-effector is fixed to the free extremity of the chain. This chain may be *serial (simple open chain)* (Figure 1.1), *tree structured* (Figure 1.2) or *closed* (Figures 1.3 and 1.4). The last two structures are termed *complex chains* since they contain at least one link with more than two joints.

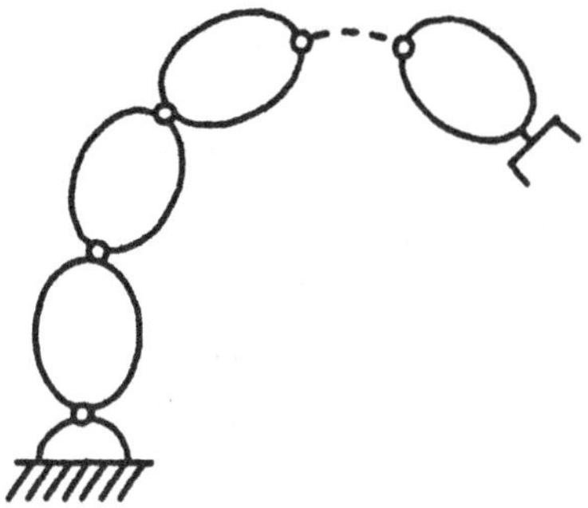

Figure 1.1. *Simple open (or serial) chain*

Serial robots with a simple open chain are the most commonly used. There are also industrial robots with closed kinematic chains, which have the advantage of being more rigid and accurate.

Figure 1.2. *Tree structured chain*

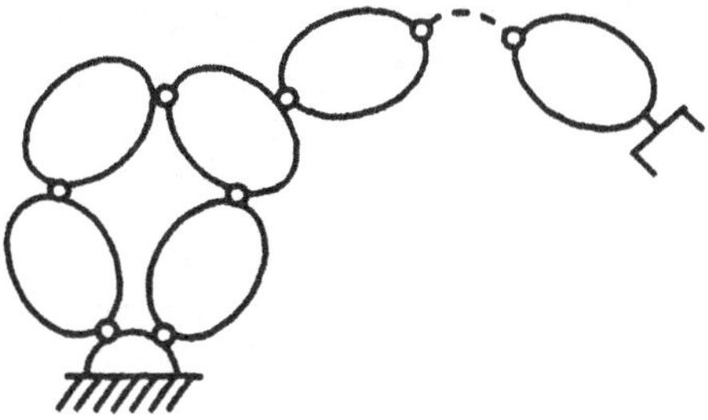

Figure 1.3. *Closed chain*

Figure 1.4 shows a specific architecture with closed chains, which is known as a *parallel robot*. In this case, the end-effector is connected to the base by several parallel chains [Inoue 85], [Fichter 86], [Reboulet 88], [Gosselin 88], [Clavel 89],

[Charentus 90], [Pierrot 91a], [Merlet 00]. The mass ratio of the payload to the robot is much higher compared to serial robots. This structure seems promising in manipulating heavy loads with high accelerations and realizing difficult assembly tasks.

Figure 1.4. *Parallel robot*

1.3. Definitions

1.3.1. *Joints*

A joint connects two successive links, thus limiting the number of degrees of freedom between them. The resulting number of degrees of freedom, m, is also called *joint mobility*, such that $0 \leq m \leq 6$.

When m = 1, which is frequently the case in robotics, the joint is either *revolute* or *prismatic*. A complex joint with several degrees of freedom can be constructed by an equivalent combination of revolute and prismatic joints. For example, a spherical joint can be obtained by using three revolute joints whose axes intersect at a point.

1.3.1.1. *Revolute joint*

This limits the motion between two links to a rotation about a common axis. The relative location between the two links is given by the angle about this axis. The revolute joint, denoted by R, is represented by the symbols shown in Figure 1.5.

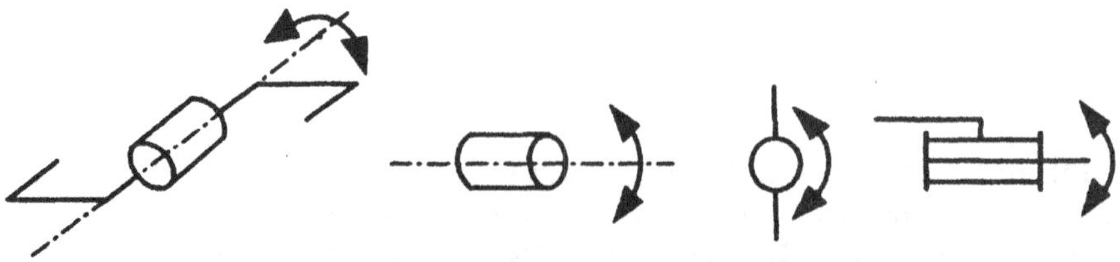

Figure 1.5. *Symbols of a revolute joint*

1.3.1.2. *Prismatic joint*

This limits the motion between two links to a translation along a common axis. The relative location between the two links is determined by the distance along this axis. The prismatic joint, denoted by P, is represented by the symbols shown in Figure 1.6.

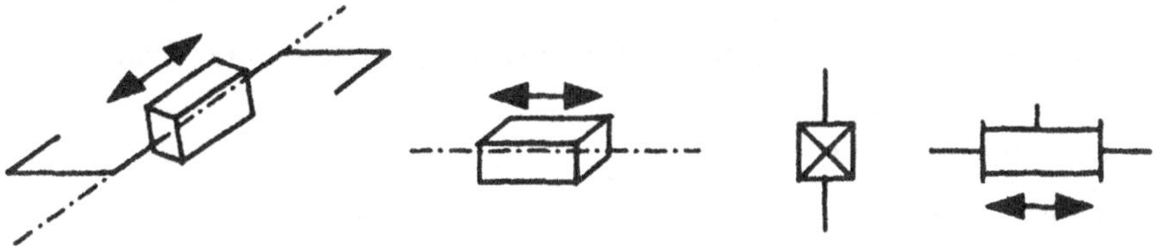

Figure 1.6. *Symbols of a prismatic joint*

1.3.2. *Joint space*

The space in which the location of all the links of a robot are represented is called *joint space*, or *configuration space*. We use the *joint variables*, $\mathbf{q} \in \Re^N$, as the coordinates of this space. Its dimension N is equal to the number of independent joints and corresponds to the number of degrees of freedom of the mechanical structure. In an open chain robot (simple or tree structured), the joint variables are generally independent, whereas a closed chain structure implies constraint relations between the joint variables.

Unless otherwise stated, we will consider that a robot with N degrees of freedom has N actuated joints.

1.3.3. *Task space*

The location, position and orientation, of the end-effector is represented in the *task space*, or *operational space*. We may consider as many task spaces as there are end-effectors. Generally, Cartesian coordinates are used to specify the position in $\Re^3$ and the rotation group SO(3) for the orientation. Thus the task space is equal to $\Re^3 \times SO(3)$. An element of the task space is represented by the vector $\mathbf{X} \in \Re^M$, where M is equal to the maximum number of independent parameters that are necessary to specify the location of the end-effector in space. Consequently, $M \leq 6$ and $M \leq N$.

1.3.4. *Redundancy*

A robot is classified as *redundant* when the number of degrees of freedom of its task space is less than the number of degrees of freedom of its joint space. This property increases the volume of the reachable workspace of the robot and enhances its performance. We will see in Chapter 6 that redundant robots can achieve a secondary objective besides the primary objective of locating and moving the end-effector with desired velocity.

Notice that a simple open chain is redundant if it contains any of the following combinations of joints:

- more than six joints;
- more than three revolute joints whose axes intersect at a point;
- more than three revolute joints with parallel axes;
- more than three prismatic joints;
- prismatic joints with parallel axes;
- revolute joints with collinear axes.

NOTES.–

- for an articulated mechanism with several end-effectors, redundancy is evaluated by comparing the number of degrees of freedom of the joint space acting on each end-effector and the number of degrees of freedom of the corresponding task space;
- another type of redundancy may occur when the number of degrees of freedom of the task is less than the number of degrees of freedom of the robot. We will discuss this case in Chapter 6.

1.3.5. *Singular configurations*

For all robots, redundant or not, it is possible that at some configurations, called *singular configurations*, the number of degrees of freedom of the end-effector becomes less than the dimension of the task space. For example, this may occur when:

- the axes of two prismatic joints become parallel;
- the axes of two revolute joints become collinear;
- the origin of the end-effector lies on a line that intersects all the joint axes.

In Chapter 5, we will present a mathematical condition to determine the number of degrees of freedom of the task space of a mechanism as well as its singular configurations.

1.4. Choosing the number of degrees of freedom of a robot

A non-redundant robot must have six degrees of freedom in order to place an arbitrary object in space. However, if the manipulated object exhibits revolution symmetry, five degrees of freedom are sufficient, since it is not necessary to specify the rotation about the revolution axis. In the same way, to locate a body in a plane, one needs only three degrees of freedom: two for positioning a point in the plane and the third to determine the orientation of the body.

From these observations, we deduce that:

- the number of degrees of freedom of a mechanism is chosen as a function of the shape of the object to be manipulated by the robot and of the class of tasks to be realized;
- a necessary but insufficient condition to have compatibility between the robot and the task is that the number of degrees of freedom of the end-effector of the robot is equal to or more than that of the task.

1.5. Architectures of robot manipulators

Without anticipating the results of the next chapters, we can say that the study of both tree structured and closed chains can be reduced to some equivalent simple open chains. Thus, the classification presented below is relevant for simple open chain architectures, but may also be generalized to the complex chains.

In order to count the possible architectures, we only consider revolute or prismatic joints whose consecutive axes are either parallel or perpendicular. Generally, with some exceptions (in particular, the last three joints of the GMF P150 and Kuka IR600 robots), the consecutive axes of currently used robots are either parallel or perpendicular. The different combinations of these four parameters yield the number of possible architectures with respect to the number of joints as shown in Table 1.1 [Delignières 87], [Chedmail 90a].

The first three joints of a robot are commonly designed in order to perform gross motion of the end-effector, and the remaining joints are used to accomplish orientation. Thus, the first three joints and the associated links constitute the shoulder or regional positioning structure. The other joints and links form the wrist.

Taking into account these considerations and the data of Table 1.1, one can count 36 possible combinations of the shoulder. Among these architectures, only 12 are mathematically distinct and non-redundant (we eliminate, *a priori*, the structures limiting the motion of the terminal point of the shoulder to linear or planar displacement, such as those having three prismatic joints with parallel axes, or three revolute joints with parallel axes). These structures are shown in Figure 1.7.

Table 1.1. *Number of possible architectures as a function of the number of degrees of freedom of the robot*

Number of degrees of freedom of the robot	Number of architectures
2	8
3	36
4	168
5	776
6	3508

A survey of industrial robots has shown that only the following five structures [Liégeois 79] are manufactured:

- anthropomorphic shoulder represented by the first RRR structure shown in Figure 1.7, like PUMA from Unimation, Acma SR400, ABB IRBx400, Comau Smart-3, Fanuc (S-xxx, Arc Mate), Kuka (KR 6 to KR 200), Reis (RV family), Staübli (RX series), etc.;
- spherical shoulder RRP: "Stanford manipulator" and Unimation robots (Series 1000, 2000, 4000);
- RPR shoulder corresponding to the first RPR structure shown in Figure 1.7: Acma-H80, Reis (RH family), etc. The association of a wrist with one revolute degree of freedom of rotation to such a shoulder can be found frequently in the industry. The resulting structure of such a robot is called SCARA (Selective Compliance Assembly Robot Arm) (Figure 1.8). It has several applications, particularly in planar assembly. SCARA, designed by Sankyo, has been manufactured by many other companies: IBM, Bosch, Adept, etc.;
- cylindrical shoulder RPP: Acma-TH8, AFMA (ROV, ROH), etc.;
- Cartesian shoulder PPP: Acma-P80, IBM-7565, Sormel-Cadratic, Olivetti-SIGMA. More recent examples: AFMA (RP, ROP series), Comau P-Mast, Reis (RL family), SEPRO, etc.

The second RRR structure of Figure 1.7, which is equivalent to a spherical joint, is generally used as a wrist. Other types of wrists are shown in Figure 1.9 [Delignières 87].

A robot, composed of a shoulder with three degrees of freedom and a spherical wrist, constitutes a classical six degree-of-freedom structure (Figure 1.10). Note that the position of the center of the spherical joint depends only on the configuration of joints 1, 2 and 3. We will see in Chapter 4 that, due to this property, the inverse geometric model, providing the joint variables for a given location of the end-effector, can be obtained analytically for such robots.

According to the survey carried out by the French Association of Industrial Robotics (AFRI) and RobAut Journal [Fages 98], the classification of robots in France (17794 robots), with respect to the number of degrees of freedom, is as follows: 4.5% of the robots have three degrees of freedom, 27% have four, 9% have five and 59.5% have six or more. As far as the architecture of the shoulder is concerned, there is a clear dominance of the RRR anthropomorphic shoulder (65.5%), followed by the Cartesian shoulder (20.5%), then the cylindrical shoulder (7%) and finally the SCARA shoulder (7%).

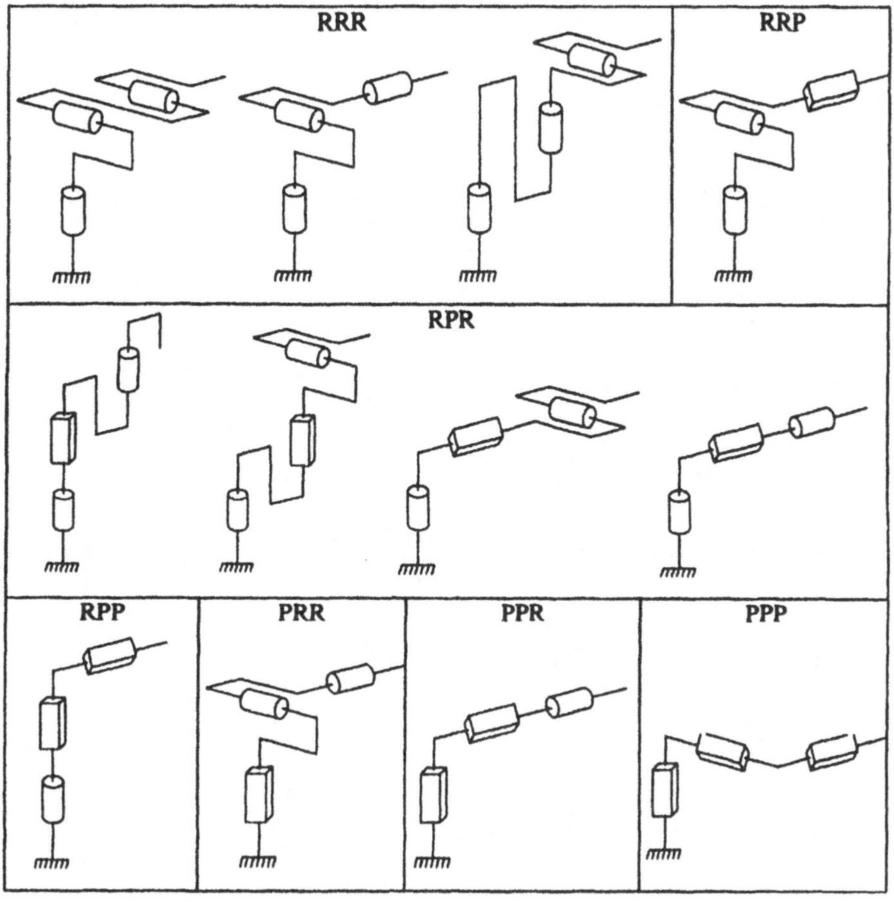

Figure 1.7. *Architectures of the shoulder (from [Milenkovic 83])*

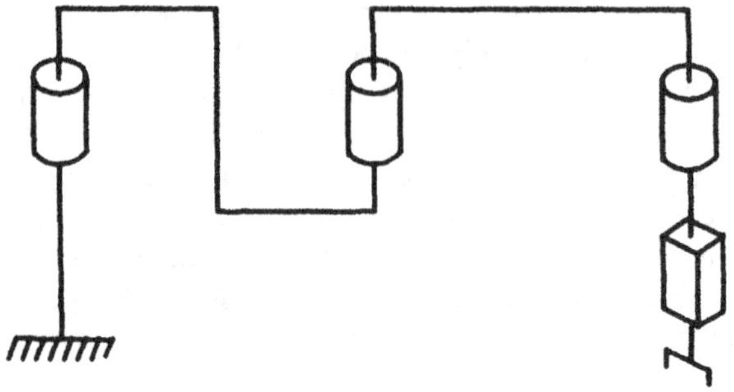

Figure 1.8. *SCARA robot*

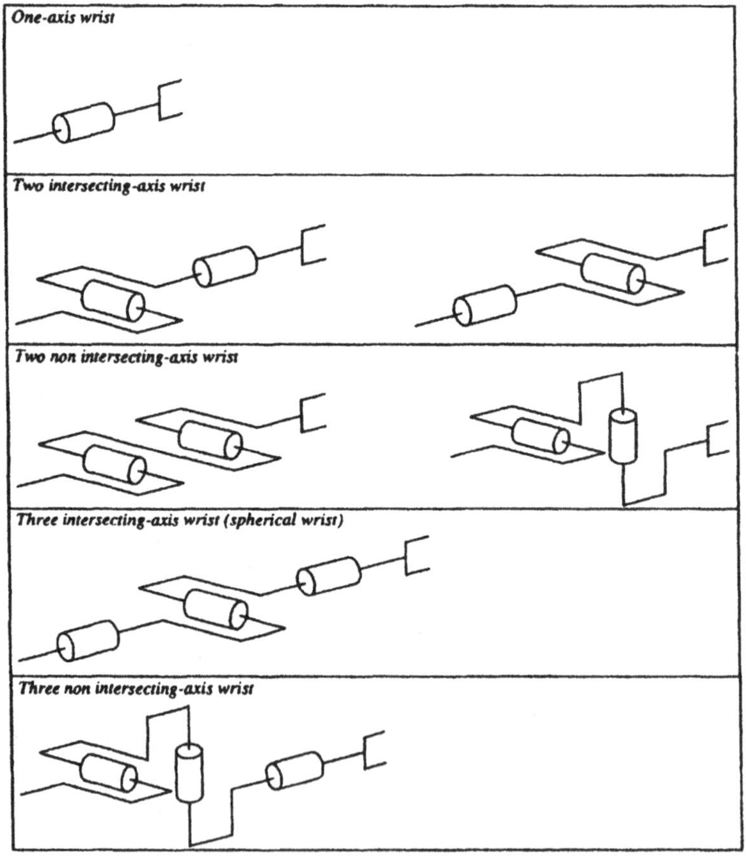

Figure 1.9. *Architectures of the wrist (from [Delignières 87])*

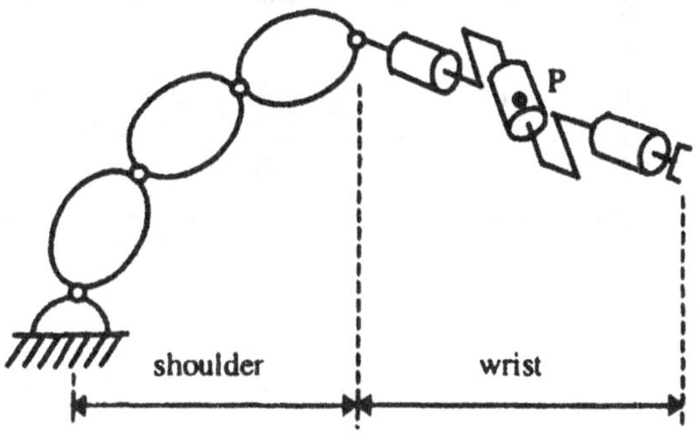

Figure 1.10. *Classical six degree-of-freedom robot*

1.6. Characteristics of a robot

The standard ISO 9946 specifies the characteristics that manufacturers of robots must provide. Here, we describe some of these characteristics that may help the user in choosing an appropriate robot with respect to a given application:

- *workspace*: defines the space that can be swept by the end-effector. Its range depends on the number of degrees of freedom, the joint limits and the length of the links;
- *payload*: maximum load carried by the robot;
- *maximum velocity and acceleration*: determine the cycle time;
- *position accuracy* (Figure 1.11): indicates the difference between a commanded position and the mean of the attained positions when visiting the commanded position several times from different initial positions;
- *position repeatability* (Figure 1.11): specifies the precision with which the robot returns to a commanded position. It is given as the distance between the mean of the attained positions and the furthermost attained position;
- *resolution*: the smallest increment of movement that can be achieved by the joint or the end-effector.

However, other characteristics must also be taken into account: technical (energy, control, programming, etc.) and commercial (price, maintenance, etc.). Thus, the selection criteria are sometimes difficult to formulate and are often contradictory. To a certain extent, the simulation and modeling tools available in Computer Aided Design (CAD) packages may help in making the best choice [Dombre 88b], [Zeghloul 91], [Chedmail 92], [Chedmail 98].

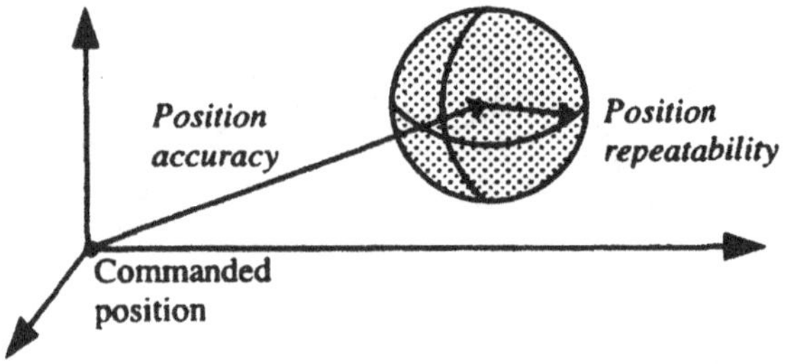

Figure 1.11. *Position accuracy and repeatability (from [Priel 90])*

1.7. Conclusion

In this chapter, we have presented the definitions of some technical terms related to the field of modeling, identification and control of robots. We will frequently come across these terms in this book and some of them will be reformulated in a more analytical or mathematical way. The figures mentioned here justify the choice of the robots that are taken as examples in the following chapters. In the next chapter, we present the transformation matrix concept, which constitutes an important mathematical tool for the modeling of robots.

Chapter 2

Transformation matrix between vectors, frames and screws

2.1. Introduction

In robotics, we assign one or more frames to each link of the robot and each object of the workcell. Thus, transformation of frames is a fundamental concept in the modeling and programming of a robot. It enables us to:

- compute the location, position and orientation of robot links relative to each other;
- describe the position and orientation of objects;
- specify the trajectory and velocity of the end-effector of a robot for a desired task;
- describe and control the forces when the robot interacts with its environment;
- implement sensory-based control using information provided by various sensors, each having its own reference frame.

In this chapter, we present a notation that allows us to describe the relationship between different frames and objects of a robotic cell. This notation, called *homogeneous transformation,* has been widely used in computer graphics [Roberts 65], [Newman 79] to compute the projections and perspective transformations of an object on a screen. Currently, this is also being used extensively in robotics [Pieper 68], [Paul 81]. We will show how the points, vectors and transformations between frames can be represented using this approach. Then, we will define the differential transformations between frames as well as the representation of velocities and forces using screws.

2.2. Homogeneous coordinates

2.2.1. *Representation of a point*

Let $({}^iP_x, {}^iP_y, {}^iP_z)$ be the Cartesian coordinates of an arbitrary point P with respect to the frame R_i, which is described by the origin O_i and the axes $\mathbf{x}_i$, $\mathbf{y}_i$, $\mathbf{z}_i$ (Figure 2.1). The homogeneous coordinates of P with respect to frame R_i are defined by $(w\,{}^iP_x, w\,{}^iP_y, w\,{}^iP_z, w)$, where w is a scaling factor. In robotics, w is taken to be equal to 1. Thus, we represent the homogeneous coordinates of P by the (4x1) column vector:

$$ {}^i\mathbf{P} = \begin{bmatrix} {}^iP_x \\ {}^iP_y \\ {}^iP_z \\ 1 \end{bmatrix} \qquad [2.1] $$

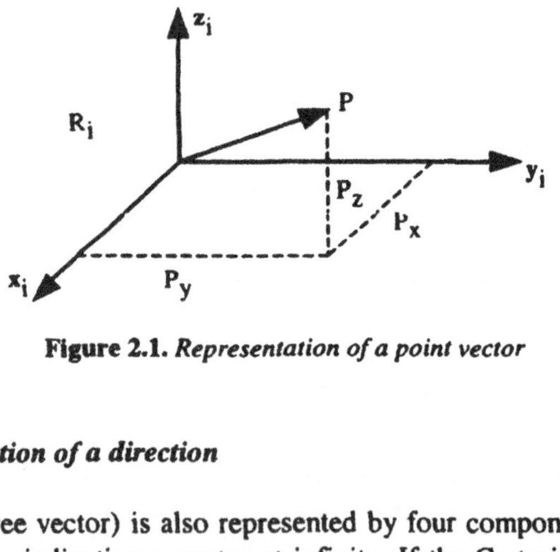

Figure 2.1. *Representation of a point vector*

2.2.2. *Representation of a direction*

A direction (free vector) is also represented by four components, but the fourth component is zero, indicating a vector at infinity. If the Cartesian coordinates of a unit vector **u** with respect to frame R_i are $({}^iu_x, {}^iu_y, {}^iu_z)$, its homogeneous coordinates will be:

$$ {}^i\mathbf{u} = \begin{bmatrix} {}^iu_x \\ {}^iu_y \\ {}^iu_z \\ 0 \end{bmatrix} \qquad [2.2] $$

2.2.3. *Representation of a plane*

The homogeneous coordinates of a plane Q, whose equation with respect to a frame R_i is ${}^i\alpha x + {}^i\beta y + {}^i\gamma z + {}^i\delta = 0$, are given by:

$$ {}^i\mathbf{Q} = [\, {}^i\alpha \quad {}^i\beta \quad {}^i\gamma \quad {}^i\delta \,] \tag{2.3} $$

If a point P lies in the plane Q, then the matrix product ${}^i\mathbf{Q}\,{}^i\mathbf{P}$ is zero:

$$ {}^i\mathbf{Q}\,{}^i\mathbf{P} = [\, {}^i\alpha \quad {}^i\beta \quad {}^i\gamma \quad {}^i\delta \,] \begin{bmatrix} {}^iP_x \\ {}^iP_y \\ {}^iP_z \\ 1 \end{bmatrix} = 0 \tag{2.4} $$

2.3. Homogeneous transformations [Paul 81]

2.3.1. *Transformation of frames*

The transformation, translation and/or rotation, of a frame R_i into frame R_j (Figure 2.2) is represented by the (4x4) homogeneous transformation matrix ${}^i\mathbf{T}_j$ such that:

$$ {}^i\mathbf{T}_j = [\, {}^i\mathbf{s}_j \quad {}^i\mathbf{n}_j \quad {}^i\mathbf{a}_j \quad {}^i\mathbf{P}_j \,] = \begin{bmatrix} s_x & n_x & a_x & P_x \\ s_y & n_y & a_y & P_y \\ s_z & n_z & a_z & P_z \\ 0 & 0 & 0 & 1 \end{bmatrix} \tag{2.5a} $$

where ${}^i\mathbf{s}_j$, ${}^i\mathbf{n}_j$ and ${}^i\mathbf{a}_j$ contain the components of the unit vectors along the $\mathbf{x}_j$, $\mathbf{y}_j$ and $\mathbf{z}_j$ axes respectively expressed in frame R_i, and where ${}^i\mathbf{P}_j$ is the vector representing the coordinates of the origin of frame R_j expressed in frame R_i.

We can also say that the matrix ${}^i\mathbf{T}_j$ defines frame R_j relative to frame R_i. Thereafter, the transformation matrix [2.5a] will occasionally be written in the form of a partitioned matrix:

$$ {}^i\mathbf{T}_j = \begin{bmatrix} {}^i\mathbf{A}_j & {}^i\mathbf{P}_j \\ 0\ 0\ 0 & 1 \end{bmatrix} = \begin{bmatrix} {}^i\mathbf{s}_j & {}^i\mathbf{n}_j & {}^i\mathbf{a}_j & {}^i\mathbf{P}_j \\ 0 & 0 & 0 & 1 \end{bmatrix} \tag{2.5b} $$

Apparently, this is in violation of the homogeneous notation since the vectors have only three components. In any case, the distinction in the representation with either three or four components will always be clear in the text.

In summary:

- the matrix ${}^{i}\mathbf{T}_j$ represents the transformation from frame R_i to frame R_j;
- the matrix ${}^{i}\mathbf{T}_j$ can be interpreted as representing the frame R_j (three orthogonal axes and an origin) with respect to frame R_i.

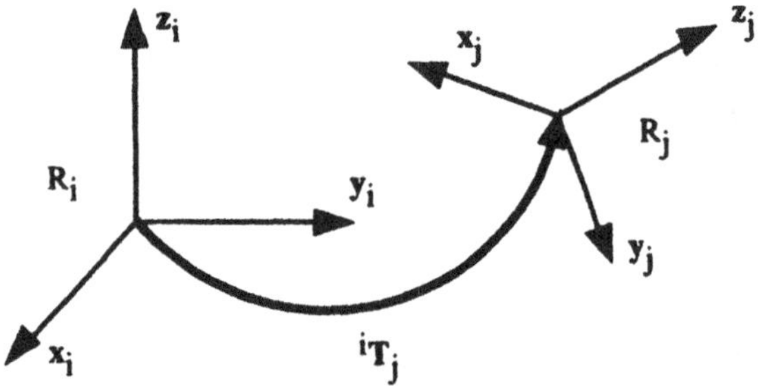

Figure 2.2. *Transformation of frames*

2.3.2. *Transformation of vectors*

Let the vector ${}^{j}\mathbf{P}$ define the homogeneous coordinates of the point P with respect to frame R_j (Figure 2.3). Thus, the homogeneous coordinates of P with respect to frame R_i can be obtained as:

$$ {}^{i}\mathbf{P} = {}^{i}(\mathbf{O_iP}) = {}^{i}\mathbf{s}_j\,{}^{j}P_x + {}^{i}\mathbf{n}_j\,{}^{j}P_y + {}^{i}\mathbf{a}_j\,{}^{j}P_z + {}^{i}\mathbf{P}_j = {}^{i}\mathbf{T}_j\,{}^{j}\mathbf{P} \qquad [2.6] $$

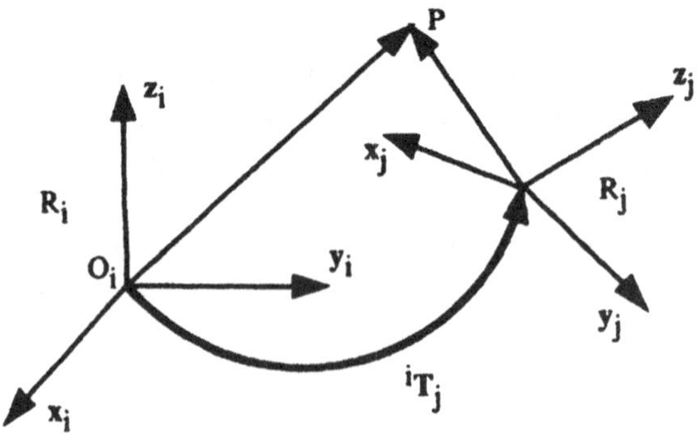

Figure 2.3. *Transformation of a vector*

Thus the matrix ${}^{i}T_j$ allows us to calculate the coordinates of a vector with respect to frame R_i in terms of its coordinates in frame R_j.

• **Example 2.1.** Deduce the matrices ${}^{i}T_j$ and ${}^{j}T_i$ from Figure 2.4. Using equation [2.5a], we directly obtain:

$$ {}^{i}T_j = \begin{bmatrix} 0 & 0 & 1 & 3 \\ 0 & 1 & 0 & 12 \\ -1 & 0 & 0 & 6 \\ 0 & 0 & 0 & 1 \end{bmatrix}, \quad {}^{j}T_i = \begin{bmatrix} 0 & 0 & -1 & 6 \\ 0 & 1 & 0 & -12 \\ 1 & 0 & 0 & -3 \\ 0 & 0 & 0 & 1 \end{bmatrix} $$

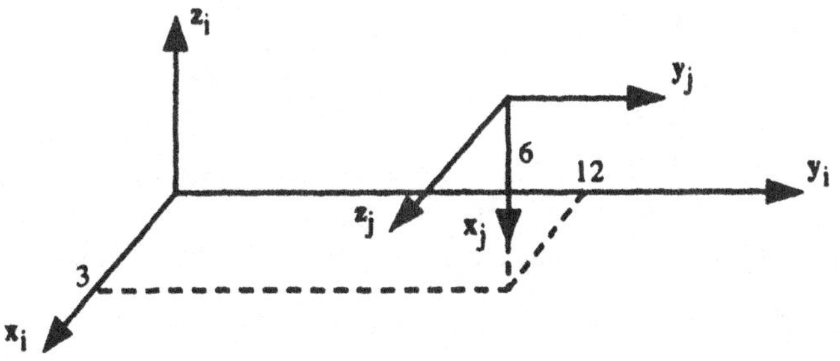

Figure 2.4. *Example 2.1*

2.3.3. *Transformation of planes*

The relative position of a point with respect to a plane is invariant with respect to the transformation applied to the set of {point, plane}. Thus:

$$ {}^{j}Q\,{}^{j}P = {}^{i}Q\,{}^{i}P = {}^{i}Q\,{}^{i}T_j\,{}^{j}P $$

leading to:

$$ {}^{j}Q = {}^{i}Q\,{}^{i}T_j \qquad [2.7] $$

2.3.4. *Transformation matrix of a pure translation*

Let **Trans**(a, b, c) be this transformation, where a, b and c denote the translation along the **x**, **y** and **z** axes respectively. Since the orientation is invariant, the transformation **Trans**(a, b, c) is expressed as (Figure 2.5):

$$
{}^i\mathbf{T}_j = \mathbf{Trans}(a, b, c) = \begin{bmatrix} 1 & 0 & 0 & a \\ 0 & 1 & 0 & b \\ 0 & 0 & 1 & c \\ 0 & 0 & 0 & 1 \end{bmatrix} \quad [2.8]
$$

From now on, we will also use the notation **Trans(u, d)** to denote a translation along an axis **u** by a value d. Thus, the matrix **Trans**(a, b, c) can be decomposed into the product of three matrices **Trans(x**, a) **Trans(y**, b) **Trans(z**, c), taking any order of the multiplication.

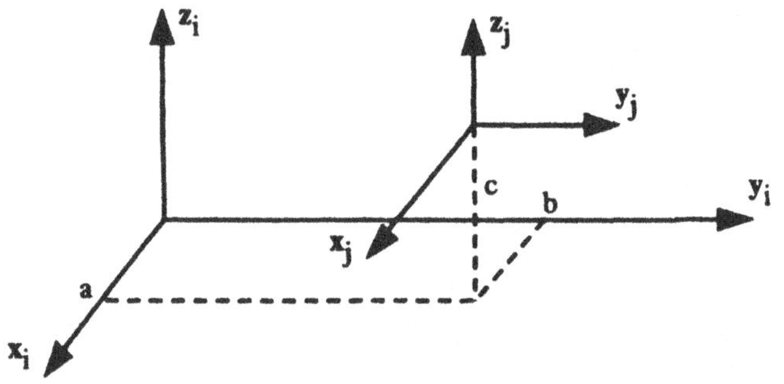

Figure 2.5. *Transformation of pure translation*

2.3.5. *Transformation matrices of a rotation about the principle axes*

2.3.5.1. *Transformation matrix of a rotation about the x axis by an angle θ*

Let **Rot(x**, θ) be this transformation. From Figure 2.6, we deduce that the components of the unit vectors ${}^i\mathbf{s}_j$, ${}^i\mathbf{n}_j$, ${}^i\mathbf{a}_j$ along the axes $\mathbf{x}_j$, $\mathbf{y}_j$ and $\mathbf{z}_j$ respectively of frame R_j expressed in frame R_i are as follows:

$$
\begin{cases} {}^i\mathbf{s}_j = [1 \quad 0 \quad 0 \quad 0]^T \\ {}^i\mathbf{n}_j = [0 \quad C\theta \quad S\theta \quad 0]^T \\ {}^i\mathbf{a}_j = [0 \quad -S\theta \quad C\theta \quad 0]^T \end{cases} \quad [2.9]
$$

where Sθ and Cθ represent sin(θ) and cos(θ) respectively, and the superscript T indicates the transpose of the vector.

$$
{}^{i}\mathbf{T}_j = \mathbf{Rot}(\mathbf{x}, \theta) = \begin{bmatrix} 1 & 0 & 0 & 0 \\ 0 & C\theta & -S\theta & 0 \\ 0 & S\theta & C\theta & 0 \\ 0 & 0 & 0 & 1 \end{bmatrix} = \begin{bmatrix} & \mathbf{rot}(\mathbf{x}, \theta) & & 0 \\ & & & 0 \\ & & & 0 \\ 0 & 0 & 0 & 1 \end{bmatrix} \qquad [2.10]
$$

where $\mathbf{rot}(\mathbf{x}, \theta)$ denotes the (3x3) orientation matrix.

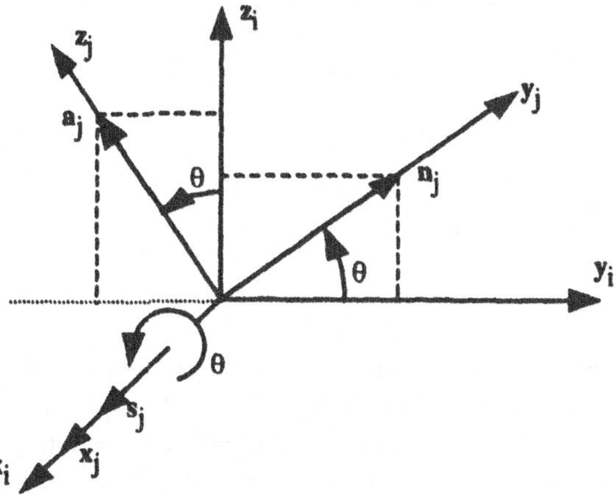

Figure 2.6. *Transformation of a pure rotation about the x-axis*

2.3.5.2. *Transformation matrix of a rotation about the y axis by an angle θ*

In the same way, we obtain:

$$
{}^{i}\mathbf{T}_j = \mathbf{Rot}(\mathbf{y}, \theta) = \begin{bmatrix} C\theta & 0 & S\theta & 0 \\ 0 & 1 & 0 & 0 \\ -S\theta & 0 & C\theta & 0 \\ 0 & 0 & 0 & 1 \end{bmatrix} = \begin{bmatrix} & \mathbf{rot}(\mathbf{y}, \theta) & & 0 \\ & & & 0 \\ & & & 0 \\ 0 & 0 & 0 & 1 \end{bmatrix} \qquad [2.11]
$$

2.3.5.3. *Transformation matrix of a rotation θ about the z axis by an angle θ*

We can also verify that:

$$
{}^{i}\mathbf{T}_j = \mathbf{Rot}(\mathbf{z},\theta) = \begin{bmatrix} C\theta & -S\theta & 0 & 0 \\ S\theta & C\theta & 0 & 0 \\ 0 & 0 & 1 & 0 \\ 0 & 0 & 0 & 1 \end{bmatrix} = \begin{bmatrix} & \mathbf{rot}(\mathbf{z},\theta) & & 0 \\ & & & 0 \\ & & & 0 \\ 0 & 0 & 0 & 1 \end{bmatrix} \quad [2.12]
$$

2.3.6. *Properties of homogeneous transformation matrices*

a) From equations [2.5], a transformation matrix can be written as:

$$
\mathbf{T} = \begin{bmatrix} s_x & n_x & a_x & P_x \\ s_y & n_y & a_y & P_y \\ s_z & n_z & a_z & P_z \\ 0 & 0 & 0 & 1 \end{bmatrix} = \begin{bmatrix} \mathbf{A} & & & \mathbf{P} \\ 0 & 0 & 0 & 1 \end{bmatrix} \quad [2.13]
$$

The matrix **A** represents the rotation whereas the column matrix **P** represents the translation. For a transformation of pure translation, $\mathbf{A} = \mathbf{I}_3$ ($\mathbf{I}_3$ represents the identity matrix of order 3), whereas $\mathbf{P} = \mathbf{0}$ for a transformation of pure rotation. The matrix **A** represents the direction cosine matrix. It contains three independent parameters (one of the vectors **s**, **n** or **a** can be deduced from the vector product of the other two, for example $\mathbf{s} = \mathbf{n}\mathrm{x}\mathbf{a}$; moreover, the dot product $\mathbf{n}.\mathbf{a}$ is zero and the magnitudes of **n** and **a** are equal to 1).

b) The matrix **A** is orthogonal, i.e. its inverse is equal to its transpose:

$$
\mathbf{A}^{-1} = \mathbf{A}^T \quad [2.14]
$$

c) The inverse of a matrix ${}^{i}\mathbf{T}_j$ defines the matrix ${}^{j}\mathbf{T}_i$.

To express the components of a vector ${}^{i}\mathbf{P}_1$ into frame R_j, we write:

$$
{}^{j}\mathbf{P}_1 = {}^{j}\mathbf{T}_i \, {}^{i}\mathbf{P}_1 \quad [2.15]
$$

If we postmultiply equation [2.6] by ${}^{i}\mathbf{T}_j^{-1}$ (inverse of ${}^{i}\mathbf{T}_j$), we obtain:

$$
{}^{i}\mathbf{T}_j^{-1} \, {}^{i}\mathbf{P}_1 = {}^{j}\mathbf{P}_1 \quad [2.16]
$$

From equations [2.15] and [2.16], we deduce that:

$$
{}^{i}\mathbf{T}_j^{-1} = {}^{j}\mathbf{T}_i \quad [2.17]
$$

d) We can easily verify that:

$$\mathbf{Rot}^{-1}(\mathbf{u}, \theta) = \mathbf{Rot}(\mathbf{u}, -\theta) = \mathbf{Rot}(-\mathbf{u}, \theta) \qquad [2.18]$$

$$\mathbf{Trans}^{-1}(\mathbf{u}, d) = \mathbf{Trans}(-\mathbf{u}, d) = \mathbf{Trans}(\mathbf{u}, -d) \qquad [2.19]$$

e) The inverse of a transformation matrix represented by equation [2.13] can be obtained as:

$$\mathbf{T}^{-1} = \begin{bmatrix} & -\mathbf{s}^T\mathbf{P} \\ \mathbf{A}^T & -\mathbf{n}^T\mathbf{P} \\ & -\mathbf{a}^T\mathbf{P} \\ 0\ 0\ 0 & 1 \end{bmatrix} = \begin{bmatrix} \mathbf{A}^T & -\mathbf{A}^T\mathbf{P} \\ 0\ 0\ 0 & 1 \end{bmatrix} \qquad [2.20]$$

f) Composition of two matrices. The multiplication of two transformation matrices gives a transformation matrix:

$$\mathbf{T}_1\,\mathbf{T}_2 = \begin{bmatrix} \mathbf{A}_1 & \mathbf{P}_1 \\ 0\ 0\ 0 & 1 \end{bmatrix}\begin{bmatrix} \mathbf{A}_2 & \mathbf{P}_2 \\ 0\ 0\ 0 & 1 \end{bmatrix} = \begin{bmatrix} \mathbf{A}_1\mathbf{A}_2 & \mathbf{A}_1\mathbf{P}_2 + \mathbf{P}_1 \\ 0\quad 0\quad 0 & 1 \end{bmatrix} \qquad [2.21]$$

Note that the matrix multiplication is non-commutative ($\mathbf{T}_1\mathbf{T}_2 \neq \mathbf{T}_2\mathbf{T}_1$).

g) If a frame R_0 is subjected to k consecutive transformations (Figure 2.7) and if each transformation i, (i = 1, ..., k), is defined with respect to the current frame R_{i-1}, then the transformation ${}^0\mathbf{T}_k$ can be deduced by multiplying all the transformation on the right as:

$$ {}^0\mathbf{T}_k = {}^0\mathbf{T}_1\ {}^1\mathbf{T}_2\ {}^2\mathbf{T}_3 \ldots {}^{k-1}\mathbf{T}_k \qquad [2.22]$$

h) If a frame R_j, defined by ${}^i\mathbf{T}_j$, undergoes a transformation $\mathbf{T}$ that is defined relative to frame R_j, then R_j will be transformed into $R_{j'}$ with ${}^i\mathbf{T}_{j'} = \mathbf{T}\ {}^i\mathbf{T}_j$ (Figure 2.8).

From the properties g and h, we deduce that:

- multiplication on the right (postmultiplication) of the transformation ${}^i\mathbf{T}_j$ indicates that the transformation is defined with respect to the current frame R_j;
- multiplication on the left (premultiplication) indicates that the transformation is defined with respect to the reference frame R_i.

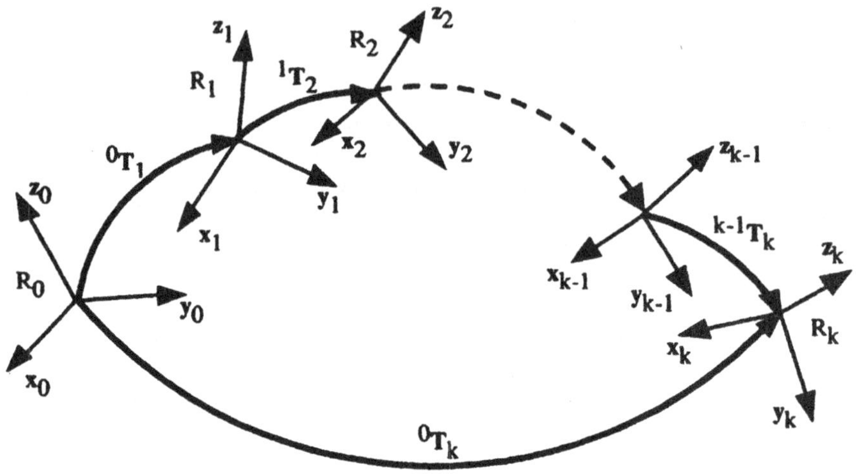

Figure 2.7. *Composition of transformations: multiplication on the right*

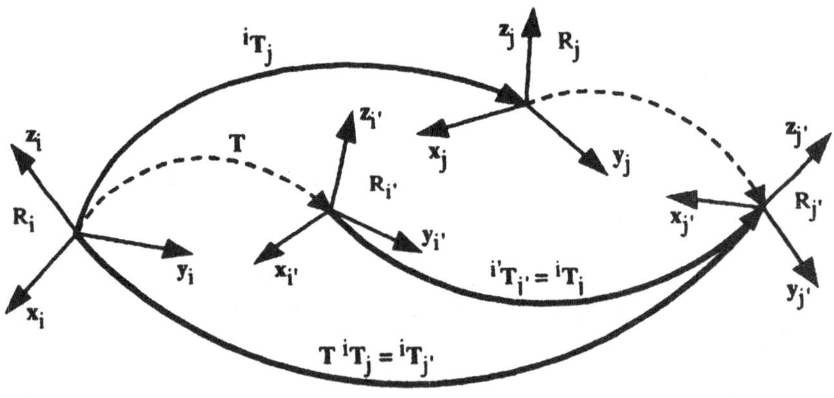

Figure 2.8. *Composition of transformations: multiplication on the left*

• **Example 2.2.** Consider the composite transformation illustrated in Figure 2.9 and defined by:

$$^{0}\mathbf{T}_2 = \mathbf{Rot}(\mathbf{x}, \frac{\pi}{6})\ \mathbf{Trans}(\mathbf{y}, d)$$

– reading $^{0}\mathbf{T}_2$ from left to right (Figure 2.9a): first, we apply the rotation; the new location of frame R_0 is denoted by frame R_1; then, the translation is defined with respect to frame R_1;

– reading ${}^{0}\mathbf{T}_2$ from right to left (Figure 2.9b): first we apply the translation, then the rotation is defined with respect to frame R_0.

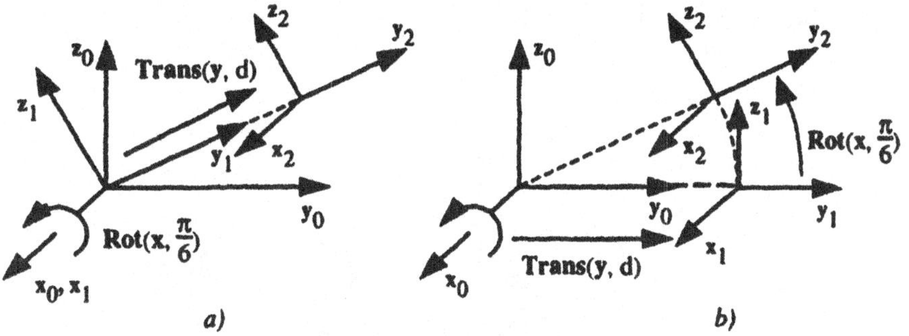

Figure 2.9. *Example 2.2*

i) Consecutive transformations about the same axis. We note the following properties:

$$\mathbf{Rot}(\mathbf{u}, \theta_1)\, \mathbf{Rot}(\mathbf{u}, \theta_2) = \mathbf{Rot}[\mathbf{u}, (\theta_1 + \theta_2)] \qquad [2.23]$$

$$\mathbf{Rot}(\mathbf{u}, \theta)\, \mathbf{Trans}(\mathbf{u}, d) = \mathbf{Trans}(\mathbf{u}, d)\, \mathbf{Rot}(\mathbf{u}, \theta) \qquad [2.24]$$

j) Decomposition of a transformation matrix. A transformation matrix can be decomposed into two transformation matrices, one represents a pure translation and the second a pure rotation:

$$\mathbf{T} = \begin{bmatrix} & \mathbf{A} & & \mathbf{P} \\ 0 & 0 & 0 & 1 \end{bmatrix} = \begin{bmatrix} & \mathbf{I}_3 & & \mathbf{P} \\ 0 & 0 & 0 & 1 \end{bmatrix} \begin{bmatrix} & \mathbf{A} & & \mathbf{0} \\ 0 & 0 & 0 & 1 \end{bmatrix} \qquad [2.25]$$

2.3.7. *Transformation matrix of a rotation about a general vector located at the origin*

Let $\mathbf{Rot}(\mathbf{u}, \theta)$ be the transformation representing a rotation of an angle θ about an axis, with unit vector $\mathbf{u} = [u_x \quad u_y \quad u_z]^T$, located at the origin of frame R_i (Figure 2.10). We define the frame R_k such that $\mathbf{z}_k$ is along the vector $\mathbf{u}$ and $\mathbf{x}_k$ is along the common normal between $\mathbf{z}_k$ and $\mathbf{z}_i$. The matrix ${}^{i}\mathbf{T}_k$ can be obtained as:

$${}^{i}\mathbf{T}_k = \mathbf{Rot}(\mathbf{z}, \alpha)\, \mathbf{Rot}(\mathbf{x}, \beta) \qquad [2.26]$$

where α is the angle between $\mathbf{x}_i$ and $\mathbf{x}_k$ about $\mathbf{z}_i$, and β is the angle between $\mathbf{z}_i$ and $\mathbf{u}$ about $\mathbf{x}_k$.

From equation [2.26], we obtain:

$$\mathbf{u} = {}^{i}\mathbf{a}_k = \begin{bmatrix} u_x \\ u_y \\ u_z \end{bmatrix} = \begin{bmatrix} S\alpha\, S\beta \\ -C\alpha\, S\beta \\ C\beta \end{bmatrix} \qquad [2.27]$$

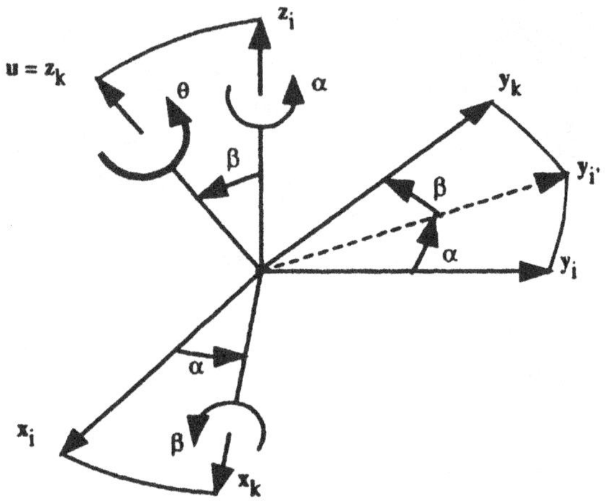

Figure 2.10. *Transformation of pure rotation about any axis*

The rotation about **u** is equivalent to the rotation about $\mathbf{z}_k$. From properties g and h of § 2.3.6, we deduce that:

$$\mathbf{Rot(u, \theta)}\ {}^{i}\mathbf{T}_k = {}^{i}\mathbf{T}_k\ \mathbf{Rot(z, \theta)} \qquad [2.28]$$

thus:

$$\begin{aligned} \mathbf{Rot(u, \theta)} &= {}^{i}\mathbf{T}_k\ \mathbf{Rot(z, \theta)}\ {}^{i}\mathbf{T}_k^{-1} \\ &= \mathbf{Rot(z, \alpha)\ Rot(x, \beta)\ Rot(z, \theta)\ Rot(x, -\beta)\ Rot(z, -\alpha)} \end{aligned} \qquad [2.29]$$

From this relation and using equation [2.27], we obtain:

$$\mathbf{Rot\ (u, \theta)} = \begin{bmatrix} & \mathbf{rot(u, \theta)} & & \begin{matrix} 0 \\ 0 \\ 0 \end{matrix} \\ 0 & 0 & 0 & 1 \end{bmatrix}$$

$$= \begin{bmatrix} u_x^2(1-C\theta)+C\theta & u_xu_y(1-C\theta)-u_zS\theta & u_xu_z(1-C\theta)+u_yS\theta & 0 \\ u_xu_y(1-C\theta)+u_zS\theta & u_y^2(1-C\theta)+C\theta & u_yu_z(1-C\theta)-u_xS\theta & 0 \\ u_xu_z(1-C\theta)-u_yS\theta & u_yu_z(1-C\theta)+u_xS\theta & u_z^2(1-C\theta)+C\theta & 0 \\ 0 & 0 & 0 & 1 \end{bmatrix} \quad [2.30]$$

We can easily remember this relation by writing it as:

$$\mathbf{rot}(\mathbf{u}, \theta) = \mathbf{u}\,\mathbf{u}^T(1 - C\theta) + \mathbf{I}_3\, C\theta + \hat{\mathbf{u}}\, S\theta \quad [2.31]$$

where $\hat{\mathbf{u}}$ indicates the skew-symmetric matrix defined by the components of the vector $\mathbf{u}$ such that:

$$\hat{\mathbf{u}} = \begin{bmatrix} 0 & -u_z & u_y \\ u_z & 0 & -u_x \\ -u_y & u_x & 0 \end{bmatrix} \quad [2.32]$$

Note that the vector product $\mathbf{u}\mathrm{x}\mathbf{V}$ is obtained by $\hat{\mathbf{u}}\,\mathbf{V}$.

2.3.8. *Equivalent angle and axis of a general rotation*

Let $\mathbf{T}$ be any arbitrary rotational transformation matrix such that:

$$\mathbf{T} = \begin{bmatrix} s_x & n_x & a_x & 0 \\ s_y & n_y & a_y & 0 \\ s_z & n_z & a_z & 0 \\ 0 & 0 & 0 & 1 \end{bmatrix} \quad [2.33]$$

We solve the following expression for $\mathbf{u}$ and θ:

$$\mathbf{Rot}(\mathbf{u}, \theta) = \mathbf{T} \qquad \text{with } 0 \le \theta \le \pi$$

Adding the diagonal terms of equations [2.30] and [2.33], we obtain:

$$C\theta = \frac{1}{2}(s_x + n_y + a_z - 1) \quad [2.34]$$

From the off-diagonal terms, we obtain:

$$\begin{cases} 2\,u_x\,S\theta = n_z - a_y \\ 2\,u_y\,S\theta = a_x - s_z \\ 2\,u_z\,S\theta = s_y - n_x \end{cases} \qquad [2.35]$$

yielding:

$$S\theta = \frac{1}{2}\sqrt{(n_z - a_y)^2 + (a_x - s_z)^2 + (s_y - n_x)^2} \qquad [2.36]$$

From equations [2.34] and [2.36], we deduce that:

$$\theta = \text{arctg}\,(S\theta/C\theta) \qquad \text{with } 0 \le \theta \le \pi \qquad [2.37]$$

u_x, u_y and u_z are calculated using equation [2.35] if $S\theta \neq 0$. When $S\theta$ is small, the elements u_x, u_y and u_z cannot be determined with good accuracy by this equation. However, in the case where $C\theta < 0$, we obtain u_x, u_y and u_z more accurately using the diagonal terms of **Rot(u, θ)** as follows:

$$u_x = \pm\sqrt{\frac{s_x - C\theta}{1 - C\theta}},\; u_y = \pm\sqrt{\frac{n_y - C\theta}{1 - C\theta}},\; u_z = \pm\sqrt{\frac{a_z - C\theta}{1 - C\theta}} \qquad [2.38]$$

From equation [2.35], we deduce that:

$$\begin{cases} u_x = \text{sign}(n_z - a_y)\sqrt{\dfrac{s_x - C\theta}{1 - C\theta}} \\ u_y = \text{sign}(a_x - s_z)\sqrt{\dfrac{n_y - C\theta}{1 - C\theta}} \\ u_z = \text{sign}(s_y - n_x)\sqrt{\dfrac{a_z - C\theta}{1 - C\theta}} \end{cases} \qquad [2.39]$$

where sign(.) indicates the sign function of the expression between brackets, thus $\text{sign}(e) = +1$ if $e \ge 0$, and $\text{sign}(e) = -1$ if $e < 0$.

• **Example 2.3.** Suppose that the location of a frame R_E, which is fixed to the end-effector of a robot, relative to the reference frame R_0 is given by the matrix **Rot(x, – π/4)**. Determine the vector ${}^E\mathbf{u}$ and the angle of rotation θ that transforms frame R_E to the location **Rot(y, π/4) Rot(z, π/2)**. We can write:

$$\mathbf{Rot}(\mathbf{x}, -\pi/4)\,\mathbf{Rot}(\mathbf{u}, \theta) = \mathbf{Rot}(\mathbf{y}, \pi/4)\,\mathbf{Rot}(\mathbf{z}, \pi/2)$$

Thus:

$$\mathbf{Rot(u, \theta)} = \mathbf{Rot(x, \pi/4)\,Rot(y, \pi/4)\,Rot(z, \pi/2)}$$

$$= \begin{bmatrix} 0 & -1/\sqrt{2} & 1/\sqrt{2} & 0 \\ 1/\sqrt{2} & -1/2 & -1/2 & 0 \\ 1/\sqrt{2} & 1/2 & 1/2 & 0 \\ 0 & 0 & 0 & 1 \end{bmatrix}$$

Using equations [2.34] and [2.36], we get: $C\theta = -\frac{1}{2}$, $S\theta = \frac{\sqrt{3}}{2}$, giving $\theta = 2\pi/3$.

Equation [2.35] yields: $u_x = \frac{1}{\sqrt{3}}$, $u_y = 0$, $u_z = \sqrt{\frac{2}{3}}$.

2.4. Kinematic screw

In this section, we will use the concept of screw to describe the velocity of a body in space.

2.4.1. *Definition of a screw*

A vector field H on $\Re^3$ is a screw if there exist a point O_i and a vector $\mathbf{\Omega}$ such that for all points O_j in $\Re^3$:

$$\mathbf{H}_j = \mathbf{H}_i + \mathbf{\Omega} \times \mathbf{O}_i\mathbf{O}_j$$

where $\mathbf{H}_j$ is the vector of H at O_j and the symbol x indicates the vector product; $\mathbf{\Omega}$ is called the vector of the *screw* of H.

Then, it is easy to prove that for every couple of points O_k and O_m:

$$\mathbf{H}_m = \mathbf{H}_k + \mathbf{\Omega} \times \mathbf{O}_k\mathbf{O}_m$$

Thus, the screw at a point O_i is well defined by the vectors $\mathbf{H}_i$ and $\mathbf{\Omega}$, which can be concatenated in a single (6x1) vector.

2.4.2. *Representation of velocity (kinematic screw)*

Since the set of velocity vectors at all the points of a body defines a screw field, the screw at a point O_i can be defined by:

- $\mathbf{V}_i$ representing the linear velocity at O_i with respect to the fixed frame R_0, such that $\mathbf{V}_i = \frac{d}{dt}(\mathbf{O}_0\mathbf{O}_i)$;
- ω_i representing the angular velocity of the body with respect to frame R_0. It constitutes the vector of the screw of the velocity vector field.

Thus, the velocity of a point O_j is calculated in terms of the velocity of the point O_i by the following equation:

$$\mathbf{V}_j = \mathbf{V}_i + \omega_i \times \mathbf{O}_i\mathbf{O}_j \qquad [2.40]$$

The components of $\mathbf{V}_i$ and ω_i can be concatenated to form the *kinematic screw* vector $\mathbb{V}_i$, i.e.:

$$\mathbb{V}_i = [\ \mathbf{V}_i^T \ \ \omega_i^T\ ]^T \qquad [2.41]$$

The kinematic screw is also called *twist* or *spatial velocity*.

2.4.3. *Transformation of screws*

Let ${}^i\mathbf{V}_i$ and ${}^i\omega_i$ be the vectors representing the kinematic screw in O_i, origin of frame R_i, expressed in frame R_i. To calculate ${}^j\mathbf{V}_j$ and ${}^j\omega_j$ representing the kinematic screw in O_j expressed in frame R_j, we first note that:

$$\omega_j = \omega_i \qquad [2.42]$$

$$\mathbf{V}_j = \mathbf{V}_i + \omega_i \times \mathbf{L}_{i,j} \qquad [2.43]$$

$\mathbf{L}_{i,j}$ being the position vector connecting O_i to O_j.

Equations [2.42] and [2.43] can be rewritten as:

$$\begin{bmatrix} \mathbf{V}_j \\ \omega_j \end{bmatrix} = \begin{bmatrix} \mathbf{I}_3 & -\hat{\mathbf{L}}_{i,j} \\ \mathbf{0}_3 & \mathbf{I}_3 \end{bmatrix} \begin{bmatrix} \mathbf{V}_i \\ \omega_i \end{bmatrix} \qquad [2.44]$$

where $\mathbf{I}_3$ and $\mathbf{0}_3$ represent the (3x3) identity matrix and zero matrix respectively. Projecting this relation in frame R_i, we obtain:

$$\begin{bmatrix} {}^i\mathbf{V}_j \\ {}^i\omega_j \end{bmatrix} = \begin{bmatrix} \mathbf{I}_3 & -{}^i\hat{\mathbf{P}}_j \\ \mathbf{0}_3 & \mathbf{I}_3 \end{bmatrix} \begin{bmatrix} {}^i\mathbf{V}_i \\ {}^i\omega_i \end{bmatrix} \qquad [2.45]$$

Since ${}^j\mathbf{V}_j = {}^j\mathbf{A}_i\,{}^i\mathbf{V}_j$ and ${}^j\omega_j = {}^j\mathbf{A}_i\,{}^i\omega_j$, equation [2.45] gives:

$${}^j\mathbb{V}_j = {}^j\mathbb{T}_i\,{}^i\mathbb{V}_i \qquad [2.46]$$

where ${}^j\mathbb{T}_i$ is the (6x6) transformation matrix between screws:

$${}^j\mathbb{T}_i = \begin{bmatrix} {}^j\mathbf{A}_i & -{}^j\mathbf{A}_i\,{}^i\hat{\mathbf{P}}_j \\ \mathbf{0}_3 & {}^j\mathbf{A}_i \end{bmatrix} \qquad [2.47]$$

The transformation matrices between screws have the following properties:

i) product:

$${}^0\mathbb{T}_j = {}^0\mathbb{T}_1\,{}^1\mathbb{T}_2 \ldots {}^{j-1}\mathbb{T}_j \qquad [2.48]$$

ii) inverse:

$${}^j\mathbb{T}_i^{-1} = \begin{bmatrix} {}^i\mathbf{A}_j & {}^i\hat{\mathbf{P}}_j\,{}^i\mathbf{A}_j \\ \mathbf{0}_3 & {}^i\mathbf{A}_j \end{bmatrix} = {}^i\mathbb{T}_j \qquad [2.49]$$

Note that equation [2.49] gives another possibility, other than equation [2.45], to define the transformation matrix between screws.

2.5. Differential translation and rotation of frames

The differential transformation of the position and orientation – or location – of a frame R_i attached to any body may be expressed by a differential translation vector $\mathbf{dP}_i$ expressing the translation of the origin of frame R_i, and of a differential rotation vector $\boldsymbol{\delta}_i$, equal to $\mathbf{u}_i\, d\theta$, representing the rotation of an angle $d\theta$ about an axis, with unit vector $\mathbf{u}_i$, passing through the origin O_i.

Given a transformation ${}^{i}\mathbf{T}_{j}$, the transformation ${}^{i}\mathbf{T}_{j} + \mathbf{d}^{i}\mathbf{T}_{j}$ can be calculated, taking into account the property h of § 2.3.6, by the premultiplication rule as:

$$ {}^{i}\mathbf{T}_{j} + \mathbf{d}^{i}\mathbf{T}_{j} = \mathbf{Trans}({}^{i}dx_{i}, {}^{i}dy_{i}, {}^{i}dz_{i})\ \mathbf{Rot}({}^{i}\mathbf{u}_{i}, d\theta)\ {}^{i}\mathbf{T}_{j} \qquad [2.50] $$

Thus, the differential of ${}^{i}\mathbf{T}_{j}$ is equal to:

$$ \mathbf{d}^{i}\mathbf{T}_{j} = [\mathbf{Trans}({}^{i}dx_{i}, {}^{i}dy_{i}, {}^{i}dz_{i})\ \mathbf{Rot}({}^{i}\mathbf{u}_{i}, d\theta) - \mathbf{I}_{4}]\ {}^{i}\mathbf{T}_{j} \qquad [2.51] $$

In the same way, the transformation ${}^{i}\mathbf{T}_{j} + \mathbf{d}^{i}\mathbf{T}_{j}$ can be calculated, using the postmultiplication rule as:

$$ {}^{i}\mathbf{T}_{j} + \mathbf{d}^{i}\mathbf{T}_{j} = {}^{i}\mathbf{T}_{j}\ \mathbf{Trans}({}^{j}dx_{j}, {}^{j}dy_{j}, {}^{j}dz_{j})\ \mathbf{Rot}({}^{j}\mathbf{u}_{j}, d\theta) \qquad [2.52] $$

and the differential of ${}^{i}\mathbf{T}_{j}$ becomes:

$$ \mathbf{d}^{i}\mathbf{T}_{j} = {}^{i}\mathbf{T}_{j}\ [\mathbf{Trans}({}^{j}dx_{j}, {}^{j}dy_{j}, {}^{j}dz_{j})\ \mathbf{Rot}({}^{j}\mathbf{u}_{j}, d\theta) - \mathbf{I}_{4}] \qquad [2.53] $$

From equations [2.51] and [2.53], the differential transformation matrix $\mathbf{\Delta}$ is defined as [Paul 81]:

$$ \mathbf{\Delta} = [\mathbf{Trans}(dx, dy, dz)\ \mathbf{Rot}(\mathbf{u}, d\theta) - \mathbf{I}_{4}] \qquad [2.54] $$

such that:

$$ \mathbf{d}^{i}\mathbf{T}_{j} = {}^{i}\mathbf{\Delta}\ {}^{i}\mathbf{T}_{j} \qquad [2.55] $$

or:

$$ \mathbf{d}^{i}\mathbf{T}_{j} = {}^{i}\mathbf{T}_{j}\ {}^{j}\mathbf{\Delta} \qquad [2.56] $$

Assuming that $d\theta$ is sufficiently small so that $S(d\theta) \approx d\theta$ and $C(d\theta) \approx 1$, the transformation matrix of a pure rotation $d\theta$ about an axis of unit vector $\mathbf{u}$ can be calculated from equations [2.30] and [2.54] as:

$$ {}^{j}\mathbf{\Delta} = \begin{bmatrix} {}^{j}\hat{\boldsymbol{\delta}}_{j} & {}^{j}\mathbf{dP}_{j} \\ 0\ \ 0\ \ 0 & 0 \end{bmatrix} = \begin{bmatrix} {}^{j}\hat{\mathbf{u}}_{j}\, d\theta & {}^{j}\mathbf{dP}_{j} \\ 0\ \ 0\ \ 0 & 0 \end{bmatrix} \qquad [2.57] $$

where $\hat{\mathbf{u}}$ and $\hat{\boldsymbol{\delta}}$ represent the skew-symmetric matrices defined by the vectors $\mathbf{u}$ and $\boldsymbol{\delta}$ respectively.

Note that the transformation matrix between screws can also be used to transform the differential translation and rotation vectors between frames:

$$\begin{bmatrix} {}^{j}\mathbf{dP}_j \\ {}^{j}\delta_j \end{bmatrix} = {}^{j}\mathbb{T}_i \begin{bmatrix} {}^{i}\mathbf{dP}_i \\ {}^{i}\delta_i \end{bmatrix} \qquad [2.58]$$

In a similar way as for the kinematic screw, we call the concatenation of $\mathbf{dP}_i$ and δ_i the *differential screw*.

• **Example 2.4.** Consider using the differential model of a robot to control its displacement. The differential model calculates the joint increments corresponding to the desired elementary displacement of frame R_n fixed to the terminal link (Figure 2.11). However, the task of the robot is often described in the tool frame R_E, which is also fixed to the terminal link. The problem is to calculate ${}^{n}\mathbf{dP}_n$ and ${}^{n}\delta_n$ in terms of ${}^{E}\mathbf{dP}_E$ and ${}^{E}\delta_E$.

Let the transformation describing the tool frame in frame R_n be:

$${}^{n}\mathbf{T}_E = \begin{bmatrix} 0 & 1 & 0 & 0 \\ -1 & 0 & 0 & 0.1 \\ 0 & 0 & 1 & -0.3 \\ 0 & 0 & 0 & 1 \end{bmatrix}$$

and that the value of the desired elementary displacement is:

$${}^{E}\mathbf{dP}_E = [\ 0 \quad 0 \quad -0.01\ ]^T,\ {}^{E}\delta_E = [\ 0 \quad -0.05 \quad 0\ ]^T$$

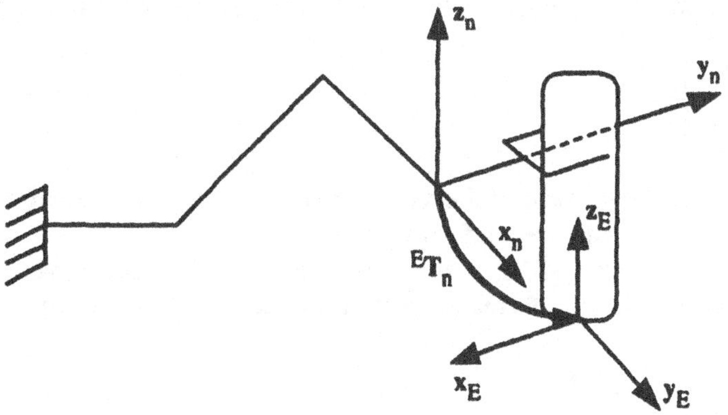

Figure 2.11. *Example 2.4*

Using equation [2.58], we obtain:

$$ {}^{n}\delta_n = {}^{n}\mathbf{A}_E \, {}^{E}\delta_E \, , \quad {}^{n}\mathbf{dP}_n = {}^{n}\mathbf{A}_E \, ({}^{E}\delta_E x \, {}^{E}\mathbf{P}_n + {}^{E}\mathbf{dP}_E) $$

The numerical application gives:

$$ {}^{n}\mathbf{dP}_n = [\ 0 \quad 0.015 \quad -0.005\ ]^T, \ {}^{n}\delta_n = [\ -0.05 \quad 0 \quad 0\ ]^T $$

In a similar way, we can evaluate the error in the location of the tool frame due to errors in the position and orientation of the terminal frame. Suppose that the position error is equal to 10 mm in all directions and that the rotation error is estimated as 0.01 radian about the **x** axis:

$$ {}^{n}\mathbf{dP}_n = [\ 0.01 \quad 0.01 \quad 0.01\ ]^T, \ {}^{n}\delta_n = [\ 0.01 \quad 0 \quad 0\ ]^T $$

The error on the tool frame is calculated by:

$$ {}^{E}\delta_E = {}^{E}\mathbf{A}_n \, {}^{n}\delta_n, \quad {}^{E}\mathbf{dP}_E = {}^{E}\mathbf{A}_n \, ({}^{n}\delta_n x \, {}^{n}\mathbf{P}_E + {}^{n}\mathbf{dP}_n) $$

which results in:

$$ {}^{E}\mathbf{dP}_E = [\ -0.013 \quad 0.01 \quad 0.011\ ]^T, \ {}^{E}\delta_E = [\ 0 \quad 0.01 \quad 0\ ]^T $$

2.6. Representation of forces (wrench)

A collection of forces and moments acting on a body can be reduced to a *wrench* $\mathbb{f}_i$ at point O_i, which is composed of a force $\mathbf{f}_i$ at O_i and a moment $\mathbf{m}_i$ about O_i:

$$ \mathbb{f}_i = \begin{bmatrix} \mathbf{f}_i \\ \mathbf{m}_i \end{bmatrix} \qquad [2.59] $$

Note that the vector field of the moments constitutes a screw where the vector of the screw is $\mathbf{f}_i$. Thus, the wrench forms a screw.

Consider a given wrench ${}^{i}\mathbb{f}_i$, expressed in frame R_i. For calculating the equivalent wrench ${}^{j}\mathbb{f}_j$, we use the transformation matrix between screws such that:

$$ \begin{bmatrix} {}^{j}\mathbf{m}_j \\ {}^{j}\mathbf{f}_j \end{bmatrix} = {}^{j}\mathbb{T}_i \begin{bmatrix} {}^{i}\mathbf{m}_i \\ {}^{i}\mathbf{f}_i \end{bmatrix} \qquad [2.60] $$

which gives:

$$^j\mathbf{f}_j = {}^j\mathbf{A}_i\, {}^i\mathbf{f}_i \qquad [2.61]$$
$$^j\mathbf{m}_j = {}^j\mathbf{A}_i\, ({}^i\mathbf{f}_i x\, {}^i\mathbf{P}_j + {}^i\mathbf{m}_i) \qquad [2.62]$$

It is often more practical to permute the order of $\mathbf{f}_i$ and $\mathbf{m}_i$. In this case, equation [2.60] becomes:

$$\begin{bmatrix} {}^j\mathbf{f}_j \\ {}^j\mathbf{m}_j \end{bmatrix} = {}^i\mathbb{T}_j^{T} \begin{bmatrix} {}^i\mathbf{f}_i \\ {}^i\mathbf{m}_i \end{bmatrix} \qquad [2.63]$$

• **Example 2.5.** Let the transformation matrix $^n\mathbf{T}_E$ describing the location of the tool frame with respect to the terminal frame be:

$$^n\mathbf{T}_E = \begin{bmatrix} 0 & 1 & 0 & 0 \\ -1 & 0 & 0 & 0.1 \\ 0 & 0 & 1 & 0.5 \\ 0 & 0 & 0 & 1 \end{bmatrix}$$

Supposing that we want to exert a wrench $^E\mathbb{f}_E$ with this tool such that $^E\mathbf{f}_E = [0\ \ 0\ \ 5]^T$ and $^E\mathbf{m}_E = [0\ \ 0\ \ 3]^T$, determine the corresponding wrench $^n\mathbb{f}_n$ at the origin O_n and referred to frame R_n. Using equations [2.61] and [2.62], it follows that:

$$^n\mathbf{f}_n = {}^n\mathbf{A}_E\, {}^E\mathbf{f}_E$$
$$^n\mathbf{m}_n = {}^n\mathbf{A}_E\, ({}^E\mathbf{f}_E x\, {}^E\mathbf{P}_n + {}^E\mathbf{m}_E)$$

The numerical application leads to:

$$^n\mathbf{f}_n = [\ 0\ \ 0\ \ 0.5\ ]^T$$
$$^n\mathbf{m}_n = [\ 0.5\ \ 0\ \ 3\ ]^T$$

2.7. Conclusion

In the first part of this chapter, we have developed the homogeneous transformation matrix. This notation constitutes the basic tool for the modeling of robots and their environment. Other techniques have been used in robotics: quaternion [Yang 66], [Castelain 86], (3x3) rotation matrices [Coiffet 81] and the Rodrigues formulation [Wang 83]. Readers interested in these techniques can consult the given references.

We have also recalled some definitions about screws, and transformation matrices between screws, as well as differential transformations. These concepts will be used extensively in this book. In the following chapter, we deal with the problem of robot modeling.

Chapter 3

Direct geometric model of serial robots

3.1. Introduction

The design and control of a robot requires the computation of some mathematical models such as:

- transformation models between the joint space (in which the configuration of the robot is defined) and the task space (in which the location of the end-effector is specified). These transformation models are very important since robots are controlled in the joint space, whereas tasks are defined in the task space. Two classes of models are considered:
  - direct and inverse geometric models, which give the location of the end-effector as a function of the joint variables of the mechanism and vice versa;
  - direct and inverse kinematic models, which give the velocity of the end-effector as a function of the joint velocities and vice versa;
- dynamic models giving the relations between the input torques or forces of the actuators and the positions, velocities and accelerations of the joints.

The automatic symbolic computation of these models has largely been addressed in the literature [Dillon 73], [Khalil 76], [Zabala 78], [Kreuzer 79], [Aldon 82], [Cesareo 84], [Megahed 84], [Murray 84], [Kircánski 85], [Burdick 86], [Izaguirre 86], [Khalil 89a]. The algorithms presented in this book have been used in the development of the software package SYMORO+ [Khalil 97], which deals with all the above-mentioned models.

The modeling of robots in a systematic and automatic way requires an adequate method for the description of their structure. Several methods and notations have been proposed [Denavit 55], [Sheth 71], [Renaud 75], [Khalil 76], [Borrel 79],

[Craig 86a]. The most popular among these is the Denavit-Hartenberg method [Denavit 55]. This method is developed for serial structures and presents ambiguities when applied to robots with closed or tree chains. For this reason, we will use the notation of Khalil and Kleinfinger [Khalil 86a], which gives a unified description for all mechanical articulated systems with a minimum number of parameters.

In this chapter, we will present the geometric description and the direct geometric model of serial robots. Tree and closed loop structures will be covered in Chapter 7.

3.2. Description of the geometry of serial robots

A serial robot is composed of a sequence of n + 1 links and n joints. The links are assumed to be perfectly rigid. The joints are either revolute or prismatic and are assumed to be ideal (no backlash, no elasticity). A complex joint can be represented by an equivalent combination of revolute and prismatic joints with zero-length massless links. The links are numbered such that link 0 constitutes the base of the robot and link n is the terminal link (Figure 3.1). Joint j connects link j to link j – 1 and its variable is denoted q_j. In order to define the relationship between the location of links, we assign a frame R_j attached to each link j, such that:

- the $\mathbf{z}_j$ axis is along the axis of joint j;
- the $\mathbf{x}_j$ axis is aligned with the common normal between $\mathbf{z}_j$ and $\mathbf{z}_{j+1}$. If $\mathbf{z}_j$ and $\mathbf{z}_{j+1}$ are parallel or collinear, the choice of $\mathbf{x}_j$ is not unique. The intersection of $\mathbf{x}_j$ and $\mathbf{z}_j$ defines the origin O_j. In the case of intersecting joint axes, the origin is at the point of intersection of the joint axes;
- the $\mathbf{y}_j$ axis is formed by the right-hand rule to complete the coordinate system $(\mathbf{x}_j, \mathbf{y}_j, \mathbf{z}_j)$.

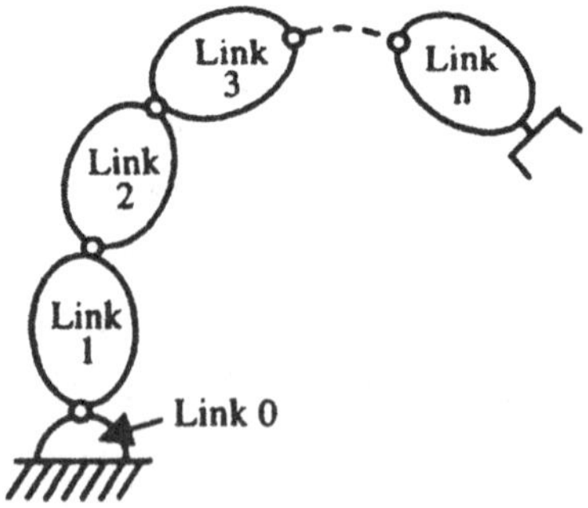

Figure 3.1. *Robot with simple open structure*

The transformation matrix from frame R_{j-1} to frame R_j is expressed as a function of the following four geometric parameters (Figure 3.2):

- α_j: the angle between $\mathbf{z}_{j-1}$ and $\mathbf{z}_j$ about $\mathbf{x}_{j-1}$;
- d_j: the distance between $\mathbf{z}_{j-1}$ and $\mathbf{z}_j$ along $\mathbf{x}_{j-1}$;
- θ_j: the angle between $\mathbf{x}_{j-1}$ and $\mathbf{x}_j$ about $\mathbf{z}_j$;
- r_j: the distance between $\mathbf{x}_{j-1}$ and $\mathbf{x}_j$ along $\mathbf{z}_j$.

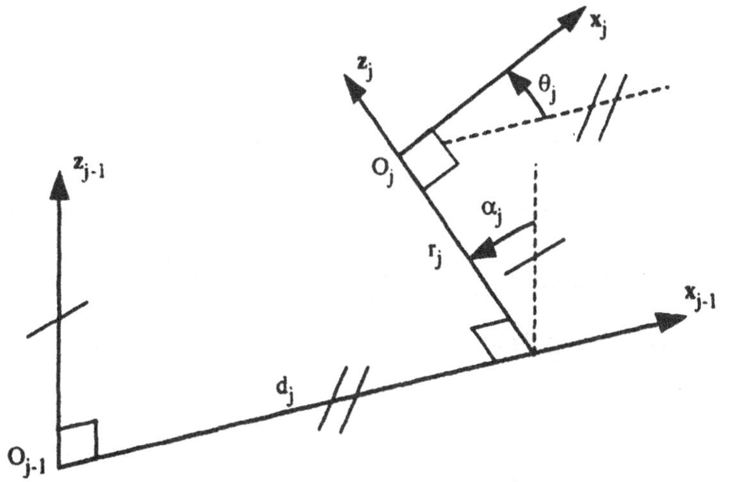

Figure 3.2. *The geometric parameters in the case of a simple open structure*

The variable of joint j, defining the relative orientation or position between links $j-1$ and j, is either θ_j or r_j, depending on whether the joint is revolute or prismatic respectively. This is defined by the relation:

$$q_j = \bar{\sigma}_j \theta_j + \sigma_j r_j \qquad [3.1a]$$

with:

- $\sigma_j = 0$ if joint j is revolute;
- $\sigma_j = 1$ if joint j is prismatic;
- $\bar{\sigma}_j = 1 - \sigma_j$.

By analogy, we define the parameter $\bar{q}_j$ by:

$$\bar{q}_j = \sigma_j \theta_j + \bar{\sigma}_j r_j \qquad [3.1b]$$

The transformation matrix defining frame R_j relative to frame R_{j-1} is given as (Figure 3.2):

$$^{j-1}\mathbf{T}_j = \mathbf{Rot}(\mathbf{x}, \alpha_j)\ \mathbf{Trans}(\mathbf{x}, d_j)\ \mathbf{Rot}(\mathbf{z}, \theta_j)\ \mathbf{Trans}(\mathbf{z}, r_j)$$

$$= \begin{bmatrix} C\theta_j & -S\theta_j & 0 & d_j \\ C\alpha_j S\theta_j & C\alpha_j C\theta_j & -S\alpha_j & -r_j S\alpha_j \\ S\alpha_j S\theta_j & S\alpha_j C\theta_j & C\alpha_j & r_j C\alpha_j \\ 0 & 0 & 0 & 1 \end{bmatrix} \qquad [3.2]$$

We note that the (3x3) rotation matrix $^{j-1}\mathbf{A}_j$ can be obtained as:

$$^{j-1}\mathbf{A}_j = \mathbf{rot}(\mathbf{x}, \alpha_j)\ \mathbf{rot}(\mathbf{z}, \theta_j) \qquad [3.3]$$

The transformation matrix defining frame R_{j-1} relative to frame R_j is given as:

$$^{j}\mathbf{T}_{j-1} = \mathbf{Trans}(\mathbf{z}, -r_j)\ \mathbf{Rot}(\mathbf{z}, -\theta_j)\ \mathbf{Trans}(\mathbf{x}, -d_j)\ \mathbf{Rot}(\mathbf{x}, -\alpha_j)$$

$$= \begin{bmatrix} & & & -d_j C\theta_j \\ & ^{j-1}\mathbf{A}_j^T & & d_j S\theta_j \\ & & & -r_j \\ 0 & 0 & 0 & 1 \end{bmatrix} \qquad [3.4]$$

NOTES.–

- the frame R_0 is chosen to be aligned with frame R_1 when $q_1 = 0$. This means that $\mathbf{z}_0$ is aligned with $\mathbf{z}_1$, whereas the origin O_0 is coincident with the origin O_1 if joint 1 is revolute, and $\mathbf{x}_0$ is parallel to $\mathbf{x}_1$ if joint 1 is prismatic. This choice makes $\alpha_1 = 0$, $d_1 = 0$ and $\bar{q}_1 = 0$;
- in a similar way, the choice of the $\mathbf{x}_n$ axis to be aligned with $\mathbf{x}_{n-1}$ when $q_n = 0$ makes $\bar{q}_n = 0$;
- if joint j is prismatic, the $\mathbf{z}_j$ axis must be taken to be parallel to the joint axis but can have any position in space. So, we place it in such a way that $d_j = 0$ or $d_{j+1} = 0$;
- if $\mathbf{z}_j$ is parallel to $\mathbf{z}_{j+1}$, we place $\mathbf{x}_j$ in such a way that $r_j = 0$ or $r_{j+1} = 0$;
- assuming that each joint is driven by an independent actuator, the vector of joint variables $\mathbf{q}$ can be obtained from the vector of encoder readings $\mathbf{q}_c$ using the relation:

$$\mathbf{q} = \mathbf{K}\,\mathbf{q}_c + \mathbf{q}_0$$

where **K** is an (n×n) constant matrix and $\mathbf{q}_0$ is an offset vector representing the robot configuration when $\mathbf{q}_c = \mathbf{0}$;

- if a chain contains two or more consecutive parallel joints, the transformation matrices between them can be reduced to one equivalent transformation matrix using the sum of the joint variables. For example, if $\alpha_{j+1} = 0$, i.e. if $\mathbf{z}_j$ and $\mathbf{z}_{j+1}$ are parallel, the transformation ${}^{j-1}\mathbf{T}_{j+1}$ is written as:

$${}^{j-1}\mathbf{T}_{j+1} = {}^{j-1}\mathbf{T}_j\,{}^{j}\mathbf{T}_{j+1} = \mathbf{Rot}(\mathbf{x}, \alpha_j)\,\mathbf{Trans}(\mathbf{x}, d_j)\,\mathbf{Rot}(\mathbf{z}, \theta_j)\,\mathbf{Trans}(\mathbf{z}, r_j)\,\mathbf{Trans}(\mathbf{x}, d_{j+1})\,\mathbf{Rot}(\mathbf{z}, \theta_{j+1})\,\mathbf{Trans}(\mathbf{z}, r_{j+1}) \quad [3.5]$$

$$= \begin{bmatrix} C(\theta_j+\theta_{j+1}) & -S(\theta_j+\theta_{j+1}) & 0 & d_j+d_{j+1}C\theta_j \\ C\alpha_j S(\theta_j+\theta_{j+1}) & C\alpha_j C(\theta_j+\theta_{j+1}) & -S\alpha_j & d_{j+1}C\alpha_j S\theta_j-(r_j+r_{j+1})S\alpha_j \\ S\alpha_j S(\theta_j+\theta_{j+1}) & S\alpha_j C(\theta_j+\theta_{j+1}) & C\alpha_j & d_{j+1}S\alpha_j S\theta_j+(r_j+r_{j+1})C\alpha_j \\ 0 & 0 & 0 & 1 \end{bmatrix}$$

and the inverse transformation has the expression:

$${}^{j+1}\mathbf{T}_{j-1} = \begin{bmatrix} & & & -d_jC(\theta_j+\theta_{j+1})-d_{j+1}C\theta_{j+1} \\ & {}^{j-1}\mathbf{A}^{T}_{j+1} & & d_jS(\theta_j+\theta_{j+1})+d_{j+1}S\theta_{j+1} \\ & & & -(r_j+r_{j+1}) \\ 0 & 0 & 0 & 1 \end{bmatrix} \quad [3.6]$$

The above expressions contain terms in $(\theta_j + \theta_{j+1})$ and $(r_j + r_{j+1})$. This result can be generalized for the case of multiple consecutive parallel axes [Kleinfinger 86a].

• **Example 3.1.** Geometric description of the Stäubli RX-90 robot (Figure 3.3a). The shoulder is of RRR type and the wrist has three revolute joints whose axes intersect at a point (Figure 3.3b). From a methodological point of view, we first place the $\mathbf{z}_j$ axes on the joint axes, then the $\mathbf{x}_j$ axes according to the previously mentioned conventions. Then, we determine the geometric parameters defining each frame R_j with respect to frame R_{j-1}. The link coordinate frames are indicated in Figure 3.3b and the geometric parameters are given in Table 3.1.

Table 3.1. *Geometric parameters of the Stäubli RX-90 robot*

j	σ_j	α_j	d_j	θ_j	r_j
1	0	0	0	θ_1	0
2	0	$\pi/2$	0	θ_2	0
3	0	0	D3	θ_3	0
4	0	$-\pi/2$	0	θ_4	RL4
5	0	$\pi/2$	0	θ_5	0
6	0	$-\pi/2$	0	θ_6	0

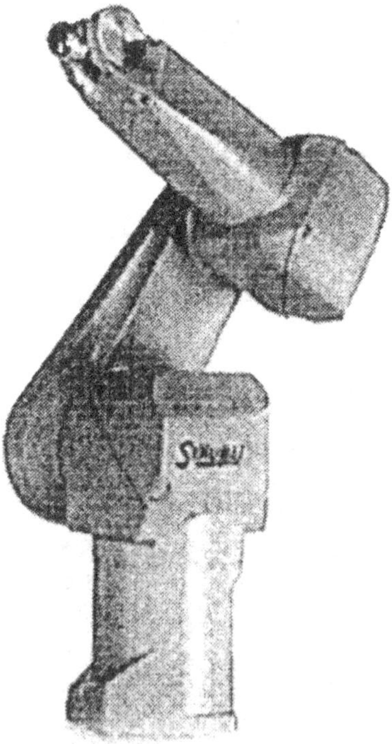

Figure 3.3a. *General view of the Stäubli RX-90 robot*
(Courtesy of Stäubli company)

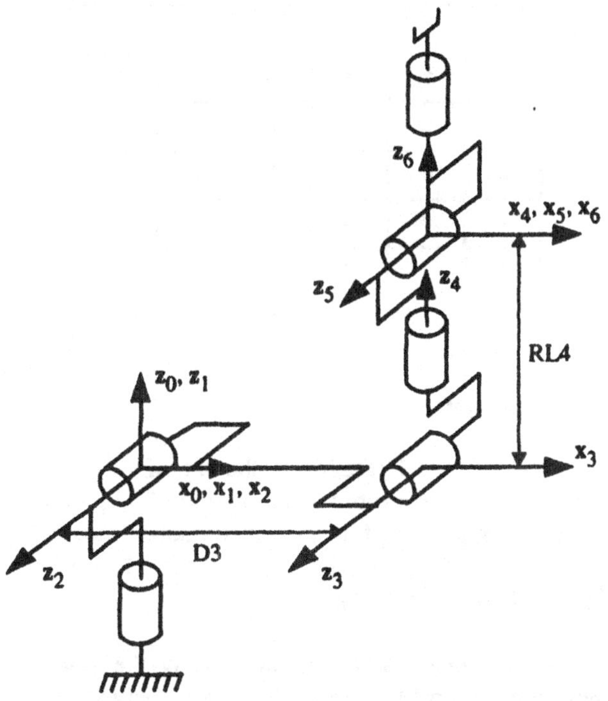

Figure 3.3b. *Link coordinate frames for the Stäubli RX-90 robot*

• **Example 3.2.** Geometric description of a SCARA robot (Figure 3.4). The geometric parameters of a four degree-of-freedom SCARA robot are given in Table 3.2.

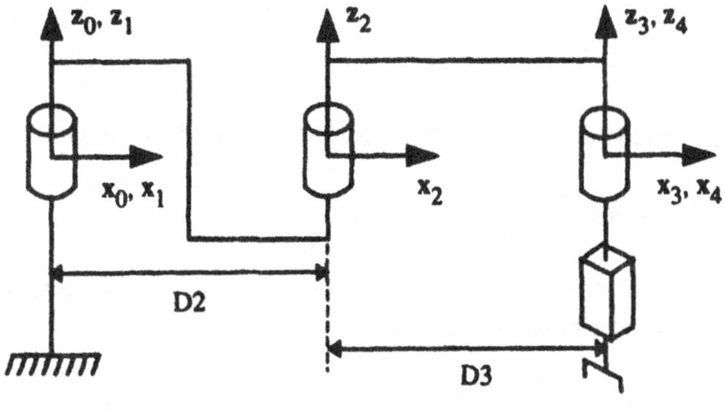

Figure 3.4. *SCARA Robot*

Table 3.2. *Geometric parameters of a SCARA robot*

j	σ_j	α_j	d_j	θ_j	r_j
1	0	0	0	θ_1	0
2	0	0	D2	θ_2	0
3	0	0	D3	θ_3	0
4	1	0	0	0	r_4

3.3. Direct geometric model

The Direct Geometric Model (DGM) is the set of relations that defines the location of the end-effector of the robot as a function of its joint coordinates. For a serial structure, it may be represented by the transformation matrix ${}^0\mathbf{T}_n$ as:

$$ {}^0\mathbf{T}_n = {}^0\mathbf{T}_1(q_1)\, {}^1\mathbf{T}_2(q_2)\ \dots\ {}^{n-1}\mathbf{T}_n(q_n) \qquad [3.7] $$

This relation can be numerically computed using the general transformation matrix ${}^{j-1}\mathbf{T}_j$ given by equation [3.2], or symbolically derived after substituting the values of the constant geometric parameters in the transformation matrices (Example 3.3). The symbolic method needs less computational operations.

The direct geometric model of a robot may also be represented by the relation:

$$ \mathbf{X} = \mathbf{f}(\mathbf{q}) \qquad [3.8] $$

where $\mathbf{q}$ is the vector of joint variables such that:

$$ \mathbf{q} = [q_1 \quad q_2 \dots q_n]^T \qquad [3.9] $$

The position and orientation of the terminal link are defined as:

$$ \mathbf{X} = [x_1 \quad x_2 \dots x_m]^T \qquad [3.10] $$

There are several possibilities of defining the vector $\mathbf{X}$ as we will see in § 3.6. For example, with the elements of the matrix ${}^0\mathbf{T}_n$:

$$ \mathbf{X} = [P_x \quad P_y \quad P_z \quad s_x \quad s_y \quad s_z \quad n_x \quad n_y \quad n_z \quad a_x \quad a_y \quad a_z]^T \qquad [3.11] $$

Taking into account that $\mathbf{s} = \mathbf{n} \times \mathbf{a}$, we can also take:

$$\mathbf{X} = [P_x \quad P_y \quad P_z \quad n_x \quad n_y \quad n_z \quad a_x \quad a_y \quad a_z]^T \qquad [3.12]$$

• **Example 3.3.** Symbolic direct geometric model of the Stäubli RX-90 robot (Figure 3.3). From Table 3.1 and using equation [3.2], we write the elementary transformation matrices $^{j-1}\mathbf{T}_j$ as:

$$^0\mathbf{T}_1 = \begin{bmatrix} C1 & -S1 & 0 & 0 \\ S1 & C1 & 0 & 0 \\ 0 & 0 & 1 & 0 \\ 0 & 0 & 0 & 1 \end{bmatrix}, \; ^1\mathbf{T}_2 = \begin{bmatrix} C2 & -S2 & 0 & 0 \\ 0 & 0 & -1 & 0 \\ S2 & C2 & 0 & 0 \\ 0 & 0 & 0 & 1 \end{bmatrix}, \; ^2\mathbf{T}_3 = \begin{bmatrix} C3 & -S3 & 0 & D3 \\ S3 & C3 & 0 & 0 \\ 0 & 0 & 1 & 0 \\ 0 & 0 & 0 & 1 \end{bmatrix}$$

Since the joint axes 2 and 3 are parallel, we can write the transformation matrix $^1\mathbf{T}_3$ using equation [3.5] as:

$$^1\mathbf{T}_3 = \begin{bmatrix} C23 & -S23 & 0 & C2D3 \\ 0 & 0 & -1 & 0 \\ S23 & C23 & 0 & S2D3 \\ 0 & 0 & 0 & 1 \end{bmatrix}$$

with $C23 = \cos(\theta_2 + \theta_3)$ and $S23 = \sin(\theta_2 + \theta_3)$.

$$^3\mathbf{T}_4 = \begin{bmatrix} C4 & -S4 & 0 & 0 \\ 0 & 0 & 1 & RL4 \\ -S4 & -C4 & 0 & 0 \\ 0 & 0 & 0 & 1 \end{bmatrix}, \; ^4\mathbf{T}_5 = \begin{bmatrix} C5 & -S5 & 0 & 0 \\ 0 & 0 & -1 & 0 \\ S5 & C5 & 0 & 0 \\ 0 & 0 & 0 & 1 \end{bmatrix}, \; ^5\mathbf{T}_6 = \begin{bmatrix} C6 & -S6 & 0 & 0 \\ 0 & 0 & 1 & 0 \\ -S6 & -C6 & 0 & 0 \\ 0 & 0 & 0 & 1 \end{bmatrix}$$

In order to compute $^0\mathbf{T}_6$, it is better to multiply the matrices $^{j-1}\mathbf{T}_j$ starting from the last transformation matrix and working back to the base, mainly for two reasons:

- the intermediate matrices $^j\mathbf{T}_6$, denoted as $\mathbf{U}_j$, will be used to obtain the inverse geometric model (Chapter 4);
- this reduces the number of operations (additions and multiplications) of the model.

We thus compute successively $\mathbf{U}_j$ for j = 5, ..., 0:

$$\mathbf{U}_5 = {}^5\mathbf{T}_6$$

$$U_4 = {}^4T_6 = {}^4T_5\, U_5 = \begin{bmatrix} C5C6 & -C5S6 & -S5 & 0 \\ S6 & C6 & 0 & 0 \\ S5C6 & -S5S6 & C5 & 0 \\ 0 & 0 & 0 & 1 \end{bmatrix}$$

$$U_3 = {}^3T_6 = {}^3T_4\, U_4 = \begin{bmatrix} C4C5C6-S4S6 & -C4C5S6-S4C6 & -C4S5 & 0 \\ S5C6 & -S5S6 & C5 & RL4 \\ -S4C5C6-C4S6 & S4C5S6-C4C6 & S4S5 & 0 \\ 0 & 0 & 0 & 1 \end{bmatrix}$$

$$U_2 = {}^2T_6 = {}^2T_3\, U_3$$

The **s**, **n**, **a**, **P** vectors of $\mathbf{U}_2$ are:

$s_x = C3(C4C5C6 - S4S6) - S3S5C6$
$s_y = S3(C4C5C6 - S4S6) + C3S5C6$
$s_z = - S4C5C6 - C4S6$
$n_x = - C3(C4C5S6 + S4C6) + S3S5S6$
$n_y = - S3(C4C5S6 + S4C6) - C3S5S6$
$n_z = S4C5S6 - C4C6$
$a_x = - C3C4S5 - S3C5$
$a_y = - S3C4S5 + C3C5$
$a_z = S4S5$
$P_x = - S3RL4 + D3$
$P_y = C3RL4$
$P_z = 0$

$$U_1 = {}^1T_6 = {}^1T_2\, U_2 = {}^1T_3\, U_3$$

The corresponding **s**, **n**, **a**, **P** vectors are:

$s_x = C23(C4C5C6 - S4S6) - S23S5C6$
$s_y = S4C5C6 + C4S6$
$s_z = S23(C4C5C6 - S4S6) + C23S5C6$
$n_x = - C23(C4C5S6 + S4C6) + S23S5S6$
$n_y = - S4C5S6 + C4C6$
$n_z = - S23(C4C5S6 + S4C6) - C23S5S6$
$a_x = - C23C4S5 - S23C5$
$a_y = - S4S5$
$a_z = - S23C4S5 + C23C5$
$P_x = - S23\, RL4 + C2D3$

$P_y = 0$
$P_z = C23\ RL4 + S2D3$

Finally:

$$U_0 = {}^0T_6 = {}^0T_1\ U_1$$

The corresponding **s**, **n**, **a**, **P** vectors are:

$s_x = C1(C23(C4C5C6 - S4S6) - S23S5C6) - S1(S4C5C6 + C4S6)$
$s_y = S1(C23(C4C5C6 - S4S6) - S23S5C6) + C1(S4C5C6 + C4S6)$
$s_z = S23(C4C5C6 - S4S6) + C23S5C6$
$n_x = C1(- C23\ (C4C5S6 + S4C6) + S23S5S6) + S1(S4C5S6 - C4C6)$
$n_y = S1(- C23\ (C4C5S6 + S4C6) + S23S5S6) - C1(S4C5S6 - C4C6)$
$n_z = - S23(C4C5S6 + S4C6) - C23S5S6$
$a_x = - C1(C23C4S5 + S23C5) + S1S4S5$
$a_y = - S1(C23C4S5 + S23C5) - C1S4S5$
$a_z = - S23C4S5 + C23C5$
$P_x = - C1(S23\ RL4 - C2D3)$
$P_y = - S1(S23\ RL4 - C2D3)$
$P_z = C23\ RL4 + S2D3$

3.4. Optimization of the computation of the direct geometric model

The control of a robot manipulator requires fast computation of its different models. An efficient method to reduce the computation time is to generate a symbolic customized model for each specific robot. To obtain this model, we expand the matrix multiplications to transform them into scalar equations. Each element of a matrix containing at least one mathematical operation is replaced by an intermediate variable. This variable is written in the output file that contains the customized model. The elements that do not contain any operation are kept without modification. We propagate the matrix obtained in the subsequent equations. Consequently, customizing eliminates multiplications by one and zero, and additions with zero. Moreover, if the robot has two or more successive revolute joints with parallel axes, it is more interesting to replace the corresponding product of matrices by a single matrix, which is calculated using equation [3.5]. We can also compute ${}^0\mathbf{s}_n$ using the vector product (${}^0\mathbf{n}_n \times {}^0\mathbf{a}_n$). In this case, the multiplication of the transformation matrices from the end-effector to the base saves the computation of the vectors ${}^j\mathbf{s}_n$ of the intermediate matrices ${}^j\mathbf{T}_n$, (j = n, …, 1).

• **Example 3.4.** Direct geometric model of the Stäubli RX-90 robot using the customized symbolic method.

a) computation of all the elements (s, n, a, P)

We denote Tijrs as the element (r, s) of the matrix $^{i}\mathbf{T}_{j}$. As in Example 3.3, the product of the matrices is carried out starting from the last transformation matrix. We obtain the following intermediate variables for the matrix $^{4}\mathbf{T}_{6}$:

T4611 = C5 C6
T4612 = – C5 S6
T4631 = S5 C6
T4632 = – S5 S6

Proceeding in the same way, the other intermediate variables are written as:

T3611 = C4 T4611 – S4 S6
T3612 = C4 T4612 – S4 C6
T3613 = – C4 S5
T3631 = – S4 T4611 – C4 S6
T3632 = – S4 T4612 – C4 C6
T3633 = S4 S5
T1314 = D3 C2
T1334 = D3 S2
T1611 = C23 T3611 – S23 T4631
T1612 = C23 T3612 – S23 T4632
T1613 = C23 T3613 – S23 C5
T1614 = – S23 RL4 + T1314
T1631 = S23 T3611 + C23 T4631
T1632 = S23 T3612 + C23 T4632
T1633 = S23 T3613 + C23 C5
T1634 = C23 RL4 + T1334
T0611 = C1 T1611 + S1 T3631
T0612 = C1 T1612 + S1 T3632
T0613 = C1 T1613 + S1 T3633
T0614 = C1 T1614
T0621 = S1 T1611 – C1 T3631
T0622 = S1 T1612 – C1 T3632
T0623 = S1 T1613 – C1 T3633
T0624 = S1 T1614
T0631 = T1631
T0632 = T1632
T0633 = T1633
T0634 = T1634

Total number of operations: 44 multiplications and 18 additions

b) computing only the columns (n, a, P)

T4612 = – C5 S6
T4632 = – S5 S6
T3612 = C4 T4612 – S4 C6
T3613 = – C4 S5
T3632 = – S4 T4612 – C4 C6
T3633 = S4 S5
T1314 = D3 C2
T1334 = D3 S2
T1612 = C23 T3612 – S23 T4632
T1613 = C23 T3613 – S23 C5
T1614 = – S23 RL4 + T1314
T1632 = S23 T3612 + C23 T4632
T1633 = S23 T3613 + C23 C5
T1634 = C23 RL4 + T1334
T0612 = C1 T1612 + S1 T3632
T0613 = C1 T1613 + S1 T3633
T0614 = C1 T1614
T0622 = S1 T1612 – C1 T3632
T0623 = S1 T1613 – C1 T3633
T0624 = S1 T1614
T0632 = T1632
T0633 = T1633
T0634 = T1634

Total number of operations: 30 multiplications and 12 additions

These equations constitute a complete direct geometric model. However, the computation of ${}^0s_6$ requires six multiplications and three additions corresponding to the vector product (${}^0\mathbf{n}_6 \times {}^0\mathbf{a}_6$).

3.5. Transformation matrix of the end-effector in the world frame

The robot is a component among others in a robotic workcell. It is generally associated with fastening devices, sensors..., and eventually with other robots. Consequently, we have to define a reference world frame R_f, which may be different than the base reference frame R_0 of the robot (Figure 3.5). The transformation matrix defining R_0 with reference to R_f will be denoted as $\mathbf{Z} = {}^f\mathbf{T}_0$.

Moreover, very often, a robot is not intended to perform a single operation at the workcell: it has interchangeable different tools. In order to facilitate the programming of the task, it is more practical to define one or more functional frames, called *tool frames* for each tool. We denote $\mathbf{E} = {}^n\mathbf{T}_E$ as the transformation matrix defining the tool frame with respect to the terminal link frame.

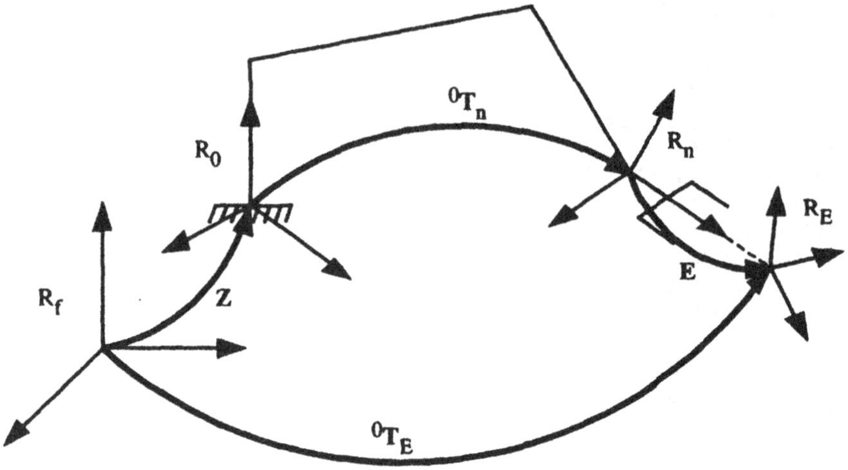

Figure 3.5. *Transformations between the end-effector and the world frame*

Thus, the transformation matrix ^fT_E can be written as:

$$^fT_E = Z\,{}^0T_n(q)\,E \quad [3.13]$$

In most programming languages, the user can specify **Z** and **E**.

3.6. Specification of the orientation

Previously, we have used the elements of the matrix ^0T_n to represent the position and orientation of the end-effector in frame R_0. This means the use of the Cartesian coordinates to describe the position:

$$^0P_n = [\ P_x\ \ P_y\ \ P_z\ ]^T \quad [3.14]$$

and the use of the direction cosine matrix for the orientation:

$$^0A_n = [\ {}^0s_n\ \ {}^0n_n\ \ {}^0a_n\ ] \quad [3.15]$$

Practically, all the robot manufacturers make use of the Cartesian coordinates for the position even though the cylindrical or spherical representations could appear to be more judicious for some structures of robots.

Other representations may be used for the orientation, for example: Euler angles for CINCINNATI-T3 robots and PUMA robots, Roll-Pitch-Yaw (RPY) angles for

ACMA robots, Euler parameters for ABB robots. In this section, we will show how to obtain the direction cosines **s**, **n**, **a** from the other representations and vice versa. Note that the orientation requires three independent parameters, thus the representation is redundant when it uses more than that.

3.6.1. *Euler angles*

The orientation of frame R_n expressed in frame R_0 is determined by specifying three angles, ϕ, θ and ψ, corresponding to a sequence of three rotations (Figure 3.6). The plane $(\mathbf{x}_n, \mathbf{y}_n)$ intersects the plane $(\mathbf{x}_0, \mathbf{y}_0)$ following the straight line ON, which is perpendicular to $\mathbf{z}_0$ and $\mathbf{z}_n$. The positive direction is given by the vector product $\mathbf{a}_0 \times \mathbf{a}_n$. The Euler angles are defined as:

- ϕ: angle between $\mathbf{x}_0$ and ON about $\mathbf{z}_0$, with $0 \le \phi < 2\pi$;
- θ : angle between $\mathbf{z}_0$ and $\mathbf{z}_n$ about ON, with $0 \le \theta \le \pi$;
- ψ: angle between ON and $\mathbf{x}_n$ about $\mathbf{z}_0$, with $0 \le \psi < 2\pi$.

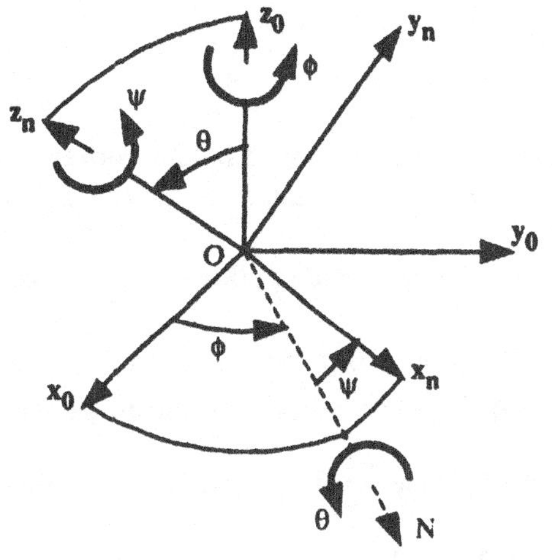

Figure 3.6. *Euler angles (z, x, z representation)*

The orientation matrix is given by:

$$
{}^0\mathbf{A}_n = \mathbf{rot}(\mathbf{z}, \phi)\ \mathbf{rot}(\mathbf{x}, \theta)\ \mathbf{rot}(\mathbf{z}, \psi)
$$

$$
= \begin{bmatrix} C\phi C\psi - S\phi C\theta S\psi & -C\phi S\psi - S\phi C\theta C\psi & S\phi S\theta \\ S\phi C\psi + C\phi C\theta S\psi & -S\phi S\psi + C\phi C\theta C\psi & -C\phi S\theta \\ S\theta S\psi & S\theta C\psi & C\theta \end{bmatrix} \quad [3.16]
$$

Inverse problem: expression of the Euler angles as functions of the direction cosines. Premultiplying equation [3.16] by **rot(z, –φ)**, we obtain [Paul 81]:

$$\mathbf{rot}(\mathbf{z}, -\phi)\ {}^{0}\mathbf{A}_{n} = \mathbf{rot}(\mathbf{x}, \theta)\ \mathbf{rot}(\mathbf{z}, \psi) \qquad [3.17]$$

Using relations [3.15] and [3.17] yields:

$$\begin{bmatrix} C\phi s_x+S\phi s_y & C\phi n_x+S\phi n_y & C\phi a_x+S\phi a_y \\ -S\phi s_x+C\phi s_y & -S\phi n_x+C\phi n_y & -S\phi a_x+C\phi a_y \\ s_z & n_z & a_z \end{bmatrix} = \begin{bmatrix} C\psi & -S\psi & 0 \\ C\theta S\psi & C\theta C\psi & -S\theta \\ S\theta S\psi & S\theta C\psi & C\theta \end{bmatrix} \qquad [3.18]$$

Equating the (1, 3) elements of both sides, we obtain:

$$C\phi\ a_x + S\phi\ a_y = 0$$

which gives:

$$\begin{cases} \phi = \text{atan2}(-a_x, a_y) \\ \phi' = \text{atan2}(a_x, -a_y) = \phi + \pi \end{cases} \qquad [3.19]$$

NOTE.– atan2 is a mathematical function (Matlab, Fortran, ...), which provides the arc tangent function from its two arguments. This function has the following characteristics:

- examining the sign of both a_x and a_y allows us to uniquely determine the angle ϕ such that $-\pi \leq \phi < \pi$;
- the accuracy of this function is uniform over its full range of definition;
- when $a_x = 0$, $a_y = 0$, $a_z = \pm 1$ the angle ϕ is undefined (singularity).

Using the (2, 3) and (3, 3) elements of equation [3.18], we obtain:

$$\theta = \text{atan2}(S\phi\ a_x - C\phi\ a_y, a_z) \qquad [3.20]$$

We proceed in a similar way to calculate ψ using the (1, 1) and (1, 2) elements:

$$\psi = \text{atan2}(-C\phi\ n_x - S\phi\ n_y, C\phi\ s_x + S\phi\ s_y) \qquad [3.21]$$

When a_x and a_y are zero, the axes z_n and z_0 are aligned, thus θ is zero or π. This situation corresponds to the singular case: the rotations ϕ and ψ are about the same axis and we can only determine their sum or difference. For example, when $a_z = 1$, we obtain:

$$^0\mathbf{A}_n = \mathbf{rot}(\mathbf{z}, \psi + \phi)$$

and from this, we deduce:

$$\psi + \phi = \text{atan2}(-n_x, n_y) \quad [3.22]$$

NOTE.– The Euler angles adopted here correspond to a (**z**, **x**, **z**) representation where the first rotation is about $\mathbf{z}_0$, followed by a rotation about the new **x** axis, followed by a last rotation about the new **z** axis. Some authors prefer the (**z**, **y**, **z**) representation [Paul 81]. A specific but interesting case can be encountered in the PUMA robot controller [Lee 83], [Dombre 88a] where an initial shift is introduced so that the orientation matrix is written as:

$$^0\mathbf{A}_n = \mathbf{rot}(\mathbf{z}, \phi)\, \mathbf{rot}(\mathbf{x}, \theta + \frac{\pi}{2})\, \mathbf{rot}(\mathbf{z}, \psi - \frac{\pi}{2}) \quad [3.23]$$

3.6.2. *Roll-Pitch-Yaw angles*

Following the convention shown in Figure 3.7, the angles ϕ , θ and ψ indicate roll, pitch and yaw respectively. If we suppose that the direction of motion (by analogy to the direction along which a ship is sailing) is along the **z** axis, the orientation matrix can be written as:

$$^0\mathbf{A}_n = \mathbf{rot}(\mathbf{z}, \phi)\, \mathbf{rot}(\mathbf{y}, \theta)\, \mathbf{rot}(\mathbf{x}, \psi)$$

$$= \begin{bmatrix} C\phi C\theta & C\phi S\theta S\psi - S\phi C\psi & C\phi S\theta C\psi + S\phi S\psi \\ S\phi C\theta & S\phi S\theta S\psi + C\phi C\psi & S\phi S\theta C\psi - C\phi S\psi \\ -S\theta & C\theta S\psi & C\theta C\psi \end{bmatrix} \quad [3.24]$$

Inverse problem: expression of the Roll-Pitch-Yaw angles as functions of the direction cosines. We use the same method discussed in the previous section. Premultiplying equation [3.24] by $\mathbf{rot}(\mathbf{z}, -\phi)$, we obtain:

$$\mathbf{rot}(\mathbf{z}, -\phi)\, ^0\mathbf{A}_n = \mathbf{rot}(\mathbf{y}, \theta)\, \mathbf{rot}(\mathbf{x}, \psi) \quad [3.25]$$

which results in:

$$\begin{bmatrix} C\phi s_x+S\phi s_y & C\phi n_x+S\phi n_y & C\phi a_x+S\phi a_y \\ -S\phi s_x+C\phi s_y & -S\phi n_x+C\phi n_y & -S\phi a_x+C\phi a_y \\ s_z & n_z & a_z \end{bmatrix} = \begin{bmatrix} C\theta & S\theta S\psi & S\theta C\psi \\ 0 & C\psi & -S\psi \\ -S\theta & C\theta S\psi & C\theta C\psi \end{bmatrix} \quad [3.26]$$

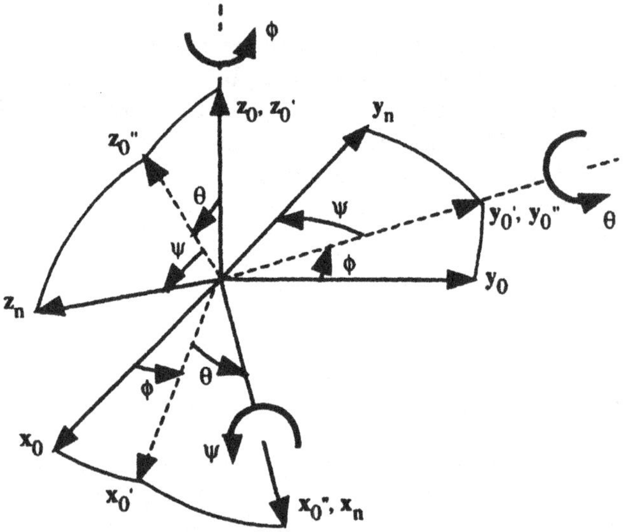

Figure 3.7. *Roll-Pitch-Yaw angles*

From the (2, 1) elements of equation [3.26], we obtain:

$$- S\phi\ s_x + C\phi\ s_y = 0$$

thus:

$$\begin{cases} \phi = \text{atan2}(s_y, s_x) \\ \phi' = \text{atan2}(-s_y, -s_x) = \phi + \pi \end{cases} \quad [3.27]$$

There is a singularity if s_x and s_y are zero ($\theta = \pm\frac{\pi}{2}$).

In the same way, from the (1, 1) and (1, 3) elements, then from the (2, 2) and (2, 3) elements, we deduce that:

$$\theta = \text{atan2}(-s_z, C\phi\ s_x + S\phi\ s_y) \quad [3.28]$$

$$\psi = \text{atan2}(S\phi a_x - C\phi\ a_y, - S\phi\ n_x + C\phi\ n_y) \quad [3.29]$$

3.6.3. *Quaternions*

The quaternions are also called *Euler parameters* or *Olinde-Rodrigues parameters*. In this representation, the orientation is expressed by four parameters that describe the orientation by a rotation of an angle θ ($0 \leq \theta \leq \pi$) about an axis of unit vector **u** (Figure 3.8). We define the quaternions as:

$$\begin{cases} Q_1 = C(\theta/2) \\ Q_2 = u_x\, S(\theta/2) \\ Q_3 = u_y\, S(\theta/2) \\ Q_4 = u_z\, S(\theta/2) \end{cases} \qquad [3.30]$$

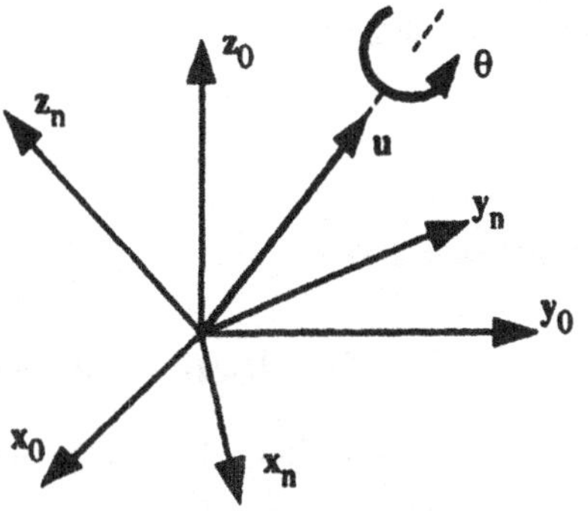

Figure 3.8. *The quaternions*

From these relations, we obtain:

$$Q_1^2 + Q_2^2 + Q_3^2 + Q_4^2 = 1 \qquad [3.31]$$

The orientation matrix ${}^0\mathbf{A}_n$ is deduced from equation [2.30], defining **rot(u**, θ**)**, after rewriting its elements as a function of Q_i. We note that:

$$C\theta = C^2(\theta/2) - S^2(\theta/2) = 2Q_1^2 - 1 \qquad [3.32]$$

and that:

$$\begin{cases} Q_2^2 = u_x^2\, S^2(\theta/2) = \frac{1}{2} u_x^2 (1 - C\theta) \\ Q_3^2 = \frac{1}{2} u_y^2 (1 - C\theta) \\ Q_4^2 = \frac{1}{2} u_z^2 (1 - C\theta) \\ Q_2\, Q_3 = \frac{1}{2} u_x\, u_y (1 - C\theta) \\ Q_2\, Q_4 = \frac{1}{2} u_x\, u_z (1 - C\theta) \\ Q_3\, Q_4 = \frac{1}{2} u_y u_z (1 - C\theta) \\ u_x\, S\theta = 2\, u_x\, S(\theta/2)\, C(\theta/2) = 2\, Q_1\, Q_2 \\ u_y\, S\theta = 2\, Q_1\, Q_3 \\ u_z\, S\theta = 2\, Q_1\, Q_4 \end{cases} \qquad [3.33]$$

Thus, the orientation matrix is given as:

$$^0\mathbf{A}_n = \begin{bmatrix} 2(Q_1^2+Q_2^2) - 1 & 2(Q_2Q_3-Q_1Q_4) & 2(Q_2Q_4+Q_1Q_3) \\ 2(Q_2Q_3+Q_1Q_4) & 2(Q_1^2+Q_3^2) - 1 & 2(Q_3Q_4-Q_1Q_2) \\ 2(Q_2Q_4-Q_1Q_3) & 2\,(Q_3Q_4+Q_1Q_2) & 2\,(Q_1^2+Q_4^2) - 1 \end{bmatrix} \qquad [3.34]$$

For more information on the algebra of quaternions, the reader can refer to [de Casteljau 87].

Inverse problem: expression of the quaternions as functions of the direction cosines. Equating the elements of the diagonals of the right sides of equations [3.34] and [3.15] leads to:

$$Q_1 = \frac{1}{2}\sqrt{s_x + n_y + a_z + 1} \qquad [3.35]$$

which is always positive. If we then subtract the (2, 2) and (3, 3) elements from the (1, 1) element, we can write after simplifying:

$$4\, Q_2^2 = s_x - n_y - a_z + 1 \qquad [3.36]$$

This expression gives the magnitude of Q_2. For determining the sign, we consider the difference of the (3, 2) and (2, 3) elements, which leads to:

$$4\,Q_1Q_2 = n_z - a_y \qquad [3.37]$$

The parameter Q_1 being always positive, the sign of Q_2 is that of $(n_z - a_y)$, which allows us to write:

$$Q_2 = \frac{1}{2}\,\text{sign}\,(n_z - a_y)\sqrt{s_x - n_y - a_z + 1} \qquad [3.38]$$

Similar reasoning for Q_3 and Q_4 gives:

$$Q_3 = \frac{1}{2}\,\text{sign}\,(a_x - s_z)\sqrt{-\,s_x + n_y - a_z + 1} \qquad [3.39]$$

$$Q_4 = \frac{1}{2}\,\text{sign}\,(s_y - n_x)\sqrt{-\,s_x - n_y + a_z + 1} \qquad [3.40]$$

These expressions exhibit no singularity.

3.7. Conclusion

In this chapter, we have shown how to calculate the direct geometric model of a serial robot. This model is unique and is given in the form of explicit equations. The description of the geometry is based on rules that have an intrinsic logic facilitating its application. This method can be generalized to tree and closed loop structures (Chapter 7). It can also be extended to systems with lumped elasticity [Khalil 00a].

We have also presented the methods that are frequently used in robotics to specify the orientation of a body in space. We have shown how to calculate the orientation matrix from these representations and inversely, how to find the parameters of these descriptions from the orientation matrix.

Having calculated the direct geometric model, in the next chapter we study the inverse geometric problem, which consists of computing the joint variables as functions of a given location of the end-effector.

Chapter 4

Inverse geometric model of serial robots

4.1. Introduction

The direct geometric model of a robot provides the location of the end-effector in terms of the joint coordinates. The problem of computing the joint variables corresponding to a specified location of the end-effector is called the inverse geometric problem. This problem is at the center of computer control algorithms for robots. It has in general a multiple solution and its complexity is highly dependent on the geometry of the robot. The model that gives all the possible solutions for this problem is called the *Inverse Geometric Model* (IGM). In this chapter, we will present three methods to obtain the IGM of serial robots. First, we present the Paul method [Paul 81], which can be used to obtain an explicit solution for robots with relatively simple geometry that have many zero distances and parallel or perpendicular joint axes. Then, we develop a variation on the Pieper method [Pieper 68], which provides the analytical solution for the IGM of six degree-of-freedom robots with three prismatic joints or three revolute joints whose axes intersect at a point. Finally, we expose the Raghavan-Roth method [Raghavan 90], which gives the IGM for six degree-of-freedom robots with general (arbitrary) geometry using, at most, a sixteen degree polynomial.

When the inverse geometric model cannot be obtained or if it is difficult to implement in real time applications, iterative numerical techniques can be used. For this purpose, several algorithms can be found in the literature [Grudić 93]. Most of these algorithms use either the Newton-Raphson-based method [Pieper 68], [Goldenberg 85] or inverse Jacobian-based methods [Pieper 68], [Whitney 69], [Fournier 80], [Featherstone 83a]. In Chapter 6, we will present the second technique.

4.2. Mathematical statement of the problem

Let ${}^{f}\mathbf{T}_{E}^{d}$ be the homogeneous transformation matrix representing the desired location of the tool frame R_E relative to the world frame. In general, we can express ${}^{f}\mathbf{T}_{E}^{d}$ in the following form (§ 3.5):

$$ {}^{f}\mathbf{T}_{E}^{d} = \mathbf{Z}\,{}^{0}\mathbf{T}_{n}(\mathbf{q})\,\mathbf{E} \tag{4.1} $$

where (Figure 3.5):

- $\mathbf{Z}$ is the transformation matrix defining the location of the robot (frame R_0) relative to the world frame;
- ${}^{0}\mathbf{T}_{n}$ is the transformation matrix of the terminal frame R_n relative to frame R_0. It is a function of the joint variable vector $\mathbf{q}$;
- $\mathbf{E}$ is the transformation matrix defining the tool frame R_E relative to R_n.

Putting all the known matrices of relation [4.1] on the left side leads to:

$$ \mathbf{U}_0 = {}^{0}\mathbf{T}_{n}(\mathbf{q}) \tag{4.2} $$

with $\mathbf{U}_0 = \mathbf{Z}^{-1}\,{}^{f}\mathbf{T}_{E}^{d}\,\mathbf{E}^{-1}$

The problem is composed of a set of twelve nonlinear equations of n unknowns. The regular case has a finite number of solutions, whereas for redundant robots or in some singular configurations we obtain an infinite number of solutions. When the desired location is outside the reachable workspace there is no solution.

We say that a robot manipulator is *solvable* [Pieper 68], [Roth 76] when it is possible to compute all the joint configurations corresponding to a given location of the end-effector. Now, all non-redundant manipulators can be considered to be solvable [Lee 88], [Raghavan 90]. The number of solutions depends on the architecture of the robot manipulator and the amplitude of the joint limits. For six degree-of-freedom robots with only revolute joints (6R), or having five revolute joints and one prismatic joint (5R1P), the maximum number of solutions is sixteen. When the robot has three revolute joints whose axes intersect at a point, the maximum number of solutions is eight. For the 3P3R robots, this number reduces to two. In all cases, it decreases when the geometric parameters take certain particular values.

Robots with less than six degrees of freedom are not able to place the end-effector frame in an arbitrary location. Thus, we only specify the task in terms of placing some elements of the tool frame (points, axes) in the world frame. Under these conditions, the matrix $\mathbf{E}$ is not completely defined, and the equation to solve is given by:

$$\mathbf{Z}^{-1}\ {}^{f}\mathbf{T}_{E}^{d} = {}^{0}\mathbf{T}_{n}(\mathbf{q})\ \mathbf{E} \quad [4.3]$$

4.3. Inverse geometric model of robots with simple geometry

For robots with simple geometry, where most of the distances (r_j and d_j) are zero and most of the angles (θ_j and α_j) are zero or $\pm\pi/2$, the inverse geometric model can be analytically obtained using the Paul method [Paul 81]. Most commercially available robots can be solved using this method.

4.3.1. *Principle*

Let us consider a robot manipulator whose transformation matrix has the expression:

$$^{0}\mathbf{T}_{n} = {}^{0}\mathbf{T}_{1}(q_1)\ {}^{1}\mathbf{T}_{2}(q_2) \ldots {}^{n-1}\mathbf{T}_{n}(q_n) \quad [4.4]$$

Let $\mathbf{U}_0$ be the desired location such that:

$$\mathbf{U}_0 = \begin{bmatrix} s_x & n_x & a_x & P_x \\ s_y & n_y & a_y & P_y \\ s_z & n_z & a_z & P_z \\ 0 & 0 & 0 & 1 \end{bmatrix} \quad [4.5]$$

The IGM is obtained by solving the following equation:

$$\mathbf{U}_0 = {}^{0}\mathbf{T}_{1}(q_1)\ {}^{1}\mathbf{T}_{2}(q_2) \ldots {}^{n-1}\mathbf{T}_{n}(q_n) \quad [4.6]$$

To find the solutions of this equation, Paul [Paul 81] proposed to move each joint variable to the left side one after the other by successively premultiplying equation [4.6] by ${}^{j}\mathbf{T}_{j-1}$, for j varying from 1 to n − 1. Then, the joint variables are determined by equating the elements of the two sides of each equation. For example, for a six degree-of-freedom robot, we proceed as follows:

- premultiply equation [4.6] by ${}^{1}\mathbf{T}_{0}$:

$$^{1}\mathbf{T}_{0}\ \mathbf{U}_0 = {}^{1}\mathbf{T}_{2}\ {}^{2}\mathbf{T}_{3}\ {}^{3}\mathbf{T}_{4}\ {}^{4}\mathbf{T}_{5}\ {}^{5}\mathbf{T}_{6} \quad [4.7]$$

The elements of the left side are constants or functions of q_1. The elements of the right side are constants or functions of $q_2, \ldots, q_6$;

- try to solve q_1 by equating the elements of the two sides of equation [4.7];
- premultiply equation [4.7] by ${}^2\mathbf{T}_1$ and try to determine q_2;
- continue the process until all the variables are solved.

In summary, the equations used to obtain all the joint variables are written as:

$$\begin{aligned}
\mathbf{U}_0 &= {}^0\mathbf{T}_1\, {}^1\mathbf{T}_2\, {}^2\mathbf{T}_3\, {}^3\mathbf{T}_4\, {}^4\mathbf{T}_5\, {}^5\mathbf{T}_6 \\
{}^1\mathbf{T}_0\, \mathbf{U}_0 &= {}^1\mathbf{T}_2\, {}^2\mathbf{T}_3\, {}^3\mathbf{T}_4\, {}^4\mathbf{T}_5\, {}^5\mathbf{T}_6 \\
{}^2\mathbf{T}_1\, \mathbf{U}_1 &= {}^2\mathbf{T}_3\, {}^3\mathbf{T}_4\, {}^4\mathbf{T}_5\, {}^5\mathbf{T}_6 \\
{}^3\mathbf{T}_2\, \mathbf{U}_2 &= {}^3\mathbf{T}_4\, {}^4\mathbf{T}_5\, {}^5\mathbf{T}_6 \\
{}^4\mathbf{T}_3\, \mathbf{U}_3 &= {}^4\mathbf{T}_5\, {}^5\mathbf{T}_6 \\
{}^5\mathbf{T}_4\, \mathbf{U}_4 &= {}^5\mathbf{T}_6
\end{aligned} \qquad [4.8]$$

with $\mathbf{U}_j = {}^j\mathbf{T}_6 = {}^j\mathbf{T}_{j-1}\, \mathbf{U}_{j-1}$

The resolution of equations [4.8] needs intuition, but the use of this method on a large number of industrial robots has shown that only few fundamental types of equations are encountered [Khalil 86b] (Table 4.1). The solutions of these equations are given in Appendix 1.

NOTES.–

- the matrices of the right side of equations [4.8] are already available when computing the direct geometric model (DGM) if the multiplication of the transformation matrices is started from the end of the robot;
- in certain cases, it may be more convenient to solve the robot by first determining q_n and ending with q_1. In this case, we postmultiply equation [4.6] by ${}^j\mathbf{T}_{j-1}$ for j varying from n to 2.

Table 4.1. *Types of equations encountered in the Paul method*

Type 1	$X\, r_i = Y$
Type 2	$X\, S\theta_i + Y\, C\theta_i = Z$
Type 3	$X1\, S\theta_i + Y1\, C\theta_i = Z1$ $X2\, S\theta_i + Y2\, C\theta_i = Z2$
Type 4	$X1\, r_j\, S\theta_i = Y1$ $X2\, r_j\, C\theta_i = Y2$
Type 5	$X1\, S\theta_i = Y1 + Z1\, r_j$ $X2\, C\theta_i = Y2 + Z2\, r_j$
Type 6	$W\, S\theta_j = X\, C\theta_i + Y\, S\theta_i + Z1$ $W\, C\theta_j = X\, S\theta_i - Y\, C\theta_i + Z2$
Type 7	$W1\, C\theta_j + W2\, S\theta_j = X\, C\theta_i + Y\, S\theta_i + Z1$ $W1\, S\theta_j - W2\, C\theta_j = X\, S\theta_i - Y\, C\theta_i + Z2$
Type 8	$X\, C\theta_i + Y\, C(\theta_i + \theta_j) = Z1$ $X\, S\theta_i + Y\, S(\theta_i + \theta_j) = Z2$

r_i: prismatic joint variable,
$S\theta_i$, $C\theta_i$: sine and cosine of a revolute joint variable θ_i.

4.3.2. *Special case: robots with a spherical wrist*

Most six degree-of-freedom industrial robots have a spherical wrist composed of three revolute joints whose axes intersect at a point (Figure 4.1). This structure is characterized by the following set of geometric parameters:

$$\begin{cases} d_5 = r_5 = d_6 = 0 \\ \sigma_4 = \sigma_5 = \sigma_6 = 0 \\ S\alpha_5 \neq 0,\ S\alpha_6 \neq 0 \text{ (non-redundant robot)} \end{cases}$$

The position of the center of the spherical joint is obtained as a function of the joint variables q_1, q_2 and q_3. This type of structure allows the decomposition of the six degree-of-freedom problem into two three degree-of-freedom problems representing a position equation and an orientation equation. The position problem, which is a function of q_1, q_2 and q_3, is first solved, then the orientation problem allows us to determine θ_4, θ_5, θ_6.

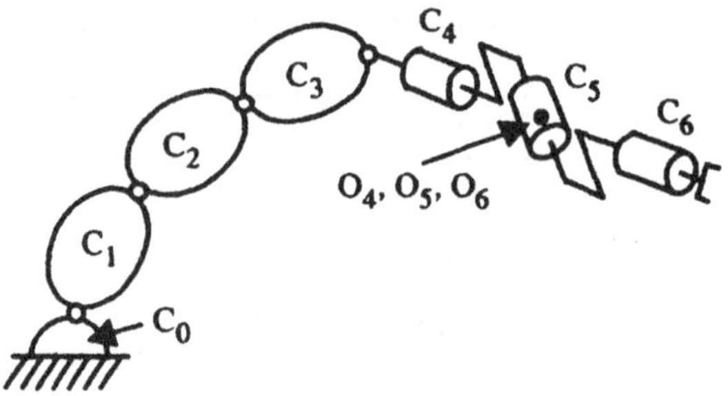

Figure 4.1. *Six degree-of-freedom robot with a spherical wrist*

4.3.2.1. *Position equation*

Since ${}^0\mathbf{P}_6 = {}^0\mathbf{P}_4$, the fourth column of the transformation matrix ${}^0\mathbf{T}_4$ is equal to the fourth column of $\mathbf{U}_0$:

$$\begin{bmatrix} Px \\ Py \\ Pz \\ 1 \end{bmatrix} = {}^0\mathbf{T}_1\,{}^1\mathbf{T}_2\,{}^2\mathbf{T}_3\,{}^3\mathbf{T}_4 \begin{bmatrix} 0 \\ 0 \\ 0 \\ 1 \end{bmatrix} \quad [4.9]$$

We obtain the variables q_1, q_2, q_3 by successively premultiplying this equation by ${}^j\mathbf{T}_0$, j = 1, 2, to isolate and determine sequentially the joint variables. The elements of the right side have already been calculated for the DGM.

4.3.2.2. *Orientation equation*

The orientation part of equation [4.2] is written as:

$$[\,\mathbf{s} \quad \mathbf{n} \quad \mathbf{a}\,] = {}^0\mathbf{A}_6(\mathbf{q})$$

yielding:

$${}^3\mathbf{A}_0(q_1, q_2, q_3)\,[\,\mathbf{s} \quad \mathbf{n} \quad \mathbf{a}\,] = {}^3\mathbf{A}_6(\theta_4, \theta_5, \theta_6)$$

which can be written as:

$$[\,\mathbf{F} \quad \mathbf{G} \quad \mathbf{H}\,] = {}^3\mathbf{A}_6(\theta_4, \theta_5, \theta_6) \quad [4.10]$$

Since q_1, q_2, q_3 have been determined, the left side elements are considered to be known. To obtain θ_4, θ_5, θ_6, we successively premultiply equation [4.10] by ${}^4\mathbf{A}_3$

then by $^5A_4$ and proceed by equating the elements of the two sides. Again, the elements of the right side have already been calculated for the DGM.

• **Example 4.1.** IGM of the Stäubli RX-90 robot. The geometric parameters are given in Table 3.1. The robot has a spherical wrist. The DGM is developed in Chapter 3.

a) Computation of $\theta_1, \theta_2, \theta_3$

i) by developing equation [4.9], we obtain:

$$\begin{bmatrix} P_x \\ P_y \\ P_z \\ 1 \end{bmatrix} = \begin{bmatrix} C1\,(-S23\,RL4 + C2\,D3) \\ S1\,(-S23\,RL4 + C2\,D3) \\ C23\,RL4 + S2\,D3 \\ 1 \end{bmatrix}$$

Note that the elements of the right side constitute the fourth column of $^0T_6$, which have already been calculated for the DGM. No variable can be determined from this equation;

ii) premultiplying the previous equation by $^1T_0$, *we obtain the left side elements as:*

$$\begin{aligned} U(1) &= C1\,P_x + S1\,P_y \\ U(2) &= -S1\,P_x + C1\,P_y \\ U(3) &= P_z \end{aligned}$$

The elements of the right side are obtained from the fourth column of $^1T_6$:

$$\begin{aligned} T(1) &= -S23\,RL4 + C2\,D3 \\ T(2) &= 0 \\ T(3) &= C23\,RL4 + S2\,D3 \end{aligned}$$

By equating U(2) and T(2), we obtain the following two solutions for θ_1:

$$\begin{cases} \theta_1 = \text{atan2}(P_y, P_x) \\ \theta'_1 = \theta_1 + \pi \end{cases}$$

iii) premultiplying by $^2\mathbf{T}_1$, *we obtain the elements of the left side as:*

$$\begin{aligned} U(1) &= C2\,(C1\,P_x + S1\,P_y) + S2\,P_z \\ U(2) &= -\,S2\,(C1\,P_x + S1\,P_y) + C2\,P_z \\ U(3) &= S1\,P_x - C1\,P_y \end{aligned}$$

The elements of the right side represent the fourth column of $^2\mathbf{T}_6$:

$$\begin{aligned} T(1) &= -\,S3\,RL4 + D3 \\ T(2) &= C3\,RL4 \\ T(3) &= 0 \end{aligned}$$

We determine θ_2 and θ_3 by considering the first two elements, which constitute a type-6 system of equations (Table 4.1). First, an equation in θ_2 is obtained:

$$X\,S2 + Y\,C2 = Z$$

with:

$$\begin{aligned} X &= -\,2P_z\,D3 \\ Y &= -\,2\,B1D3 \\ Z &= (RL4)^2 - (D3)^2 - (P_z)^2 - (B1)^2 \\ B1 &= P_x\,C1 + P_y\,S1 \end{aligned}$$

from which we deduce that:

$$\begin{cases} C2 = \dfrac{YZ - \varepsilon X\sqrt{X^2 + Y^2 - Z^2}}{X^2 + Y^2} \\ S2 = \dfrac{XZ + \varepsilon Y\sqrt{X^2 + Y^2 - Z^2}}{X^2 + Y^2} \end{cases} \qquad \text{with } \varepsilon = \pm 1$$

This gives two solutions of the following form:

$$\theta_2 = \text{atan2}(S2, C2)$$

θ_2 being known, we obtain:

$$\theta_3 = \text{atan2}(S3, C3)$$

with:

$$\begin{cases} S3 = \dfrac{-Pz\,S2 - B1C2 + D3}{RL4} \\ C3 = \dfrac{-B1S2 + Pz\,C2}{RL4} \end{cases}$$

b) Computation of θ_4, θ_5, θ_6

Once the variables θ_1, θ_2, θ_3 are determined, we define the (3×3) orientation matrix ${}^3\mathbf{A}_6$ as follows:

$$[\ \mathbf{F}\quad \mathbf{G}\quad \mathbf{H}\ ] = {}^3\mathbf{A}_0\,[\ \mathbf{s}\quad \mathbf{n}\quad \mathbf{a}\ ]$$

The elements of **F** are written as:

$$\begin{aligned} U(1,1) &= C23\,(C1\,s_x + S1\,s_y) + S23\,s_z \\ U(2,1) &= -\,S23\,(C1\,s_x + S1\,s_y) + C23\,s_z \\ U(3,1) &= S1\,s_x - C1\,s_y \end{aligned}$$

The elements of **G** and **H** are obtained from **F** by replacing (s_x, s_y, s_z) by (n_x, n_y, n_z) and (a_x, a_y, a_z) respectively.

i) equating the elements of $[\ \mathbf{F}\quad \mathbf{G}\quad \mathbf{H}\ ] = {}^3\mathbf{A}_6$

The elements of ${}^3\mathbf{A}_6$ are obtained from ${}^3\mathbf{T}_6$, which is calculated for the DGM:

$${}^3\mathbf{A}_6 = \begin{bmatrix} C6C5C4-S6S4 & -S6C5C4-C6S4 & -S5C4 \\ C6S5 & -S6S5 & C5 \\ -C6C5S4-S6C4 & S6C5S4-C6C4 & S5S4 \end{bmatrix}$$

We can determine θ_5 from the (2, 3) elements using an arccosine function. But this solution is not retained, considering that another one using an atan2 function may appear in the next equations;

ii) equating the elements of ${}^4\mathbf{A}_3\,[\ \mathbf{F}\quad \mathbf{G}\quad \mathbf{H}\ ] = {}^4\mathbf{A}_6$

The elements of the first column of the left side are written as:

$$\begin{aligned} U(1,1) &= C4\,F_x - S4\,F_z \\ U(2,1) &= -\,C4\,F_z - S4\,F_x \\ U(3,1) &= F_y \end{aligned}$$

The elements of the second and third columns are obtained by replacing (F_x, F_y, F_z) with (G_x, G_y, G_z) and (H_x, H_y, H_z) respectively. The elements of ${}^4A_6$ are obtained from ${}^4T_6$, which has already been calculated for the DGM:

$$
{}^4A_6 = \begin{bmatrix} C6C5 & -S6C5 & -S5 \\ S6 & C6 & 0 \\ C6S5 & -S6S5 & C5 \end{bmatrix}
$$

From the (2, 3) elements, we obtain a type-2 equation in θ_4:

$$
-C4\,H_z - S4\,H_x = 0
$$

which gives two solutions:

$$
\begin{cases} \theta_4 = \text{atan2}(H_z, -H_x) \\ \theta'_4 = \theta_4 + \pi \end{cases}
$$

From the (1, 3) and (3, 3) elements, we obtain a type-3 system of equations in θ_5:

$$
\begin{aligned} -S5 &= C4\,H_x - S4\,H_z \\ C5 &= H_y \end{aligned}
$$

whose solution is:

$$
\theta_5 = \text{atan2}(S5, C5)
$$

Finally, by considering the (2, 1) and (2, 2) elements, we obtain a type-3 system of equations in θ_6:

$$
\begin{aligned} S6 &= -C4\,F_z - S4\,F_x \\ C6 &= -C4\,G_z - S4\,G_x \end{aligned}
$$

whose solution is:

$$
\theta_6 = \text{atan2}(S6, C6)
$$

NOTES.– By examining the IGM solution of the Stäubli RX-90 robot, it can be observed that:

a) The robot has the following singular positions:

i) shoulder singularity: takes place when the point O_6 lies on the z_0 axis (Figure 4.2a). Thus $P_x = P_y = 0$, which corresponds to $S23RL4 - C2D3 = 0$. In this case, both the two arguments of the atan2 function used to determine θ_1 are zero, thus leaving θ_1 undetermined. We are free to choose any value for θ_1, but frequently the current value is assigned. This means that one can always find a solution, but when leaving this configuration, a small change in the desired location may require a significant variation in θ_1, impossible to realize due to the speed and acceleration limits of the actuator;

ii) wrist singularity: takes place when $C23(C1a_x + S1a_y) + S23a_z = H_x = 0$ and $(S1ax - C1ay) = H_z = 0$. The two arguments of the atan2 function used to determine θ_4 are zero. From the (2, 3) element of ${}^3A_6$, we notice that in this case $C\theta_5 = \pm 1$. Thus, the axes of joints 4 and 6 are collinear and it is the sum $\theta_4 \pm \theta_6$ that can be determined (Figure 4.2b). For example, when $\theta_5 = 0$, the orientation equation becomes:

$$[\,F \quad G \quad H\,] = {}^3A_6 = \begin{bmatrix} C46 & -S46 & 0 \\ 0 & 0 & 1 \\ -S46 & -C46 & 0 \end{bmatrix}$$

Thus, $\theta 4 + \theta 6 = \text{atan2}(-G_x, -G_z)$. We can arbitrarily assign θ_4 to its current value and calculate the corresponding θ_6. We can also calculate the values of θ_4 and θ_6 for which the joints 4 and 6 move away from their limits;

iii) elbow singularity: occurs when $C3 = 0$. This singularity will be discussed in Chapter 6. It does not affect the inverse geometric model computation (Figure 4.2c).

b) The above-mentioned singularities are classified as first order singularities. Singularities of higher order may occur when several singularities of first order take place simultaneously.

c) Number of solutions: in the regular case, the Stäubli RX-90 robot has eight solutions for the IGM (product of the number of possible solutions for each joint). Some of these configurations may not be accessible because of the joint limits.

4.3.3. *Inverse geometric model of robots with more than six degrees of freedom*

A robot with more than six degrees of freedom is redundant and its inverse geometric problem has an infinite number of solutions. To obtain a closed form solution, (n – 6) additional relations are needed. Two strategies are possible:

- arbitrarily fixing (n – 6) joint values to reduce the problem to six unknowns. The selection of the fixed joints is determined by the task specifications and the robot structure;

- introducing (n – 6) additional relations describing the redundancy, as is done in certain seven degree-of-freedom robots [Hollerbach 84b].

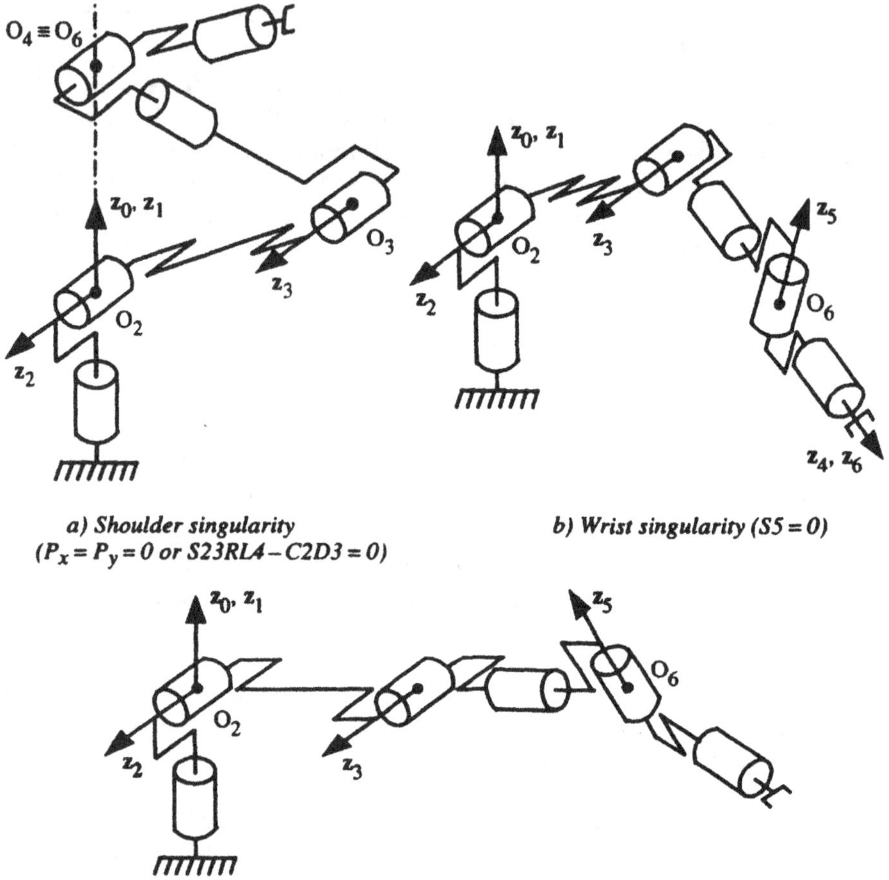

a) Shoulder singularity
($P_x = P_y = 0$ or $S23RL4 - C2D3 = 0$)

b) Wrist singularity ($S5 = 0$)

c) Elbow singularity ($C3 = 0$)

Figure 4.2. *Singular positions of the Stäubli RX-90 robot*

4.3.4. *Inverse geometric model of robots with less than six degrees of freedom*

When the robot has less than six degrees of freedom, the end-effector frame R_E cannot be placed at an arbitrary location except if certain elements of ${}^0\mathbf{T}_E{}^d$ have specific values to compensate for the missing degrees of freedom. Otherwise, instead of realizing frame-to-frame contact, we consider tasks with less degrees of freedom such as point-to-point contact, or (point-axis) to (point-axis) contact [Manaoui 85].

In the next example, we will study this problem for the four degree-of-freedom SCARA robot whose geometric parameters are given in Table 3.2.

• **Example 4.2.** IGM of the SCARA robot (Figure 3.4).

i) frame-to-frame contact

In this case, the system of equations to be solved is given by equation [4.2] and U_0 is defined by equation [4.5]:

$$U_0 = {}^0T_4 = \begin{bmatrix} C123 & -S123 & 0 & C12D3+C1D2 \\ S123 & C123 & 0 & S12D3+S1D2 \\ 0 & 0 & 1 & r_4 \\ 0 & 0 & 0 & 1 \end{bmatrix}$$

Examining the elements of this matrix reveals that frame-to-frame contact is possible if the third column of U_0 is equal to $[0 \quad 0 \quad 1 \quad 0]^T$. This implies two independent conditions, which compensate for the missing degrees of freedom. By equating the (3, 4) elements, we obtain:

$$r_4 = P_z$$

The (1, 4) and (2, 4) elements give a type-8 system of equations in θ_1 and θ_2 with the following solution:

$$\theta_2 = \text{atan2}(\pm\sqrt{1-(C2)^2}, C2)$$
$$\theta_1 = \text{atan2}(S1, C1)$$

with:

$$C2 = \frac{D^2-(D2)^2-(D3)^2}{2\,D2\,D3} \qquad D^2 = P_x{}^2 + P_y{}^2$$

$$S1 = \frac{B1\,P_y - B2\,P_x}{D^2} \qquad C1 = \frac{B1\,P_x + B2\,P_y}{D^2}$$

$$B1 = D2 + D3\,C2, \quad B2 = D3\,S2$$

After determining θ_1 and θ_2, we obtain θ_3 as:

$$\theta_3 = \text{atan2}(s_y, s_x) - \theta 2 - \theta 1$$

ii) (point-axis) to (point-axis) contact

Let us suppose that the tool is defined by an axis of unit vector $\mathbf{a}_E$, passing by O_E such that:

$$
\begin{aligned}
{}^4\mathbf{P}_E &= [Q_x \quad Q_y \quad Q_z]^T \\
{}^4\mathbf{a}_E &= [W_x \quad W_y \quad W_z]^T
\end{aligned}
$$

The task consists of placing the point O_E at a point of the environment while aligning the axis $\mathbf{a}_E$ with an axis of the environment, which are defined by:

$$
\begin{aligned}
{}^0\mathbf{P}_E^d &= [P_x \quad P_y \quad P_z]^T \\
{}^0\mathbf{a}_E^d &= [a_x \quad a_y \quad a_z]^T
\end{aligned}
$$

The system to be solved is written as:

$$
\begin{bmatrix} - & - & a_x & P_x \\ - & - & a_y & P_y \\ - & - & a_z & P_z \\ - & - & 0 & 1 \end{bmatrix} = {}^0\mathbf{T}_4 \begin{bmatrix} - & - & W_x & Q_x \\ - & - & W_y & Q_y \\ - & - & W_z & Q_z \\ - & - & 0 & 1 \end{bmatrix}
$$

After simplifying, we obtain:

$$
\begin{bmatrix} P_x \\ P_y \\ P_z \end{bmatrix} = \begin{bmatrix} Q_x C123 - Q_y S123 + C12D3 + C1D2 \\ Q_x S123 + Q_y C123 + S12D3 + S1D2 \\ Q_z + r4 \end{bmatrix}
$$

$$
\begin{bmatrix} a_x \\ a_y \\ a_z \end{bmatrix} = \begin{bmatrix} W_x C123 - W_y S123 \\ W_x S123 + W_y C123 \\ W_z \end{bmatrix}
$$

Thus, we deduce that the condition $a_z = W_z$ must be satisfied to realize the task. The IGM solutions are obtained in the following way:

- from the a_x and a_y equations, we obtain $(\theta_1 + \theta_2 + \theta_3)$ by solving a type-3 system (Appendix 1):

$$
\theta_1 + \theta_2 + \theta_3 = \text{atan2}(S123, C123)
$$

$$
\text{with } S123 = \frac{a_y W_x - a_x W_y}{W_x^2 + W_y^2} \text{ and } C123 = \frac{a_x W_x + a_y W_y}{W_x^2 + W_y^2} \text{ if } (W_x^2 + W_y^2) \neq 0;
$$

- when $W_x = W_y = 0$, the axis of the end-effector is vertical and its orientation cannot be changed. Any value for θ_3 may be taken;
- from P_x and P_y equations, we obtain θ_1 and θ_2 by solving a type-8 system of equations;
- finally, from the third element of the position equation, we obtain $r_4 = P_z - Q_z$.

In summary, the task of a SCARA robot can be described in one of the following ways:

- placing the tool frame onto a specified frame provided that the third column of the matrix ${}^0T_4^d = {}^0T_E^d\, E^{-1} = [0\ \ 0\ \ 1\ \ 0]^T$, in order to satisfy that $\mathbf{z}_4$ is vertical;
- placing an axis and a point of the tool frame respectively onto an axis and a point of the environment provided that $a_z = W_z$. The obvious particular case is to locate a horizontal axis of the end-effector frame in a horizontal plane ($a_z = W_z = 0$).

4.4. Inverse geometric model of decoupled six degree-of-freedom robots

4.4.1. *Introduction*

The IGM of a six degree-of-freedom decoupled robot can be computed by solving two sub-problems, each having three unknowns [Pieper 68]. Two classes of structures are considered:

a) robots having a spherical joint given by one of the following four combinations: XXX(RRR), (RRR)XXX, XX(RRR)X, X(RRR)XX, where (RRR) denotes a spherical joint and X denotes either a revolute (R) or a prismatic (P) joint. Consequently, each combination results in eight structures;

b) robots having three revolute and three prismatic joints as given by one of the following 20 combinations: PPPRRR, PPRPRR, PPRRPR, ...

In this section, we present the inverse geometric model of these structures using two general equations [Khalil 90c], [Bennis 91a]. These equations make use of the six types of equations shown in Table 4.2. The first three types have already been used in the Paul method (Table 4.1). The explicit solution of a type-10 equation can be obtained symbolically using software packages like Maple or Mathematica. In general, however, the numerical solution is more accurate. We note that a type-11 equation can be transformed into type-10 using the half-angle transformation by writing $C\theta_i$ and $S\theta_i$ as:

$$C\theta_i = \frac{1 - t^2}{1 + t^2} \text{ and } S\theta_i = \frac{2t}{1 + t^2} \text{ with } t = \tan\frac{\theta_i}{2}$$

Table 4.2. *Types of equations encountered in the Pieper method*

Type 1	$X\, r_i = Y$
Type 2	$X\, C\theta_i + Y\, S\theta_i = Z$
Type 3	$X1\, S\theta_i + Y1\, C\theta_i = Z1$ $X2\, S\theta_i + Y2\, C\theta_i = Z2$
Type 9	$a_2\, r_i^2 + a_1\, r_i + a_0 = 0$
Type 10	$a_4\, r_i^4 + a_3\, r_i^3 + a_2\, r_i^2 + a_1\, r_i + a_0 = 0$
Type 11	$a_4\, S\theta_i^2 + a_3\, C\theta_i\, S\theta_i + a_2\, C\theta_i + a_1\, S\theta_i + a_0 = 0$

4.4.2. *Inverse geometric model of six degree-of-freedom robots having a spherical joint*

In this case, equation [4.6] is decoupled into two equations, each containing three variables:

- a position equation, which is a function of the joint variables that do not belong to the spherical joint;
- an orientation equation, which is a function of the joint variables of the spherical joint.

4.4.2.1. *General solution of the position equation*

The revolute joint axes $m-1$, m and $m+1$ ($2 \le m \le 5$) form a spherical joint if:

$$\begin{cases} d_m = r_m = d_{m+1} = 0 \\ S\alpha_m \neq 0 \\ S\alpha_{m+1} \neq 0 \end{cases}$$

The position of the center of the spherical joint, O_m or O_{m-1}, is independent of the joint variables θ_{m-1}, θ_m and θ_{m+1}. Thus, we can show (Figure 4.3) that the position of O_m relative to frame R_{m-2} is given by:

$$^{m-2}\mathbf{T}_{m+1}\,\mathbf{Trans}(z,-r_{m+1})\mathbf{p}_0 = \begin{bmatrix} ^{m-2}\mathbf{P}_{m-1} \\ 1 \end{bmatrix} = \begin{bmatrix} d_{m-1} \\ -r_{m-1}S\alpha_{m-1} \\ r_{m-1}C\alpha_{m-1} \\ 1 \end{bmatrix} \qquad [4.11]$$

where $\mathbf{p}_0 = [\,0 \;\; 0 \;\; 0 \;\; 1\,]^T$ and ${}^{m-2}\mathbf{P}_{m-1}$ is obtained using equation [3.2].

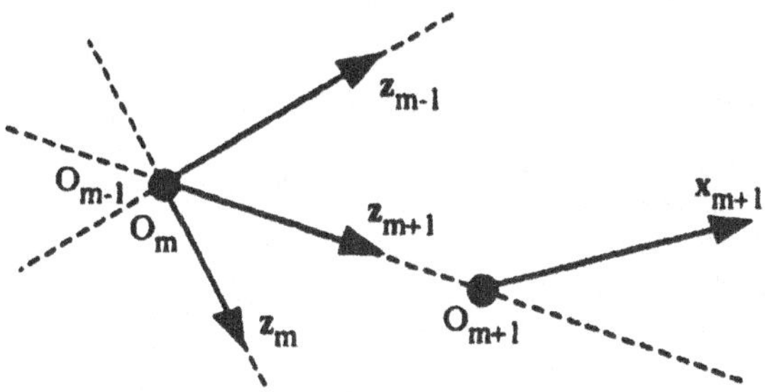

Figure 4.3. *Axes of a spherical joint*

To obtain the position equation, we write equation [4.6] in the following form:

$$ {}^0\mathbf{T}_{m-2}\, {}^{m-2}\mathbf{T}_{m+1}\, {}^{m+1}\mathbf{T}_6 = \mathbf{U}_0 \qquad [4.12] $$

Postmultiplying this relation by ${}^6\mathbf{T}_{m+1}$ **Trans**($\mathbf{z}$, $-r_{m+1}$) $\mathbf{p}_0$ and using equation [4.11], we obtain:

$$ {}^0\mathbf{T}_{m-2}\begin{bmatrix} {}^{m-2}\mathbf{P}_{m-1} \\ 1 \end{bmatrix} = \mathbf{U}_0\, {}^6\mathbf{T}_{m+1}\, \mathbf{Trans}(\mathbf{z}, -r_{m+1})\, \mathbf{p}_0 \qquad [4.13] $$

Equation [4.13] can be written in the general form:

$$ \mathbf{Rot}(\mathbf{z}, \theta_i)\, \mathbf{Trans}(\mathbf{z}, r_i)\, \mathbf{Rot}(\mathbf{x}, \alpha_j)\, \mathbf{Trans}(\mathbf{x}, d_j)\, \mathbf{Rot}(\mathbf{z}, \theta_j)\, \mathbf{Trans}(\mathbf{z}, r_j)\begin{bmatrix} \mathbf{f}(q_k) \\ 1 \end{bmatrix} = \begin{bmatrix} \mathbf{g} \\ 1 \end{bmatrix} \qquad [4.14] $$

where:

- the subscripts i, j and k represent the joints that do not belong to the spherical joint; i and j represent two successive joints;
- the vector **f** is a function of the joint variable q_k;
- the vector **g** is a constant.

By combining the parameters $\bar{q}_i$ and $\bar{q}_j$ with **g** and **f** respectively, equation [4.14] becomes:

$$ \mathbf{Rot/Trans}(\mathbf{z}, q_i)\, \mathbf{Rot}(\mathbf{x}, \alpha_j)\, \mathbf{Trans}(\mathbf{x}, d_j)\, \mathbf{Rot/Trans}(\mathbf{z}, q_j)\begin{bmatrix} \mathbf{F}(q_k) \\ 1 \end{bmatrix} = \begin{bmatrix} \mathbf{G} \\ 1 \end{bmatrix} \qquad [4.15] $$

with:

- $\mathbf{Rot/Trans}(\mathbf{z}, q_i) = \mathbf{Rot}(\mathbf{z}, \theta_i)$ if $q_i = \theta_i$
  $= \mathbf{Trans}(\mathbf{z}, r_i)$ if $q_i = r_i$
- $\begin{bmatrix} \mathbf{F}(q_k) \\ 1 \end{bmatrix} = \begin{bmatrix} F_x \\ F_y \\ F_z \\ 1 \end{bmatrix} = \mathbf{Rot/Trans}(\mathbf{z}, \bar{q}_j) \begin{bmatrix} \mathbf{f}(q_k) \\ 1 \end{bmatrix}$
- $\begin{bmatrix} \mathbf{G} \\ 1 \end{bmatrix} = \begin{bmatrix} G_x \\ G_y \\ G_z \\ 1 \end{bmatrix} = \mathbf{Rot/Trans}(\mathbf{z}, -\bar{q}_i) \begin{bmatrix} \mathbf{g} \\ 1 \end{bmatrix}$
- $\mathbf{Rot/Trans}(\mathbf{z}, \bar{q}_i) = \mathbf{Trans}(\mathbf{z}, r_i)$ if joint i is revolute
  $= \mathbf{Rot}(\mathbf{z}, \theta_i)$ if joint i is prismatic

The components of **G** are constants and those of **F** are functions of the joint variable q_k. We note that if joint k is revolute, then:

$$\|\mathbf{F}\|^2 = a\,C\theta_k + b\,S\theta_k + c \qquad [4.16]$$

where a, b and c are constants.

Table 4.3 shows the equations that are used to obtain the joint variable q_k according to the types of joints i and j (columns 1 and 2). The variables q_i and q_j are then computed using equation [4.15]. Table 4.4 indicates the type of the obtained equations and the maximum number of solutions for each structure; the last column of the table indicates the order in which we calculate them. In Example 4.3, we will develop the solution for the case where joints i and j are revolute. We note that the maximum number of solutions for q_i, q_j and q_k is four.

NOTE.– The assignment of i, j and k for the joints that do not belong to the spherical joint is not unique. In order to get a simple solution for q_k, this assignment can be changed using the concept of the inverse robot (presented in Appendix 2). For instance, if the spherical joint is composed of the joints 4, 5 and 6, we can take $i = 1$, $j = 2$, $k = 3$. But we can also take $i = 3$, $j = 2$, $k = 1$ by using the concept of the inverse robot. We can easily verify that the second choice is more interesting if these joints are revolute and $S\alpha_2 \neq 0$, $d_2 \neq 0$ but $d_3 = 0$ or $S\alpha_3 = 0$.

Table 4.3. *Solutions of q_k and types of equations*

i	j	Conditions	Equations for q_k	Type θ_k	Type r_k
R	R	$S\alpha_j = 0$	$C\alpha_j F_z(q_k) = G_z$	2	1
		$d_j = 0$	$\|F\|^2 = \|G\|^2$	2	9
		$d_j \neq 0$ and $S\alpha_j \neq 0$	$\left[\frac{\|F\|^2-\|G\|^2-d_j^2}{2d_j}\right]^2 + \left[\frac{F_z - C\alpha_j G_z}{S\alpha_j}\right]^2 = G_x^2 + G_y^2$	11	10
R	P	$C\alpha_j = 0$	$F_y(q_k) = S\alpha_j G_z$	2	1
		$C\alpha_j \neq 0$	$(F_x+d_j)^2 + \left[\frac{F_y - S\alpha_j G_z}{C\alpha_j}\right]^2 = G_x^2 + G_y^2$	11	9
P	R	$C\alpha_j = 0$	$G_y = - S\alpha_j F_z(q_k)$	2	1
		$C\alpha_j \neq 0$	$(G_x-d_j)^2 + \left[\frac{G_y + S\alpha_j F_z}{C\alpha_j}\right]^2 = F_x^2 + F_y^2$	11	9
P	P		$F_x + d_j = G_x$	2	1

Table 4.4. *Type of equations and maximum number of solutions for q_i, q_j and q_k*

			Type / Number of solutions				
i	j	Conditions	θ_k	r_k	q_i	q_j	Order
R	R	$S\alpha_j = 0$	2 / 2	1 / 1	3 / 1	2 / 2	θ_j then θ_i
		$d_j = 0$	2 / 2	9 / 2	3 / 1	3 / 2	θ_j then θ_i
		$d_j \neq 0$ and $S\alpha_j \neq 0$	11 / 4	10 / 4	3 / 1	3 / 1	θ_i then θ_j
R	P	$C\alpha_j = 0$	2 / 2	1 / 1	2 / 2	1 / 1	θ_i then r_j
		$C\alpha_j \neq 0$	11 / 4	9 / 2	3 / 1	1 / 1	θ_i then r_j
P	R	$C\alpha_j = 0$	2 / 2	1 / 1	1 / 1	2 / 2	θ_j then r_i
		$C\alpha_j \neq 0$	11 / 4	9 / 2	1 / 1	3 / 1	θ_j then r_i
P	P		2 / 2	1 / 1	1 / 1	1 / 1	r_j then r_i

• **Example 4.3.** Solving q_k when joints i and j are revolute. In this case, equation [4.15] is written as:

$$\mathbf{Rot}(\mathbf{z}, \theta_i)\ \mathbf{Rot}(\mathbf{x}, \alpha_j)\ \mathbf{Trans}(\mathbf{x}, d_j)\ \mathbf{Rot}(\mathbf{z}, \theta_j) \begin{bmatrix} \mathbf{F}(q_k) \\ 1 \end{bmatrix} = \begin{bmatrix} \mathbf{G} \\ 1 \end{bmatrix} \tag{4.17}$$

Postmultiplying equation [4.17] by $\mathbf{Rot}(\mathbf{z}, -\theta_i)$, we obtain:

$$\begin{bmatrix} C\theta_j & -S\theta_j & 0 & d_j \\ C\alpha_j S\theta_j & C\alpha_j C\theta_j & -S\alpha_j & 0 \\ S\alpha_j S\theta_j & S\alpha_j C\theta_j & C\alpha_j & 0 \\ 0 & 0 & 0 & 1 \end{bmatrix} \begin{bmatrix} F_x \\ F_y \\ F_z \\ 1 \end{bmatrix} = \begin{bmatrix} C\theta_i & S\theta_i & 0 & 0 \\ -S\theta_i & C\theta_i & 0 & 0 \\ 0 & 0 & 1 & 0 \\ 0 & 0 & 0 & 1 \end{bmatrix} \begin{bmatrix} G_x \\ G_y \\ G_z \\ 1 \end{bmatrix} \tag{4.18}$$

Expanding equation [4.18] gives:

$$C\theta_j F_x - S\theta_j F_y + d_j = C\theta_i G_x + S\theta_i G_y \tag{4.18a}$$
$$C\alpha_j S\theta_j F_x + C\alpha_j C\theta_j F_y - S\alpha_j F_z = -S\theta_i G_x + C\theta_i G_y \tag{4.18b}$$
$$S\alpha_j S\theta_j F_x + S\alpha_j C\theta_j F_y + C\alpha_j F_z = G_z \tag{4.18c}$$

Three cases are considered depending on the values of the geometric parameters α_j and d_j:

a) $S\alpha_j = 0$ (thus $C\alpha_j = \pm 1$), $d_j \neq 0$. Equation [4.18c] can be written as:

$$C\alpha_j F_z(q_k) = G_z \tag{4.19}$$

We thus deduce that:

- if $q_k = \theta_k$, equation [4.19] is of type 2 in θ_k;
- if $q_k = r_k$, equation [4.19] is of type 1 in r_k.

Having determined q_k, the components of **F** are considered to be known. Adding the squares of equations [4.18a] and [4.18b] eliminates θ_i and gives a type-2 equation in θ_j:

$$F_x^2 + F_y^2 + d_j^2 + 2\, d_j (C\theta_j F_x - S\theta_j F_y) = G_x^2 + G_y^2 \tag{4.20}$$

After obtaining θ_j, equations [4.18a] and [4.18b] give a system of type-3 equations in θ_i.

b) $d_j = 0$ and $S\alpha_j \neq 0$. Adding the squares of equations [4.18] gives:

$$\|\mathbf{F}\|^2 = \|\mathbf{G}\|^2 \qquad [4.21]$$

Note that $\|\mathbf{F}\|^2$ is a function of q_k whereas $\|\mathbf{G}\|^2$ is a constant:

- if $q_k = \theta_k$, equation [4.21] is of type 2 in θ_k;
- if $q_k = r_k$, equation [4.21] is of type 9 in r_k.

Having obtained q_k and $\mathbf{F}$, equation [4.18c] gives θ_j using the type-2 equation. Finally, equations [4.18a] and [4.18b] give a system of type-3 equations in θ_i.

c) $d_j \neq 0$ and $S\alpha_j \neq 0$. Writing equation [4.17] in the form:

$$\begin{bmatrix} \mathbf{F}(q_k) \\ 1 \end{bmatrix} = \mathbf{Rot}(\mathbf{z}, -\theta_j)\ \mathbf{Trans}(\mathbf{x}, -d_j)\ \mathbf{Rot}(\mathbf{x}, -\alpha_j)\ \mathbf{Rot}(\mathbf{z}, -\theta_i) \begin{bmatrix} \mathbf{G} \\ 1 \end{bmatrix} \qquad [4.22]$$

after expanding, we obtain the third component as:

$$F_z = S\alpha_j\, S\theta_i\, G_x - S\alpha_j\, C\theta_i\, G_y + C\alpha_j\, G_z \qquad [4.23a]$$

Adding the squares of the components of equation [4.22] eliminates θ_j:

$$\|\mathbf{G}\|^2 + d_j^2 - 2\, d_j\, (C\theta_i\, G_x - S\theta_i\, G_y) = \|\mathbf{F}\|^2 \qquad [4.23b]$$

By eliminating θ_i from equations [4.23], we obtain:

$$\left[\frac{\|\mathbf{F}\|^2 - \|\mathbf{G}\|^2 - d_j^2}{2\, d_j}\right]^2 + \left[\frac{F_z - C\alpha_j\, G_z}{S\alpha_j}\right]^2 = G_x^2 + G_y^2 \qquad [4.24]$$

Here, we distinguish two cases:

- if $q_k = \theta_k$, equation [4.24] is of type 11 in θ_k;
- if $q_k = r_k$, equation [4.24] is of type 10 in r_k.

Knowing θ_k, equations [4.23a] and [4.23b] give a system of type-3 equations in θ_i. Finally, equations [4.18a] and [4.18b] are of type 3 in θ_j.

• **Example 4.4.** The variables θ_1, θ_2, θ_3 for the Stäubli RX-90 robot can be determined with the following equations using the Pieper method:

- equation for θ_3: $-2D3\ RL4\ S3 = (P_x)^2 + (P_y)^2 + (P_z)^2 - (D3)^2 - (RL4)^2$
- equation for θ_2: $(-RL4\ S3 + D3)\ S2 + (RL4\ C3)\ C2 = P_z$
- equations for θ_1: $[(-RL4\ S3 + D3)\ C2 - RL4\ C3\ S2]\ C1 = P_x$

  $[(-RL4\ S3 + D3)\ C2 - RL4\ C3\ S2]\ S1 = P_y$

4.4.2.2. General solution of the orientation equation

The spherical joint variables θ_{m-1}, θ_m and θ_{m+1} are determined from the orientation equation, which is deduced from equation [4.2] as:

$$^{0}\mathbf{A}_{m-2}\,{}^{m-2}\mathbf{A}_{m+1}\,{}^{m+1}\mathbf{A}_6 = [\,\mathbf{s}\quad \mathbf{n}\quad \mathbf{a}\,] \qquad [4.25]$$

The matrices $^{0}\mathbf{A}_{m-2}$ and $^{m+1}\mathbf{A}_6$ are functions of the variables that have already been obtained. Using equation [3.3] and after rearranging, equation [4.25] becomes:

$$\mathbf{rot}(\mathbf{z}, \theta_{m-1})\,\mathbf{rot}(\mathbf{x}, \alpha_m)\,\mathbf{rot}(\mathbf{z}, \theta_m)\,\mathbf{rot}(\mathbf{x}, \alpha_{m+1})\,\mathbf{rot}(\mathbf{z}, \theta_{m+1}) = [\,\mathbf{S}\quad \mathbf{N}\quad \mathbf{A}\,] \qquad [4.26]$$

with $[\,\mathbf{S}\quad \mathbf{N}\quad \mathbf{A}\,] = \mathbf{rot}(\mathbf{x}, -\alpha_{m-1})\,{}^{m-2}\mathbf{A}_0\,[\,\mathbf{s}\quad \mathbf{n}\quad \mathbf{a}\,]\,{}^{6}\mathbf{A}_{m+1}$

The left side of equation [4.26] is a function of the joint variables θ_{m-1}, θ_m and θ_{m+1} whereas the right side is known. Since $\mathbf{rot}(\mathbf{z}, \theta)$ defines a rotation about the axis $\mathbf{z}_0 = [\,0\quad 0\quad 1\,]^T$, then $\mathbf{z}_0$ is invariant with this rotation, which results in:

$$\mathbf{rot}(\mathbf{z}, \theta)\,\mathbf{z}_0 = \mathbf{z}_0 \text{ and } \mathbf{z}_0^T\,\mathbf{rot}(\mathbf{z}, \theta) = \mathbf{z}_0^T \qquad [4.27]$$

i) determination of θ_m

To eliminate θ_{m-1}, we premultiply equation [4.26] by $\mathbf{z}_0^T$ and postmultiply it by $\mathbf{z}_0$:

$$\mathbf{z}_0^T\,\mathbf{rot}(\mathbf{x}, \alpha_m)\,\mathbf{rot}(\mathbf{z}, \theta_m)\,\mathbf{rot}(\mathbf{x}, \alpha_{m+1})\,\mathbf{z}_0 = \mathbf{z}_0^T\,[\,\mathbf{S}\quad \mathbf{N}\quad \mathbf{A}\,]\,\mathbf{z}_0 \qquad [4.28]$$

thus, we obtain:

$$S\alpha_m\,S\alpha_{m+1}\,C\theta_m + C\alpha_m\,C\alpha_{m+1} = A_z$$

Equation [4.28] is of type 2 in θ_m and gives two solutions (Appendix 1);

ii) determination of θ_{m-1}

Having obtained θ_m, let us write:

$$[\,\mathbf{S1}\quad \mathbf{N1}\quad \mathbf{A1}\,] = \mathbf{rot}(\mathbf{x}, \alpha_m)\,\mathbf{rot}(\mathbf{z}, \theta_m)\,\mathbf{rot}(\mathbf{x}, \alpha_{m+1}) \qquad [4.29]$$

Postmultiplying equation [4.26] by $\mathbf{z}_0$ and using equation [4.29] gives:

$$\mathbf{rot}(\mathbf{z}, \theta_{m-1})\,\mathbf{A1} = \mathbf{A} \qquad [4.30]$$

The first two elements of [4.30] give a type-3 system of equations in θ_{m-1};

iii) determination of θ_{m+1}

By premultiplying equation [4.26] by $\mathbf{z}_0^T$ and using equation [4.29], we obtain:

$$[\ S1_z\ \ N1_z\ \ A1_z\ ]\ \mathbf{rot}(\mathbf{z}, \theta_{m+1}) = [\ S_z\ \ N_z\ \ A_z\ ] \quad [4.31]$$

This gives a type-3 system of equations in θ_{m+1}.

These equations yield two solutions for the spherical joint variables. Thus, the maximum number of solutions of the IGM for a six degree-of-freedom robot with a spherical joint is eight.

4.4.3. *Inverse geometric model of robots with three prismatic joints*

The IGM of this class of robots is obtained by solving firstly the three revolute joint variables using the orientation equation. After this, the prismatic joint variables are obtained using the position equation. The number of solutions for the IGM of such robots is two.

4.4.3.1. *Solution of the orientation equation*

Let the revolute joints be denoted by i, j and k. The orientation equation can be deduced from equations [4.2] and [3.3] as:

$$\mathbf{rot}(\mathbf{z}, \theta_i)\ [\ \mathbf{S1}\ \ \mathbf{N1}\ \ \mathbf{A1}\ ]\ \mathbf{rot}(\mathbf{z},\theta_j)\ [\ \mathbf{S2}\ \ \mathbf{N2}\ \ \mathbf{A2}\ ]\ \mathbf{rot}(\mathbf{z},\theta_k) = [\ \mathbf{S3}\ \ \mathbf{N3}\ \ \mathbf{A3}\ ] \quad [4.32]$$

where the orientation matrices $[\ \mathbf{Si}\ \ \mathbf{Ni}\ \ \mathbf{Ai}\ ]$, for $\mathbf{i} = 1, 2, 3$, are known. The solution of equation [4.32] is similar to that of § 4.4.2.2 and gives two solutions.

4.4.3.2. *Solution of the position equation*

Let the prismatic joints be denoted by i', j' and k'. The revolute joint variables being determined, the position equation is written as:

$$\mathbf{Trans}(\mathbf{z}, r_{i'})\ \mathbf{T1}\ \mathbf{Trans}(\mathbf{z}, r_{j'})\ \mathbf{T2}\ \mathbf{Trans}(\mathbf{z}, r_{k'}) = \mathbf{T3} \quad [4.33]$$

with $\mathbf{Ti} = \begin{bmatrix} \mathbf{Si'} & \mathbf{Ni'} & \mathbf{Ai'} & \mathbf{Pi'} \\ 0 & 0 & 0 & 1 \end{bmatrix}$

The matrices **Ti**, for **i** = **1, 2, 3**, are known. The previous equation gives a system of three linear equations in $r_{i'}$, $r_{j'}$ and $r_{k'}$.

4.5. Inverse geometric model of general robots

The Raghavan-Roth method [Raghavan 90] gives the solution to the inverse geometric problem for six degree-of-freedom robots with general geometry (the geometric parameters may have arbitrary real values). In this method, we first compute all possible solutions of one variable q_i using a polynomial equation, which is called the *characteristic polynomial*. Then, the other variables are uniquely derived for each q_i. This method is based on the *dyalitic elimination* technique presented in Appendix 3.

In order to illustrate this method, we consider the 6R robot and rewrite equation [4.2] as follows:

$$^0\mathbf{T}_1\,{}^1\mathbf{T}_2\,{}^2\mathbf{T}_3\,{}^3\mathbf{T}_4 = \mathbf{U}_0\,{}^6\mathbf{T}_5\,{}^5\mathbf{T}_4 \qquad [4.34]$$

The left and right sides of equation [4.34] represent the transformation of frame R_4 relative to frame R_0 using two distinct paths. The joint variables appearing in the elements of the previous equation are:

$$\begin{bmatrix} \theta_1,\theta_2,\theta_3,\theta_4 & \theta_1,\theta_2,\theta_3,\theta_4 & \theta_1,\theta_2,\theta_3 & \theta_1,\theta_2,\theta_3 \\ \theta_1,\theta_2,\theta_3,\theta_4 & \theta_1,\theta_2,\theta_3,\theta_4 & \theta_1,\theta_2,\theta_3 & \theta_1,\theta_2,\theta_3 \\ \theta_1,\theta_2,\theta_3,\theta_4 & \theta_1,\theta_2,\theta_3,\theta_4 & \theta_1,\theta_2,\theta_3 & \theta_1,\theta_2,\theta_3 \\ 0 & 0 & 0 & 1 \end{bmatrix} = \begin{bmatrix} \theta_5,\theta_6 & \theta_5,\theta_6 & \theta_5,\theta_6 & \theta_5,\theta_6 \\ \theta_5,\theta_6 & \theta_5,\theta_6 & \theta_5,\theta_6 & \theta_5,\theta_6 \\ \theta_5,\theta_6 & \theta_5,\theta_6 & \theta_5,\theta_6 & \theta_5,\theta_6 \\ 0 & 0 & 0 & 1 \end{bmatrix}$$

From this equation, we observe that the third and fourth columns of the left side are independent of θ_4. This is due to the fact that the elements of the third and fourth columns of the transformation matrix $^{i-1}\mathbf{T}_i$ are independent of θ_i (see equation [3.2]). From equation [4.34], we can thus establish the following equations devoid of θ_4:

$$\mathbf{a}_l = \mathbf{a}_r \qquad [4.35a]$$

$$\mathbf{P}_l = \mathbf{P}_r \qquad [4.35b]$$

where the vectors **a** and **P** contain the first three elements of the third and fourth columns of equation [4.34] respectively, and the subscripts "l" and "r" indicate the left and right sides respectively. Equations [4.35] give six scalar equations.

It is now necessary to eliminate four of the five remaining variables to obtain a polynomial equation in one variable. This requires the use of the following additional equations:

$$(\mathbf{a}^T \mathbf{P})_l = (\mathbf{a}^T \mathbf{P})_r \quad [4.36a]$$

$$(\mathbf{P}^T \mathbf{P})_l = (\mathbf{P}^T \mathbf{P})_r \quad [4.36b]$$

$$(\mathbf{a}\mathrm{x}\mathbf{P})_l = (\mathbf{a}\mathrm{x}\mathbf{P})_r \quad [4.36c]$$

$$[\mathbf{a}(\mathbf{P}^T \mathbf{P}) - 2\mathbf{P}(\mathbf{a}^T \mathbf{P})]_l = [\mathbf{a}(\mathbf{P}^T \mathbf{P}) - 2\mathbf{P}(\mathbf{a}^T \mathbf{P})]_r \quad [4.36d]$$

Equations [4.36a] and [4.36b] are scalar, whereas equations [4.36c] and [4.36d] are vectors. They do not contain $\sin^2(.)$, $\cos^2(.)$ or $\sin(.)\cos(.)$. We thus have fourteen scalar equations that may be written in the following matrix form:

$$\mathbf{A}\,\mathbf{X1} = \mathbf{B}\,\mathbf{Y} \quad [4.37]$$

where:

- $\mathbf{X1} = [\ S2S3 \quad S2C3 \quad C2S3 \quad C2C3 \quad S2 \quad C2 \quad S3 \quad C3 \quad 1\ ]^T$ [4.38]
- $\mathbf{Y} = [\ S5S6 \quad S5C6 \quad C5S6 \quad C5C6 \quad S5 \quad C5 \quad S6 \quad C6\ ]^T$ [4.39]
- **A**: (14x9) matrix whose elements are linear combinations of S1 and C1;
- **B**: (14x8) matrix whose elements are constants.

To eliminate θ5 and θ6, we select eight scalar equations out of equation [4.37]. The system [4.37] will be partitioned as:

$$\begin{bmatrix} \mathbf{A1} \\ \mathbf{A2} \end{bmatrix} \mathbf{X1} = \begin{bmatrix} \mathbf{B1} \\ \mathbf{B2} \end{bmatrix} \mathbf{Y} \quad [4.40]$$

where **A1 X1 = B1 Y** gives six equations, and **A2 X1 = B2 Y** represents the remaining eight equations. By eliminating **Y**, we obtain the following system of equations:

$$\mathbf{D}\,\mathbf{X1} = \mathbf{0}_{6x1} \quad [4.41]$$

where $\mathbf{D} = [\mathbf{A1} - \mathbf{B1}\,\mathbf{B2}^{-1}\,\mathbf{A2}]$ is a (6x9) matrix whose elements are functions of S1 and C1.

Using the half-angle transformation for the sine and cosine functions in equation [4.41] ($Ci = \frac{1 - x_i^2}{1 + x_i^2}$ and $Si = \frac{2x_i}{1 + x_i^2}$ with $x_i = \tan\frac{\theta_i}{2}$ for i = 1, 2, 3) yields the new homogeneous system of equations:

$$\mathbf{E}\,\mathbf{X2} = \mathbf{0}_{6x1} \qquad [4.42]$$

where **E** is a (6x9) matrix whose elements are quadratic functions of x_1, and:

$$\mathbf{X2} = [\; x_2^2x_3^2 \;\; x_2^2x_3 \;\; x_2^2 \;\; x_2x_3^2 \;\; x_2x_3 \;\; x_2 \;\; x_3^2 \;\; x_3 \;\; 1 \;]^T \qquad [4.43]$$

Thus, we have a system of six equations with nine unknowns. We now eliminate x_2 and x_3 dyalitically (see Appendix 3). Multiplying equation [4.42] by x_2, we obtain six additional equations with only three new unknowns:

$$\mathbf{E}\,\mathbf{X3} = \mathbf{0}_{6x1} \qquad [4.44]$$

with $\mathbf{X3} = [\; x_2^3x_3^2 \;\; x_2^3x_3 \;\; x_2^3 \;\; x_2^2x_3^2 \;\; x_2^2x_3 \;\; x_2^2 \;\; x_2x_3^2 \;\; x_2x_3 \;\; x_2 \;]^T$.

Combining equations [4.42] and [4.44], we obtain a system of twelve homogeneous equations:

$$\mathbf{S}\,\mathbf{X} = \mathbf{0}_{12x1} \qquad [4.45]$$

where:

$$\mathbf{X} = [\; x_2^3x_3^2 \;\; x_2^3x_3 \;\; x_2^3 \;\; x_2^2x_3^2 \;\; x_2^2x_3 \;\; x_2^2 \;\; x_2x_3^2 \;\; x_2x_3 \;\; x_2 \;\; x_3^2 \;\; x_3 \;\; 1 \;]^T \qquad [4.46]$$

and **S** is a (12x12) matrix whose elements are quadratic functions of x_1 and has the following form:

$$\mathbf{S} = \begin{bmatrix} \mathbf{E} & \mathbf{0}_{6x3} \\ \mathbf{0}_{6x3} & \mathbf{E} \end{bmatrix} \qquad [4.47]$$

In order that equation [4.45] has a non-trivial solution, the determinant of the matrix **S** must be zero. The characteristic polynomial of equation [4.47], which gives the solution for x_1, can be obtained from:

$$\det(\mathbf{S}) = 0 \qquad [4.48]$$

It can be shown that this determinant, which is a polynomial of degree 24, has $(1+x_1^2)^4$ as a common factor [Raghavan 90]. Thus, equation [4.48] is written as:

$$\det(\mathbf{S}) = f(x_1)\,(1+x_1^2)^4 = 0 \qquad [4.49]$$

The polynomial $f(x_1)$ is of degree sixteen and represents the characteristic polynomial of the robot. The real roots of this polynomial give all the solutions for

θ_1. For each value of θ_1, we can calculate the matrix S. The variables θ_2 and θ_3 are uniquely determined by solving the linear system of equation [4.45]. By substituting θ_1, θ_2 and θ_3 in equation [4.37] and using eight equations, we can calculate θ_5 and θ_6. Finally, we consider the following equation to calculate θ_4:

$$ {}^4T_3 = {}^4T_6\, U_0\, {}^0T_3 \qquad [4.50] $$

By using the (1, 1) and (2, 1) elements, we obtain θ_4 using an atan2 function.

The same method can also be applied to six degree-of-freedom robots having prismatic joints. It this case, Si and Ci have to be replaced by r_i^2 and r_i in X1 and Y respectively, i being the prismatic joint.

NOTE.– Equation [4.34] is a particular form of equation [4.2] that can be written in several other forms [Mavroidis 93], for example:

$$ {}^4T_5\, {}^5T_6\, {}^6T_7\, {}^0T_1 = {}^4T_3\, {}^3T_2\, {}^2T_1 \qquad [4.51a] $$
$$ {}^5T_6\, {}^6T_7\, {}^0T_1\, {}^1T_2 = {}^5T_4\, {}^4T_3\, {}^3T_2 \qquad [4.51b] $$
$$ {}^6T_7\, {}^0T_1\, {}^1T_2\, {}^2T_3 = {}^6T_5\, {}^5T_4\, {}^4T_3 \qquad [4.51c] $$
$$ {}^0T_1\, {}^1T_2\, {}^2T_3\, {}^3T_4 = {}^7T_6\, {}^6T_5\, {}^5T_4 \qquad [4.51d] $$
$$ {}^1T_2\, {}^2T_3\, {}^3T_4\, {}^4T_5 = {}^1T_0\, {}^7T_6\, {}^6T_5 \qquad [4.51e] $$
$$ {}^2T_3\, {}^3T_4\, {}^4T_5\, {}^5T_6 = {}^2T_1\, {}^1T_0\, {}^7T_6 \qquad [4.51f] $$

with ${}^7T_6 = U_0$ and ${}^6T_7 = U_0^{-1}$

The selection of the starting equation not only defines the variable of the characteristic equation but also the degree of the corresponding polynomial. For specific values of the geometric parameters, certain columns of the matrix S become dependent and it is necessary to either change the selected variables and columns [Khalil 94b], [Murareci 97] or choose another starting equation [Mavroidis 93].

When the robot is in a singular configuration, the rows of the matrix S are linearly dependent. In this case, it is not possible to find a solution. In fact, this method has proved the maximum number of solutions that can be obtained for the inverse geometric problem of serial robots, but it is hardly usable to develop a general numerical method to treat any robot architecture.

4.6. Conclusion

In this chapter, we have presented three methods for calculating the inverse geometric model. The Paul method is applicable to a large number of structures with particular geometrical parameters where most of the distances are zero and most of the angles are zero or $\pm\pi/2$. The Pieper method gives the solution for the six degree-of-freedom robots having three prismatic joints or three revolute joints whose axes intersect at a point. Finally, the general method provides the solution for the IGM of six degree-of-freedom robots with general geometry.

The analytical solution, as compared to the differential methods discussed in the next chapter, is useful for obtaining all the solutions of the inverse geometric model. Some of them may be eliminated because they do not satisfy the joint limits. Generally, the selected solution is left to the robot's user and depends on the task specifications: to avoid collisions between the robot and its environment; to ensure the continuity of the trajectory as required in certain tasks prohibiting configuration changes (machining, welding,...); to avoid as much as possible the singular configurations that may induce control problems (namely discontinuity of velocity), etc.

Chapter 5

Direct kinematic model of serial robots

5.1. Introduction

The direct kinematic model of a robot manipulator gives the velocity of the end-effector $\dot{\mathbf{X}}$ in terms of the joint velocities $\dot{\mathbf{q}}$. It is written as:

$$\dot{\mathbf{X}} = \mathbf{J}(\mathbf{q})\,\dot{\mathbf{q}} \qquad [5.1]$$

where $\mathbf{J}(\mathbf{q})$ denotes the (mxn) Jacobian matrix.

The same Jacobian matrix also appears in the direct differential model, which provides the differential displacement of the end-effector $\mathbf{dX}$ in terms of the differential variation of the joint variables $\mathbf{dq}$:

$$\mathbf{dX} = \mathbf{J}(\mathbf{q})\,\mathbf{dq} \qquad [5.2]$$

The Jacobian matrix has multiple applications in robotics [Whitney 69], [Paul 81]. The most obvious is the use of its inverse to numerically compute a solution for the inverse geometric model, i.e. to compute the joint variables $\mathbf{q}$ corresponding to a given location of the end-effector $\mathbf{X}$ (Chapter 6). The transpose of the Jacobian matrix is used in the static model to compute the necessary joint forces and torques to exert specified forces and moments on the environment. The Jacobian matrix is also used to determine the singularities and to analyze the reachable workspace of robots [Borrel 86], [Wenger 89].

In this chapter, we will present the computation of the Jacobian matrix and expose its different applications for serial robots. The kinematic model of complex chain robots will be studied in Chapter 7.

5.2. Computation of the Jacobian matrix from the direct geometric model

The Jacobian matrix can be obtained by differentiating the DGM, $\mathbf{X} = \mathbf{f}(\mathbf{q})$, using the partial derivative $\frac{\partial \mathbf{f}}{\partial \mathbf{q}}$ such that:

$$J_{ij} = \frac{\partial f_i(\mathbf{q})}{\partial q_j} \quad \text{for } i = 1, \ldots, m \text{ and } j = 1, \ldots, n \qquad [5.3]$$

where J_{ij} is the (i, j) element of the Jacobian matrix **J**.

This method is convenient for simple robots having a reduced number of degrees of freedom as shown in the following example. The computation of the basic Jacobian matrix, also known as kinematic Jacobian matrix, is more practical for a general n degree-of-freedom robot. It is presented in § 5.3.

• **Example 5.1.** Let us consider the three degree-of-freedom planar robot presented in Figure 5.1. Let us denote the link lengths by L1, L2 and L3.

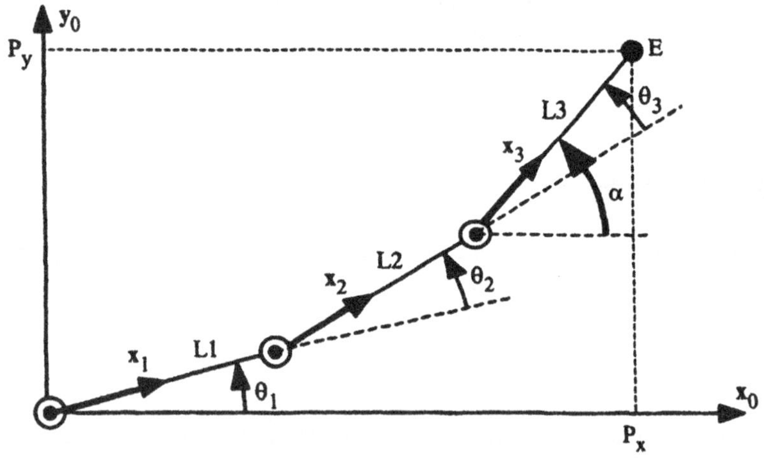

Figure 5.1. *Example of a three degree-of-freedom planar robot*

The task coordinates, defined as the position coordinates (P_x, P_y) of the terminal point E and the angle α giving the orientation of the third link relative to frame R_0, are such that:

$$P_x = C1\,L1 + C12\,L2 + C123\,L3$$
$$P_y = S1\,L1 + S12\,L2 + S123\,L3$$

$$\alpha = \theta_1 + \theta_2 + \theta_3$$

where C1 = $\cos(\theta_1)$, S1 = $\sin(\theta_1)$, C12 = $\cos(\theta_1+\theta_2)$, S12 = $\sin(\theta_1+\theta_2)$, C123 = $\cos(\theta_1+\theta_2+\theta_3)$ and S123 = $\sin(\theta_1+\theta_2+\theta_3)$.

The Jacobian matrix is obtained by differentiating these expressions with respect to θ_1, θ_2 and θ_3:

$$\mathbf{J} = \begin{bmatrix} -S1L1-S12L2-S123L3 & -S12L2-S123L3 & -S123L3 \\ C1L1+C12L2+C123L3 & C12L2+C123L3 & C123L3 \\ 1 & 1 & 1 \end{bmatrix}$$

5.3. Basic Jacobian matrix

In this section, we present a direct method to compute the Jacobian matrix of a serial mechanism without differentiating the DGM. The Jacobian matrix obtained is called the *basic Jacobian matrix*, or *kinematic Jacobian matrix*. It relates the kinematic screw of frame R_n to the joint velocities $\dot{\mathbf{q}}$:

$$\mathbb{V}_n = \begin{bmatrix} \mathbf{V}_n \\ \boldsymbol{\omega}_n \end{bmatrix} = \mathbf{J}_n \, \dot{\mathbf{q}} \qquad [5.4a]$$

where $\mathbf{V}_n$ and $\boldsymbol{\omega}_n$ are the linear and angular velocities of frame R_n respectively. We note that $\mathbf{V}_n$ is the derivative of the position vector $\mathbf{P}_n$ with respect to time, while $\boldsymbol{\omega}_n$ is not the derivative of any orientation vector.

The basic Jacobian matrix also gives the relationship between the differential translation and rotation vectors ($\mathbf{dP}_n$, $\boldsymbol{\delta}_n$) of frame R_n in terms of the differential joint variables $\mathbf{dq}$:

$$\begin{bmatrix} \mathbf{dP}_n \\ \boldsymbol{\delta}_n \end{bmatrix} = \mathbf{J}_n \, \mathbf{dq} \qquad [5.4b]$$

We will show in § 5.11 that the Jacobian giving the end-effector velocity $\dot{\mathbf{X}}$, for any task coordinate representation, can be deduced from the basic Jacobian $\mathbf{J}_n$.

5.3.1. *Computation of the basic Jacobian matrix*

The velocity $\dot{q}_k$ of joint k produces the linear and angular velocities ($\mathbf{V}_{k,n}$ and $\omega_{k,n}$ respectively) at the terminal frame R_n. Two cases are considered:

- if joint k is prismatic ($\sigma_k = 1$, Figure 5.2):

$$\begin{cases} \mathbf{V}_{k,n} = \mathbf{a}_k \, \dot{q}_k \\ \omega_{k,n} = \mathbf{0} \end{cases} \qquad [5.5]$$

where $\mathbf{a}_k$ is the unit vector along the $\mathbf{z}_k$ axis;

- if joint k is revolute ($\sigma_k = 0$, Figure 5.3):

$$\begin{cases} \mathbf{V}_{k,n} = \mathbf{a}_k \dot{q}_k \times \mathbf{L}_{k,n} = (\mathbf{a}_k \times \mathbf{L}_{k,n})\dot{q}_k \\ \omega_{k,n} = \mathbf{a}_k \dot{q}_k \end{cases} \qquad [5.6]$$

where $\mathbf{L}_{k,n}$ denotes the position vector connecting O_k to O_n.

Thus, $\mathbf{V}_{k,n}$ and $\omega_{k,n}$ can be written in the following general form:

$$\begin{cases} \mathbf{V}_{k,n} = [\sigma_k \mathbf{a}_k + \bar{\sigma}_k (\mathbf{a}_k \times \mathbf{L}_{k,n})] \dot{q}_k \\ \omega_{k,n} = \bar{\sigma}_k \mathbf{a}_k \dot{q}_k \end{cases} \qquad [5.7]$$

The linear and angular velocities of the terminal frame can be written as:

$$\begin{cases} \mathbf{V}_n = \sum_{k=1}^{n} \mathbf{V}_{k,n} = \sum_{k=1}^{n} [\sigma_k \mathbf{a}_k + \bar{\sigma}_k (\mathbf{a}_k \times \mathbf{L}_{k,n})] \dot{q}_k \\ \omega_n = \sum_{k=1}^{n} \omega_{k,n} = \sum_{k=1}^{n} \bar{\sigma}_k \mathbf{a}_k \dot{q}_k \end{cases} \qquad [5.8]$$

Writing equation [5.8] in matrix form and using equation [5.4], we deduce that:

$$\mathbf{J}_n = \begin{bmatrix} \sigma_1 \mathbf{a}_1 + \bar{\sigma}_1 (\mathbf{a}_1 \times \mathbf{L}_{1,n}) & \ldots & \sigma_n \mathbf{a}_n + \bar{\sigma}_n (\mathbf{a}_n \times \mathbf{L}_{n,n}) \\ \bar{\sigma}_1 \mathbf{a}_1 & \ldots & \bar{\sigma}_n \mathbf{a}_n \end{bmatrix} \qquad [5.9]$$

Referring the vectors of $\mathbf{J}_n$ with respect to frame R_i, we obtain the (6xn) Jacobian matrix ${}^i\mathbf{J}_n$ such that:

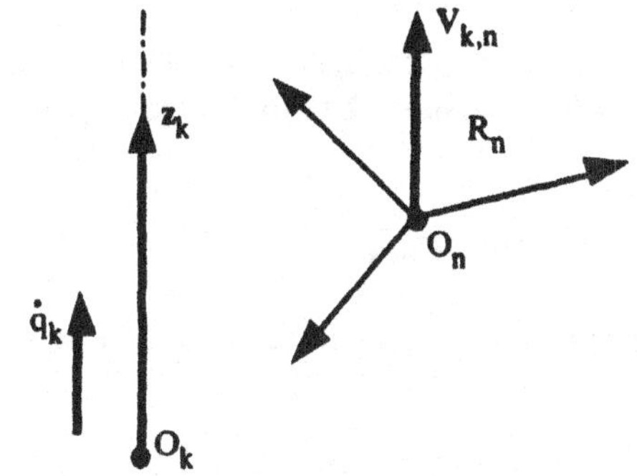

Figure 5.2. *Case of a prismatic joint*

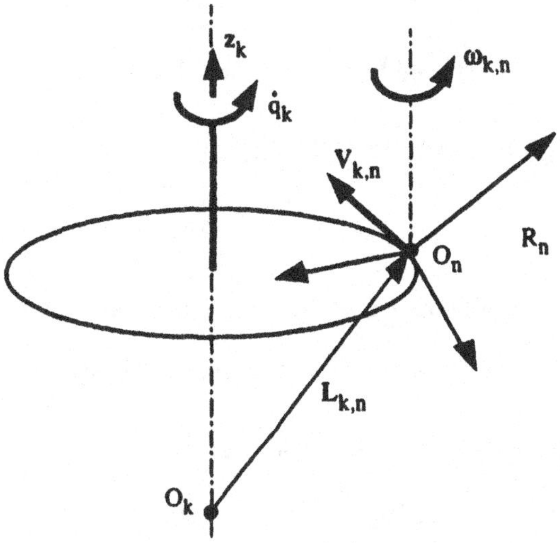

Figure 5.3. *Case of a revolute joint*

$$^{i}\mathbb{V}_n = {}^{i}\mathbf{J}_n \, \dot{\mathbf{q}} \qquad [5.10]$$

In general, we calculate $\mathbf{V}_n$ and $\boldsymbol{\omega}_n$ in frame R_n or frame R_0. The corresponding Jacobian matrix is denoted by $^{n}\mathbf{J}_n$ or $^{0}\mathbf{J}_n$ respectively. These matrices can also be computed using any matrix $^{i}\mathbf{J}_n$, for i = 0, ..., n, thanks to the following expression:

$$^{s}\mathbf{J}_n = \begin{bmatrix} ^{s}\mathbf{A}_i & \mathbf{0}_3 \\ \mathbf{0}_3 & ^{s}\mathbf{A}_i \end{bmatrix} {}^{i}\mathbf{J}_n \qquad [5.11]$$

where ${}^{s}\mathbf{A}_i$ is the (3x3) orientation matrix of frame R_i relative to frame R_s.

In general, we obtain the simplest matrix ${}^{i}\mathbf{J}_n$ when i = integer (n/2). We note that the matrices ${}^{i}\mathbf{J}_n$, for i = 0, ..., n, have the same singular positions.

5.3.2. *Computation of the matrix* ${}^{i}J_n$

Since the vector product $\mathbf{a}_k \times \mathbf{L}_{k,n}$ can be computed by $\hat{\mathbf{a}}_k \mathbf{L}_{k,n}$, the kth column of ${}^{i}\mathbf{J}_n$, denoted as ${}^{i}\mathbf{j}_{n;k}$, becomes:

$$
{}^{i}\mathbf{j}_{n;k} = \begin{bmatrix} \sigma_k\, {}^{i}\mathbf{a}_k + \bar{\sigma}_k\, {}^{i}\mathbf{A}_k\, {}^{k}\hat{\mathbf{a}}_k\, {}^{k}\mathbf{L}_{k,n} \\ \bar{\sigma}_k\, {}^{i}\mathbf{a}_k \end{bmatrix}
$$

Since ${}^{k}\mathbf{a}_k = [0 \quad 0 \quad 1]^T$ and ${}^{k}\mathbf{L}_{k,n} = {}^{k}\mathbf{P}_n$, we obtain:

$$
{}^{i}\mathbf{j}_{n;k} = \begin{bmatrix} \sigma_k\, {}^{i}\mathbf{a}_k + \bar{\sigma}_k\, (-{}^{k}P_{ny}\, {}^{i}\mathbf{s}_k + {}^{k}P_{nx}\, {}^{i}\mathbf{n}_k) \\ \bar{\sigma}_k\, {}^{i}\mathbf{a}_k \end{bmatrix} \qquad [5.12]
$$

where ${}^{k}P_{nx}$ and ${}^{k}P_{ny}$ denote the x and y components of the vector ${}^{k}\mathbf{P}_n$ respectively.

From this expression, we obtain the kth column of ${}^{n}\mathbf{J}_n$ as:

$$
{}^{n}\mathbf{j}_{n;k} = \begin{bmatrix} \sigma_k\, {}^{n}\mathbf{a}_k + \bar{\sigma}_k\, (-{}^{k}P_{ny}\, {}^{n}\mathbf{s}_k + {}^{k}P_{nx}\, {}^{n}\mathbf{n}_k) \\ \bar{\sigma}_k\, {}^{n}\mathbf{a}_k \end{bmatrix} \qquad [5.13]
$$

The column ${}^{n}\mathbf{j}_{n;k}$ is computed from the elements of the matrix ${}^{k}\mathbf{T}_n$ resulting from the DGM.

In a similar way, the kth column of ${}^{i}\mathbf{J}_n$ is also written as:

$$
{}^{i}\mathbf{j}_{n;k} = \begin{bmatrix} \sigma_k\, {}^{i}\mathbf{a}_k + \bar{\sigma}_k\, {}^{i}\hat{\mathbf{a}}_k\, ({}^{i}\mathbf{P}_n - {}^{i}\mathbf{P}_k) \\ \bar{\sigma}_k\, {}^{i}\mathbf{a}_k \end{bmatrix} \qquad [5.14]
$$

which gives for i = 0:

$$ {}^0\mathbf{j}_{n;k} = \begin{bmatrix} \sigma_k\, {}^0\mathbf{a}_k + \bar{\sigma}_k\, {}^0\hat{\mathbf{a}}_k\, ({}^0\mathbf{P}_n - {}^0\mathbf{P}_k) \\ \bar{\sigma}_k\, {}^0\mathbf{a}_k \end{bmatrix} \qquad [5.15] $$

In this case, we need to compute the matrices ${}^0\mathbf{T}_k$ for k = 1, …, n.

NOTE. – To find the Jacobian ${}^E\mathbf{J}_E$ defining the velocity of the tool frame R_E, we can either make use of equation [5.9] after replacing $\mathbf{L}_{k,n}$ by $\mathbf{L}_{k,E}$, or compute ${}^E\mathbb{V}_E$ as a function of ${}^n\mathbb{V}_n$, and deduce ${}^E\mathbf{J}_E$. From § 2.4.3, we can see that:

$$ {}^E\mathbb{V}_E = {}^E\mathbb{T}_n\, {}^n\mathbb{V}_n = {}^E\mathbb{T}_n\, {}^n\mathbf{J}_n\, \dot{\mathbf{q}} $$

where ${}^E\mathbb{T}_n$ is the (6x6) screw transformation matrix defined in equation [2.47]. Consequently, we deduce that:

$$ {}^E\mathbf{J}_E = \begin{bmatrix} {}^E\mathbf{A}_n & -{}^E\mathbf{A}_n\, {}^n\hat{\mathbf{P}}_E \\ \mathbf{0}_3 & {}^E\mathbf{P}_n \end{bmatrix} {}^n\mathbf{J}_n = {}^E\mathbb{T}_n\, {}^n\mathbf{J}_n \qquad [5.16] $$

• **Example 5.2.** Compute the Jacobian matrix ${}^6\mathbf{J}_6$ of the Stäubli RX-90 robot. Using equation [5.13] and the matrices ${}^k\mathbf{T}_6$ resulting from the DGM, we obtain:

J (1,1) = (– C6C5S4 – S6C4)(S23RL4 – C2D3)
J (2,1) = (S6C5S4 – C6C4)(S23RL4 – C2D3)
J (3,1) = S5S4(S23RL4 – C2D3)
J (4,1) = (C6C5C4 – S6S4)S23 + C6S5C23
J (5,1) = (– S6C5C4 – C6S4)S23 – S6S5C23
J (6,1) = – S5C4S23 + C5C23
J (1,2) = (– C6C5C4 + S6S4)(RL4 – S3D3) + C6S5C3D3
J (2,2) = (S6C5C4 + C6S4)(RL4 – S3D3) – S6S5C3D3
J (3,2) = S5C4(RL4 – S3D3) + C5C3D3
J (4,2) = – C6C5S4 – S6C4
J (5,2) = S6C5S4 – C6C4
J (6,2) = S5S4
J (1,3) = (– C6C5C4 + S6S4)RL4
J (2,3) = (S6C5C4 + C6S4)RL4
J (3,3) = S5C4RL4
J (4,3) = – C6C5S4 – S6C4
J (5,3) = S6C5S4 – C6C4
J (6,3) = S5S4
J (1,4) = 0
J (2,4) = 0
J (3,4) = 0
J (4,4) = C6S5
J (5,4) = – S6S5

J (6,4) = C5
J (1,5) = 0
J (2,5) = 0
J (3,5) = 0
J (4,5) = – S6
J (5,5) = – C6
J (6,5) = 0
J (1,6) = 0
J (2,6) = 0
J (3,6) = 0
J (4,6) = 0
J (5,6) = 0
J (6,6) = 1

• **Example 5.3.** Determine the Jacobian matrix $^3\mathbf{J}_6$ of the Stäubli RX-90 robot. The column k of the matrix $^3\mathbf{J}_6$ for a revolute joint is obtained from equation [5.12] as:

$$^3\mathbf{j}_{6;k} = \begin{bmatrix} -{}^kP_{6y}\,{}^3\mathbf{s}_k + {}^kP_{6x}\,{}^3\mathbf{n}_k \\ {}^3\mathbf{a}_k \end{bmatrix}$$

The elements $^kP_{6y}$ and $^kP_{6x}$ are obtained from the DGM. The vectors $^3\mathbf{s}_k$, $^3\mathbf{n}_k$ and $^3\mathbf{a}_k$, for k = 2, 3, 4 and 6, are deduced from the matrices $^3\mathbf{A}_2$, $^3\mathbf{A}_3$, $^3\mathbf{A}_4$ and $^3\mathbf{A}_6$, which are also computed for the DGM. The additional matrices to be computed are $^3\mathbf{A}_1$ and $^3\mathbf{A}_5$. Finally, we obtain:

$$^3\mathbf{J}_6 = \begin{bmatrix} 0 & -RL4+S3D3 & -RL4 & 0 & 0 & 0 \\ 0 & C3D3 & 0 & 0 & 0 & 0 \\ S23\,RL4-C2D3 & 0 & 0 & 0 & 0 & 0 \\ S23 & 0 & 0 & 0 & S4 & -S5C4 \\ C23 & 0 & 0 & 1 & 0 & C5 \\ 0 & 1 & 1 & 0 & C4 & S5S4 \end{bmatrix}$$

5.4. Decomposition of the Jacobian matrix into three matrices

We have shown in equation [5.11] that the matrix $^s\mathbf{J}_n$ could be decomposed into two matrices; the first is always of full-rank and the second contains simple elements. Renaud [Renaud 80b] has shown that the Jacobian matrix could also be decomposed into three matrices: the first two are always of full-rank and their inverse is straightforward; the third is of the same rank as $^s\mathbf{J}_n$, but contains simpler elements.

Figure 5.4 illustrates the principle of the proposed method: the influence of the joint velocities is not calculated at the level of the terminal frame R_n but at the level of an intermediate frame R_j. Therefore, we define the Jacobian matrix $J_{n,j}$ as:

$$J_{n,j} = \begin{bmatrix} \sigma_1 a_1 + \bar{\sigma}_1 (a_1 \times L_{1,j}) & \cdots & \sigma_n a_n + \bar{\sigma}_n (a_n \times L_{n,j}) \\ \bar{\sigma}_1 a_1 & \cdots & \bar{\sigma}_n a_n \end{bmatrix} \quad [5.17]$$

The matrix $J_{n,j}$ is equivalent to the Jacobian matrix defining the velocity of a frame fixed to link n and aligned instantaneously with frame R_j. We can compute J_n from $J_{n,j}$ using the expression:

$$J_n = \begin{bmatrix} I_3 & -\hat{L}_{j,n} \\ 0_3 & I_3 \end{bmatrix} J_{n,j} \quad [5.18]$$

By projecting the elements of this equation into frame R_i, we obtain:

$$^iJ_n = \begin{bmatrix} I_3 & -{}^i\hat{L}_{j,n} \\ 0_3 & I_3 \end{bmatrix} {}^iJ_{n,j} \quad [5.19]$$

with:

$$^iL_{j,n} = {}^iA_j\, {}^jP_n \quad [5.20]$$

The kth column of $^iJ_{n,j}$, deduced from equation [5.17], is expressed in frame R_i as:

$$^ij_{n,j;k} = \begin{bmatrix} \sigma_k\, {}^ia_k + \bar{\sigma}_k (-{}^kP_{jy}\, {}^is_k + {}^kP_{jx}\, {}^in_k) \\ \bar{\sigma}_k\, {}^ia_k \end{bmatrix} \quad [5.21]$$

We note that $^iJ_n = {}^iJ_{n,n}$. Thus, the matrix ^sJ_n can be expressed by the multiplication of the following three matrices where the first two are of full-rank:

$$^sJ_n = \begin{bmatrix} {}^sA_i & 0_3 \\ 0_3 & {}^sA_i \end{bmatrix} \begin{bmatrix} I_3 & -{}^i\hat{L}_{j,n} \\ 0_3 & I_3 \end{bmatrix} {}^iJ_{n,j} \quad [5.22]$$

In general, the shift frame R_j and the projection frame R_i leading to a simple matrix $^iJ_{n,j}$ are chosen such that i = integer (n/2) and j = i + 1.

Thus, for a six degree-of-freedom robot, the simplest Jacobian matrix is ${}^3\mathbf{J}_{6,4}$. If the robot has a spherical wrist, the vector $\mathbf{L}_{4,6}$ is zero and consequently ${}^3\mathbf{J}_{6,4} = {}^3\mathbf{J}_6$.

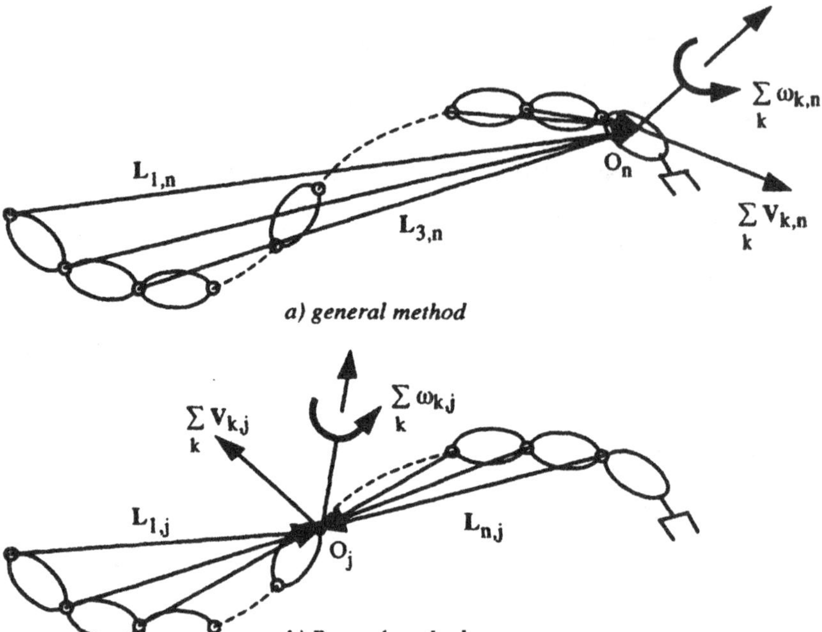

b) Renaud method

Figure 5.4. *Principle of Renaud method*

5.5. Efficient computation of the end-effector velocity

Having calculated $\mathbf{J}_n$, the linear and angular velocities $\mathbf{V}_n$ and $\boldsymbol{\omega}_n$ of frame R_n can be obtained from equation [5.4a]. However, in order to reduce the computational cost, it is more efficient, as we will see in Chapter 9, to use the following recursive equations for j = 1, ..., n:

$$\begin{cases} {}^j\boldsymbol{\omega}_{j-1} = {}^j\mathbf{A}_{j-1}\, {}^{j-1}\boldsymbol{\omega}_{j-1} \\ {}^j\boldsymbol{\omega}_j = {}^j\boldsymbol{\omega}_{j-1} + \bar{\sigma}_j\, \dot{q}_j\, {}^j\mathbf{a}_j \\ {}^j\mathbf{V}_j = {}^j\mathbf{A}_{j-1}\,({}^{j-1}\mathbf{V}_{j-1} + {}^{j-1}\boldsymbol{\omega}_{j-1} \times {}^{j-1}\mathbf{P}_j) + \sigma_j\, \dot{q}_j\, {}^j\mathbf{a}_j \end{cases} \qquad [5.23]$$

where ${}^j\mathbf{a}_j$ is the unit vector $[0 \;\; 0 \;\; 1]^T$. We initialize the algorithm by the linear and angular velocities of the robot base ($\mathbf{V}_0$ and $\boldsymbol{\omega}_0$ respectively), which are zero if the base is fixed.

5.6. Dimension of the task space of a robot

At a given joint configuration $\mathbf{q}$, the rank r of the Jacobian matrix ${}^i\mathbf{J}_n$ – hereafter written as $\mathbf{J}$ to simplify the notation – corresponds to the number of degrees of freedom of the end-effector. It defines the dimension of the accessible task space at this configuration. The number of degrees of freedom of the task space of a robot, M, is equal to the maximum rank, r_{max}, which the Jacobian matrix can have at all possible configurations. Noting the number of degrees of freedom of the robot as N (equal to n for serial robots), the following cases are considered [Gorla 84]:

- if N = M, the robot is non-redundant and has just the number of joints required to provide M degrees of freedom to the end-effector;
- if N > M, the robot is redundant of order (N – M). It has more joints than required to provide M degrees of freedom to the end-effector;
- if r < M, the Jacobian matrix is rank deficient. The robot is at a singular configuration of order (M – r). At this configuration, the robot cannot generate an end-effector velocity along some directions of the task space, which are known as *degenerate directions*. When the matrix $\mathbf{J}$ is square, the singularities are obtained by the zeros of $\det(\mathbf{J}) = 0$, where $\det(\mathbf{J})$ indicates the determinant of the Jacobian matrix of the robot. They correspond to the zeros of $\det(\mathbf{J}\,\mathbf{J}^T) = 0$ for redundant robots.

• **Example 5.4.** Computation of the singularities of the Stäubli RX-90 robot. Noting that the Jacobian matrix ${}^3\mathbf{J}_6$ (obtained in Example 5.3) has the following particular form:

$$ {}^3\mathbf{J}_6 = \begin{bmatrix} \mathbf{A} & \mathbf{0}_3 \\ \mathbf{B} & \mathbf{C} \end{bmatrix} $$

we obtain $\det({}^3\mathbf{J}_6) = \det(\mathbf{A})\det(\mathbf{C}) = -\mathrm{C3\ D3\ RL4\ S5\ (S23\ RL4 - C2\ D3)}$.

The maximum rank is $r_{max} = 6$. The robot is not redundant because it has six degrees of freedom. However, this rank drops to five in the following three singular configurations:

$$ \begin{cases} \mathrm{C3} = 0 \\ \mathrm{S23\ RL4 - C2\ D3} = 0 \\ \mathrm{S5} = 0 \end{cases} $$

- when C3 = 0, the robot is fully extended (Figure 4.2c) or fully folded. In this case, the origin O_6 is located on the boundary of its workspace: this elbow singularity has not been deduced from the inverse geometric model (§ 4.3.2,

Example 4.1). In this configuration, where the second row of ${}^3\mathbf{J}_6$ is zero, the robot cannot generate linear velocity for O_6 along the direction O_6O_2;

- the singularity S23 RL4 – C2 D3 = 0 (Figure 4.2a), already deduced from the inverse geometric model, corresponds to a configuration in which O_6 is located on the $\mathbf{z}_0$ axis (shoulder singularity). In this configuration, where $P_x = P_y = 0$, the third row of ${}^3\mathbf{J}_6$ is zero. The robot cannot generate velocity for O_6 along the normal to the plane containing the points O_2, O_3 and O_6;
- for S5 = 0 (Figure 4.2b), the axes of the joints θ_4 and θ_6 are aligned, resulting in the loss of one degree of freedom of the robot. We notice that the columns 4 and 6 of ${}^3\mathbf{J}_6$ are identical. In this configuration, the robot cannot generate rotational velocity for frame R_6 about the normal to the plane containing the axes $\mathbf{z}_4$, $\mathbf{z}_5$ and $\mathbf{z}_6$. This wrist singularity has already been deduced from the inverse geometric model.

5.7. Analysis of the robot workspace

The analysis of the workspace is very important for the design, selection and programming of robots.

5.7.1. *Workspace*

Let $\mathbf{q} = [q_1, \ldots, q_n]$ be an element of the joint space and let $\mathbf{X} = [x_1, \ldots, x_m]$ be the corresponding element in the task space, such that:

$$\mathbf{X} = \mathbf{f}(\mathbf{q}) \qquad [5.24]$$

The joint domain $\mathbf{Q}$ is defined as the set of all reachable configurations taking into account the joint limits:

$$\mathbf{Q} = \{\mathbf{q} \mid q_{imin} \leq q_i \leq q_{imax}, \forall\, i = 1, \ldots, n\} \qquad [5.25]$$

The image of $\mathbf{Q}$ by the direct geometric model DGM defines the workspace W of the robot:

$$W = \mathbf{f}(\mathbf{Q}) \qquad [5.26]$$

Thus, the workspace W is the set of positions and orientations reachable by the robot end-effector. Its geometry depends on the robot architecture. Its boundaries are defined by the singularities and the joint limits. However, when there is an obstacle in the robot workspace, additional boundaries limiting the reachable zones will appear [Wenger 89].

For robots with two joints, the workspace is easy to obtain and can be visualized in a plane. For a three degree-of-freedom positioning shoulder, the workspace can be represented by a generic planar cross section of W. This cross section contains the axis of the first joint if it is revolute, whereas it is perpendicular to the axis of the first joint if it is prismatic. The whole workspace is obtained from the generic cross section by rotating it about (or translating it along) the first joint axis. However, if there are obstacles or joint limits, the generic planar section is not sufficient for a complete analysis of the workspace.

In general, the workspace is a 6-dimensional space, which is difficult to handle. However, we can study its projection in the 3-dimensional position space.

5.7.2. *Singularity branches*

The *singularity branches* are the connected components of the set of singular configurations of Q. Since the singularities are always independent of the first joint, we can represent them in the joint space excluding the first joint. They are represented by surfaces of Q. However, for some particular cases, they can be reduced to subspaces of fewer dimensions (curves or points for example), which do not have a boundary in Q.

For the two degree-of-freedom planar robot with revolute joints shown in Figure 5.5, the determinant of the Jacobian matrix is equal to L1 L2 S2. The singularity branches, assuming unlimited joint ranges, are defined by the lines $\theta_2 = 0$ and $\theta_2 = \pm\pi$ (Figure 5.6). The corresponding workspace is presented in Figure 5.7.

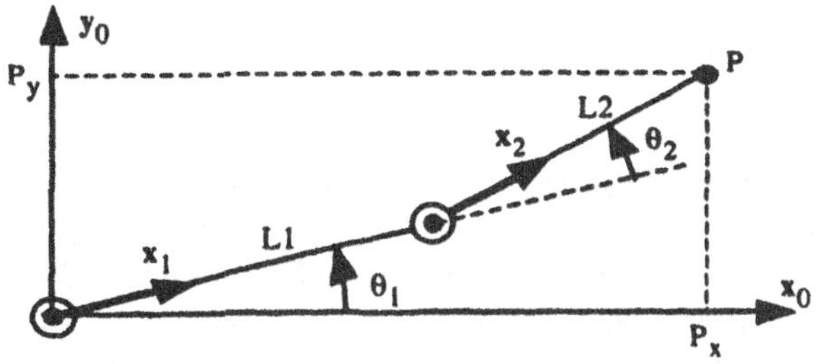

Figure 5.5. *Two degree-of-freedom planar robot*

For the Stäubli RX-90 robot, the joint space is partitioned by three singularity surfaces C3 = 0, S23 RL4 − C2 D3 = 0 and S5 = 0. Figure 5.8 shows these surfaces in the $(\theta_2, \theta_3, \theta_5)$ space and in the (θ_2, θ_3) plane.

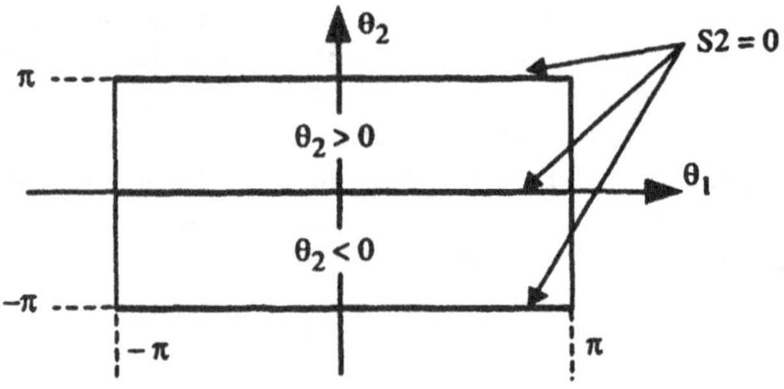

Figure 5.6. *Singularity branches of the planar robot with unlimited joints*

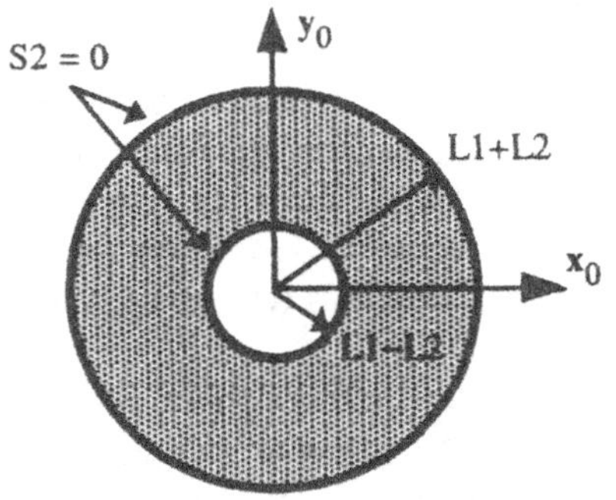

Figure 5.7. *Workspace of the planar robot with unlimited joints (L1>L2)*

5.7.3. *Jacobian surfaces*

Mapping the singularities into the workspace generally leads to surfaces (or subspaces with fewer dimensions) called *Jacobian surfaces*. These surfaces divide W into regions where the number of solutions of the IGM is constant and even [Roth 76], [Kholi 85], [Burdick 88]. In the presence of joint limits, additional boundaries appear in W, which define new regions in which the number of solutions of the IGM is always constant but not necessarily even [Spanos 85]. The Jacobian surfaces can be defined as the set of points in W where the IGM has at least two identical solutions [Kholi 87], [Spanos 85]. When the robot has three identical solutions for a point of the Jacobian surface, the robot is said to be *cuspidal* [El Omri 96].

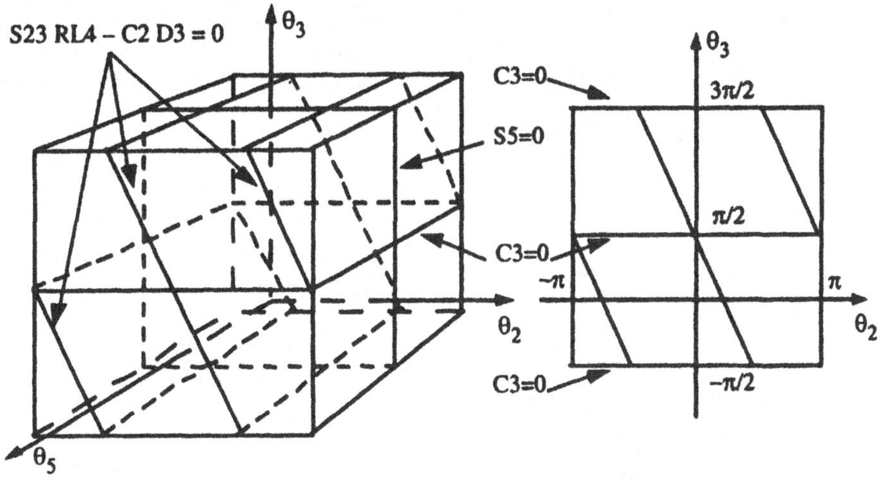

Figure 5.8. *Singularity branches of the Stäubli RX-90 robot*

In the case of a three degree-of-freedom robot, if the Jacobian surfaces are subspaces of fewer dimensions (for example a curve or a set of isolated points), the IGM for these points has an infinite number of solutions.

For the two degree-of-freedom planar robot shown in Figure 5.5, the Jacobian surfaces correspond to the singular configurations "extended arm" and "folded arm". They are represented by the circles with radii L1 + L2 and L1 – L2 respectively (Figure 5.7).

For the anthropomorphic shoulder of the Stäubli RX-90 robot, the Jacobian surfaces in the position workspace are of two types (Figure 5.9). The first is associated with the singular configurations where the point O_6 lies on the axis of the first joint. Their reciprocal mapping in **Q** give the singularity surfaces defined by S23RL4 – C2D3 = 0. For any point of these configurations, the IGM has an infinite number of solutions since θ_1 can be chosen arbitrarily. The other type of Jacobian surface corresponds to the singular configuration C3 = 0, and is represented by the surfaces of the spheres whose center is O_0, with radii D3 + RL4 ("extended arm" configuration) and D3 – RL4 ("folded arm" configuration) defining the external and internal boundaries of the workspace respectively. For the Stäubli RX-90 robot, the internal sphere is reduced to a point because D3 = RL4.

5.7.4. *Concept of aspect*

The concept of aspect has been introduced by Borrel [Borrel 86]. The aspects are the connected regions of the joint space inside which no minor of order M extracted from the Jacobian matrix **J** is zero, except if this minor is zero everywhere in the joint domain. For a non-redundant robot manipulator, the only minor of order M is the Jacobian matrix itself. Therefore, the aspects are limited by the singularity branches and the joint limits (Figures 5.6 and 5.8). Consequently, they represent the maximum singularity-free regions.

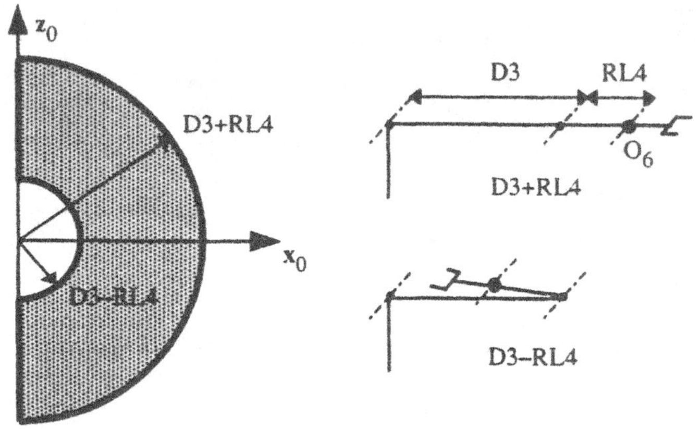

Figure 5.9. *Generic section of the workspace of an anthropomorphic shoulder with unlimited joints*

For a long time, it has been thought that the aspects also represent the uniqueness domains of the IGM solutions. Although this is indeed the case for most industrial robots with simple architectures, which are classified as non-cuspidal robots [El Omri 96], the IGM of cuspidal robots can have several solutions in the same aspect. Thus, a cuspidal robot can move from one IGM solution to another without encountering a singularity. Figure 5.10 shows a cuspidal robot with three revolute joints whose successive axes are perpendicular. The inverse geometric solution of the point X (Figure 5.11a) whose coordinates are $P_x = 2.5$, $P_y = 0$, $Pz = 0$ is given by the following four configurations (in degrees):

$$\mathbf{q}^{(1)} = [\,-101.52 \;\; -158.19 \;\; 104.88\,]^T,\; \mathbf{q}^{(2)} = [\,-50.92 \;\; -46.17 \;\; 141.16\,]^T$$
$$\mathbf{q}^{(3)} = [\,-164.56 \;\; -170.02 \;\; -12.89\,]^T,\; \mathbf{q}^{(4)} = [\,10.13 \;\; -22.33 \;\; -106.28\,]^T$$

The joint space of this robot is divided into two aspects (Figure 5.11a). We notice that the configurations $\mathbf{q}^{(2)}$ and $\mathbf{q}^{(3)}$ are located in the same aspect whereas $\mathbf{q}^{(1)}$ and $\mathbf{q}^{(4)}$ fall in the other aspect.

For cuspidal robots, the uniqueness domains of the IGM in the joint space are separated by the *characteristic surfaces* [Wenger 92], which are defined as the mapping of the Jacobian surfaces in the joint space using the IGM. Figure 5.11b shows the singularities and the characteristic surfaces of the shoulder structure of Figure 5.10.

There is no general simple rule to identify the architectures of non-cuspidal robots. However, Table 5.1 gives a list of non-cuspidal shoulders as presented in [Wenger 93], [Wenger 98].

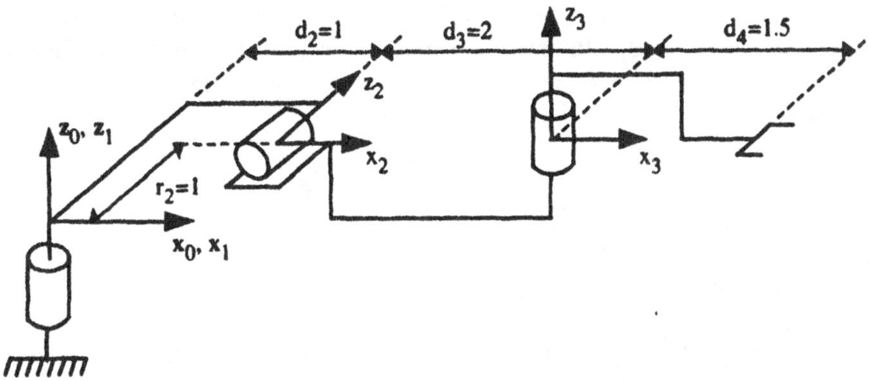

Figure 5.10. *Example of a cuspidal shoulder [Wenger 92]*

Table 5.1. *Non-cuspidal shoulders*

PPP	RPP	PRP	PPR	RRR	PRR	RPR	RRP
all	all	all	all	$s\alpha_2=0$ $s\alpha_3=0$ $d_2=0$ $d_3=0$ ($c\alpha_2=0$, $r_2=0$ and $r_3=0$)	$c\alpha_2=0$ $s\alpha_3=0$ $d_3=0$ ($s\alpha_2=0$ and $r_3=0$) ($s\alpha_2=0$ and $c\alpha_3=0$)	$c\alpha_2=0$ $c\alpha_3=0$ $s2=0$ $d_3+d_2c2=0$	$s\alpha_2=0$ $c\alpha_3=0$ $d_2=0$ ($c\alpha_2=0$, $s\alpha_3=0$ and $r_2=0$) $d_3+d_4c3=0$

5.7.5. *t-connected subspaces*

The t-connected subspaces are the regions in the workspace where any continuous trajectory can be followed by the robot end-effector. These subspaces are the mapping of the uniqueness domains in W using the DGM. For the non-cuspidal robots, the largest t-connected subspaces are the mapping of the aspects (and more generally of the free connected regions of the aspects when the environment is cluttered with obstacles [Wenger 89]). We do not present here the definition of the t-connected subspaces for the cuspidal robots. The interested reader can refer to [El Omri 96].

For the two degree-of-freedom planar robot shown in Figure 5.5, the straight line S2 = 0 separates the joint space domain into two aspects (Figure 5.12a) corresponding to the two solutions of the IGM, $\theta_2 > 0$ and $\theta_2 < 0$.

The mapping of these aspects in the workspace is identical if the joint ranges are equal to 2π. Figure 5.12b shows, for certain joint limits θ_{imax} and θ_{imin}, the t-connected regions: the hatched and non-hatched zones represent the mapping of the aspects $\theta_2 > 0$ and $\theta_2 < 0$ respectively. The trajectory PP' is located in the region mapped by the aspect $\theta_2 < 0$: thus it can only be realized if the initial configuration of the robot is $\theta_2 < 0$. Otherwise, one of the joints reaches its limit before arriving at the final position.

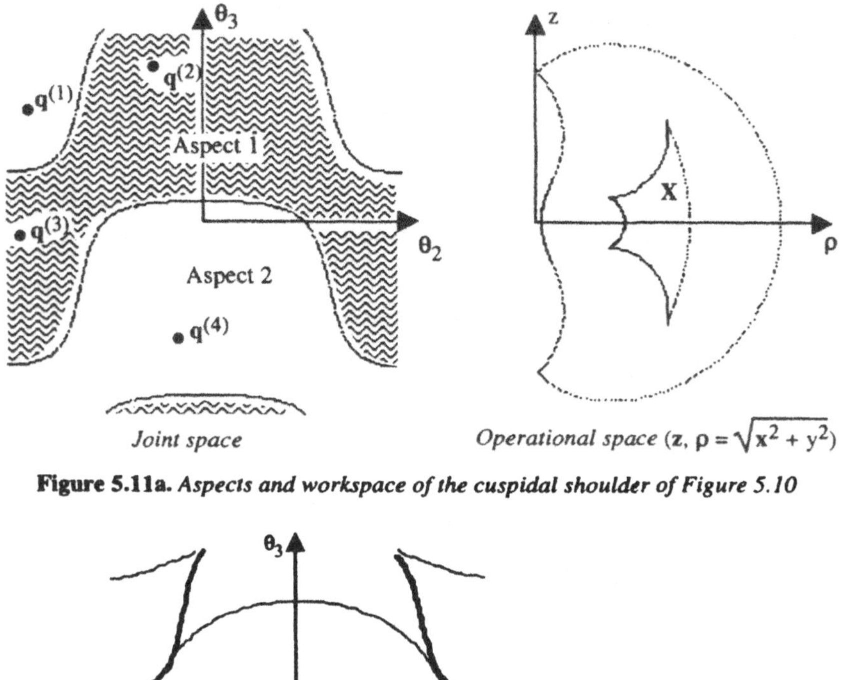

Figure 5.11a. *Aspects and workspace of the cuspidal shoulder of Figure 5.10*

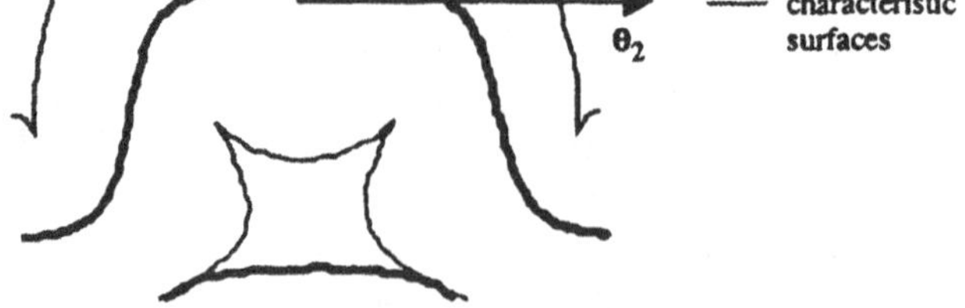

Figure 5.11b. *Singularity branches and characteristic surfaces of the cuspidal shoulder*

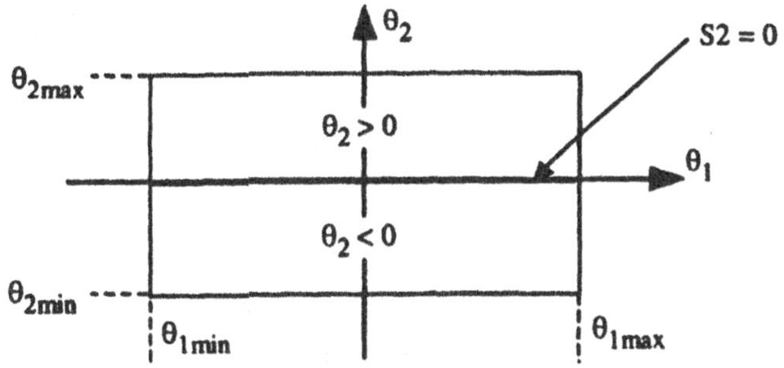

Figure 5.12a. *Aspects in the presence of joint limits*

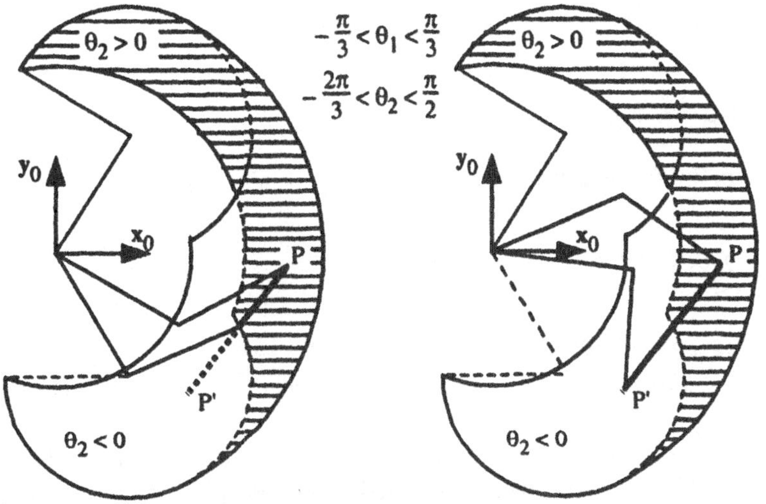

Figure 5.12b. *t-connected regions in the workspace*

5.8. Velocity transmission between joint space and task space

5.8.1. *Singular value decomposition*

At a given configuration, the (mxn) matrix **J** represents a linear mapping of the joint space velocities into the task space velocities. For simplicity, we write the basic Jacobian matrix $\mathbf{J}_n$ as **J**. When the end-effector coordinates are independent, we have n = N and m = M.

The singular value decomposition (SVD) theory states that for any (mxn) matrix **J** of rank r [Lawson 74], [Dongarra 79], [Klema 80], there exist orthogonal matrices **U** and **V** of dimensions (mxm) and (nxn) respectively such that:

$$\mathbf{J} = \mathbf{U}\,\boldsymbol{\Sigma}\,\mathbf{V}^T \qquad [5.27]$$

The (mxn) matrix $\boldsymbol{\Sigma}$ has the following form:

$$\boldsymbol{\Sigma} = \begin{bmatrix} \mathbf{S}_{rxr} & \mathbf{0}_{rx(n-r)} \\ \mathbf{0}_{(m-r)xr} & \mathbf{0}_{(m-r)x(n-r)} \end{bmatrix} \qquad [5.28]$$

S is an (rxr) diagonal matrix, formed by the non-zero singular values of **J**, which are arranged in decreasing order such that $\sigma_1 \geq \sigma_2 \geq \ldots \geq \sigma_r$. The singular values of **J** are the square roots of the eigenvalues of the matrix $\mathbf{J}^T\mathbf{J}$ if $n \geq m$ (or $\mathbf{J}\,\mathbf{J}^T$ if $n \leq m$). The columns of **V** are the eigenvectors of and are called *right singular vectors* or *input vectors* of **J**. The columns of **U** are the eigenvectors of $\mathbf{J}\,\mathbf{J}^T$ and are called *left singular vectors* or *output vectors.*

Using equation [5.27], the kinematic model becomes:

$$\dot{\mathbf{X}} = \mathbf{U}\,\boldsymbol{\Sigma}\,\mathbf{V}^T\dot{\mathbf{q}} \qquad [5.29]$$

Since $\sigma_i = 0$ for $i > r$, we can write:

$$\dot{\mathbf{X}} = \sum_{i=1}^{r} \sigma_i\,\mathbf{U}_i\,\mathbf{V}_i^T\,\dot{\mathbf{q}} \qquad [5.30]$$

From equation [5.30], we deduce that (Figure 5.13):

- the vectors $\mathbf{V}_1, \ldots, \mathbf{V}_r$ form an orthonormal basis for the subspace of $\dot{\mathbf{q}}$ generating an end-effector velocity;
- the vectors $\mathbf{V}_{r+1}, \ldots, \mathbf{V}_n$ form an orthonormal basis for the subspace of $\dot{\mathbf{q}}$ giving $\dot{\mathbf{X}} = 0$. In other words, they define the null space of **J**, denoted by $\mathcal{N}(\mathbf{J})$;
- the vectors $\mathbf{U}_1, \ldots, \mathbf{U}_r$ form an orthonormal basis for the set of the achievable end-effector velocities $\dot{\mathbf{X}}$. Hence, they define the range space of **J**, denoted by $\mathcal{R}(\mathbf{J})$;
- the vectors $\mathbf{U}_{r+1}, \ldots, \mathbf{U}_m$ form an orthonormal basis for the subspace composed of the set of $\dot{\mathbf{X}}$ that cannot be generated by the robot. In other words, they define the complement of the range space, denoted by $\mathcal{R}(\mathbf{J})^\perp$;
- the singular values represent the velocity transmission ratio from the joint space to the task space. In fact, multiplying equation [5.30] by $\mathbf{U}_i^T$ yields:

$$\mathbf{U}_i^T\,\dot{\mathbf{X}} = \sigma_i\,\mathbf{V}_i^T\,\dot{\mathbf{q}} \quad \text{for } i = 1, \ldots, r \qquad [5.31]$$

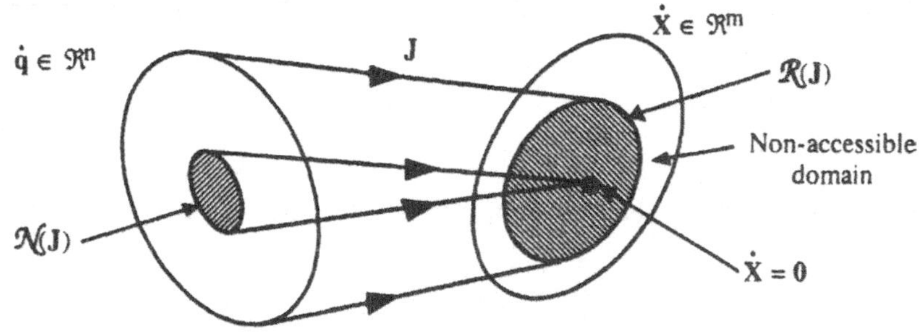

Figure 5.13. *Null space and range space of J (from [Asada 86])*

– since $J^T = V\,\Sigma\,U^T$, we deduce that:

$$\Re^m = \mathcal{R}(J) + \mathcal{R}(J)^{\perp} = \mathcal{R}(J) + \mathcal{N}(J^T)$$
$$\Re^n = \mathcal{R}(J^T) + \mathcal{N}(J)$$

5.8.2. *Velocity ellipsoid: velocity transmission performance*

The velocity transmission performance of a mechanism can be evaluated through the kinematic model [5.1]. Let us suppose that the joint velocities are limited such that:

$$-\dot{q}_{max} \leq \dot{q} \leq \dot{q}_{max} \qquad [5.32]$$

At a given configuration **q**, the task space velocity satisfying these conditions belongs to:

$$\dot{X}_{min} \leq \dot{X} \leq \dot{X}_{max} \qquad [5.33]$$

with:

$$\dot{X}_{max} = \max(J(q)\,\dot{q}) \qquad [5.34]$$

$$\dot{X}_{min} = \min(J(q)\,\dot{q}) \qquad [5.35]$$

Thus, the set of possible joint velocities (equation [5.32]) can be represented geometrically by a hyper-parallelepiped in the joint space. Equation [5.33] can also be represented by a hyper-parallelepiped in the task space. In this section, we develop another common approach to studying the velocity transmission between the joint space and the task space using an analytical ellipsoidal representation.

Let us consider the joint velocities contained in the unit sphere of the joint velocity space, such that [Yoshikawa 84b]:

$$\dot{\mathbf{q}}^T \dot{\mathbf{q}} \leq 1 \qquad [5.36]$$

We can show that the corresponding velocities in the task space are defined by the ellipsoid:

$$\dot{\mathbf{X}}^T (\mathbf{J}\,\mathbf{J}^T)^{-1} \dot{\mathbf{X}} \leq 1 \qquad [5.37]$$

The velocity ellipsoid is a useful tool for analyzing the velocity transmission performance of a robot at a given configuration. It is called the *manipulability ellipsoid*. The principal axes of the ellipsoid are given by the vectors $\mathbf{U}_1, \ldots, \mathbf{U}_m$, which are the eigenvectors of $\mathbf{J}\,\mathbf{J}^T$. The lengths of the principal axes are determined by the singular values $\sigma_1, \ldots, \sigma_m$ of $\mathbf{J}$. The optimum direction to generate velocity is along the major axis where the transmission ratio is maximum. Conversely, the velocity is most accurately controlled along the minor axis. Figure 5.14 shows the velocity ellipsoid for a 2R planar mechanism.

The volume of the velocity ellipsoid of a robot gives a measurement of its capacity to generate velocity. Consequently, we define the velocity manipulability of a robot as:

$$w(\mathbf{q}) = \sqrt{\det[\mathbf{J}(\mathbf{q})\,\mathbf{J}^T(\mathbf{q})]} \qquad [5.38]$$

For a non-redundant robot, this expression becomes:

$$w(\mathbf{q}) = |\det[\mathbf{J}(\mathbf{q})]| \qquad [5.39]$$

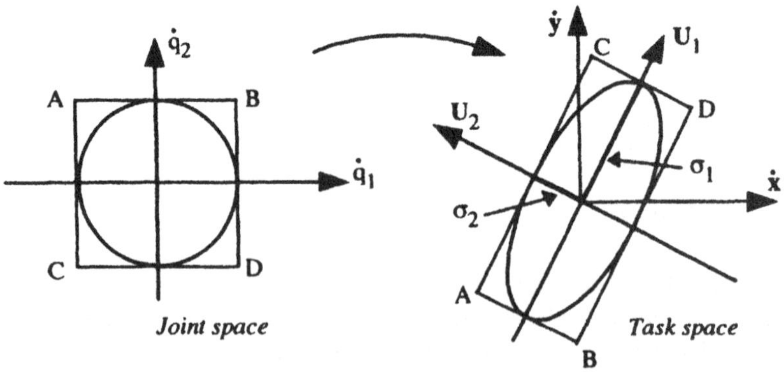

Figure 5.14. *Velocity ellipsoid for a two degree-of-freedom planar robot*

5.9. Static model

In this section, we establish the static model, which provides the joint torques (for revolute joints) or forces (for prismatic joints) corresponding to the wrench (forces and moments) exerted by the end-effector on the environment. We also discuss the duality between the kinematic model and the static model.

5.9.1. *Representation of a wrench*

Let us recall (§ 2.6) that a wrench $\mathbb{f}_i$ is represented by the screw, which is composed of a force $\mathbf{f}_i$ and a moment $\mathbf{m}_i$:

$$\mathbb{f}_i = \begin{bmatrix} \mathbf{f}_i \\ \mathbf{m}_i \end{bmatrix} \quad [5.40]$$

We assume, unless otherwise stated, that the moment is defined about the point O_i, origin of frame R_i. Let the static wrench $\mathbb{f}_{en}$ to be exerted on the environment be defined as:

$$\mathbb{f}_{en} = \begin{bmatrix} \mathbf{f}_{en} \\ \mathbf{m}_{en} \end{bmatrix} = [\; f_x \;\; f_y \;\; f_z \;\; m_x \;\; m_y \;\; m_z \;]^T \quad [5.41]$$

The subscript n indicates that the wrench is expressed at the origin O_n of frame R_n.

5.9.2. *Mapping of an external wrench into joint torques*

To compute the joint torques and forces Γ_e of a serial robot such that its end-effector can exert a static wrench $\mathbb{f}_{en}$, we make use of the principle of virtual work, which states that:

$$\Gamma_{en}^T \, \mathbf{dq}^* = \mathbb{f}_{en}^T \begin{bmatrix} \mathbf{dP}_n^* \\ \boldsymbol{\delta}_n^* \end{bmatrix} \quad [5.42]$$

where the superscript (*) indicates virtual displacements.

Substituting $\mathbf{dP}_n^*$ and $\boldsymbol{\delta}_n^*$ from equation [5.4b] gives:

$$\Gamma_e = \mathbf{J}_n^T \, \mathbb{f}_{en}$$ [5.43]

We can use either the Jacobian matrix ${}^n\mathbf{J}_n$ or ${}^0\mathbf{J}_n$ depending on whether the wrench $\mathbb{f}_{en}$ is referred to frame R_n or frame R_0 respectively.

5.9.3. *Velocity-force duality*

The Jacobian matrix appearing in the static model (equation [5.43]) is the same as that used in the differential or kinematic model. By analogy with the velocity transmission analysis (§ 5.8.1), we deduce the following results (Figure 5.15) [Asada 86]:

- the torques of the actuators are uniquely determined for an arbitrary wrench $\mathbb{f}$; the range space of $\mathbf{J}^T$, denoted as $\mathcal{R}(\mathbf{J}^T)$, is the set of Γ balancing the static wrench $\mathbb{f}$ according to equation [5.43];
- for a zero Γ, the corresponding static wrench can be non-zero; we thus define the null space of $\mathbf{J}^T$, $\mathcal{N}(\mathbf{J}^T)$, as the set of static wrenches that do not require actuator torques in order to be balanced. In this case, the endpoint wrench is borne by the structure of the robot. Note that the null space of $\mathbf{J}^T$, $\mathcal{N}(\mathbf{J}^T)$, which is the orthogonal complement of $\mathcal{R}(\mathbf{J})$, also represents the set of directions along which the robot cannot generate velocity;
- some joint torques Γ cannot be compensated by $\mathbb{f}$. These torques correspond to the vectors of the null space $\mathcal{N}(\mathbf{J})$, orthogonal complement of the space $\mathcal{R}(\mathbf{J}^T)$.

The basis of these spaces can be defined using the columns of the matrices **U** and **V** of the singular value decomposition of **J** as indicated for the velocity case (§ 5.8.1).

Analogously, we can study the force transmission performance using a force manipulability ellipsoid, which corresponds to the set of achievable wrench in the task space $\Re^m$ corresponding to the constraint $\Gamma^T \Gamma \leq 1$. Thus, the force ellipsoid is defined by $\mathbb{f}^T \, \mathbf{J} \, \mathbf{J}^T \, \mathbb{f} \leq 1$. Consequently, we can deduce that the velocity ellipsoid (equation [5.37]) and the force ellipsoid have the same principal axes but the axis lengths are reciprocal (Figure 5.16). This means that the optimum direction for generating velocity is the optimum direction for controlling force. Similarly, the optimal direction for exerting force is also the optimum direction for controlling velocity.

From the control point of view, this behavior makes sense: the velocity is controlled most accurately in the direction where the robot can resist large force disturbances, and force is most accurately controlled in the direction where the robot can rapidly adapt its motion.

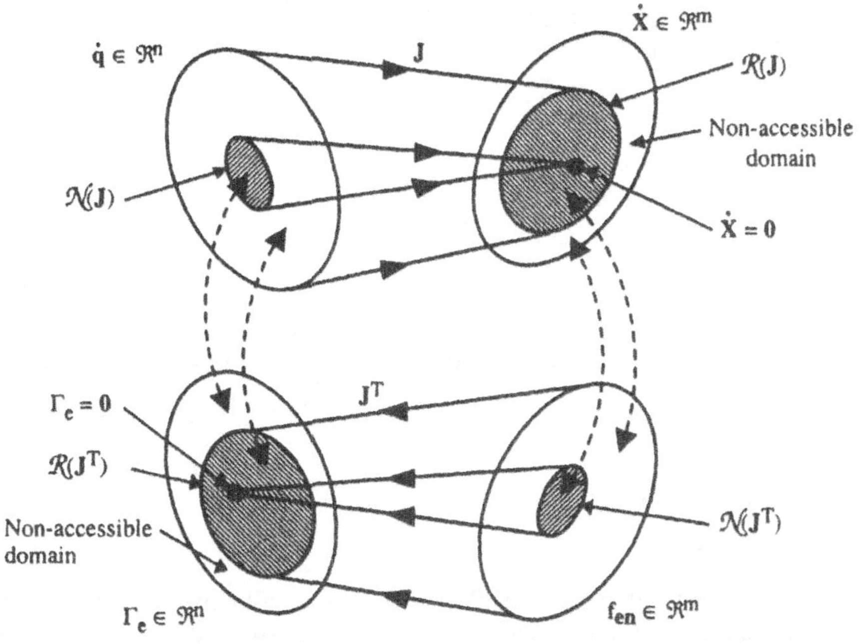

Figure 5.15. *Velocity-force duality (from [Asada 86])*

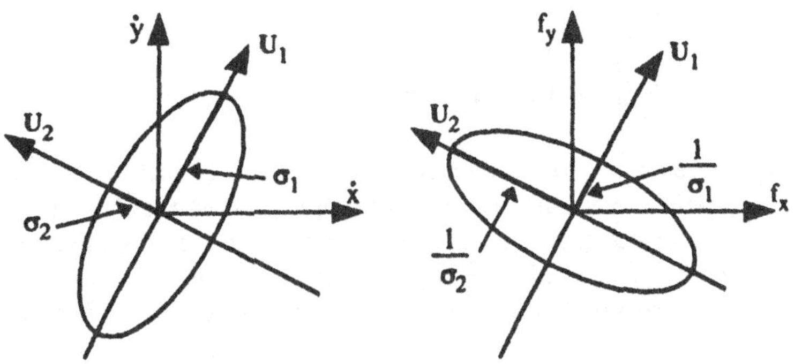

Figure 5.16. *Velocity and force ellipsoids*

5.10. Second order kinematic model

The second order kinematic model allows us to compute the acceleration of the end-effector in terms of positions, velocities and accelerations of the joints. By differentiating equation [5.1] with respect to time, we obtain the following expression:

$$\ddot{\mathbf{X}} = \mathbf{J}\,\ddot{\mathbf{q}} + \dot{\mathbf{J}}\,\dot{\mathbf{q}} \qquad [5.44]$$

where:

$$\dot{\mathbf{J}}(\mathbf{q}, \dot{\mathbf{q}}) = \frac{d}{dt}\mathbf{J}(\mathbf{q}) \qquad [5.45]$$

Using the basic Jacobian matrix, the second order kinematic model can be written as:

$$\begin{bmatrix} \dot{\mathbf{V}}_n \\ \dot{\boldsymbol{\omega}}_n \end{bmatrix} = \mathbf{J}_n\,\ddot{\mathbf{q}} + \dot{\mathbf{J}}_n\,\dot{\mathbf{q}} \qquad [5.46]$$

However, it is most efficient from the computational cost point of view to obtain $\dot{\mathbf{V}}_n$ and $\dot{\boldsymbol{\omega}}_n$ from the following recursive equations, for j = 1, ..., n, which will be developed in Chapter 9:

$$\begin{cases} {}^j\dot{\boldsymbol{\omega}}_j = {}^j\mathbf{A}_{j-1}\,{}^{j-1}\dot{\boldsymbol{\omega}}_{j-1} + \bar{\sigma}_j\,(\ddot{q}_j\,{}^j\mathbf{a}_j + {}^j\boldsymbol{\omega}_{j-1} \times \dot{q}_j\,{}^j\mathbf{a}_j) \\ {}^j\mathbf{U}_j = {}^j\hat{\dot{\boldsymbol{\omega}}}_j + {}^j\hat{\boldsymbol{\omega}}_j\,{}^j\hat{\boldsymbol{\omega}}_j \\ {}^j\dot{\mathbf{V}}_j = {}^j\mathbf{A}_{j-1}\,({}^{j-1}\dot{\mathbf{V}}_{j-1} + {}^{j-1}\mathbf{U}_{j-1}\,{}^{j-1}\mathbf{P}_j) + \sigma_j\,(\ddot{q}_j\,{}^j\mathbf{a}_j + 2\,{}^j\boldsymbol{\omega}_{j-1} \times \dot{q}_j\,{}^j\mathbf{a}_j) \end{cases} \qquad [5.47]$$

The angular velocities ${}^j\boldsymbol{\omega}_{j-1}$ and ${}^j\boldsymbol{\omega}_j$ are calculated using equation [5.23].

In certain applications, such as the control in the task space (§ 14.4.3), we need to compute the vector $\dot{\mathbf{J}}\,\dot{\mathbf{q}}$. Instead of taking the derivative of $\mathbf{J}$ with respect to time and multiplying by $\dot{\mathbf{q}}$, it is more efficient to make use of the recursive equations [5.47] with $\ddot{\mathbf{q}}$ equal to zero in order to leave out the terms involving $\ddot{\mathbf{q}}$ [Khalil 87a].

5.11. Kinematic model associated with the task coordinate representation

Let $\mathbf{X} = \begin{bmatrix} \mathbf{X}_p \\ \mathbf{X}_r \end{bmatrix}$ be any representation of the location of frame R_n relative to frame R_0, where $\mathbf{X}_p$ and $\mathbf{X}_r$ denote the position and orientation vectors respectively. The relationships between the velocities $\dot{\mathbf{X}}_p$ and $\dot{\mathbf{X}}_r$ and the velocities ${}^0\mathbf{V}_n$ and ${}^0\boldsymbol{\omega}_n$ of frame R_n are given as:

$$\begin{cases} \dot{\mathbf{X}}_p = \boldsymbol{\Omega}_p\, {}^0\mathbf{V}_n \\ \dot{\mathbf{X}}_r = \boldsymbol{\Omega}_r\, {}^0\boldsymbol{\omega}_n \end{cases} \qquad [5.48]$$

Similar relations can be derived to express the differential vectors $\mathbf{dX}_p$ and $\mathbf{dX}_r$ as functions of the vectors ${}^0\mathbf{dP}_n$ and ${}^0\boldsymbol{\delta}_n$:

$$\begin{cases} \mathbf{dX}_p = \boldsymbol{\Omega}_p\, {}^0\mathbf{dP}_n \\ \mathbf{dX}_r = \boldsymbol{\Omega}_r\, {}^0\boldsymbol{\delta}_n \end{cases} \qquad [5.49]$$

In matrix form, equation [5.48] becomes:

$$\begin{bmatrix} \dot{\mathbf{X}}_p \\ \dot{\mathbf{X}}_r \end{bmatrix} = \begin{bmatrix} \boldsymbol{\Omega}_p & \mathbf{0}_3 \\ \mathbf{0}_3 & \boldsymbol{\Omega}_r \end{bmatrix} \begin{bmatrix} {}^0\mathbf{V}_n \\ {}^0\boldsymbol{\omega}_n \end{bmatrix} = \boldsymbol{\Omega} \begin{bmatrix} {}^0\mathbf{V}_n \\ {}^0\boldsymbol{\omega}_n \end{bmatrix} \qquad [5.50]$$

Using equation [5.4a], we deduce that:

$$\begin{bmatrix} \dot{\mathbf{X}}_p \\ \dot{\mathbf{X}}_r \end{bmatrix} = \boldsymbol{\Omega}\, {}^0\mathbf{J}_n\, \dot{\mathbf{q}} = \mathbf{J}_x\, \dot{\mathbf{q}} \qquad [5.51]$$

with:

$$\mathbf{J}_x = \boldsymbol{\Omega}\, {}^0\mathbf{J}_n \qquad [5.52]$$

The matrix $\boldsymbol{\Omega}_p$ is equal to $\mathbf{I}_3$ when the position of frame R_n is described by the Cartesian coordinates.

In this section, we show how to calculate $\boldsymbol{\Omega}_r$ and $\boldsymbol{\Omega}_r^{-1}$ for different orientation representations. These expressions are necessary for establishing the kinematic model corresponding to the representation at hand. When the orientation description is not redundant, the inverse of $\mathbf{\Omega}$ can be written as:

$$\mathbf{\Omega}^{-1} = \begin{bmatrix} \mathbf{I}_3 & \mathbf{0}_3 \\ \mathbf{0}_3 & \mathbf{\Omega}_r^{-1} \end{bmatrix} \quad [5.53]$$

If the description of the orientation is redundant, which is the case with the direction cosines and the quaternions (Euler parameters), the matrices $\mathbf{\Omega}_r$, and consequently $\mathbf{\Omega}$, are rectangular. We then use the so-called left inverse, which is a particular case of the pseudoinverse (Appendix 4). The left inverse is defined by:

$$\mathbf{\Omega}^{+} = \begin{bmatrix} \mathbf{I}_3 & \mathbf{0}_3 \\ \mathbf{0}_3 & \mathbf{\Omega}_r^{+} \end{bmatrix} \quad [5.54]$$

with:

$$\begin{cases} \mathbf{\Omega}^{+} = (\mathbf{\Omega}^T \mathbf{\Omega})^{-1} \mathbf{\Omega}^T \\ \mathbf{\Omega}^{+} \mathbf{\Omega} = \mathbf{I}_6 \end{cases} \quad [5.55]$$

Such a matrix exists if $\mathbf{\Omega}$ is of rank 6, which means that $\mathbf{\Omega}_r$ is of rank 3.

5.11.1. *Direction cosines*

The velocity of the vectors **s**, **n**, **a** are given by:

$$\begin{cases} {}^0\dot{\mathbf{s}}_n = {}^0\boldsymbol{\omega}_n \times {}^0\mathbf{s}_n \\ {}^0\dot{\mathbf{n}}_n = {}^0\boldsymbol{\omega}_n \times {}^0\mathbf{n}_n \\ {}^0\dot{\mathbf{a}}_n = {}^0\boldsymbol{\omega}_n \times {}^0\mathbf{a}_n \end{cases} \quad [5.56]$$

Using the vector product operator defined in [2.32], equations [5.56] can be written in the following matrix form [Khatib 80]:

$$\dot{\mathbf{X}}_r = \begin{bmatrix} {}^0\dot{\mathbf{s}}_n \\ {}^0\dot{\mathbf{n}}_n \\ {}^0\dot{\mathbf{a}}_n \end{bmatrix} = \begin{bmatrix} -{}^0\hat{\mathbf{s}}_n \\ -{}^0\hat{\mathbf{n}}_n \\ -{}^0\hat{\mathbf{a}}_n \end{bmatrix} {}^0\boldsymbol{\omega}_n = \mathbf{\Omega}_{CD}\, {}^0\boldsymbol{\omega}_n \quad [5.57]$$

where $\mathbf{\Omega}_{CD}$ is a (9x3) matrix. To calculate $\mathbf{\Omega}_{CD}^{+}$, we use the fact that:

$$\boldsymbol{\Omega}_{CD}^{T}\,\boldsymbol{\Omega}_{CD} = 2\,\mathbf{I}_3 \quad [5.58]$$

Using equation [5.55] and taking into account that the matrices $\hat{\mathbf{s}}$, $\hat{\mathbf{n}}$, $\hat{\mathbf{a}}$ are skew-symmetric, we obtain:

$$\boldsymbol{\Omega}_{CD}^{+} = \frac{1}{2}\boldsymbol{\Omega}_{CD}^{T} = \frac{1}{2}\left[\ {}^{0}\hat{\mathbf{s}}_n \quad {}^{0}\hat{\mathbf{n}}_n \quad {}^{0}\hat{\mathbf{a}}_n\ \right] \quad [5.59]$$

5.11.2. *Euler angles*

We deduce from § 3.6.1 that ϕ is the rotation angle about $\mathbf{z}_0 = [\,0 \;\; 0 \;\; 1\,]^T$, θ is the rotation angle about the current $\mathbf{x}$ axis (after applying $\mathbf{rot}(\mathbf{z}, \phi)$) whose unit vector with respect to R_0 is $[\,C\phi \;\; S\phi \;\; 0\,]^T$, and ψ is the rotation angle about the current $\mathbf{z}$ axis (after applying $\mathbf{rot}(\mathbf{z}, \phi)\ \mathbf{rot}(\mathbf{x}, \theta)$) whose unit vector components with respect to R_0 are $[\,S\phi S\theta \;\; -C\phi S\theta \;\; C\theta\,]^T$. Thus, the velocity of frame R_n relative to frame R_0 is given by:

$${}^{0}\omega_n = \begin{bmatrix} 0 \\ 0 \\ 1 \end{bmatrix}\dot{\phi} + \begin{bmatrix} C\phi \\ S\phi \\ 0 \end{bmatrix}\dot{\theta} + \begin{bmatrix} S\phi S\theta \\ -C\phi S\theta \\ C\theta \end{bmatrix}\dot{\psi} \quad [5.60]$$

thus:

$${}^{0}\omega_n = \begin{bmatrix} 0 & C\phi & S\phi S\theta \\ 0 & S\phi & -C\phi S\theta \\ 1 & 0 & C\theta \end{bmatrix}\begin{bmatrix} \dot{\phi} \\ \dot{\theta} \\ \dot{\psi} \end{bmatrix} \quad [5.61]$$

which we identify with:

$${}^{0}\omega_n = \boldsymbol{\Omega}_{Eul}^{-1}\,\dot{\mathbf{X}}_r = \boldsymbol{\Omega}_{Eul}^{-1}\begin{bmatrix} \dot{\phi} \\ \dot{\theta} \\ \dot{\psi} \end{bmatrix} \quad [5.62]$$

By taking the inverse of $\boldsymbol{\Omega}_{Eul}^{-1}$, we obtain:

$$\Omega_{Eul} = \begin{bmatrix} -S\phi\,\text{cotg}\theta & C\phi\,\text{cotg}\theta & 1 \\ C\phi & S\phi & 0 \\ S\phi/S\theta & -C\phi/S\theta & 0 \end{bmatrix} \qquad [5.63]$$

Ω_{Eul} is singular when $S\theta = 0$, as already obtained in § 3.6.1.

5.11.3. *Roll-Pitch-Yaw angles*

Similarly, we can write:

$$^0\omega_n = \begin{bmatrix} 0 & -S\phi & C\phi C\theta \\ 0 & C\phi & S\phi C\theta \\ 1 & 0 & -S\theta \end{bmatrix} \begin{bmatrix} \dot{\phi} \\ \dot{\theta} \\ \dot{\psi} \end{bmatrix} = \Omega_{RTL}^{-1} \begin{bmatrix} \dot{\phi} \\ \dot{\theta} \\ \dot{\psi} \end{bmatrix} \qquad [5.64]$$

from which we obtain:

$$\Omega_{RTL} = \begin{bmatrix} C\phi\,\text{tg}\theta & S\phi\,\text{tg}\theta & 1 \\ -S\phi & C\phi & 0 \\ C\phi/C\theta & S\phi/C\theta & 0 \end{bmatrix} \qquad [5.65]$$

This matrix is singular when $C\theta = 0$, as already obtained in § 3.6.2.

5.11.4. *Quaternions*

Differentiating equation [3.34] with respect to time and equating the diagonal elements with those of equation [5.56] leads to the following equation:

$$\begin{cases} 2(Q_1\dot{Q}_1 + Q_2\dot{Q}_2) = (Q_2Q_4 - Q_1Q_3)\omega_y - (Q_2Q_3 + Q_1Q_4)\omega_z \\ 2(Q_1\dot{Q}_1 + Q_3\dot{Q}_3) = (Q_2Q_3 - Q_1Q_4)\omega_z - (Q_3Q_4 + Q_1Q_2)\omega_x \\ 2(Q_1\dot{Q}_1 + Q_4\dot{Q}_4) = (Q_3Q_4 - Q_1Q_2)\omega_x - (Q_2Q_4 + Q_1Q_3)\omega_y \end{cases} \qquad [5.66]$$

By differentiating equation [3.31] with respect to time, we obtain:

$$Q_1\,\dot{Q}_1 + Q_2\,\dot{Q}_2 + Q_3\,\dot{Q}_3 + Q_4\,\dot{Q}_4 = 0 \qquad [5.67]$$

From equations [5.66] and [5.67], we deduce that:

$$\dot{X}_r = \dot{Q} = [\dot{Q}_1 \quad \dot{Q}_2 \quad \dot{Q}_3 \quad \dot{Q}_4]^T = \Omega_Q \, {}^0\omega_n \qquad [5.68]$$

with:

$$\Omega_Q = \frac{1}{2}\begin{bmatrix} -Q_2 & -Q_3 & -Q_4 \\ Q_1 & Q_4 & -Q_3 \\ -Q_4 & Q_1 & Q_2 \\ Q_3 & -Q_2 & Q_1 \end{bmatrix} \qquad [5.69]$$

To obtain the inverse relationship, we use the left inverse. While taking into account that $\Omega_Q^T \Omega_Q = \frac{1}{4}$, we obtain:

$$\Omega_Q^+ = 4\,\Omega_Q^T \qquad [5.70]$$

We note that, since the integration of the angular velocity ${}^0\omega_n$ does not yield an orientation representation, equation [5.69] can be used to obtain $\dot{Q}$ whose integration gives the orientation by the Quaternion representation.

5.12. Conclusion

In this chapter, we have shown how to obtain the kinematic model of a robot manipulator using the basic Jacobian matrix. This model allows us to compute the linear and angular velocities of the end-effector in terms of the joint velocities. The Jacobian matrix can be decomposed into two or three matrices containing simpler terms.

Then, we have shown how to use the Jacobian matrix to analyze the workspace and the velocity space of a robot. We have also demonstrated how to use the Jacobian matrix to obtain the static model and we have highlighted the duality of this model with the kinematic model. Finally, the kinematic models associated with the various representations of the task coordinates have been established.

The kinematic model can also be used to find a numerical solution to the inverse geometric problem for a general robot. The necessary tool to obtain this solution is the inverse kinematic model, which is the topic of the next chapter.

Chapter 6

Inverse kinematic model of serial robots

6.1. Introduction

The inverse kinematic model gives the joint velocities $\dot{\mathbf{q}}$ for a desired end-effector velocity $\dot{\mathbf{X}}$. This model is equivalent to the inverse differential model, which determines the differential variation of the joint variables $\mathbf{dq}$ corresponding to a given differential displacement of the end-effector coordinates $\mathbf{dX}$. We obtain the inverse kinematic model by solving a system of linear equations analytically or numerically. The analytical solutions, whenever they exist, offer much lower computational complexity than the numerical solutions, but all the singular cases must be considered separately on a case by case basis [Chevallereau 87]. Thus, the computational complexity of numerical methods is compensated by its generality in handling the regular, singular and redundant cases in a unified way.

In this chapter, we present the techniques used to develop an inverse kinematic model for the regular, singular and redundant cases. The analytical solution is developed for the regular case. The numerical methods presented for the other cases are based essentially on the pseudoinverse of the Jacobian matrix. Finally, we show how to take advantage of redundancy in the inverse kinematic problem using a minimum description of tasks. We assume that the reader is familiar with the techniques of solving linear equations, which are exposed in Appendix 4.

6.2. General form of the kinematic model

From equations [5.22] and [5.50], whatever the method used to describe the end-effector coordinates, the direct kinematic model can be expressed as:

$$\dot{\mathbf{X}} = \begin{bmatrix} \mathbf{\Omega}_p & \mathbf{0}_3 \\ \mathbf{0}_3 & \mathbf{\Omega}_r \end{bmatrix} \begin{bmatrix} {}^0\mathbf{A}_i & \mathbf{0}_3 \\ \mathbf{0}_3 & {}^0\mathbf{A}_i \end{bmatrix} \begin{bmatrix} \mathbf{I}_3 & -{}^i\hat{\mathbf{L}}_{j,n} \\ \mathbf{0}_3 & \mathbf{I}_3 \end{bmatrix} {}^i\mathbf{J}_{n,j}\, \dot{\mathbf{q}} \qquad [6.1]$$

or in compact form as:

$$\dot{\mathbf{X}} = {}^0\mathbf{J}_x\, \dot{\mathbf{q}} \qquad [6.2]$$

Equation [6.1] can be written as:

$${}^i\dot{\mathbf{X}}_{n,j} = {}^i\mathbf{J}_{n,j}\, \dot{\mathbf{q}} \qquad [6.3]$$

with:

$${}^i\dot{\mathbf{X}}_{n,j} = \begin{bmatrix} \mathbf{I}_3 & {}^i\hat{\mathbf{L}}_{j,n} \\ \mathbf{0}_3 & \mathbf{I}_3 \end{bmatrix} \begin{bmatrix} {}^i\mathbf{A}_0 & \mathbf{0}_3 \\ \mathbf{0}_3 & {}^i\mathbf{A}_0 \end{bmatrix} \begin{bmatrix} \mathbf{\Omega}_p^{-1} & \mathbf{0}_3 \\ \mathbf{0}_3 & \mathbf{\Omega}_r^{+} \end{bmatrix} \dot{\mathbf{X}} \qquad [6.4]$$

We find in § 5.11 the expression of the pseudoinverse $\mathbf{\Omega}_r^{+}$ for different representations of the orientation, while $\mathbf{\Omega}_p^{-1} = \mathbf{I}_3$ if the Cartesian coordinates are used to describe the position.

Since the elements of ${}^i\mathbf{J}_{n,j}$ are simpler than those of ${}^0\mathbf{J}_x$, equation [6.3] is more appropriate for developing an analytical solution to the inverse kinematic problem. To simplify the notation, we will use the following form for both equations [6.2] and [6.3]:

$$\dot{\mathbf{X}} = \mathbf{J}\, \dot{\mathbf{q}} \qquad [6.5]$$

NOTE.– If n < 6, we cannot use the Jacobian matrix ${}^i\mathbf{J}_{n,j}$ systematically. The singularities of this matrix do not take into account the corresponding particular choice of the task coordinates [Borrel 86].

6.3. Inverse kinematic model for a regular case

In this case, the Jacobian matrix **J** is square and of full rank. Thus, it is possible to move the end-effector with finite velocity in any desired direction of the task space. The joint velocities can be evaluated using one of the following methods.

6.3.1. *First method*

We compute $\mathbf{J}^{-1}$, the inverse of $\mathbf{J}$, either numerically or analytically. Then, the joint velocity vector $\dot{\mathbf{q}}$ is obtained as:

$$\dot{\mathbf{q}} = \mathbf{J}^{-1}\,\dot{\mathbf{X}} \qquad [6.6]$$

If the matrix $\mathbf{J}$ has the following form:

$$\mathbf{J} = \begin{bmatrix} \mathbf{A} & \mathbf{0} \\ \mathbf{B} & \mathbf{C} \end{bmatrix} \qquad [6.7]$$

the matrices $\mathbf{A}$ and $\mathbf{C}$ being square and invertible, it is easy to show that:

$$\mathbf{J}^{-1} = \begin{bmatrix} \mathbf{A}^{-1} & \mathbf{0} \\ -\mathbf{C}^{-1}\mathbf{B}\mathbf{A}^{-1} & \mathbf{C}^{-1} \end{bmatrix} \qquad [6.8]$$

Consequently, the inverse of $\mathbf{J}$ reduces to the inverse of two matrices of smaller dimension. For a six degree-of-freedom robot with a spherical wrist, the general form of $\mathbf{J}$ is given by equation [6.7] where $\mathbf{A}$ and $\mathbf{C}$ are (3x3) matrices [Gorla 84].

6.3.2. *Second method*

In this method, instead of solving a linear system of n equations in n unknowns, the problem is reduced to solving two linear systems of equations of lower dimensions. In general, this technique requires less computational complexity. Let us take for example a six degree-of-freedom robot with a spherical wrist whose Jacobian matrix (see Example 5.3) can be written as:

$$\begin{bmatrix} \dot{\mathbf{X}}_a \\ \dot{\mathbf{X}}_b \end{bmatrix} = \begin{bmatrix} \mathbf{A} & \mathbf{0}_3 \\ \mathbf{B} & \mathbf{C} \end{bmatrix} \begin{bmatrix} \dot{\mathbf{q}}_a \\ \dot{\mathbf{q}}_b \end{bmatrix} \qquad [6.9]$$

$\mathbf{A}$ and $\mathbf{C}$ being (3x3) regular square matrices.

The solution $\dot{\mathbf{q}}$ is given by:

$$\begin{cases} \dot{\mathbf{q}}_a = \mathbf{A}^{-1}\,\dot{\mathbf{X}}_a \\ \dot{\mathbf{q}}_b = \mathbf{C}^{-1}\,[\dot{\mathbf{X}}_b - \mathbf{B}\,\dot{\mathbf{q}}_a] \end{cases} \qquad [6.10]$$

which, *a priori*, is simpler than that obtained by the first method.

• **Example 6.1.** Calculate the inverse kinematic model of the Stäubli RX-90 robot. The Jacobian ${}^3\mathbf{J}_6$ has been computed in Example 5.3. We develop the solutions according to equations [6.8] and [6.10].

i) first method. The inverses of **A** and **C** are respectively:

$$\mathbf{A}^{-1} = \begin{bmatrix} 0 & 0 & V1 \\ 0 & V3 & 0 \\ -1/RL4 & V2V3/RL4 & 0 \end{bmatrix}$$

$$\mathbf{C}^{-1} = \begin{bmatrix} V4 & 1 & -V5 \\ S4 & 0 & C4 \\ -C4/S5 & 0 & S4/S5 \end{bmatrix}$$

with:

$$V1 = \frac{1}{S23RL4 - C2D3}$$
$$V2 = -RL4 + S3D3$$
$$V3 = \frac{1}{C3D3}$$
$$V4 = C4\ \text{cotg}5$$
$$V5 = S4\ \text{cotg}5$$

Using equation [6.8], we obtain:

$${}^3\mathbf{J}_6^{-1} = \begin{bmatrix} 0 & 0 & V1 & 0 & 0 & 0 \\ 0 & V3 & 0 & 0 & 0 & 0 \\ -1/RL4 & V2V3/RL4 & 0 & 0 & 0 & 0 \\ -S4C5V7 & V5V6 & V8 & V4 & 1 & -V5 \\ C4/RL4 & -C4V6 & -S23S4V1 & S4 & 0 & C4 \\ S4V7 & -S4V6/S5 & S23C4V1/S5 & -C4/S5 & 0 & S4/S5 \end{bmatrix}$$

with:

$$V6 = \frac{S3}{C3RL4}$$
$$V7 = \frac{1}{S5RL4}$$
$$V8 = (-S23V4 - C23)V1$$

The computation of $\dot{\mathbf{q}}$ by equation [6.8] needs 18 additions, 47 multiplications/divisions and 8 sine/cosine functions;

ii) second method. We calculate successively $\dot{\mathbf{q}}_a$ and $\dot{\mathbf{q}}_b$:

$$\dot{\mathbf{q}}_a = \begin{bmatrix} \dot{q}_1 \\ \dot{q}_2 \\ \dot{q}_3 \end{bmatrix} = \begin{bmatrix} V1\dot{X}_3 \\ V3\dot{X}_2 \\ (-\dot{X}_1 + V2V3\dot{X}_2) / RL4 \end{bmatrix}$$

$$\dot{\mathbf{X}}_b - \mathbf{B}\,\dot{\mathbf{q}}_a = \begin{bmatrix} \dot{X}_{4'} \\ \dot{X}_{5'} \\ \dot{X}_{6'} \end{bmatrix} = \begin{bmatrix} \dot{X}_4 - S23\,\dot{q}_1 \\ \dot{X}_5 - C23\,\dot{q}_1 \\ \dot{X}_6 - \dot{q}_2 - \dot{q}_3 \end{bmatrix}$$

$$\dot{\mathbf{q}}_b = \begin{bmatrix} \dot{q}_4 \\ \dot{q}_5 \\ \dot{q}_6 \end{bmatrix} = \mathbf{C}^{-1} \begin{bmatrix} \dot{X}_{4'} \\ \dot{X}_{5'} \\ \dot{X}_{6'} \end{bmatrix} = \begin{bmatrix} C4\,\text{cotg}5\,\dot{X}_{4'} + \dot{X}_{5'} - S4\,\text{cotg}5\,\dot{X}_{6'} \\ S4\,\dot{X}_{4'} + C4\,\dot{X}_{6'} \\ (-C4\,\dot{X}_{4'} + S4\,\dot{X}_{6'}) / S5 \end{bmatrix}$$

This solution requires 12 additions, 22 multiplications/divisions and 8 sine/cosine functions.

6.4. Solution in the neighborhood of singularities

When the robot is non-redundant, the singular configurations are the roots of $\det(\mathbf{J}) = 0$. In the redundant case, they are given by the roots of $\det(\mathbf{J}\mathbf{J}^T) = 0$. Thus, singularities are identified by the rank deficiency of the matrix $\mathbf{J}$, which physically represents the inability of the robot to generate an arbitrary velocity in the task space. The neighborhood of a singular position is more precisely detected by using the singular values. In fact, the decrease of one or several singular values is generally more significant to indicate the vicinity of a singular configuration than that of examining the value of the determinant. In the neighborhood of these configurations, the use of the classical inverse of the Jacobian matrix will give excessive joint velocities. Since such high velocities are physically unrealizable, we cannot obtain an accurate motion.

The redundancy can be exploited to design robots that avoid singularities [Hollerbach 84b], [Luh 85a]. However, robots with revolute joints will have unavoidable singularities [Baillieul 84]. In § 6.5, we will see that redundancy may be exploited to go away from avoidable singularities [Baillieul 84]. An avoidable singularity is a singular configuration where the corresponding tool location can be reached with a different non-singular configuration.

6.4.1. *Use of the pseudoinverse*

The most widely proposed methods for solving the inverse kinematic problem near singularities involve the use of the pseudoinverse $\mathbf{J}^+$ of the matrix $\mathbf{J}$ (Appendix 4):

$$\dot{\mathbf{q}} = \mathbf{J}^+ \dot{\mathbf{X}} \qquad [6.11]$$

This solution, proposed by Whitney [Whitney 69], minimizes $\|\dot{\mathbf{q}}\|^2$ and $\|\dot{\mathbf{X}} - \mathbf{J}\dot{\mathbf{q}}\|^2$. Depending on $\dot{\mathbf{X}}$, the following cases are distinguished:

- $\dot{\mathbf{X}}$ belongs to $\mathcal{R}(\mathbf{J})$, representing the range space of $\mathbf{J}$: equation [6.11] gives an exact solution with zero error even though the inverse Jacobian $\mathbf{J}^{-1}$ is not defined;
- $\dot{\mathbf{X}}$ belongs to the subspace of the degenerated directions $\mathcal{R}(\mathbf{J})^{\perp}$: there are no joint velocities that can generate this velocity. In this case, the solution [6.11] gives $\dot{\mathbf{q}} = \mathbf{0}$. If the next desired velocity is also defined along this direction, the robot is blocked and it is necessary to define strategies to release it [Chevallereau 88];
- $\dot{\mathbf{X}}$ belongs to both $\mathcal{R}(\mathbf{J})$ and $\mathcal{R}(\mathbf{J})^{\perp}$: the solution [6.11] gives $\dot{\mathbf{q}}$, which only realizes the components belonging to $\mathcal{R}(\mathbf{J})$.

A major shortcoming of this method is that it produces discontinuous joint velocities near singularities [Wampler 86]. This can be seen by expressing the joint velocity solution in terms of singular value decomposition (§ 5.8.1). In fact, far from singularities, the joint velocities are given by:

$$\dot{\mathbf{q}} = \sum_{i=1}^{m} \frac{1}{\sigma_i} \mathbf{V}_i \, \mathbf{U}_i^T \dot{\mathbf{X}} \qquad [6.12]$$

While approaching a singularity, σ_{min} becomes small, leading to high joint velocities. At singularity, the smallest singular value σ_{min} becomes zero, consequently, it is not taken into account any more. The summation in equation [6.12] is carried out up to $m-1$, and the joint velocity $\dot{\mathbf{q}}$ decreases significantly.

NOTE.– Both $\|\dot{\mathbf{q}}\|^2$ and $\|\dot{\mathbf{X}}-\mathbf{J}\dot{\mathbf{q}}\|^2$ may contain elements with different units. However, using radians for the angles and meters for the distances gives good results for industrial robots of common size (1 to 2 meters reach).

6.4.2. *Use of the damped pseudoinverse*

A general approach to solving the problem of discontinuity of the pseudoinverse solution at a singular configuration is to use the damped least-squares method, which is known as the Levenberg-Marquardt stabilization method [Wampler 86], [Nakamura 87]. This solution minimizes the following expression:

$$\|\dot{\mathbf{X}} - \mathbf{J}\dot{\mathbf{q}}\|^2 + \alpha^2 \|\dot{\mathbf{q}}\|^2 \qquad [6.13]$$

where α is a constant.

This new criterion means that the end-effector tracking error is weighted against the norm of joint velocity by using the factor α , also known as the *damping factor*. This solution is typically obtained as the least-squares solution of the following system:

$$\begin{bmatrix} \mathbf{J} \\ \alpha \mathbf{I}_n \end{bmatrix} \dot{\mathbf{q}} = \begin{bmatrix} \dot{\mathbf{X}} \\ \mathbf{0}_{n\times 1} \end{bmatrix} \qquad [6.14]$$

which is given as:

$$\dot{\mathbf{q}}_a = [\mathbf{J}^T \mathbf{J} + \alpha^2 \mathbf{I}_n]^{-1} \mathbf{J}^T \dot{\mathbf{X}} \qquad [6.15]$$

When n > m, the following equivalent relation is easier to compute [Maciejewski 88]:

$$\dot{\mathbf{q}}_a = \mathbf{J}^T [\mathbf{J} \mathbf{J}^T + \alpha^2 \mathbf{I}_m]^{-1} \dot{\mathbf{X}} \qquad [6.16]$$

Using the singular value decomposition, the solution is written as:

$$\dot{\mathbf{q}}_a = \sum_{i=1}^{m} \frac{\sigma_i}{\sigma_i^2+\alpha^2} \mathbf{V}_i \, \mathbf{U}_i^T \dot{\mathbf{X}} \qquad [6.17]$$

If $\sigma_i >> \alpha$, then $\frac{\sigma_i}{\sigma_i^2+\alpha^2} \approx \frac{1}{\sigma_i}$. If $\sigma_i << \alpha$, then $\frac{\sigma_i}{\sigma_i^2+\alpha^2} \approx \frac{\sigma_i}{\alpha^2}$. The error due to the damping factor α in the joint coordinates is expressed as:

$$\dot{\mathbf{e}}_q = \dot{\mathbf{q}} - \dot{\mathbf{q}}_a = \sum_{i=1}^{m} \frac{\alpha^2}{(\sigma_i^2+\alpha^2)\sigma_i} \mathbf{V}_i \mathbf{U}_i^T \dot{\mathbf{X}} \qquad [6.18]$$

The error in $\dot{\mathbf{X}}$ is obtained as:

$$\dot{\mathbf{e}}_x = \mathbf{J} \dot{\mathbf{e}}_q = \sum_{i=1}^{m} \frac{\alpha^2}{\sigma_i^2+\alpha^2} \mathbf{U}_i \mathbf{U}_i^T \dot{\mathbf{X}} \qquad [6.19]$$

The damping factor α limits the norm of the solution. However, at positions far away from singularities, no damping is needed. Thus, a trade-off must be found between the precision of the solution and the possibility of its realization.

Wampler [Wampler 86] proposes to use a fixed damping factor $\alpha = 0.003$, while Nakamura [Nakamura 86] suggests the computation of the damping factor as a function of the manipulability w (equation [5.38]) as follows:

$$\begin{cases} \alpha = \alpha_0 \left(1 - \frac{w}{w_0}\right)^2 & \text{if } w < w_0 \\ \alpha = 0 & \text{if } w \geq w_0 \end{cases} \qquad [6.20]$$

where α_0 is a positive constant and w_0 is a threshold, which defines the boundary of the neighborhood of singular points.

A more appropriate solution can be obtained by adjusting the value of α as a function of the smallest singular value σ_{min}, which is the exact measure of the neighborhood of a singular position. Maciejewski and Klein [Maciejewski 88] propose to compute the damping factor as follows:

$$\begin{cases} \alpha = \varepsilon^2 - \sigma_{min}^2 & \text{if } \sigma_{min} \leq \varepsilon \\ \alpha = 0 & \text{if } \sigma_{min} > \varepsilon \end{cases} \qquad [6.21]$$

where ε is a constant.

In [Maciejewski 88], we find an efficient method to estimate σ_{min}. In the damping least-squares method, the robot can stay blocked in a singular configuration if the desired velocity is along the degenerated directions, i.e. when (equations [5.30] and [5.31]):

$$\dot{X} = \sum_{i=r+1}^{m} U_i (U_i^T \dot{X}) \qquad [6.22]$$

where $r < m$ gives the rank of **J**.

6.4.3. *Other approaches for controlling motion near singularities*

The kinematic model, which is a first order linearization, does not give an exact solution respecting the actuator constraints in the neighborhood of singularities. Some authors [Nielsen 91], [Chevallereau 98] have used the IGM or a kinematic model of higher order to determine the joint variables corresponding to a Cartesian motion passing through a singularity. Recently, it has been shown [Lloyd 96] that the end-effector could move along any specified path using a suitable time law.

To show the efficiency of such techniques, let us consider the case of a two degree-of-freedom planar robot in the singular configuration "extended arm". Let us suppose that we want to move the terminal point towards the origin along the **x**-axis (Figure 6.1a) (which is a degenerated direction for the kinematic model). It is easy to deduce from the kinematic model that a constant velocity motion along this direction is not feasible. However, a motion with a constant end-effector acceleration and a zero initial velocity can be proved realizable (Figure 6.1b) by developing the IGM up to the second order [Nielsen 91] or by using the second-order kinematic model [Chevallereau 98].

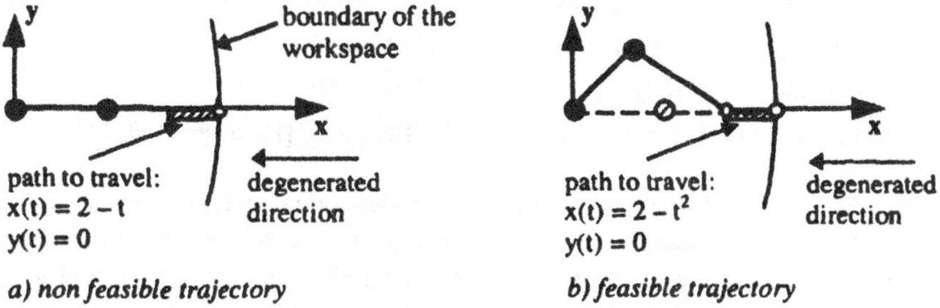

Figure 6.1. *Displacement along a degenerated direction*

In addition, Egeland and Spangelo [Egeland 91] showed that, in certain cases, a non-feasible path could become realizable after carrying out a specific motion in the null space of **J**. This motion does not modify the end-effector coordinates but it modifies the degenerated direction. Let us illustrate this method for the two degree-of-freedom planar robot with identical link lengths. From the initial configuration "folded arm" of Figure 6.2a, it is not possible to track a trajectory along the x direction. However, after a $\pi/2$ rotation of the first joint, which does not modify the terminal point coordinates but modifies the degenerated direction (Figure 6.2b), we can produce a velocity along the **x**-axis by using the kinematic model.

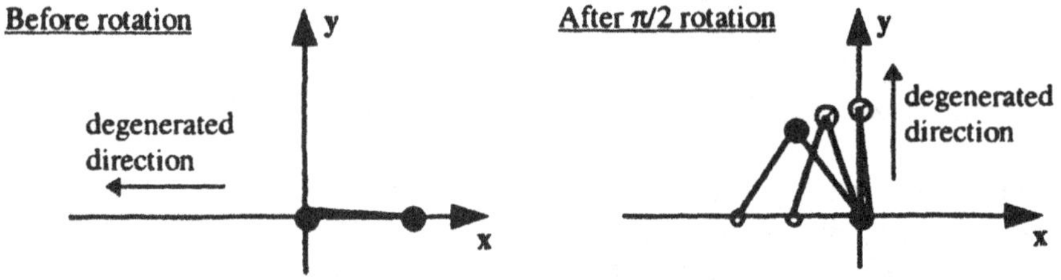

a) non feasible trajectory *b) feasible trajectory*

Figure 6.2. *Motion in the null space of J*

6.5. Inverse kinematic model of redundant robots

A robot manipulator is redundant when its number of degrees of freedom N is greater than the dimension of the workspace M. The difference (N – M) represents the degree of redundancy. In this case, the inverse kinematic model gives an infinite number of solutions. Consequently, secondary performance criteria can be optimized, such as:

- minimizing the norm of the joint velocities [Whitney 69];
- avoiding obstacles [Maciejewski 85], [Baillieul 86];
- avoiding singular configurations [Yoshikawa 84a];
- avoiding joint limits [Fournier 80], [Klein 84];
- minimizing driving joint torques [Baillieul 84], [Hollerbach 85].

When the end-effector coordinates are independent, we have n = N and m = M. For a redundant mechanism, the Jacobian **J** is represented by an (mxn) matrix, with n > m. In the following sections, we present several approaches to solving the inverse kinematic problem of redundant robots.

6.5.1. *Extended Jacobian*

In this approach, we add n – m secondary linearly independent equations to the end-effector coordinates **X** [Baillieul 85], [Chang 86], [Nenchev 92]. These equations can represent either physical constraints on the robot or constraints related to the environment. They are written in the following general form:

$$\mathbf{X}_c = \mathbf{h}(\mathbf{q}) \qquad [6.23]$$

In this expression, $\mathbf{X}_c$ is an $((n-m)\,x1)$ vector whose elements are functions of $\mathbf{q}$. Differentiating equation [6.23] with respect to time gives:

$$\dot{\mathbf{X}}_c = \mathbf{J}_h\, \dot{\mathbf{q}} \qquad [6.24]$$

where $\mathbf{J}_h = \partial\mathbf{h}(\mathbf{q})/\partial\mathbf{q}$ is the $((n-m)xn)$ Jacobian matrix of $\mathbf{h}(\mathbf{q})$. Combining this equation with the kinematic model, we obtain an (nxn) extended Jacobian matrix $\mathbf{J}_a$ and a new velocity vector $\dot{\mathbf{X}}_a$ such that:

$$\dot{\mathbf{X}}_a = \mathbf{J}_a\, \dot{\mathbf{q}} \qquad [6.25]$$

with $\dot{\mathbf{X}}_a = \begin{bmatrix} \dot{\mathbf{X}} \\ \dot{\mathbf{X}}_c \end{bmatrix}$ and $\mathbf{J}_a = \begin{bmatrix} \mathbf{J} \\ \mathbf{J}_h \end{bmatrix}$.

If the extended Jacobian $\mathbf{J}_a$ is not singular, a unique solution for the joint velocity $\dot{\mathbf{q}}$ is obtained by inverting $\mathbf{J}_a$. We can use this technique to optimize the desired objective function $\phi(\mathbf{q})$ by taking $\mathbf{h}(\mathbf{q})$ such that:

$$h_i(\mathbf{q}) = 0 = (\eta_i)^T\, \nabla\phi \quad \text{for } i = 1, \ldots, n-m \qquad [6.26]$$

where the $(nx1)$ vectors η_i, for $i = 1, \ldots, n-m$, form a basis for the null space of $\mathbf{J}$, and $\nabla\phi$ is the gradient of ϕ.

Since the calculation of the basis of the null space of the Jacobian matrix must be carried out analytically, this method can be used only for systems with a small degree of redundancy. A solution to this problem can be found in [Klein 95].

The extended Jacobian method presents the following disadvantages:

- the choice of the $(n-m)$ additional relationships is not a trivial matter;
- the extended Jacobian $\mathbf{J}_a$ may be singular even though the end-effector Jacobian is of full rank. These configurations are called *artificial singularities* or *algorithmic singularities*.

A desirable property of this method is that it yields cyclic behavior, meaning that a closed path in the task space is always tracked by a closed path in the joint space. This is important because it allows one to judge the suitability of a trajectory after executing one cycle.

6.5.2. *Jacobian pseudoinverse*

The vast majority of research in the control of redundant robots has involved the resolution through the use of the pseudoinverse $\mathbf{J}^+$ of the Jacobian matrix:

$$\dot{\mathbf{q}} = \mathbf{J}^+ \dot{\mathbf{X}} \qquad [6.27]$$

This solution minimizes $\|\dot{\mathbf{q}}\|^2$. Because of this minimization property, the early hope of researchers [Whitney 69] was that singularities would automatically be avoided. It has been proved that, without modification, this approach does not avoid singularity [Baillieul 85]. Moreover, Klein and Huang [Klein 83] have pointed out that it does not produce cyclic behavior, which is a serious practical problem.

For these reasons, we generally add to the pseudoinverse solution another component belonging to the null space of the Jacobian, in order to realize the secondary objective function.

6.5.3. *Weighted pseudoinverse*

Since each joint has different limits and even different units, it may be interesting to weight the contribution of each joint in the objective function differently. This can be achieved by the use of the weighted pseudoinverse, which minimizes a criteria C such that:

$$C = \dot{\mathbf{q}}^T \mathbf{E} \, \dot{\mathbf{q}} \qquad [6.28]$$

When $\mathbf{J}$ is of full rank, the solution is given by:

$$\dot{\mathbf{q}} = \mathbf{J}_E{}^+ \dot{\mathbf{X}} \qquad [6.29]$$

with:

$$\mathbf{J}_E{}^+ = \mathbf{E}^{-1} \mathbf{J}^T (\mathbf{J} \, \mathbf{E}^{-1} \mathbf{J}^T)^{-1} \qquad [6.30]$$

Benoit et al. [Benoit 75] propose to take for $\mathbf{E}$ the inertia matrix of the robot (Chapter 9) in order to minimize the kinetic energy. Konstantinov et al. [Konstantinov 81] have used the weighted pseudoinverse to avoid the joint limits.

6.5.4. *Jacobian pseudoinverse with an optimization term*

One of the advantages of the pseudoinverse solution is the possibility to utilize the null space to optimize another objective function (beside that of $\|\dot{\mathbf{q}}\|^2$). In fact, the general solution of the linear system [6.5] is written as (Appendix 4):

$$\dot{\mathbf{q}} = \mathbf{J}^+ \dot{\mathbf{X}} + (\mathbf{I}_n - \mathbf{J}^+ \mathbf{J})\, \mathbf{Z} \qquad [6.31]$$

where $\mathbf{Z}$ is an arbitrary (nx1) vector in the $\dot{\mathbf{q}}$ space.

The second term on the right belongs to the null space of $\mathbf{J}$. It corresponds to a self-motion of the joints that does not move the end-effector. This term, which is called *homogeneous solution* or *optimization term*, can be used to optimize a desired function $\phi(\mathbf{q})$. In fact, taking $\mathbf{Z} = \alpha\nabla\phi$ where $\nabla\phi$ is the gradient of this function with respect to $\mathbf{q}$, minimizes the function $\phi(\mathbf{q})$ when $\alpha < 0$ and maximizes it when $\alpha > 0$. Equation [6.31] is rewritten as:

$$\dot{\mathbf{q}} = \mathbf{J}^+ \dot{\mathbf{X}} + \alpha(\mathbf{I}_n - \mathbf{J}^+ \mathbf{J})\, \nabla\phi \qquad [6.32]$$

with:

$$\nabla\phi = [\frac{\partial\phi}{\partial q_1} \;\cdots\; \frac{\partial\phi}{\partial q_n}]^T \qquad [6.33]$$

The value of α allows us to realize a trade-off between the minimization of $\|\dot{\mathbf{q}}\|^2$ and the optimization of $\phi(\mathbf{q})$. In the following sections, we present two examples of desired objective functions.

6.5.4.1. *Avoiding joint limits*

A practical solution to control a redundant robot is to keep the joint variables away from their limits $\mathbf{q}_{max}$ and $\mathbf{q}_{min}$. Let:

$$\mathbf{q}_{moy} = \frac{1}{2}(\mathbf{q}_{max} + \mathbf{q}_{min}) \qquad [6.34]$$

where $\mathbf{q}_{moy}$ is the mean value of the joint positions, and:

$$\Delta\mathbf{q} = \mathbf{q}_{max} - \mathbf{q}_{min} \qquad [6.35]$$

A possible scalar function, whose minimization generates a motion away from the joint limits, can be expressed in the following quadratic form [Fournier 80]:

$$\phi(\mathbf{q}) = \sum_{i=1}^{n} \left[\frac{q_i - q_{imoy}}{\Delta q_i}\right]^2 \qquad [6.36]$$

The division by Δq_i allows us to weight the contribution of each joint in $\phi(\mathbf{q})$ such that it varies between 0 and 1. The ith element of the vector $\mathbf{Z}$ is written as (with $\alpha<0$):

$$Z_i = \frac{\alpha \, \partial\phi(\mathbf{q})}{\partial q_i} = \frac{2\alpha\,(q_i - q_{imoy})}{\Delta q_i^2} \qquad [6.37]$$

NOTE.– If the mean position of a joint corresponds to a singular configuration, it is recommended to replace the corresponding value of q_{imoy} by another value.

About the criterion [6.36], Klein [Klein 84] pointed out that the quadratic form, used generally to solve optimization problems, does not always give the best solution to the desired objectives. To avoid joint limits in particular, the following form is more suitable:

$$\phi = \max \frac{|q_i - q_{imoy}|}{|\Delta q_i|} \quad \text{for } i = 1, \ldots, n \qquad [6.38]$$

Introducing this criterion in equation [6.32] is however not as easy as the quadratic criterion. A solution consists of approximating the criterion [6.38] by a p-norm function defined as [Klein 83]:

$$\|\mathbf{q} - \mathbf{q}_{moy}\|_p = \left[\sum_{i=1}^{n} |q_i - q_{imoy}|^p\right]^{1/p} \qquad [6.39]$$

When p tends towards infinity, the corresponding p-norm meets the criterion [6.38]. However, sufficient approximation can be achieved by taking p = 6.

6.5.4.2. *Increasing the manipulability*

In § 5.8.2, we showed that the manipulability $w(\mathbf{q})$ of a robot manipulator (equation [5.38]) could be used as a measure of the ability of the mechanism to move its end-effector. At a singular point, w is minimum and is zero. In order to improve the manipulability of a structure, we can choose to maximize a scalar function ϕ such that:

$$\phi(\mathbf{q}) = \det[\mathbf{J}(\mathbf{q})\,\mathbf{J}^T(\mathbf{q})] \qquad [6.40]$$

We calculate $\mathbf{Z}$ as indicated previously with $\alpha > 0$. Maximizing ϕ moves the robot away from the singular configurations.

NOTE.- Certain singular configurations are unavoidable [Baillieul 84]. This is the case if there is no other configuration that can yield the same end-effector location. For the three degree-of-freedom planar robot of Example 6.1, the unavoidable singularities correspond to the configurations where it is fully stretched out or folded up (Figure 6.3). The other singularities are avoidable and the robot can find other configurations to achieve them (Figure 6.4).

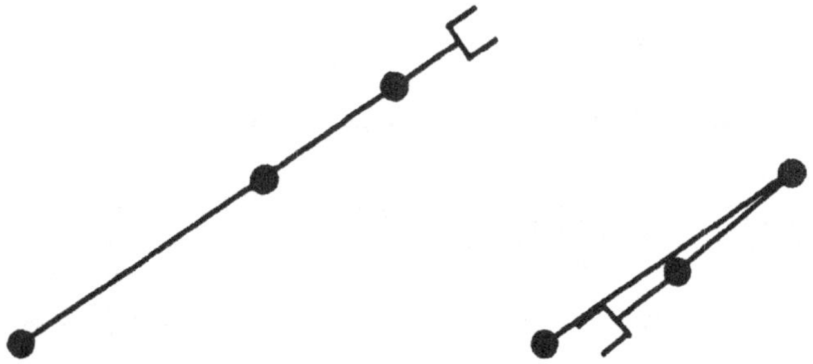

Figure 6.3. *Unavoidable singularities of a three degree-of-freedom planar robot*

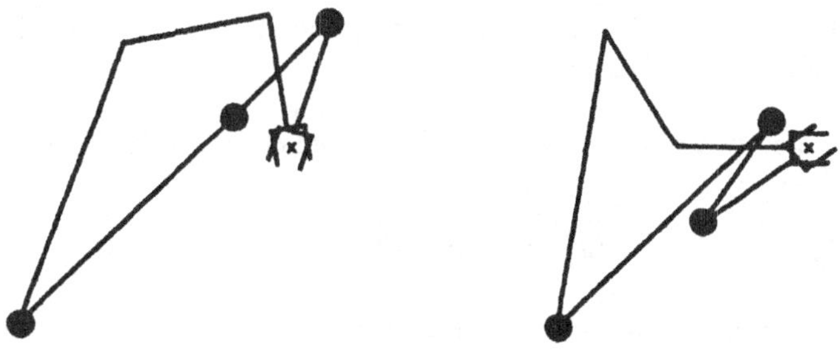

Figure 6.4. *Avoidable singularities of a three degree-of-freedom planar robot*

6.5.5. *Task-priority concept*

To solve the inverse kinematic model of redundant robots, Nakamura [Nakamura 87] introduced the concept of task priority, where a required task is divided into a primary task $\mathbf{X}_1$ of higher priority and a secondary task $\mathbf{X}_2$ of lower priority. These tasks are described by the following relationships:

$$\mathbf{X}_1 = \mathbf{f}_1(\mathbf{q}) \qquad [6.41]$$

$$\mathbf{X}_2 = \mathbf{f}_2(\mathbf{q}) \qquad [6.42]$$

Let m_1 and m_2 be the dimensions of $\mathbf{X}_1$ and $\mathbf{X}_2$ respectively. Differentiating equations [6.41] and [6.42] with respect to time gives:

$$\dot{\mathbf{X}}_1 = \mathbf{J}_1 \dot{\mathbf{q}} \qquad [6.43]$$

$$\dot{\mathbf{X}}_2 = \mathbf{J}_2 \dot{\mathbf{q}} \qquad [6.44]$$

where $\mathbf{J}_i = \partial \mathbf{f}_i(\mathbf{q})/\partial \mathbf{q}$ is the ($m_i$xn) Jacobian matrix of the task $\mathbf{X}_i$. Using the pseudoinverse, the general solution of equation [6.43] is given by:

$$\dot{\mathbf{q}} = \mathbf{J}_1^+ \dot{\mathbf{X}}_1 + (\mathbf{I}_n - \mathbf{J}_1^+ \mathbf{J}_1)\, \mathbf{Z}_1 \qquad [6.45]$$

Substituting equation [6.45] into equation [6.44] yields:

$$\mathbf{J}_2 (\mathbf{I}_n - \mathbf{J}_1^+ \mathbf{J}_1)\, \mathbf{Z}_1 = \dot{\mathbf{X}}_2 - \mathbf{J}_2 \mathbf{J}_1^+ \dot{\mathbf{X}}_1 \qquad [6.46]$$

From this equation, the vector $\mathbf{Z}_1$ can be determined by using the pseudoinverse:

$$\mathbf{Z}_1 = \mathbf{J}_3^+ [\dot{\mathbf{X}}_2 - \mathbf{J}_2 \mathbf{J}_1^+ \dot{\mathbf{X}}_1] + (\mathbf{I}_n - \mathbf{J}_3^+ \mathbf{J}_3)\, \mathbf{Z}_2 \qquad [6.47]$$

where $\mathbf{J}_3 = \mathbf{J}_2 (\mathbf{I}_n - \mathbf{J}_1^+ \mathbf{J}_1)$ is an ($m_2$xn) matrix and $\mathbf{Z}_2$ is an arbitrary (nx1) vector chosen to satisfy the optimization criterion.

The joint velocity $\dot{\mathbf{q}}$ of the robot is obtained from equations [6.45] and [6.47]:

$$\dot{\mathbf{q}} = \mathbf{J}_1^+ \dot{\mathbf{X}}_1 + (\mathbf{I}_n - \mathbf{J}_1^+ \mathbf{J}_1) \{\mathbf{J}_3^+ [\dot{\mathbf{X}}_2 - \mathbf{J}_2 \mathbf{J}_1^+ \dot{\mathbf{X}}_1] + (\mathbf{I}_n - \mathbf{J}_3^+ \mathbf{J}_3)\, \mathbf{Z}_2\} \qquad [6.48]$$

The interpretation of this method is illustrated in Figure 6.5.

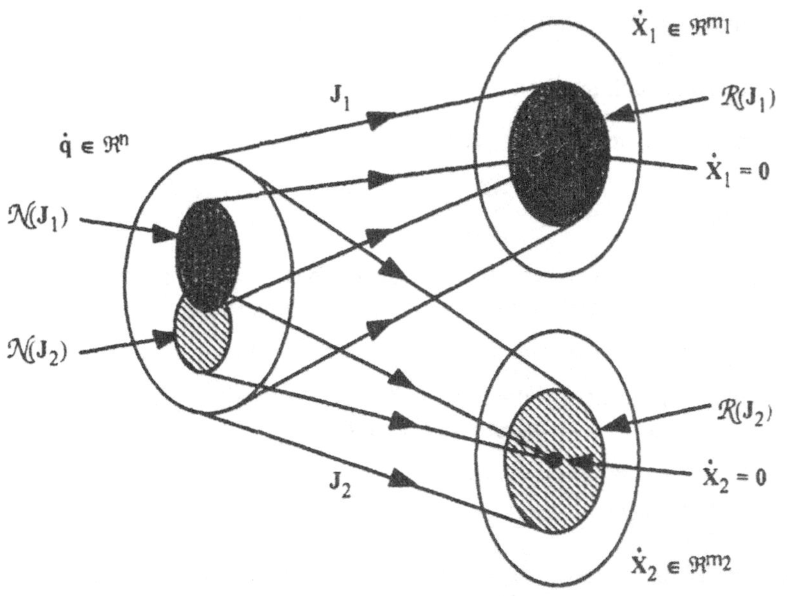

Figure 6.5. *Null space and range space of tasks X_1 and X_2 (from [Nakamura 87])*

6.6. Numerical calculation of the inverse geometric problem

When it is not possible to find a closed-form solution to the inverse geometric problem, we can use the differential model to compute an iterative numerical solution. To obtain the joint positions $\mathbf{q}^d$ corresponding to a desired location ${}^0\mathbf{T}_n^d$ of the terminal link, we proceed as follows:

- initialize $\mathbf{q}^c$ by the current joint configuration or by any random value within the joint domain of the robot;
- calculate the location of the terminal frame ${}^0\mathbf{T}_n^c$ corresponding to $\mathbf{q}^c$ using the direct geometric model;
- calculate the vectors of position error $\mathbf{dX}_p$ and rotation error $\mathbf{dX}_r$, representing the difference between the desired location ${}^0\mathbf{T}_n^d$ and the current location ${}^0\mathbf{T}_n^c$. Note that $\mathbf{dX}_p = \mathbf{dP}_n = \mathbf{P}_n^d - \mathbf{P}_n^c$ and $\mathbf{dX}_r = \mathbf{u}\,\alpha$, where the angle α and the unit vector $\mathbf{u}$ are obtained by solving the equation (§ 2.3.8): ${}^0\mathbf{A}_n^d = \mathbf{rot}(\mathbf{u}, \alpha)\,{}^0\mathbf{A}_n^c$, which can be written as ${}^0\mathbf{A}_n^d\,({}^0\mathbf{A}_n^c)^T = \mathbf{rot}(\mathbf{u}, \alpha)$;
- if $\mathbf{dX}_p$ and $\mathbf{dX}_r$ are sufficiently small, then $\mathbf{q}^d = \mathbf{q}^c$ and stop the calculation;

- to remain in the validity domain of the differential model, which is a first order expansion, we must introduce thresholds S_p and S_r on $\mathbf{dX_p}$ and $\mathbf{dX_r}$ respectively such that:

  - if $||\mathbf{dX_p}|| > S_p$, then $\mathbf{dX_p} = \dfrac{\mathbf{dX_p}}{||\mathbf{dX_p}||} S_p$

  - if $||\mathbf{dX_r}|| > S_r$, then $\mathbf{dX_r} = \dfrac{\mathbf{dX_r}}{||\mathbf{dX_r}||} S_r$

  The values 0.2 meter and 0.2 radian for these thresholds are acceptable for most of the industrial robots in view of their dimensions;
- calculate the Jacobian matrix ${}^0\mathbf{J}_n(\mathbf{q^c})$ denoted as $\mathbf{J}$;
- calculate the joint variation $\mathbf{dq} = \mathbf{J}^+ \, \mathbf{dX}$. An optimization term in the null space of $\mathbf{J}$ can also be taken into account;
- update the current joint configuration: $\mathbf{q^c} = \mathbf{q^c} + \mathbf{dq}$;
- return to the second step.

This algorithm converges rapidly and can be executed in real time. If it does not converge within a relatively large number of iterations, we have to restart the calculation using a new random value $\mathbf{q^c}$; if no convergence occurs for many different values of $\mathbf{q^c}$, it can be stated that there is no solution.

6.7. Minimum description of tasks [Fournier 80], [Dombre 81]

In current robot controllers, the desired trajectory of the end-effector is described by a sequence of frames. However, in many industrial applications, it is not necessary to completely specify the location of the end-effector frame and the task could be described by a reduced number of coordinates. For example:

- when the manipulated object is symmetric: for a spherical object, it is not necessary to specify the orientation; likewise, the rotation of a cylindrical object about its axis can be left free;
- releasing an object into a container: if the end-effector is already above the container, only an approach distance has to be specified; the task is thus described by a translational component;
- transferring objects from one point to another with arbitrary orientation; the task can be described by three translational components;
- placing a cylindrical object on a conveyor: the only orientation constraint is that the principal axis of the cylinder is horizontal; if the end-effector is already above the conveyor, the task could be described by two components (one vertical translation and one rotation).

When the number of components of a task is less than the number of degrees of freedom of the robot, the robot is redundant with respect to the task. Consequently, an infinite number of solutions can be obtained to realize such tasks. This redundancy can be exploited to satisfy secondary optimization criteria (§ 6.5).

6.7.1. *Principle of the description*

The proposed description of task is minimal in the sense that it only constrains the degrees of freedom of the task that have a functional role. The formulation is based on the use of the contact conditions between usual surfaces (plane, cylinder, sphere) that describe usual mechanical joints (or pairing) (Table 6.1 and Figure 6.6). To these six joints, we add the composite revolute and prismatic joints, which have one degree of mobility (Figure 6.7), and the fixed rigid pairing, which has no degree of freedom.

The description of a task is realized by a sequence of virtual mechanical joints. The choice of a type of joint is dictated by the local constraints associated with the task.

Table 6.1. *Simple mechanical joints*

	Plane	**Cylinder**	**Sphere**
Plane	Plane contact	Line contact	Point contact
Cylinder		Cylindrical joint	Cylindrical groove joint
Sphere			Spherical joint

A practical description of the mechanical joint formulation consists of specifying the task in terms of contact between two simple geometric entities (point, line, plane), one belonging to the robot, the other to the environment [Dombre 85]. A spherical joint, for example, is specified by matching two points. In the same way, the revolute and prismatic joints will be specified with two simultaneous combinations of geometric elements. The choice is not unique: a revolute joint for example can be achieved either by a line-to-line contact and a point-to-plane contact simultaneously or by a line-to-line contact and a point-to-point contact.

This geometric description is particularly convenient for graphic programming of tasks. Figure 6.8 shows the example of a peg-in-hole assembly, realized with the CAD/CAM software package CATIA [Catia] in which this formulation was implemented for robotic application. The different steps are as follows:

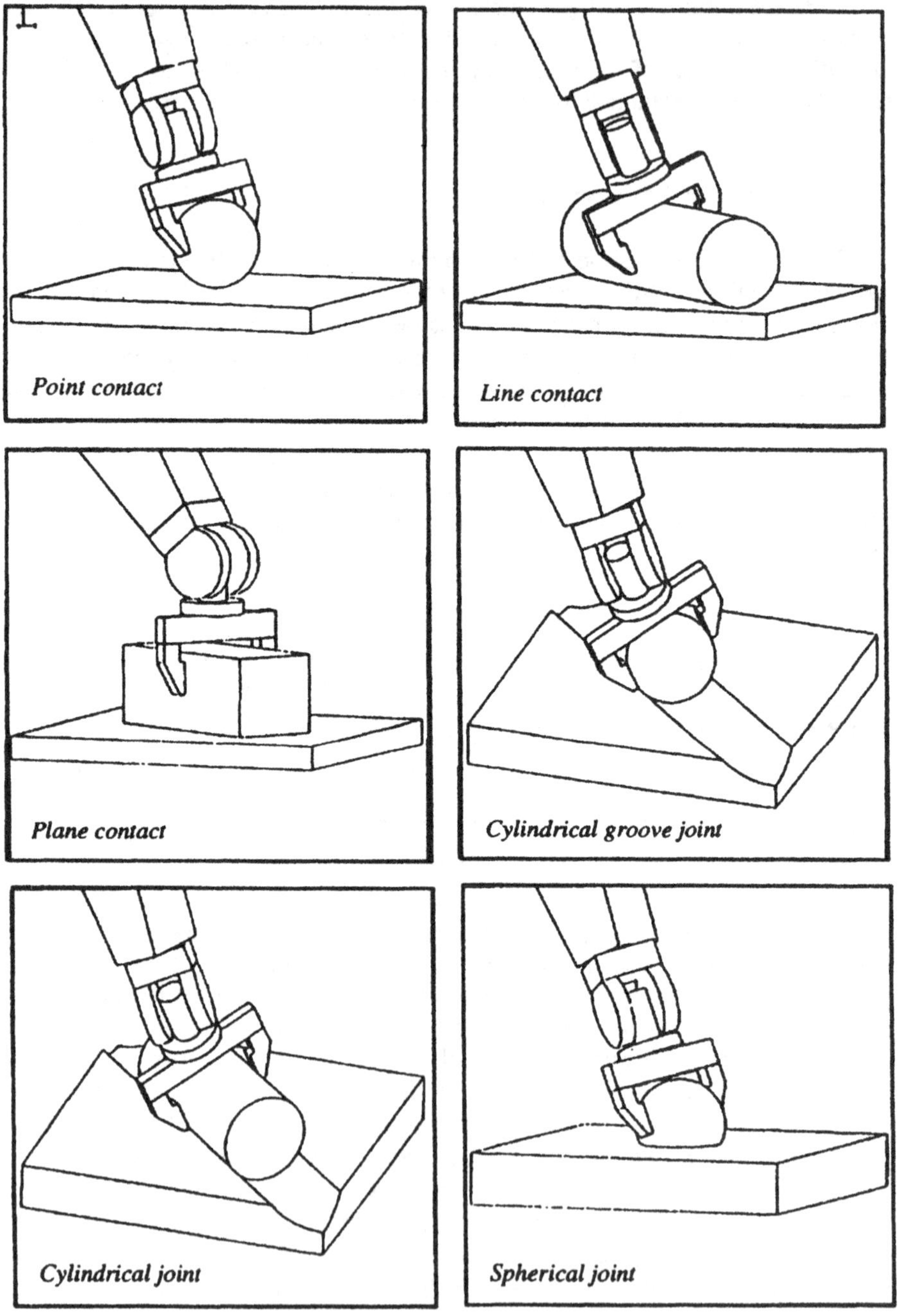

Figure 6.6. *Simple mechanical joints*

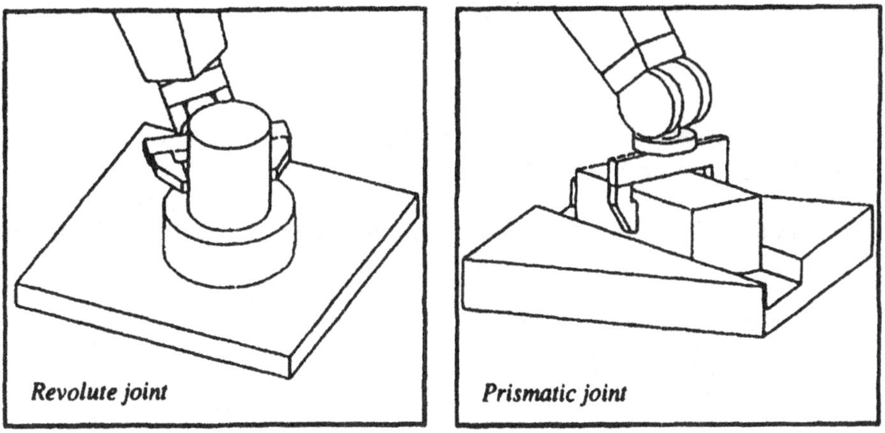

Figure 6.7. *Revolute and prismatic joints*

1) definition of a point-to-point contact (spherical joint) by selecting a point of the robot and a point of the environment (Figure 6.8a); after execution, the cylinder is positioned with an arbitrary orientation above the assembly site (Figure 6.8b);

2) definition of a line-to-line contact (cylindrical contact) by selecting a line of the robot and a line of the environment (Figure 6.8b); after execution, the axes of the hole and the peg are aligned (Figure 6.8c);

3) definition of a revolute joint by selecting a point and a line of the robot, and a point and a line of the environment (Figure 6.8c); after execution, the assembly task is completed (Figure 6.8d).

6.7.2. *Differential models associated with the minimum description of tasks*

To implement these types of tasks, we write the differential model of the location of frame R_E in the following form:

$$\begin{bmatrix} {}^0\mathbf{dP}_E \\ {}^0\boldsymbol{\delta}_E \end{bmatrix} = \begin{bmatrix} {}^0\mathbf{A}_n & \mathbf{0}_3 \\ \mathbf{0}_3 & {}^0\mathbf{A}_n \end{bmatrix} \begin{bmatrix} {}^n\mathbf{dP}_E \\ {}^n\boldsymbol{\delta}_E \end{bmatrix} = \begin{bmatrix} {}^0\mathbf{A}_n & \mathbf{0}_3 \\ \mathbf{0}_3 & {}^0\mathbf{A}_n \end{bmatrix} \begin{bmatrix} \mathbf{I}_3 & -{}^n\hat{\mathbf{P}}_E \\ \mathbf{0}_3 & \mathbf{I}_3 \end{bmatrix} \begin{bmatrix} {}^n\mathbf{dP}_n \\ {}^n\boldsymbol{\delta}_n \end{bmatrix}$$

$$= \begin{bmatrix} {}^0\mathbf{A}_n & -{}^0\mathbf{A}_n\,{}^n\hat{\mathbf{P}}_E \\ \mathbf{0}_3 & {}^0\mathbf{A}_n \end{bmatrix} {}^n\mathbf{J}_n\,\mathbf{dq} \qquad [6.49]$$

where ${}^n\mathbf{P}_E$ defines the origin of frame R_E referred to frame R_n.

The differential model of a virtual joint can be written as:

$$\mathbf{dX} = \mathbf{H}\,{}^{n}\mathbf{J}_n\,\mathbf{dq} \qquad [6.50]$$

where ${}^{n}\mathbf{J}_n$ and $\mathbf{H}$ are (6xn) and (cx6) matrices respectively, and c indicates the number of constraint equations of the task.

We will show in the following section how to determine **H** for the virtual joints [Dombre 81].

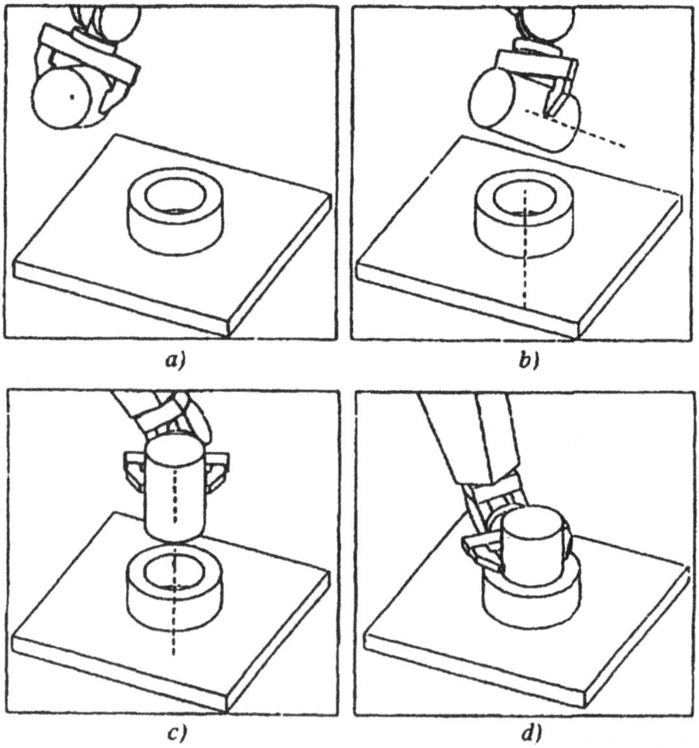

Figure 6.8. *Graphic programming of an assembly task with a minimum description*

6.7.2.1. *Point contact (point on plane)*

This joint drives a point O_E of the tool on any position on a plane Q (Figure 6.9). Let **N** be the unit vector normal to the plane Q and let O_D be an arbitrary point of Q. The necessary global displacement to realize the point contact is expressed in frame R_0 by:

$$r = {}^{0}N^{T} [{}^{0}P_{D} - {}^{0}P_{E}] \qquad [6.51]$$

where ${}^{0}P_{D}$ and ${}^{0}P_{E}$ define the coordinates of the points O_D and O_E in frame R_0.

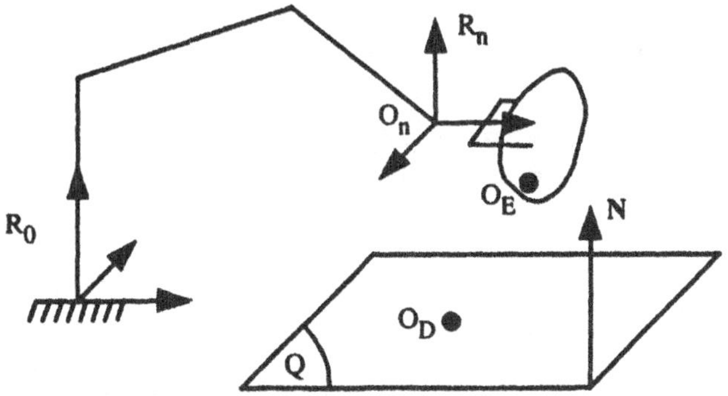

Figure 6.9. *Realization of point contact*

The displacement r is realized by a sequence of elementary displacements along a single direction such that (equation [6.49]):

$$dX = dr = {}^{0}N^{T}\,{}^{0}dP_{E} = [\ {}^{0}N^{T}\,{}^{0}A_{n} \quad -{}^{0}N^{T}\,{}^{0}A_{n}\,{}^{n}\hat{P}_{E}\ ]\ {}^{n}J_{n}\,dq$$
$$= {}^{0}N^{T}\,{}^{0}A_{n}\,[\ I_3 \quad -{}^{n}\hat{P}_{E}\ ]\ {}^{n}J_{n}\,dq \qquad [6.52]$$

Expression [6.52] constitutes the differential model of the point contact. The matrix **H** is given by the row vector ${}^{0}N^{T}\,{}^{0}A_{n}\,[\ I_3 \quad -{}^{n}\hat{P}_{E}\ ]$.

6.7.2.2. *Line contact (line on plane)*

The equations of a line contact are derived from Figure 6.10. The line U_E is driven on plane Q without constraining its orientation in the plane. We can realize this joint by simultaneously carrying out a rotation and a translation [Dombre 88a]. However, it is more judicious to avoid the calculation of an angle by defining the task as driving two points O_{E1} and O_{E2} of U_E on plane Q. The joint is thus equivalent to two point contact. The corresponding differential model is written as:

$$dX = \begin{bmatrix} dr_1 \\ dr_2 \end{bmatrix} = \begin{bmatrix} {}^{0}N^{T}\,{}^{0}A_{n} & -{}^{0}N^{T}\,{}^{0}A_{n}\,{}^{n}\hat{P}_{E1} \\ {}^{0}N^{T}\,{}^{0}A_{n} & -{}^{0}N^{T}\,{}^{0}A_{n}\,{}^{n}\hat{P}_{E2} \end{bmatrix} {}^{n}J_{n}\,dq \qquad [6.53]$$

where **H** is a (2xn) matrix.

We can generalize this approach for the other joints where the jth row of **H** takes the following general form:

$$\mathbf{H}_j = {}^0\mathbf{N}_j^T \, {}^0\mathbf{A}_n \left[\ \mathbf{I}_3 \quad -{}^n\hat{\mathbf{P}}_{Ej} \ \right] \qquad [6.54]$$

where ${}^0\mathbf{N}_j$ denotes the unit vector of the normal to the plane of the jth point contact, and ${}^n\mathbf{P}_{E_j}$ is the vector of the coordinates of the tool point O_{Ej} with respect to frame R_n.

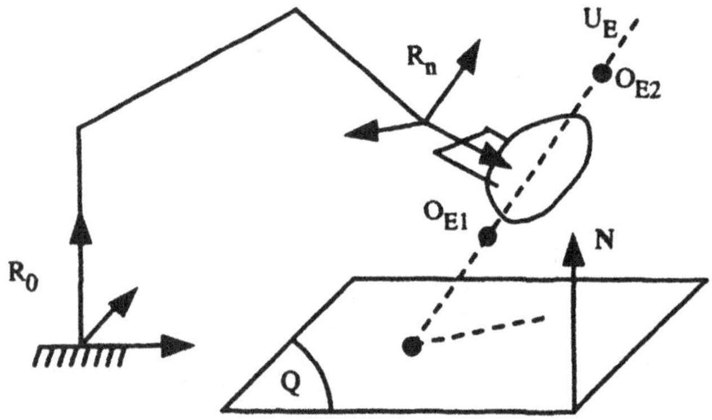

Figure 6.10. *Realization of line contact*

6.7.2.3. *Planar contact (plane on plane)*

This joint drives a plane Q_E attached to the tool on a plane Q_D of the environment, without orientation or position constraints (Figure 6.11). We select three non-aligned points O_{E1}, O_{E2} and O_{E3} in Q_E, then we carry out three simultaneous point contacts.

6.7.2.4. *Cylindrical groove joint (point on line)*

The cylindrical groove joint drives a point O_E of the tool on a line U_D of the environment. This is done by simultaneously realizing two point contacts of O_E on two arbitrary orthogonal planes Q_{D1} and Q_{D2} whose intersection is the line U_D (Figure 6.12).

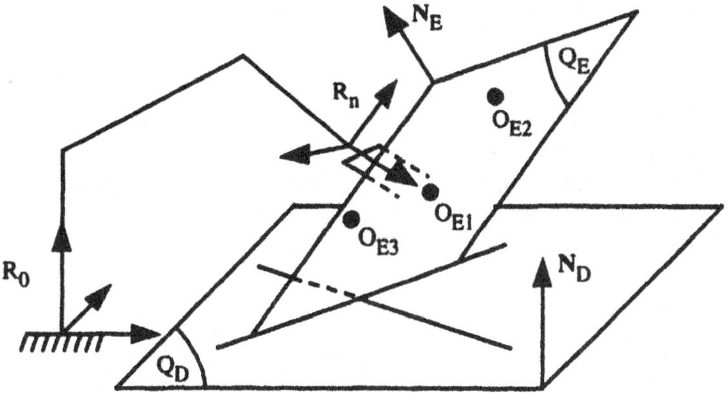

Figure 6.11. *Realization of a plane contact*

6.7.2.5. *Cylindrical joint (line on line)*

The task consists of aligning two lines U_E and U_D without position or orientation constraints along and about these lines (Figure 6.12). We define two arbitrary orthogonal planes Q_{D1} and Q_{D2} whose intersection is the line U_D and whose normals are N_{D1} and N_{D2} respectively. To realize a cylindrical joint, any two distinct points O_{E1} and O_{E2} of the line U_E are driven simultaneously on the planes Q_{D1} and Q_{D2}. In other words, the cylindrical joint corresponds to four point contacts.

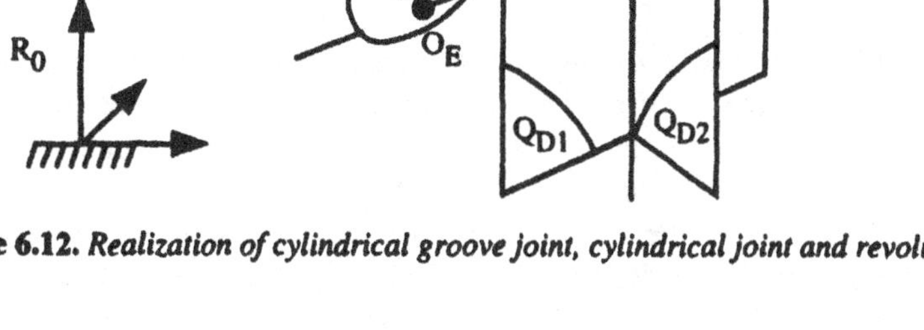

Figure 6.12. *Realization of cylindrical groove joint, cylindrical joint and revolute joint*

6.7.2.6. *Spherical joint (point on point)*

A spherical joint drives a point O_E of the tool on a point O_D of the environment without constraining the orientation of the tool. The task can be realized by three point contacts that drive O_E simultaneously on the planes Q_{D1}, Q_{D2} and Q_{D3}, which are parallel to the planes (y_0, z_0), (x_0, z_0), (x_0, y_0) and pass through the point O_D. The required displacements r_1, r_2 and r_3 are the components of the vector $\mathbf{O_EO_D}$ along the axes of frame R_0. The task is defined as:

$$\mathbf{dX} = \begin{bmatrix} dr_1 \\ dr_2 \\ dr_3 \end{bmatrix} = [\ {}^0\mathbf{A}_n \quad -{}^0\mathbf{A}_n{}^n\hat{\mathbf{P}}_E\ ]\ {}^n\mathbf{J}_n\,\mathbf{dq} \qquad [6.55]$$

6.7.2.7. *Revolute joint (line-point on line-point)*

A revolute joint (Figure 6.12) consists of aligning a line U_E of the tool with a line U_D of the environment while simultaneously driving a point O_E of U_E on a plane Q_D normal to U_D (not represented in the figure). Let O_{E1} and O_{E2} be any two distinct points on U_E. Similar to the cylindrical groove joint, let us consider that Q_{D1} and Q_{D2} are two arbitrary orthogonal planes whose intersection is the line U_D. The joint is thus equivalent to the simultaneous realization of five point contacts:

- driving the point O_{E1} on the planes Q_{D1} and Q_{D2};
- driving the point O_{E2} on the planes Q_{D1} and Q_{D2};
- driving the point O_E on the plane Q_D.

In practice, it is more convenient to describe the revolute joint by a line-to-line contact and a point-to-point contact. This choice leads to seven equations, and the rank of the matrix $\mathbf{H}\,\mathbf{J}$ is five.

6.7.2.8. *Prismatic joint (plane-plane on plane-plane)*

A prismatic joint consists of aligning two lines of the tool with two geometrically compatible lines of the environment, and making a translation along an arbitrary axis. To simplify, we consider that the two lines are perpendicular and the displacement is carried out along one of these lines.

Let U_{E1} and U_{E2} be the two lines of the tool and let U_{D1} and U_{D2} be two compatible lines of the environment (Figure 6.13). Let us suppose that the free translation is along the line U_{D1}. Let Q_{D1a} and Q_{D1b} be two arbitrary orthogonal planes whose intersection is U_{D1}, and O_{E1a} and O_{E1b} be any two distinct points on the line U_{E1}. We realize the prismatic joint by five point contacts:

- driving the point O_{E1a} on the planes Q_{D1a} and Q_{D2b};
- driving the point O_{E1b} on the planes Q_{D1a} and Q_{D2b};
- driving any point of U_{E2}, that is not the intersection of U_{E1} and U_{E2}, on the plane formed by the lines U_{D1} and U_{D2}.

Similar to the revolute joint case, it may be more convenient for the user to specify a prismatic joint using two plane-to-plane contacts. In this case, the number of equations is six.

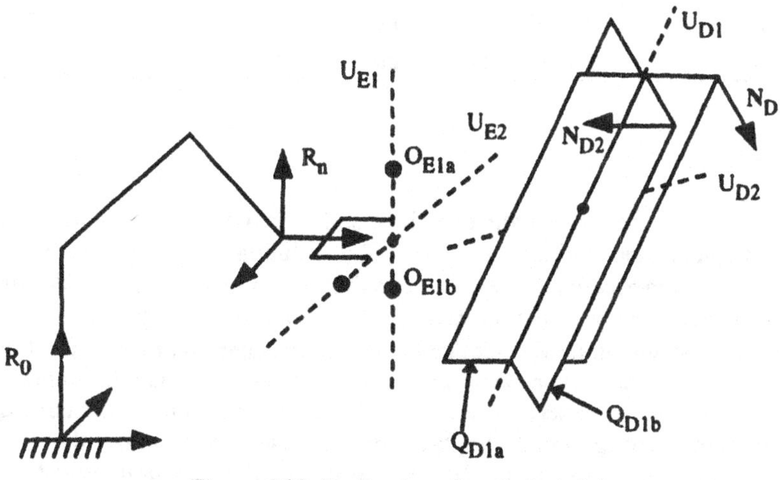

Figure 6.13. *Realization of a prismatic joint*

NOTES.–

- for the fixed rigid pairing, we use the complete description of $\mathbf{dX} = \mathbf{J}\ \mathbf{dq}$;
- Table 6.2 summarizes the specification of each virtual mechanical joint as well as the number of necessary equations.

Table 6.2. *Equivalence between virtual mechanical joints and geometric specification*

Type of joint	Elements of the tool	Elements of the environment	Number of independent equations	Total number of equations
Point contact	Point	Plane	1	1
Line contact	Line	Plane	2	2
Plane contact	Plane	Plane	3	3
Cylindrical groove	Point	Line	2	2
Spherical	Point	Point	3	3
Cylindrical	Line	Line	4	4
Revolute	Line-Point	Line-Point	5	7
Prismatic	Plane-Plane	Plane-Plane	5	6

6.8. Conclusion

In this chapter, we have studied the inverse kinematic model by considering the regular, singular and redundant cases. The solution may be obtained either analytically or numerically. The analytical solution can be used for simple robots in regular configurations, whereas the numerical methods are more general.

We have also shown how to reduce the functional degrees of freedom of the task using a description method based on the virtual mechanical joints formulation.

The redundancy, whether it is a built-in feature of the robot or the consequence of a minimum description of the task, can be used to optimize the trajectory generation of the mechanism. In this respect, the solution based on the pseudoinverse method proves to be very powerful. It allows us to realize secondary optimization functions such as keeping the joints away from their limits or improving the manipulability.

Chapter 7

Geometric and kinematic models of complex chain robots

7.1. Introduction

In this chapter, we develop a method to describe the geometry of complex robots with tree or closed chain structures. This method constitutes the extension of the notation presented in Chapter 3 for serial robots [Khalil 86a]. We also present the computation of the direct and inverse geometric models of such mechanisms. Finally, we establish their direct and inverse kinematic models. The results are illustrated using the AKR-3000 robot, which contains two closed loops, and the Acma SR400 robot, which contains a parallelogram closed loop.

7.2. Description of tree structured robots

A tree structured robot is composed of n mobile links and n joints. The links are assumed to be perfectly rigid. The joints are either revolute or prismatic and assumed to be ideal (no backlash, no elasticity). A complex joint can be represented by an equivalent combination of revolute and prismatic joints with zero-length massless links.

The links are numbered consecutively from the base to the terminal links. Thus, link 0 is the fixed base and link n is one of the terminal links (Figure 7.1). Joint j connects link j to link a(j), where a(j) denotes the link antecedent to link j, and consequently $a(j) < j$. We define a main branch as the set of links on the path between the base and a terminal link. Thus, a tree structure has as many main branches as the number of terminal links.

The topology of the system is defined by a(j) for j = 1, ..., n. In order to compute the relationship between the locations of the links, we attach a frame R_i to each link i such that:

- $\mathbf{z}_i$ is along the axis of joint i;
- $\mathbf{x}_i$ is taken along the common normal between $\mathbf{z}_i$ and one of the succeeding joint axes, which are fixed on link i. Figure 7.2 shows the case where links j and k are articulated on link i.

Two cases are considered for computing the transformation matrix ${}^i\mathbf{T}_j$, which defines the location of frame R_j relative to frame R_i with i = a(j):

1) if $\mathbf{x}_i$ is along the common normal between $\mathbf{z}_i$ and $\mathbf{z}_j$, then ${}^i\mathbf{T}_j$ is the same as the transformation matrix between two consecutive frames of serial structure. It is obtained as a function of the four geometric parameters (α_j, d_j, θ_j, r_j) as defined in § 3.2 (equation [3.2]):

$$
{}^i\mathbf{T}_j = \mathbf{Rot}(\mathbf{x}, \alpha_j)\, \mathbf{Trans}(\mathbf{x}, d_j)\, \mathbf{Rot}(\mathbf{z}, \theta_j)\, \mathbf{Trans}(\mathbf{z}, r_j)
$$

$$
= \begin{bmatrix} C\theta_j & -S\theta_j & 0 & d_j \\ C\alpha_j S\theta_j & C\alpha_j C\theta_j & -S\alpha_j & -r_j S\alpha_j \\ S\alpha_j S\theta_j & S\alpha_j C\theta_j & C\alpha_j & r_j C\alpha_j \\ 0 & 0 & 0 & 1 \end{bmatrix} \qquad [7.1]
$$

2) if $\mathbf{x}_i$ is not along the common normal between $\mathbf{z}_i$ and $\mathbf{z}_j$, then the matrix ${}^i\mathbf{T}_j$ must be defined using six geometric parameters. This case is illustrated in Figure 7.2, where $\mathbf{x}_i$ is along the common normal between $\mathbf{z}_i$ and $\mathbf{z}_k$. To obtain the six parameters defining frame R_j relative to frame R_i, we define $\mathbf{u}_j$ as the common normal between $\mathbf{z}_i$ and $\mathbf{z}_j$. The transformation from frame R_i to frame R_j can be obtained as a function of the six geometric parameters (γ_j , b_j, α_j, d_j, θ_j, r_j) where:
   - γ_j is the angle between $\mathbf{x}_i$ and $\mathbf{u}_j$ about $\mathbf{z}_i$;
   - b_j is the distance between $\mathbf{x}_i$ and $\mathbf{u}_j$ along $\mathbf{z}_i$.

The parameters γ_j and b_j permit to define $\mathbf{u}_j$ with respect to $\mathbf{x}_i$, whereas the classical parameters α_j, d_j, θ_j, r_j permit to define frame R_j with respect to the intermediate frame whose $\mathbf{x}$ axis is along $\mathbf{u}_j$ and $\mathbf{z}$ axis is along $\mathbf{z}_j$.

The transformation matrix ${}^i\mathbf{T}_j$ is obtained as:

$$
{}^i\mathbf{T}_j = \mathbf{Rot}(\mathbf{z}, \gamma_j)\, \mathbf{Trans}(\mathbf{z}, b_j)\, \mathbf{Rot}(\mathbf{x}, \alpha_j)\, \mathbf{Trans}(\mathbf{x}, d_j)\, \mathbf{Rot}(\mathbf{z}, \theta_j)\, \mathbf{Trans}(\mathbf{z}, r_j) \qquad [7.2]
$$

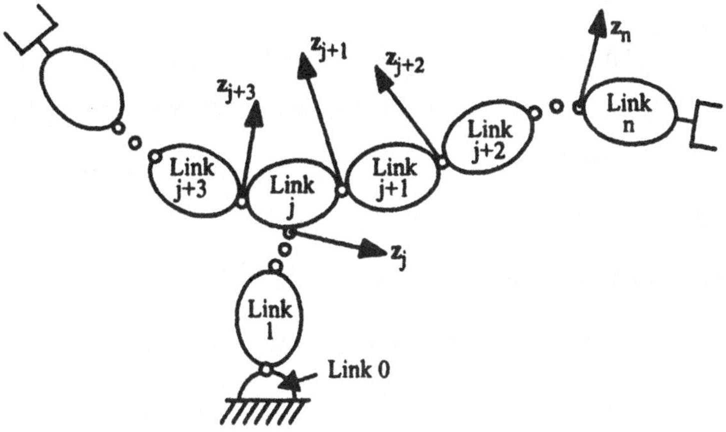

Figure 7.1. *Notations for a tree structured robot*

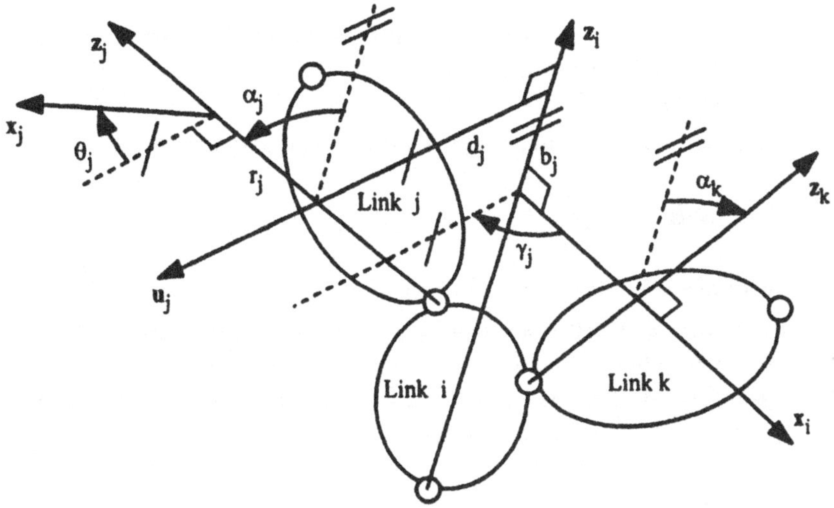

Figure 7.2. *Geometric parameters for a link with more than two joints*

After development, we obtain:

$$^{i}\mathbf{T}_j = \begin{bmatrix} C\gamma_j C\theta_j - S\gamma_j C\alpha_j S\theta_j & -C\gamma_j S\theta_j - S\gamma_j C\alpha_j C\theta_j & S\gamma_j S\alpha_j & d_j C\gamma_j + r_j S\gamma_j S\alpha_j \\ S\gamma_j C\theta_j + C\gamma_j C\alpha_j S\theta_j & -S\gamma_j S\theta_j + C\gamma_j C\alpha_j C\theta_j & -C\gamma_j S\alpha_j & d_j S\gamma_j - r_j C\gamma_j S\alpha_j \\ S\alpha_j S\theta_j & S\alpha_j C\theta_j & C\alpha_j & r_j C\alpha_j + b_j \\ 0 & 0 & 0 & 1 \end{bmatrix} \quad [7.3]$$

The inverse transformation $^{j}\mathbf{T}_i$ is expressed by:

$$^{j}\mathbf{T}_i = \begin{bmatrix} & & & -b_j S\alpha_j S\theta_j - d_j C\theta_j \\ & {}^{i}\mathbf{A}_j^{T} & & -b_j S\alpha_j C\theta_j + d_j S\theta_j \\ & & & -b_j C\alpha_j - r_j \\ 0 & 0 & 0 & 1 \end{bmatrix} \quad [7.4]$$

NOTES.–

- equation [7.3] represents the general form of the transformation matrix. The special case of serial robots (equation [7.1]) can be obtained from it by setting $b_j = 0$ and $\gamma_j = 0$;
- as for the serial structure, the joint variable q_j is given by:

$$q_j = \bar{\sigma}_j \, \theta_j + \sigma_j \, r_j \quad [7.5]$$

  where $\sigma_j = 0$ if joint j is revolute, $\sigma_j = 1$ if joint j is prismatic and $\bar{\sigma}_j = 1 - \sigma_j$;
- we set $\sigma_j = 2$ to define a frame R_j with constant position and orientation with respect to frame a(j). In this case, q_j and $\bar{\sigma}_j$ are not defined;
- the definition of frame R_0 and the frames fixed to the terminal links can be made as in the case of serial robots.

7.3. Description of robots with closed chains

A closed chain structure consists of a set of rigid links connected to each other with joints where at least one closed loop exists. This structure enhances both the accuracy and the load-carrying capacity of the robot. The system is composed of L joints and n + 1 links, where link 0 is the fixed base and L > n. It may contain several terminal links. The number of independent closed loops is equal to:

$$B = L - n \quad [7.6]$$

The joints are either active or passive. The N active joints are provided with actuators. We assume that the number of actuated joints is equal to the number of degrees of freedom of the robot. Thus, the position and orientation of all the links can be determined as a function of the active joint variables.

We introduce the parameter μ_j such that:

- $\mu_j = 1$ if joint j is actuated (active joint);
- $\mu_j = 0$ if joint j is non-actuated (passive joint).

To determine the geometric parameters of a mechanism with closed chains, we proceed as follows:

a) construct an equivalent tree structure having n joints by virtually cutting each closed chain at one of its passive joints. Since a closed loop contains several passive joints, select the joint to be cut in such a way that the difference between the number of links of the two branches from the root of the loop[1] to the links connected to the cut joint is as small as possible. This choice reduces the computational complexity of the dynamic model [Kleinfinger 86a]. The geometric parameters of the equivalent tree structure are determined as described in the previous section;

b) number the cut joints from $n+1$ to L. For each cut joint k, assign a frame R_k fixed on one of the links connected to this joint, for instance link j. The $\mathbf{z}_k$ axis is taken along the axis of joint k, and the $\mathbf{x}_k$ axis is aligned with the common normal between $\mathbf{z}_k$ and $\mathbf{z}_j$ (Figure 7.3). Let $i = a(k)$ where link i denotes the other link of joint k. The transformation matrix from frame R_i to frame R_k can be obtained as a function of the usual six (or four) geometric parameters γ_k, b_k, α_k, d_k, θ_k, r_k, where q_k is equal to θ_k or r_k;

c) since frame R_k is fixed on link j, the transformation matrix between frames R_j and R_k is constant. To avoid any confusion, this transformation will be denoted by ${}^jT_{k+B}$, with $j = a(k+B)$. The geometric parameters defining this transformation will have as a subscript $k+B$. Note that frame R_{k+B} is aligned with frame R_k, and that both r_{k+B} and θ_{k+B} are zero.

In summary, the geometric description of a structure with closed loops is defined by an equivalent tree structure that is obtained by cutting each closed loop at one of its joints and by adding two frames at each cut joint. The total number of frames is equal to $n+2B$ and the geometric parameters of the last B frames are constants.

[1] The root of a loop is the first common link when going from the links of the cut joint to the base of the robot.

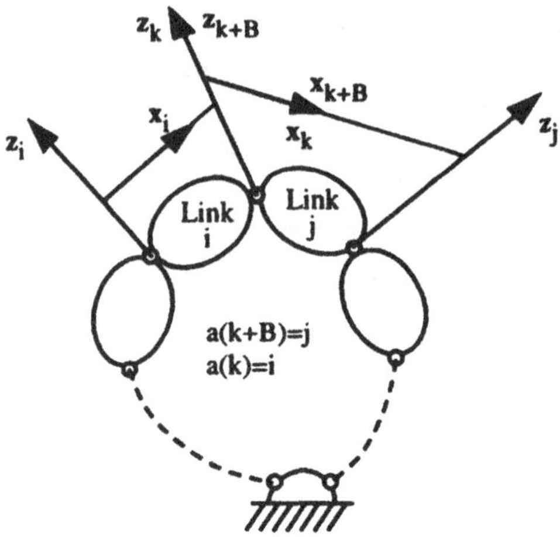

Figure 7.3. *Frames of a cut joint*

The (Lx1) joint variable vector $\mathbf{q}$ is written as:

$$\mathbf{q} = \begin{bmatrix} \mathbf{q}_a \\ \mathbf{q}_p \\ \mathbf{q}_c \end{bmatrix} \qquad [7.7]$$

with:

- $\mathbf{q}_a$: vector containing the N active joint variables;
- $\mathbf{q}_p$: vector containing the p=n−N passive joint variables of the equivalent tree structure;
- $\mathbf{q}_c$: vector containing the B variables of the cut joints. When a cut joint has several degrees of freedom (spherical, universal, ...), we can consider all of its joint variables to be belonging to $\mathbf{q}_c$.

Only the N active variables $\mathbf{q}_a$ are independent. Thus, there are c=L−N independent constraint equations between the joint variables $\mathbf{q}$. These relations form the geometric constraint equations satisfying the closure of the loops. Since R_k and R_{k+B} are aligned, the geometric constraint equations for each loop can be written as:

$$^{k+B}\mathbf{T}_j \ldots {}^{i}\mathbf{T}_k = \mathbf{I}_4 \qquad [7.8]$$

where $\mathbf{I}_4$ is the (4x4) identity matrix.

For a spatial loop, the maximum number of independent geometric constraint equations is six, while for a planar loop this number reduces to three. The geometric constraint equations can be obtained from relation [7.8]. They are represented by the nonlinear equation:

$$\boldsymbol{\phi}(\mathbf{q}) = \begin{bmatrix} \phi_1(\mathbf{q}) \\ \phi_2(\mathbf{q}) \\ \dots \\ \phi_{(L-N)}(\mathbf{q}) \end{bmatrix} = \mathbf{0}_{(L-N)x1} \quad [7.9]$$

To determine the locations of all the links of the closed chain structure, we have to compute the passive joint variables in terms of the active joint variables. For simple mechanisms, equation [7.9] may be solved in an analytical closed-form such that:

$$\mathbf{q}_p = \mathbf{g}_p(\mathbf{q}_a) \quad [7.10a]$$
$$\mathbf{q}_c = \mathbf{g}_c(\mathbf{q}_a, \mathbf{q}_p) \quad [7.10b]$$

Otherwise, numerical methods based on the inverse differential model can be used (§ 7.9) [Uicker 69], [Wang 91].

• **Example 7.1.** Description of the geometry of the Acma SR400 robot. This mechanism has six degrees of freedom, eight moving links and nine revolute joints. It contains a parallelogram closed loop. Joints 3, 8 and 9 are passive. The equivalent tree structure is obtained by cutting the loop at joint 9, which connects link 3 and link 8. The link coordinate frames are shown in Figure 7.4. The geometric parameters are given in Table 7.1.

• **Example 7.2.** Description of the AKR-3000 painting robot. This six degree-of freedom robot has 12 joints and 10 links. It contains two independent closed loops. Figure 7.5 shows the link coordinate frames. The first loop is cut at the joint connecting links 5 and 7, and the second loop is cut at the joint connecting links 2 and 6. Joints 1, 5, 6, 8, 9 and 10 are active, while joints 2, 3, 4, 7, 11 and 12 are passive. The geometric parameters are given in Table 7.2.

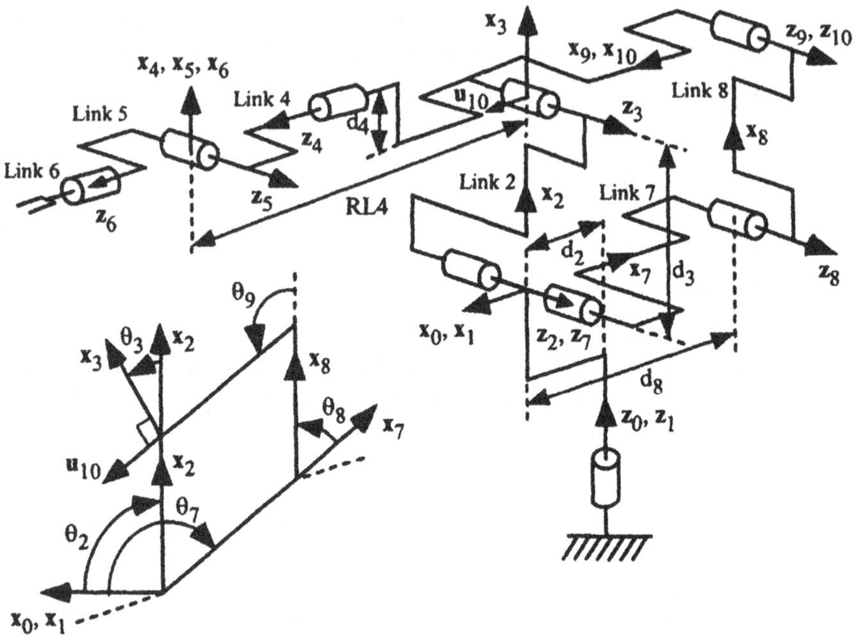

Figure 7.4. *Acma SR400 robot*

Table 7.1. *Geometric parameters of the Acma SR400 robot*

j	a(j)	μ_j	σ_j	γ_j	b_j	α_j	d_j	θ_j	r_j
1	0	1	0	0	0	0	0	θ1	0
2	1	1	0	0	0	$-\pi/2$	d2	θ_2	0
3	2	0	0	0	0	0	d3	θ_3	0
4	3	1	0	0	0	$-\pi/2$	d_4	θ_4	RL4
5	4	1	0	0	0	$\pi/2$	0	θ_5	0
6	5	1	0	0	0	$-\pi/2$	0	θ_6	0
7	1	1	0	0	0	$-\pi/2$	d2	θ_7	0
8	7	0	0	0	0	0	d_8	θ_8	0
9	8	0	0	0	0	0	$d_9=d_3$	θ_9	0
10	3	0	2	$\pi/2$	0	0	$d_{10}=-d_8$	0	0

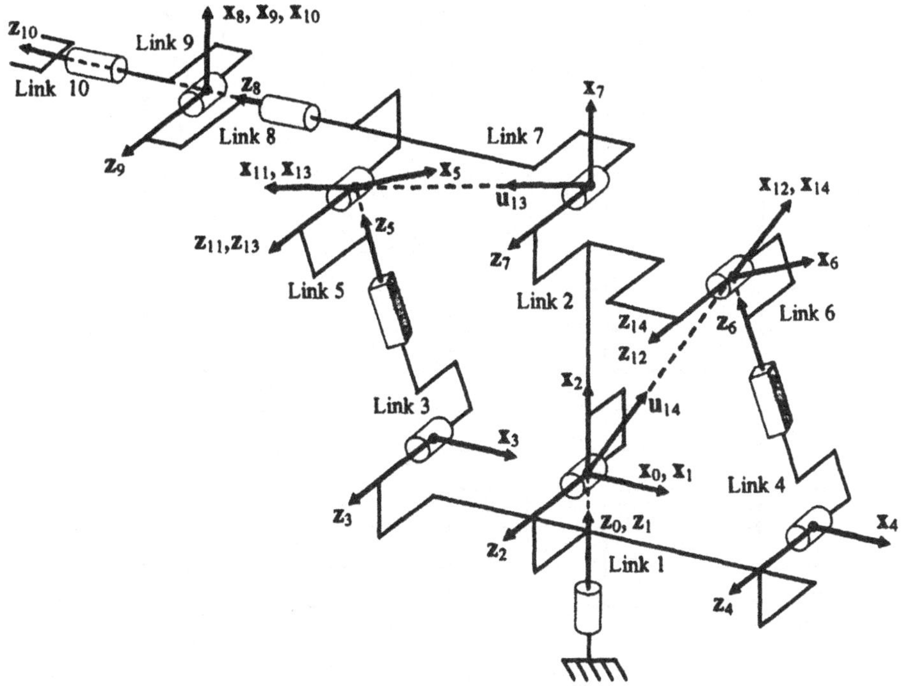

Figure 7.5. *AKR-3000 robot*

7.4. Direct geometric model of tree structured robots

We have shown in Chapter 3 that the DGM of a serial robot is obtained from the transformation matrix ${}^0\mathbf{T}_n$ giving the location of the terminal link n relative to frame R_0. The extension to tree structured robots is straightforward. The transformation matrix ${}^0\mathbf{T}_k$ specifying the location of the terminal link k relative to frame R_0 is obtained by multiplying the transformation matrices along the main branch connecting this terminal link to the base:

$$ {}^0\mathbf{T}_k = {}^0\mathbf{T}_i \cdots {}^{a(a(k))}\mathbf{T}_{a(k)}\, {}^{a(k)}\mathbf{T}_k \qquad [7.11] $$

Table 7.2. *Geometric parameters of the AKR-3000 robot*

j	a(j)	μ_j	σ_j	γ_j	b_j	α_j	d_j	θ_j	r_j
1	0	1	0	0	0	0	0	θ_1	0
2	1	0	0	0	0	$\pi/2$	0	θ_2	0
3	1	0	0	0	0	$\pi/2$	d_3	θ_3	0
4	1	0	0	0	0	$\pi/2$	d_4	θ_4	0
5	3	1	1	0	0	$-\pi/2$	0	0	r_5
6	4	1	1	0	0	$-\pi/2$	0	0	r_6
7	2	0	0	0	0	0	d_7	θ_7	0
8	7	1	0	0	0	$-\pi/2$	0	θ_8	r_8
9	8	1	0	0	0	$\pi/2$	0	θ_9	0
10	9	1	0	0	0	$-\pi/2$	0	θ_{10}	0
11	5	0	0	0	0	$\pi/2$	0	θ_{11}	0
12	6	0	0	0	0	$\pi/2$	0	θ_{12}	0
13	7	0	2	γ_{13}	0	0	d_{13}	0	0
14	2	0	2	γ_{14}	0	0	d_{14}	0	0

7.5. Direct geometric model of robots with closed chains

For a robot with closed chains, the DGM gives the location of the terminal(s) link(s) as a function of the active joint variables. The location of a terminal link k relative to the base ${}^0\mathbf{T}_k$ is obtained, as usual, by multiplying the transformation matrices along the direct shortest path between the base and the terminal link as given by equation [7.11].

If the matrix ${}^0\mathbf{T}_k$ contains passive joint variables, we have to compute these variables in terms of the active joint variables. This implies solution of the geometric constraint equations [7.8] as developed in § 7.7.

• **Example 7.3.** Direct geometric model of the Acma SR400 robot (Figure 7.4). The location of the terminal link relative to frame R_0 is obtained as:

$$ {}^0\mathbf{T}_6 = {}^0\mathbf{T}_1\, {}^1\mathbf{T}_2\, {}^2\mathbf{T}_3\, {}^3\mathbf{T}_4\, {}^4\mathbf{T}_5\, {}^5\mathbf{T}_6 $$

Since joint 3 is passive, we have to express q_3 in terms of the active variable q_7. This equation will be developed in Example 7.6.

• **Example 7.4.** Direct geometric model of the AKR-3000 robot (Figure 7.5). The transformation matrix through the direct path between the terminal link 10 and the base is written as:

$$^{0}\mathbf{T}_{10} = {}^{0}\mathbf{T}_{1}\,{}^{1}\mathbf{T}_{2}\,{}^{2}\mathbf{T}_{7}\,{}^{7}\mathbf{T}_{8}\,{}^{8}\mathbf{T}_{9}\,{}^{9}\mathbf{T}_{10}$$

The computation of the passive joint variables θ_2 and θ_7 in terms of the active joint variables r_5 and r_6 is developed in Example 7.5.

7.6. Inverse geometric model of closed chain robots

The IGM of a robot with a closed chain structure gives the active joint variables as a function of the location of the end-effector.

We first determine the joint variables of the direct path between the base and the end-effector. This problem can be solved using the approaches developed for serial robots in Chapter 4. Then, we solve the geometric constraint equations of the loop to compute the passive joint variables belonging to this path in terms of the active joint variables (§ 7.7). To use the methods of Chapter 4, we have to define the link frames such that the geometric parameters b_j and γ_j of the frames of the direct path of the terminal link are zero. Otherwise, they can be eliminated by grouping them with the parameters r_i and θ_i, for $i = a(j)$, respectively. This can be proved by developing the elements of two consecutive transformation matrices $^{a(i)}\mathbf{T}_i$ and $^{i}\mathbf{T}_j$:

$$^{a(i)}\mathbf{T}_i\,{}^{i}\mathbf{T}_j = \mathbf{Rot}(\mathbf{x}, \alpha_i)\,\mathbf{Trans}(\mathbf{x}, d_i)\,\mathbf{Rot}(\mathbf{z}, \theta_i)\,\mathbf{Trans}(\mathbf{z}, r_i)\,\mathbf{Rot}(\mathbf{z}, \gamma_j)$$
$$\mathbf{Trans}(\mathbf{z}, b_j)\,\mathbf{Rot}(\mathbf{x}, \alpha_j)\,\mathbf{Trans}(\mathbf{x}, d_j)\,\mathbf{Rot}(\mathbf{z}, \theta_j)\,\mathbf{Trans}(\mathbf{z}, r_j)$$

This equation can be rewritten as:

$$^{a(i)}\mathbf{T}_i\,{}^{i}\mathbf{T}_j = \mathbf{Rot}(\mathbf{x}, \alpha_i)\,\mathbf{Trans}(\mathbf{x}, d_i)\,\mathbf{Rot}(\mathbf{z}, \theta_i')\,\mathbf{Trans}(\mathbf{z}, r_i')\,\mathbf{Rot}(\mathbf{x}, \alpha_j)$$
$$\mathbf{Trans}(\mathbf{x}, d_j)\,\mathbf{Rot}(\mathbf{z}, \theta_j)\,\mathbf{Trans}(\mathbf{z}, r_j) \qquad [7.12]$$

with $r_i' = r_i + b_j$ and $\theta_i' = \theta_i + \gamma_j$.

7.7. Resolution of the geometric constraint equations of a simple loop

7.7.1. *Introduction*

The computation of the geometric and dynamic models of robots with closed loop structure requires the resolution of the geometric constraint equations of the loops [7.8]. The objective is to compute the passive joint variables in terms of the active joint variables. Equation [7.8] constitutes a system of twelve nonlinear equations with up to six independent unknowns. Thus, a closed loop can have at most six passive joints, and a planar loop can have at most three passive joints. The problem of solving the geometric constraint equations is similar to that of the inverse geometric model of serial robots (Chapter 4). In this section, we develop an analytical method to obtain the solution for simple loops having three passive joints [Bennis 93], which is the case for most industrial robots with closed chains. This method can be extended to loops with four passive joints. We assume that the structure is compatible with the closure constraints of the loops, otherwise there would be no solution. This hypothesis will be developed in § 7.10 and will be verified when solving the constraint equations.

7.7.2. *General principle*

Let the three passive joints of the loop be denoted by i, j and k. We can obtain two systems of equations, one by using the position elements and the other by using the orientation elements of equation [7.8].

The general case is when a passive joint is situated between two active joints. Thus, equation [7.8] can be written as:

$$^{k}\mathbf{T}_{a(i)}\,^{a(i)}\mathbf{T}_{i}(q_i)\,^{i}\mathbf{T}_{a(j)}\,^{a(j)}\mathbf{T}_{j}(q_j)\,^{j}\mathbf{T}_{a(k)}\,^{a(k)}\mathbf{T}_{k}(q_k) = \mathbf{I}_4 \qquad [7.13a]$$

The transformation matrices $^{k}\mathbf{T}_{a(i)}$, $^{i}\mathbf{T}_{a(j)}$ and $^{j}\mathbf{T}_{a(k)}$ are functions of the supposed known active joint variables. The matrices $^{a(i)}\mathbf{T}_{i}$, $^{a(j)}\mathbf{T}_{j}$ and $^{a(k)}\mathbf{T}_{k}$ are functions of the unknown variables q_i, q_j and q_k respectively.

Equation [7.13a] can also be written in the following forms:

$$^{j}\mathbf{T}_{a(k)}\,^{a(k)}\mathbf{T}_{k}\,^{k}\mathbf{T}_{a(i)}\,^{a(i)}\mathbf{T}_{i}\,^{i}\mathbf{T}_{a(j)}\,^{a(j)}\mathbf{T}_{j} = \mathbf{I}_4 \qquad [7.13b]$$

$$^{a(k)}\mathbf{T}_{k}\,^{k}\mathbf{T}_{a(i)}\,^{a(i)}\mathbf{T}_{i}\,^{i}\mathbf{T}_{a(j)}\,^{a(j)}\mathbf{T}_{j}\,^{j}\mathbf{T}_{a(k)} = \mathbf{I}_4 \qquad [7.13c]$$

$$^{i}\mathbf{T}_{a(j)}\,^{a(j)}\mathbf{T}_{j}\,^{j}\mathbf{T}_{a(k)}\,^{a(k)}\mathbf{T}_{k}\,^{k}\mathbf{T}_{a(i)}\,^{a(i)}\mathbf{T}_{i} = \mathbf{I}_4 \qquad [7.13d]$$

We can rewrite equation [7.13a] such that the first transformation matrix contains the variable of joint i, namely θ_i or r_i.

i) position equation

We can obtain the position equation of the loop by postmultiplying equation [7.13a] by the vector $\mathbf{p}_0 = [\,0 \quad 0 \quad 0 \quad 1\,]^T$, and premultiplying it by **Rot(x, $-\alpha_i$) Trans(x, $-d_i$) Rot(z, $-\gamma_i$) Trans(z, $-b_i$)** $^{a(i)}\mathbf{T}_k$. This equation depends on r_k and not on θ_k, and can be rewritten as:

$$\mathbf{Rot}(\mathbf{z}, \theta_i)\ \mathbf{Trans}(\mathbf{z}, r_i)\ {}^{i}\mathbf{T}_{a(j)}\ \mathbf{Rot}(\mathbf{x}, \alpha_j)\ \mathbf{Trans}(\mathbf{x}, d_j)$$
$$\mathbf{Rot}(\mathbf{z}, \theta_j)\ \mathbf{Trans}(\mathbf{z}, r_j) \begin{bmatrix} \mathbf{f}(r_k) \\ 1 \end{bmatrix} = \begin{bmatrix} \mathbf{g} \\ 1 \end{bmatrix} \qquad [7.14]$$

where $\mathbf{g}$ is known.

ii) orientation equation

The orientation equation can be obtained by premultiplying relation [7.13a] by $\mathbf{Rot}(\mathbf{x}, -\alpha_j)\ \mathbf{Rot}(\mathbf{z}, -\gamma_i)\ {}^{a(i)}\mathbf{T}_k$ while only keeping the orientation terms:

$$\mathbf{rot}(\mathbf{z},\theta_i)\ [\ \mathbf{S}_1\ \ \mathbf{N}_1\ \ \mathbf{A}_1\ ]\ \mathbf{rot}(\mathbf{z}, \theta_j)\ [\ \mathbf{S}_2\ \ \mathbf{N}_2\ \ \mathbf{A}_2\ ]\ \mathbf{rot}(\mathbf{z}, \theta_k) = [\ \mathbf{S}_3\ \ \mathbf{N}_3\ \ \mathbf{A}_3\ ] \qquad [7.15]$$

where the (3x1) vectors $\mathbf{S}_i$, $\mathbf{N}_i$ and $\mathbf{A}_i$, for i = 1, 2, 3, are functions of the active joint variables.

Equations [7.14] and [7.15] have the same form as the position and orientation equations of serial robots (§ 4.4). However, as they are functions of the same variables, they must be resolved simultaneously [Bennis 91a]. An application of this method is given in Example 7.5 for the AKR-3000 robot.

• **Example 7.5.** Resolution of the loop constraint equations of the AKR-3000 robot. From Table 7.2, we deduce that:

$$\mathbf{q}_a = [\ \theta_1\ \ r_5\ \ r_6\ \ \theta_8\ \ \theta_9\ \ \theta_{10}\ ]^T$$
$$\mathbf{q}_p = [\ \theta_2\ \ \theta_3\ \ \theta_4\ \ \theta_7\ ]^T$$
$$\mathbf{q}_c = [\ \theta_{11}\ \ \theta_{12}\ ]^T$$

a) Equation of the loop composed of links 1, 2, 4 and 6

Frames R_{12} and R_{14} are placed on the cut joint between links 2 and 6. The loop contains three revolute passive joints with parallel axes and one prismatic active joint. We need to calculate the passive variables θ_2, θ_4 and θ_{12} in terms of the active variable r_6. For convenience, let us group γ_{14} with θ_2 such that:

$$\theta_2' = \theta_2 + \gamma_{14} \qquad [7.16]$$

The geometric constraint equation of the loop can be written as:

$$ {}^{1}\mathbf{T}_4\ {}^{4}\mathbf{T}_6\ {}^{6}\mathbf{T}_{12} = {}^{1}\mathbf{T}_2\ {}^{2}\mathbf{T}_{14} \qquad [7.17]$$

i) orientation equation

$$^{1}A_4\, ^{4}A_6\, ^{6}A_{12} = \, ^{1}A_2\, ^{2}A_{14} \qquad [7.18]$$

Since the axes of the passive joints are parallel, equation [7.18] gives:

$$\mathbf{rot}(z, \theta_4)\, \mathbf{rot}(z, \theta_{12}) = \mathbf{rot}(z, \theta_2') \qquad [7.19]$$

We deduce that:

$$\theta_4 + \theta_{12} = \theta_2 + \gamma_{14} \qquad [7.20]$$

ii) position equation

We choose to eliminate the passive joint variable θ_{12}. Premultiplying equation [7.17] by $^{4}\mathbf{T}_1$ and postmultiplying it by the vector $\mathbf{p}_0 = [\,0 \;\; 0 \;\; 0 \;\; 1\,]^T$ leads to:

$$^{4}\mathbf{T}_6\, ^{6}\mathbf{T}_{12}\, \mathbf{p}_0 = \, ^{4}\mathbf{T}_1\, ^{1}\mathbf{T}_2\, ^{2}\mathbf{T}_{14}\, \mathbf{p}_0 \qquad [7.21]$$

which gives:

$$\mathbf{Rot}(z, -\theta_4) \begin{bmatrix} \mathbf{F}(\theta_2') \\ 1 \end{bmatrix} = \begin{bmatrix} \mathbf{G}(r_6) \\ 1 \end{bmatrix} \qquad [7.22]$$

with:

$$\begin{bmatrix} \mathbf{F}(\theta_2') \\ 1 \end{bmatrix} = \mathbf{Trans}(x, -d_4)\, \mathbf{Rot}(z, \theta_2')\, ^{2}\mathbf{T}_{14}\, \mathbf{p}_0 = \begin{bmatrix} C\theta_2' d_{14} - d_4 \\ S\theta_2' d_{14} \\ 0 \\ 1 \end{bmatrix} \qquad [7.23]$$

$$\begin{bmatrix} \mathbf{G}(r_6) \\ 1 \end{bmatrix} = \, ^{4}\mathbf{T}_6\, ^{6}\mathbf{T}_{12}\, \mathbf{p}_0 = \begin{bmatrix} 0 \\ r_6 \\ 0 \\ 1 \end{bmatrix} \qquad [7.24]$$

θ_2' can be obtained in terms of r_6 by developing the expression:

$$\| \mathbf{F} \|^2 = \| \mathbf{G} \|^2 \qquad [7.25]$$

which leads to:

$$d_{14}^2 + d_4^2 - 2\, d_4\, d_{14}\, C(\theta_2') = r_6^2 \qquad [7.26]$$

Having determined θ_2' from equation [7.26], the components of **F** are known. The variable θ_4 can be obtained from equation [7.22] as:

$$\begin{cases} -S\theta_4\, r_6 = F_x \\ C\theta_4\, r_6 = F_y \end{cases} \quad [7.27]$$

where $\mathbf{F} = [F_x \quad F_y \quad F_z]^T$.

The variable θ_{12} can be determined from equation [7.20].

b) Equation of the loop composed of links 1, 2, 7, and 3

Frames R_{11} and R_{13} are placed on the cut joint between links 5 and 7. The loop contains four revolute passive joints with parallel axes and one prismatic active joint, but the passive variable θ_2 has already been obtained from the first loop. Thus, the three unknowns θ_{11}, θ_3 and θ_7 have to be computed in terms of the active variable r_5 and the variable θ_2. To simplify the development, let us group γ_{13} with θ_7 such that:

$$\theta_7' = \theta_7 + \gamma_{13} \quad [7.28]$$

The loop equation is written as:

$$^1\mathbf{T}_3\, ^3\mathbf{T}_5\, ^5\mathbf{T}_{11} = \, ^1\mathbf{T}_2\, ^2\mathbf{T}_7\, ^7\mathbf{T}_{13} \quad [7.29]$$

i) orientation equation

By proceeding as for the first loop, we obtain the equation:

$$\theta_2 + \theta_7' = \theta_3 + \theta_{11} \quad [7.30]$$

ii) position equation

We choose to eliminate the passive variable θ_3 from equation [7.29], which can be rewritten as:

$$^2\mathbf{T}_7\, ^7\mathbf{T}_{13}\, ^{13}\mathbf{T}_{11}\, ^{11}\mathbf{T}_5\, ^5\mathbf{T}_3\, \mathbf{p}_0 = \, ^2\mathbf{T}_1\, ^1\mathbf{T}_3\, \mathbf{p}_0 \quad [7.31]$$

where the vector $^1\mathbf{T}_3\, \mathbf{p}_0$ is devoid of θ_3. Equation [7.31] becomes:

$$\mathbf{Rot}(\mathbf{z}, \theta_7') \begin{bmatrix} \mathbf{F}(\theta_{11}, r_5) \\ 1 \end{bmatrix} = \begin{bmatrix} \mathbf{G}(\theta_2) \\ 1 \end{bmatrix} \quad [7.32]$$

with:

$$\begin{bmatrix} \mathbf{F} \\ 1 \end{bmatrix} = \mathbf{Trans}(\mathbf{x}, d_{13})\ \mathbf{Rot}(\mathbf{z}, -\theta_{11})\ \mathbf{Rot}(\mathbf{x}, -\alpha_{11})\ {}^{5}\mathbf{T}_{3}\ \mathbf{p}_{0} = \begin{bmatrix} -S\theta_{11}r_5+d_{13} \\ -S\theta_{11}r_5 \\ 0 \\ 1 \end{bmatrix} \quad [7.33]$$

$$\begin{bmatrix} \mathbf{G} \\ 1 \end{bmatrix} = \mathbf{Trans}(\mathbf{x}, -d_7)\ {}^{2}\mathbf{T}_{1}\ {}^{1}\mathbf{T}_{3}\ \mathbf{p}_{0} = \begin{bmatrix} C\theta_2 d_3-d_7 \\ -S\theta_2 d_3 \\ 0 \\ 1 \end{bmatrix} \quad [7.34]$$

The variable θ_{11} can be obtained in terms of the variables r_5 and θ_2 by developing the equality $\| \mathbf{F} \|^2 = \| \mathbf{G} \|^2$:

$$r_5^2 + d_{13}^2 - 2d_{13}\, r_5\, S\theta_{11} = d_7^2 + d_3^2 - 2d_3\, d_7\, C\theta_2 \quad [7.35]$$

Having determined θ_{11}, we calculate θ_7' thanks to equation [7.32]:

$$\begin{cases} C\theta_7'\, F_x - S\theta_7'\, F_y = G_x \\ S\theta_7'\, F_x + C\theta_7'\, F_y = G_y \end{cases} \quad [7.36]$$

The variable θ_3 can be determined from equation [7.30].

7.7.3. *Particular case of a parallelogram loop*

A parallelogram loop has three revolute passive joints and one revolute active joint. In this section, we propose a specific method, which is simpler than the general approach exposed previously. We use the fact that the orientation matrix between the frames of two parallel links is constant. Thus, if links k_1, k_2, k_3 and k_4 constitute a parallelogram, where links k_1 and k_2 are parallel to links k_3 and k_4 respectively (Figure 7.6), then [Bennis 91a]:

$$^{k1}\mathbf{A}_{k3} = \mathbf{rot}(\mathbf{u}, \theta_{c1}) = \text{constant} \quad [7.37a]$$

$$^{k2}\mathbf{A}_{k4} = \mathbf{rot}(\mathbf{u}, \theta_{c2}) = \text{constant} \quad [7.37b]$$

$$^{k}\mathbf{A}_{k+B} = {}^{k}\mathbf{A}_{k3}\, {}^{k3}\mathbf{A}_{k2}\, {}^{k2}\mathbf{A}_{k4}\, {}^{k4}\mathbf{A}_{k+B} = \mathbf{I}_3 \quad [7.37c]$$

The constants θ_{c1} and θ_{c2} can be obtained from a particular configuration of the robot. Equation [7.37c] allows us to compute the cut joint variable in terms of the other joint variables.

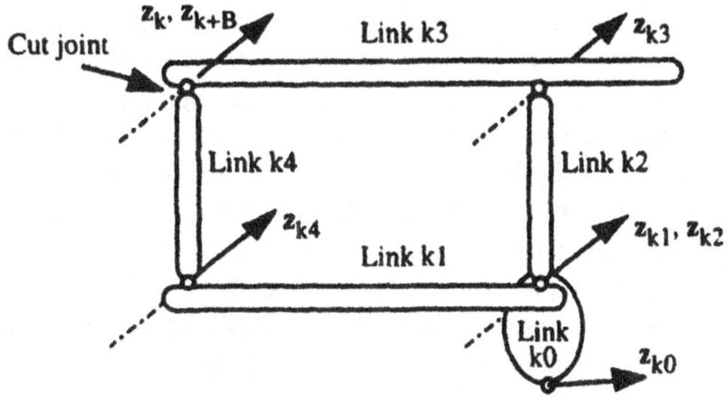

Figure 7.6. *Parallelogram closed loop*

• **Example 7.6.** Resolution of the geometric constraint equations of the Acma SR400 robot. From Table 7.1, we deduce that:

$$\mathbf{q_a} = [\ \theta_1\ \ \theta_2\ \ \theta_4\ \ \theta_5\ \ \theta_6\ \ \theta_7\ ]^T$$
$$\mathbf{q_p} = [\ \theta_3\ \ \theta_8\ ]^T$$
$$\mathbf{q_c} = [\ \theta_9\ ]$$

The loop composed of links 7, 2, 3 and 8 forms a parallelogram (Figure 7.4). Joints 3, 8 and 9 are passive, whereas joint 7 is active. In this loop, links 2 and 3 are parallel to links 8 and 7 respectively. Using equations [7.37] we obtain:

$$^7\mathbf{A}_3 = {}^7\mathbf{A}_1\,{}^1\mathbf{A}_2\,{}^2\mathbf{A}_3 = \mathbf{rot}(\mathbf{z}, -\theta_7+\theta_2+\theta_3) = \mathbf{rot}(\mathbf{z}, \pi/2) \quad [7.38]$$
$$^2\mathbf{A}_8 = {}^2\mathbf{A}_1\,{}^1\mathbf{A}_7\,{}^7\mathbf{A}_8 = \mathbf{rot}(\mathbf{z}, -\theta_2)\,\mathbf{rot}(\mathbf{z}, \theta_7)\,\mathbf{rot}(\mathbf{z}, \theta_8) = \mathbf{rot}(\mathbf{z}, 0) \quad [7.39]$$
$$^9\mathbf{A}_8\,{}^8\mathbf{A}_7\,{}^7\mathbf{A}_1\,{}^1\mathbf{A}_2\,{}^2\mathbf{A}_3\,{}^3\mathbf{A}_{10} = \mathbf{I}_3 \quad [7.40]$$

From these equations we deduce that:

$$\theta_3 = \pi/2 - \theta_2 + \theta_7 \quad [7.41]$$
$$\theta_8 = \theta_2 - \theta_7 \quad [7.42]$$
$$\theta_9 = \pi/2 + \theta_3 = \pi - \theta_2 + \theta_7 \quad [7.43]$$

7.8. Kinematic model of complex chain robots

The kinematic model provides the velocity of the terminal link corresponding to the specified velocities of the active joints. Since the joints of a tree structured robot are actuated and independent, the kinematic model for these robots can be obtained by applying the techniques developed for serial robots to each main branch. For robots with closed chains, we first compute the kinematic model of the direct chain between the end-effector and the base by proceeding as for a serial robot. Then, we compute the passive joint velocities in terms of the active joint velocities. The solution can be obtained either by differentiating the geometric constraint equations, or by resolving the kinematic constraint equations of the loops.

The kinematic constraint equations can be obtained by equating the velocities of frames R_k and R_{k+B} associated with each cut joint. They can be computed using the Jacobian matrix of the two branches of each loop as follows:

$$\begin{bmatrix} \mathbf{V}_k \\ \omega_k \end{bmatrix} = \mathbf{J}_k \, \dot{\mathbf{q}}_{b1} = \mathbf{J}_{k+B} \, \dot{\mathbf{q}}_{b2} \qquad [7.44]$$

where $\dot{\mathbf{q}}_{b1}$ and $\dot{\mathbf{q}}_{b2}$ are the joint velocities along each branch of the loop.

Equation [7.44] can be rewritten as:

$$\mathbf{J}_k \, \dot{\mathbf{q}}_{b1} - \mathbf{J}_{k+B} \, \dot{\mathbf{q}}_{b2} = \mathbf{0} \qquad [7.45]$$

Using equation [5.9] and taking into account Figure 7.7 leads to:

$$\mathbf{J}_k = \begin{bmatrix} \sigma_e \mathbf{a}_e + \bar{\sigma}_e(\mathbf{a}_e \times \mathbf{L}_{e,k}) & \cdots & \sigma_k \mathbf{a}_k + \bar{\sigma}_k(\mathbf{a}_k \times \mathbf{L}_{k,k}) \\ \bar{\sigma}_e \mathbf{a}_e & \cdots & \bar{\sigma}_k \mathbf{a}_k \end{bmatrix} \qquad [7.46]$$

where e indicates the first link, after the root of the loop, of the branch leading to frame R_k, and:

$$\mathbf{J}_{k+B} = \begin{bmatrix} \sigma_d \mathbf{a}_d + \bar{\sigma}_d(\mathbf{a}_d \times \mathbf{L}_{d,k+B}) & \cdots & \sigma_j \mathbf{a}_j + \bar{\sigma}_j(\mathbf{a}_j \times \mathbf{L}_{j,k+B}) \\ \bar{\sigma}_d \mathbf{a}_d & \cdots & \bar{\sigma}_j \mathbf{a}_j \end{bmatrix} \qquad [7.47]$$

where d indicates the first link, after the root of the loop, of the branch leading to frame R_{k+B}.

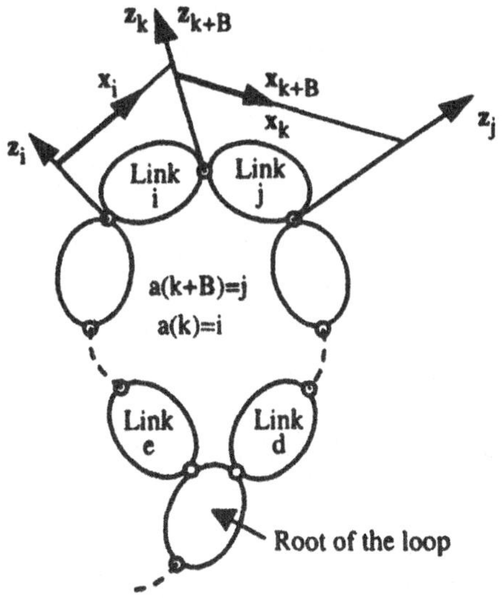

Figure 7.7. *General notations for a closed loop*

The elements of the Jacobian matrices J_k and J_{k+B} are generally computed with respect to frames R_k and R_{k+B} respectively, or with respect to the frame of the root of the loop. By combining the constraint equations of all the loops, and after eliminating the possible zero rows, the kinematic constraint equation can be written as [Zghaib 92]:

$$\mathbf{J}\,\dot{\mathbf{q}} = \mathbf{0} \qquad [7.48]$$

which can be developed as:

$$\begin{bmatrix} \mathbf{W}_a & \mathbf{W}_p & \mathbf{0} \\ \mathbf{W}_{ac} & \mathbf{W}_{pc} & \mathbf{W}_c \end{bmatrix} \begin{bmatrix} \dot{\mathbf{q}}_a \\ \dot{\mathbf{q}}_p \\ \dot{\mathbf{q}}_c \end{bmatrix} = \mathbf{0} \qquad [7.49]$$

with:

- $\dot{\mathbf{q}}_a$: (Nx1) vector of the active joint velocities;
- $\dot{\mathbf{q}}_p$: (px1) vector of the passive joint velocities of the equivalent tree structure, where p=n−N;
- $\dot{\mathbf{q}}_c$: (Bx1) vector of the cut joint velocities;
- the dimensions of the matrices are the following: $\mathbf{W}_a$ (pxN), $\mathbf{W}_p$ (pxp), $\mathbf{W}_{ac}$ (BxN), $\mathbf{W}_{pc}$ (Bxp), $\mathbf{W}_c$ (BxB).

When the cut joint is complex with several degrees of freedom (spherical, universal, ...), we can consider the corresponding joint velocities to be belonging to $\dot{\mathbf{q}}_c$. From the first row of equation [7.49], we obtain:

$$\mathbf{W}_p\,\dot{\mathbf{q}}_p = -\,\mathbf{W}_a\,\dot{\mathbf{q}}_a \qquad [7.50]$$

If the system is compatible with the loop constraints, the rank of $\mathbf{W}_p$ will be equal to p (outside the possible singular positions). We deduce that:

$$\dot{\mathbf{q}}_p = -\,\mathbf{W}_p^{-1}\,\mathbf{W}_a\,\dot{\mathbf{q}}_a = \mathbf{W}\,\dot{\mathbf{q}}_a \qquad [7.51]$$

By differentiating equation [7.49] with respect to time, we obtain the acceleration constraint equation:

$$\begin{bmatrix} \mathbf{W}_a & \mathbf{W}_p & \mathbf{0} \\ \mathbf{W}_{ac} & \mathbf{W}_{pc} & \mathbf{W}_c \end{bmatrix} \begin{bmatrix} \ddot{\mathbf{q}}_a \\ \ddot{\mathbf{q}}_p \\ \ddot{\mathbf{q}}_c \end{bmatrix} + \begin{bmatrix} \mathbf{\Psi} \\ \mathbf{\Phi} \end{bmatrix} = \mathbf{0} \qquad [7.52]$$

where $\mathbf{\Psi}$ and $\mathbf{\Phi}$ represent the vector $\dot{\mathbf{J}}\,\dot{\mathbf{q}}$.

The components of the vectors $\mathbf{\Psi}$ and $\mathbf{\Phi}$ of the loop of the cut joint k can be determined with the recursive equations giving the terminal accelerations ${}^k\dot{\mathbf{V}}_k$ and ${}^k\dot{\boldsymbol{\omega}}_k$ relative to the root of the loop when setting $\ddot{\mathbf{q}}_{b1} = 0$ and $\ddot{\mathbf{q}}_{b2} = 0$ (§ 5.10). We obtain:

$$\begin{bmatrix} {}^k\dot{\mathbf{V}}_k(\ddot{\mathbf{q}}_{b1}=0) \\ {}^k\dot{\boldsymbol{\omega}}_k(\ddot{\mathbf{q}}_{b1}=0) \end{bmatrix} - \begin{bmatrix} {}^{k+B}\dot{\mathbf{V}}_{k+B}(\ddot{\mathbf{q}}_{b2}=0) \\ {}^{k+B}\dot{\boldsymbol{\omega}}_{k+B}(\ddot{\mathbf{q}}_{b2}=0) \end{bmatrix} = {}^k\dot{\mathbf{J}}_k\,\dot{\mathbf{q}}_{b1} - {}^{k+B}\dot{\mathbf{J}}_{k+B}\,\dot{\mathbf{q}}_{b2} \qquad [7.53]$$

The vectors $\mathbf{\Psi}$ and $\mathbf{\Phi}$ are determined by applying equation [7.53] to all the loops, k = 1, ..., L, and by grouping them in the same order as those of the velocity constraint equation [7.49]. The computation of equation [7.53] can be carried out using the efficient recursive equations [5.47] after replacing $j-1$ by $a(j)$.

• **Example 7.7.** Calculation of the kinematic constraint equations of the Acma SR400 robot described in Example 7.1. By differentiating with respect to time the geometric constraint equations developed in Example 7. 6, we obtain:

$$\begin{cases} \dot{\theta}_3 = -\dot{\theta}_2 + \dot{\theta}_7 \\ \dot{\theta}_8 = \dot{\theta}_2 - \dot{\theta}_7 \\ \dot{\theta}_9 = -\dot{\theta}_2 + \dot{\theta}_7 \end{cases}$$

The acceleration constraint equations can be obtained by differentiating the kinematic constraint equations.

• **Example 7.8.** Computation of the kinematic constraint equations of the AKR-3000 robot described in Example 7.2. We recall that:

$$\dot{\mathbf{q}}_a = [\dot{\theta}_1 \;\; \dot{r}_5 \;\; \dot{r}_6 \;\; \dot{\theta}_8 \;\; \dot{\theta}_9 \;\; \dot{\theta}_{10}]^T$$

$$\dot{\mathbf{q}}_p = [\dot{\theta}_2 \;\; \dot{\theta}_3 \;\; \dot{\theta}_4 \;\; \dot{\theta}_7]^T$$

$$\dot{\mathbf{q}}_c = [\dot{\theta}_{11} \;\; \dot{\theta}_{12}]^T$$

The kinematic constraint equations can be obtained either by projecting the Jacobian matrices in the terminal frame R_k of each loop or in the frame fixed to the root of each loop. We present these two cases in the following.

i) projection of the Jacobian matrices in the base frame of the loop.

The kinematic constraint equation of the first loop is written as:

$$^1\mathbf{J}_{11}\, \dot{\mathbf{q}}_{b1,1} - {}^1\mathbf{J}_{13}\, \dot{\mathbf{q}}_{b2,1} = \mathbf{0}$$

with $\dot{\mathbf{q}}_{b1,1} = [\dot{\theta}_3 \quad \dot{r}_5 \quad \dot{\theta}_{11}]^T$ and $\dot{\mathbf{q}}_{b2,1} = [\dot{\theta}_2 \quad \dot{\theta}_7]^T$, giving the following non-trivial equations:

$$-r_5\, C3\, \dot{\theta}_3 - S3\, \dot{r}_5 + (d_7\, S2 - d_{13}\, S\alpha)\, \dot{\theta}_2 + d_{13}\, S\alpha\, \dot{\theta}_7 = 0$$

$$-r_5\, S3\, \dot{\theta}_3 + C3\, \dot{r}_5 - (d_7\, C2 + d_{13}\, C\alpha)\, \dot{\theta}_2 - d_{13}\, C\alpha\, \dot{\theta}_7 = 0$$

$$-\dot{\theta}_3 - \dot{\theta}_{11} + \dot{\theta}_2 + \dot{\theta}_7 = 0$$

with $\alpha = \gamma_{13} + \theta_2 + \theta_7$.

The kinematic constraint equation of the second loop is written as:

$$^1\mathbf{J}_{12}\, \dot{\mathbf{q}}_{b1,2} - {}^1\mathbf{J}_{14}\, \dot{\mathbf{q}}_{b2,2} = \mathbf{0}$$

with $\dot{q}_{b1,2} = [\ \dot{\theta}_4 \quad \dot{r}_6 \quad \dot{\theta}_{12}]^T$ and $\dot{q}_{b2,2} = \dot{\theta}_2$, giving the following non-trivial equations:

$$-r_6\,C4\,\dot{\theta}_4 - S4\,\dot{r}_6 + d_{14}\,S(\gamma_{14}+\theta_2)\,\dot{\theta}_2 = 0$$

$$-r_6\,S4\,\dot{\theta}_4 + C4\,\dot{r}_6 - d_{14}\,C(\gamma_{14}+\theta_2)\,\dot{\theta}_2 = 0$$

$$-\dot{\theta}_4 - \dot{\theta}_{12} + \dot{\theta}_2 = 0$$

These six equations can easily be put in the form [7.49].

ii) projection of the Jacobian matrices in the terminal frame of the loop.

The kinematic constraint equation of the first loop is written as:

$$^{11}\mathbf{J}_{11}\,\dot{\mathbf{q}}_{b1,1} - {}^{13}\mathbf{J}_{13}\,\dot{\mathbf{q}}_{b2,1} = \mathbf{0}$$

After developing, we obtain the following non-trivial equations:

$$-r_5\,C\theta_{11}\,\dot{\theta}_3 + S\theta_{11}\,\dot{r}_5 - d_7\,S(\gamma_{13}+\theta_7)\,\dot{\theta}_2 = 0$$

$$r_5\,S\theta_{11}\,\dot{\theta}_3 + C\theta_{11}\,\dot{r}_5 - [d_{13} + d_7\,C(\gamma_{13}+\theta_7)]\,\dot{\theta}_2 - d_{13}\,\dot{\theta}_7 = 0$$

$$-\dot{\theta}_3 - \dot{\theta}_{11} + \dot{\theta}_2 + \dot{\theta}_7 = 0$$

The kinematic constraint equation of the second loop is written as:

$$^{12}\mathbf{J}_{12}\,\dot{\mathbf{q}}_{b1,2} - {}^{14}\mathbf{J}_{14}\,\dot{\mathbf{q}}_{b2,2} = \mathbf{0}$$

After developing, we obtain the following non-trivial equations:

$$-r_6\,C\theta_{12}\,\dot{\theta}_4 + S\theta_{12}\,\dot{r}_6 + d_{14}\,\dot{\theta}_2 = 0$$

$$r_6\,S\theta_{12}\,\dot{\theta}_4 + C\theta_{12}\,\dot{r}_6 = 0$$

$$-\dot{\theta}_4 - \dot{\theta}_{12} + \dot{\theta}_2 = 0$$

We note that the equations of the second solution are less complicated, but they have the disadvantage of being functions of the cut joint variables.

7.9. Numerical calculation of $\mathbf{q}_p$ and $\mathbf{q}_c$ in terms of $\mathbf{q}_a$

Based on equation [7.45], we derive the following differential model, which can be used to numerically compute the variables $\mathbf{q}_p$ and $\mathbf{q}_c$ for a given $\mathbf{q}_a$:

$$\begin{bmatrix} {}^{b_1}\mathbf{J}_{n+1+B} & -{}^{b_1}\mathbf{J}_{n+1} \\ \cdots & \cdots \\ {}^{b_B}\mathbf{J}_{L+B} & -{}^{b_B}\mathbf{J}_{L} \end{bmatrix} \begin{bmatrix} \mathbf{dq}_p \\ \mathbf{dq}_c \end{bmatrix} = \begin{bmatrix} {}^{b_1}\mathbf{dX}^{n+1} \\ \cdots \\ {}^{b_B}\mathbf{dX}^{L} \end{bmatrix} \qquad [7.54]$$

where $\mathbf{dX}^k$ corresponds to the vector of position and orientation errors between frame R_k and frame R_{k+B}, and b_j denotes the frame of the root of loop j.

The left side terms of equation [7.45] are deduced from equation [7.54] by combining the equations of all the loops, eliminating the trivial rows, and by eliminating the columns corresponding to $\mathbf{dq}_a$ since $\mathbf{q}_a$ is given.

To calculate $\mathbf{q}_p$ and $\mathbf{q}_c$ we use the following algorithm:

1) from the current configuration of $\mathbf{q}_p$ and $\mathbf{q}_c$ (which can be initialized by random values within the joint domain), compute ${}^{b_j}\mathbf{T}_k(\mathbf{q})$ and ${}^{b_j}\mathbf{T}_{k+B}(\mathbf{q})$ for k = n+1, ..., L;

2) compute the vector of position and orientation errors of all the loops:

$$\mathbf{dX} = \begin{bmatrix} \mathbf{dX}^{n+1} \\ \cdots \\ \mathbf{dX}^{L} \end{bmatrix}$$

where $\mathbf{dX}^k$ corresponds to the difference between the transformation matrices ${}^{b_j}\mathbf{T}_k(\mathbf{q})$ and ${}^{b_j}\mathbf{T}_{k+B}(\mathbf{q})$ as detailed in § 6.6;

3) – if $\mathbf{dX}$ is sufficiently small, then the desired $\mathbf{q}_p$ and $\mathbf{q}_c$ are equal to their current values. Stop the calculation;

– if $\|\mathbf{dX}\| > S$, then set $\mathbf{dX} = \dfrac{\mathbf{dX}}{\|\mathbf{dX}\|} S$, where S is a fixed threshold value;

4) calculate $\mathbf{dq}_p$ and $\mathbf{dq}_c$ corresponding to $\mathbf{dX}$ by solving equation [7.54]. Update the variables $\mathbf{q}_p$ and $\mathbf{q}_c$ using the equation:

$$\begin{cases} \mathbf{q}_p = \mathbf{q}_p + \mathbf{dq}_p \\ \mathbf{q}_c = \mathbf{q}_c + \mathbf{dq}_c \end{cases} \qquad [7.55]$$

5) return to the first step.

7.10. Number of degrees of freedom of robots with closed chains

The number of degrees of freedom N of a mechanism is equal to the number of independent joint variables required to specify the location of all the links with respect to the fixed base. Thus N is equal to the difference between the number of joints L and the number of independent geometric constraint equations c :

$$N = L - c \qquad [7.56]$$

The problem consists in determining c. As mentioned in § 7.7.1, this number is, at most, six for a spatial loop and three for a planar loop.

The following simple formula is true for most robot structures:

$$N = L - \sum_{j=1}^{B} c_j \qquad [7.57]$$

where c_j is the number of independent geometric constraint equations of the loop j, in general six for a spatial loop and three for a planar loop. The application of equation [7.57] for the SR400 robot and AKR robot gives N=6.

Several formulas like [7.57] have been proposed to systematically determine the number of degrees of freedom of a mechanism. These formulas are based on the number of links and joints and their degrees of freedom but do not take into account the geometric constraints that some mechanisms possess. Consequently, they may provide an erroneous result [Sheth 71]. For example, these formulas do not work with the Bennett mechanism [Bennett 03]. For this reason, an exact solution is obtained by analyzing the geometric and kinematic constraints using the mechanism theory techniques [Le Borzec 75], [Hervé 78], or by studying the rank of the Jacobian matrix as given in the following.

From equation [7.49] we deduce that $\dot{\mathbf{q}}$ belongs to the null space of **J**. Therefore, at a given configuration, the number of degrees of freedom is equal to the dimension of the null space of **J**. Consequently:

$$N = \min_{\mathbf{q}} (\dim (\mathcal{N}(\mathbf{J}))) \qquad [7.58]$$

where $\mathcal{N}(\mathbf{J})$ is the null space of the matrix **J**.

Thus, the number of independent constraints c is given by the maximum value of the rank of the matrix **J** (equation 7.49) [Angeles 88], [Gosselin 90]:

$$c = \max_{\mathbf{q}} (\text{rank}\, \mathbf{J}(\mathbf{q})) \qquad [7.59]$$

7.11. Classification of singular positions

The general kinematic equation of a closed chain robot is given by [Gosselin 90]:

$$\mathbf{J}_1(\mathbf{q})\,\dot{\mathbf{q}}_a = \mathbf{J}_2(\mathbf{q})\,\dot{\mathbf{X}} \qquad [7.60]$$

where $\dot{\mathbf{X}}$ is the velocity of the end-effector.

Input/output singularities occur in the configurations where $\mathbf{J}_1$ and/or $\mathbf{J}_2$ are singular. Three kinds of singularities are encountered:

i) $\mathbf{J}_1$ *is singular:* in this configuration, we can find the non-zero vector $\dot{\mathbf{q}}_a \neq \mathbf{0}$ for which $\dot{\mathbf{X}} = \mathbf{0}$. The terminal link loses one or more degrees of freedom and it cannot generate motion in some directions. This kind of singularity is the only one that occurs in serial robots. It is called *serial singularity*;

ii) $\mathbf{J}_2$ *is singular:* in this case, the structure is not in static equilibrium. Thus, $\dot{\mathbf{X}} \neq \mathbf{0}$ even though the actuated joints are locked ($\dot{\mathbf{q}}_a = \mathbf{0}$). This kind of singularity takes place in parallel robots (Chapter 8) and is known as *parallel singularity*. To avoid such singularity, redundant actuators may be used;

iii) $\mathbf{J}_1$ *and* $\mathbf{J}_2$ *are singular:* in this case, the structure has both serial and parallel singularities simultaneously. Thus, some links may move even though $\dot{\mathbf{q}}_a = \mathbf{0}$ and motions in some directions of the terminal link are unrealizable.

From equation [7.49], we can deduce another type of singularity that occurs when $\mathbf{W}_p$ is singular, giving $\dot{\mathbf{q}}_p \neq \mathbf{0}$ while $\dot{\mathbf{q}}_a = \mathbf{0}$. This singularity is termed *internal singularity*.

7.12. Conclusion

The method of description presented in this chapter allows extension of the results obtained for serial robots to complex chain robots. In fact, a serial robot can be considered as a special case of a tree structured robot. We showed that the computation of the direct and inverse geometric models of a closed chain robot could be formulated as the calculation of these models for a serial structure together with the resolution of the geometric constraint equations of the closure of the loops.

The geometric constraint equations of the loops can be solved using the methods for computing the IGM exposed in Chapter 4. We also presented a general analytical method for loops with less than four passive joints.

We also showed how to establish the kinematic model of such structures and how to obtain the kinematic constraint equations using the Jacobian matrix. The problem of the determination of the number of degrees of freedom of a closed chain robot has finally been addressed. For a survey of the manipulability, which we did not cover here, the reader is referred to [Bicchi 98].

The following chapter deals with the geometric and kinematic modeling of parallel robots, which require a specific approach for the description and the modeling.

Chapter 8

Introduction to geometric and kinematic modeling of parallel robots

8.1. Introduction

Parallel architectures were originally proposed in the context of tire-testing machines and flight simulators [Gough 56], [Stewart 65]. Since then, they have been used in other applications requiring manipulation of heavy loads with high accelerations such as vehicle driving simulators or the riding simulator developed for the French National Riding School.

Recently, these kind of structures have attracted considerable interest in various manufacturing applications due to their inherent characteristics, as compared with those of serial robots, which include high structural rigidity and better dynamic performances. This concept is currently used in designing new generations of high speed machine tools.

This chapter deals with the geometric and kinematic modeling of such robots. It is shown that the closed-form solution of the inverse geometric model is straightforward for a six degree-of-freedom parallel robot. The explicit formulation of the direct geometric model is usually more complicated since it can have up to 40 solutions [Husty 96]. Similarly, the computation of the inverse kinematic model is easier than the computation of the direct kinematic model.

8.2. Parallel robot definition

A parallel robot is composed of a *mobile platform* connected to a *fixed base* by a set of identical parallel kinematic chains, which are called *legs*. The end-effector is fixed to the mobile platform. A parallel robot is said to be *fully parallel* when the number of legs is greater or equal to the number of degrees of freedom of the mobile platform, each parallel chain having a single actuator [Gosselin 88]. For example,

Figures 8.1 and 8.2 show a six degree-of freedom fully parallel robot where the mobile platform and the base are linked together by six legs. The desired mobile platform location can be obtained by changing the leg lengths using actuated prismatic joints. This architecture has been used by Gough in 1947 to design tire-testing machines and has inspired Stewart [Stewart 65] to design a flight simulator. It is known as the Gough-Stewart parallel robot.

Note that a *hybrid parallel robot* is formed by a series of several parallel robots. The advantage of such a structure is to increase the workspace of the terminal platform. Figure 8.3 shows the Logabex hybrid robot [Charentus 90], which is composed of four units of parallel structures.

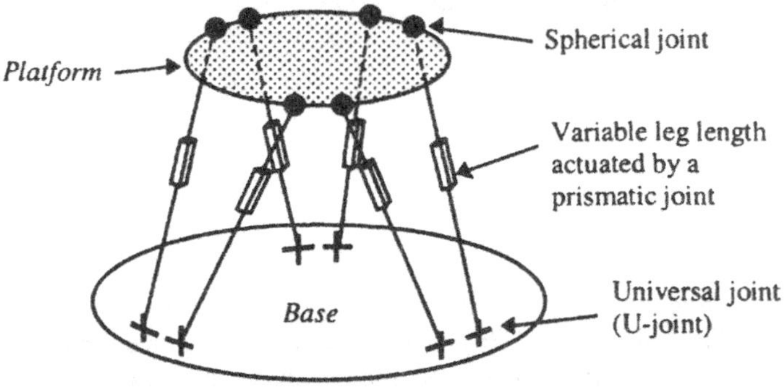

Figure 8.1. *A six degree-of-freedom parallel robot*

8.3. Comparing performance of serial and parallel robots

The main criteria for comparing performance of serial and parallel robots are the workspace, the ratio between the payload and the robot mass, the accuracy, and the dynamic behavior:

i) workspace. The main drawback of a parallel robot is its comparatively small workspace. It is determined by the intersection of the workspaces of all the parallel kinematic chains;

ii) payload - robot mass ratio. In a serial architecture, the end-effector and the manipulated object are located at the extremity of the mechanical chain. Consequently, each actuator must have the necessary power to move not only the object, but also the links and actuators in between. This leads to a poor payload - robot mass ratio.

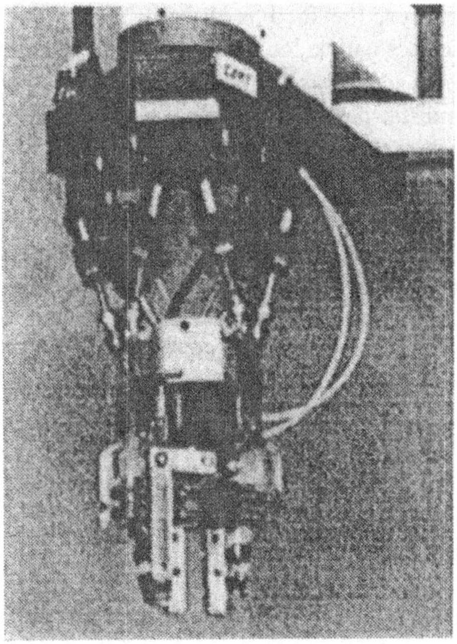

Figure 8.2. *The SPACE-1 parallel robot from CERT (courtesy of CERT)*

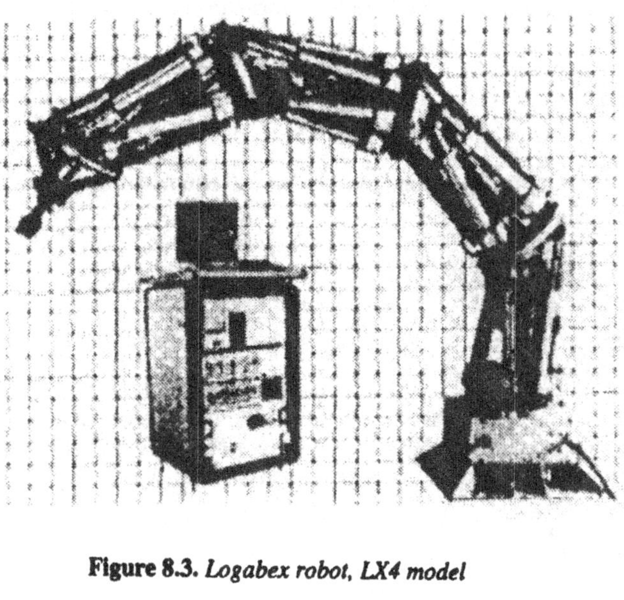

Figure 8.3. *Logabex robot, LX4 model*

In parallel structures, the load is directly supported by all the actuators. Besides, the actuators can be located either on or close to the base. Therefore, the links between the mobile platform and the base can be lightened considerably, and the payload - robot mass ratio is much higher, generally with a factor of at least 10;

iii) accuracy and repeatability. Serial robots accumulate errors from one joint to the next, since defaults like clearance, friction, flexibility, etc. also act in a serial manner. Moreover, the influence of a joint default on the end-effector location is larger when the joint is close to the robot base.
Parallel robots do not present this drawback and their architecture provides a remarkable rigidity even with light connecting links;

iv) dynamic behavior. Considering their high payload - robot mass ratio and their reduced coupling effect between joints, parallel robots have better dynamic performance.

8.4. Number of degrees of freedom

A parallel manipulator is a complex closed loop structure. We recall that computing the number of degrees of freedom of a closed structure using classical formulas (§ 7.10) that do not take into account the geometric constraints may give wrong results. Nevertheless, the number of degrees of freedom of a parallel robot N can be determined using the following simple relationship, which is derived from equation [7.57] and proved to be true for a large number of architectures:

$$N = \sum_{i=1}^{L} m_i - \sum_{j=1}^{B} c_j \qquad [8.1]$$

with:

- B: number of independent closed loops, equal to $(n_c - 1)$, where n_c is the number of parallel chains;
- m_i: mobility of joint i;
- L: total number of joints;
- c_j: number of constraints of the jth loop. In general, $c_j = 3$ for a planar loop, $c_j = 6$ for a spatial loop.

If the kinematic chains between the base and the mobile platform are identical and if the loops have the same number of constraints, equation [8.1] becomes:

$$N = n_c\, d - c_j\, B \qquad [8.2]$$

where d is the sum of the degrees of freedom of the joints in a chain and c_j indicates the number of constraints per loop.

For non-redundant robots, N gives the number of degrees of freedom of the mobile platform with respect to the base.

8.5. Parallel robot architectures

Practically, two classes of parallel robots may be distinguished: planar robots and spatial robots. In this section, we present both classes as well as a specific family of spatial robots: the *Delta robots*.

8.5.1. *Planar parallel robots*

A planar robot is composed of a mobile platform with three degrees of freedom with respect to the base: two translations and one rotation about the normal to the plane of the mobile platform (Figure 8.4). In accordance with the definition of the fully parallel robot, the mobile platform is connected to the base by three legs, each including an actuated joint.

The number of independent loops B of the planar robots of Figure 8.4 is two. If we assume that the three legs are identical, the application of equation [8.2] leads to:

$$N = 3d - 6 \qquad [8.3]$$

From equation [8.3], we deduce that, for N = 3, the number of degrees of freedom d of a leg must be three (two passive and one active).

Let A_1, A_2 and A_3 be the connection points of the base with the legs and B_1, B_2 and B_3 be the connection points of the legs with the mobile platform. To control the position and the orientation of the mobile platform, we have to change the length of the A_iB_i legs. The following three architectures are possible for the legs:

- R-P-R architecture (Figure 8.4a), which is actuated by the prismatic joint;
- P-R-R architecture (Figure 8.4b), which is actuated by the prismatic joint;
- R-R-R architecture (Figure 8.4c), which is actuated by the revolute joint close to the base.

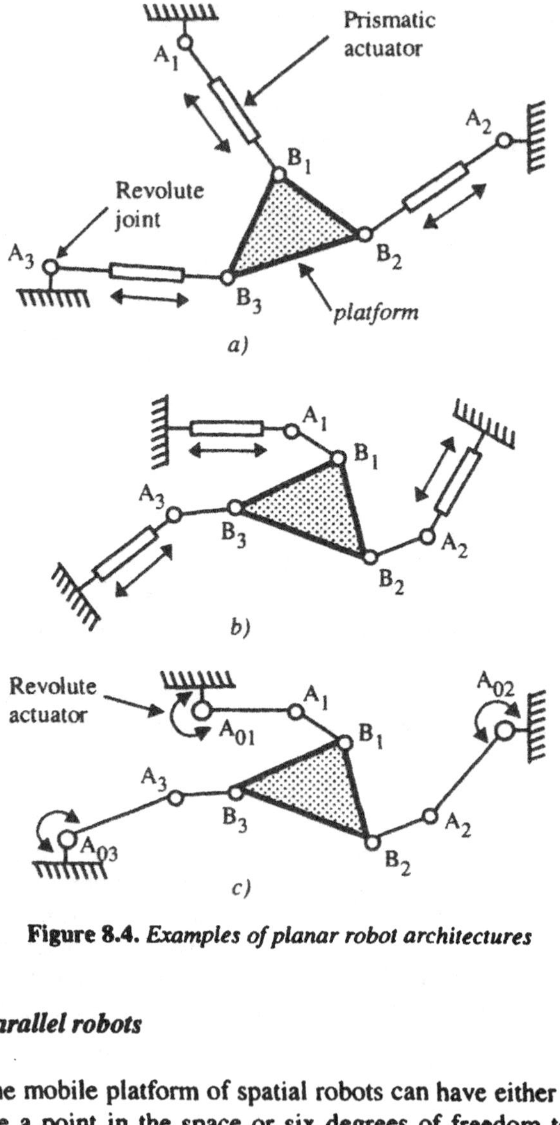

Figure 8.4. *Examples of planar robot architectures*

8.5.2. *Spatial parallel robots*

In general, the mobile platform of spatial robots can have either three degrees of freedom to place a point in the space or six degrees of freedom to place the end-effector at any arbitrary location. To determine the type of joints of each leg, we proceed as for planar robots while assuming that the legs are identical.

8.5.2.1. *Three degree-of-freedom spatial robots*

The mobile platform is connected to the base by three legs. There are two independent loops, and equation [8.2] yields:

$$N = 3\,d - 12 \qquad [8.4]$$

Since N = 3, the number of degrees of freedom d of each leg must be five (four passive and one active). The passive degrees of freedom can be distributed according to the combinations (0,4), (1,3) or (2,2).

The (1,3) combination is the most commonly used. It is composed of a revolute joint at one end and a spherical joint (RRR) at the other. Figure 8.5a depicts such an example, where the leg length is actuated by a prismatic joint, giving for each leg an R-P-(RRR) architecture. Figure 8.5b gives an example of the (0,4) combination, where the four degree-of-freedom joint is constructed with two universal joints (RR). Each leg is actuated by a revolute joint fixed on the base and presents an R-(RR)-(RR) architecture.

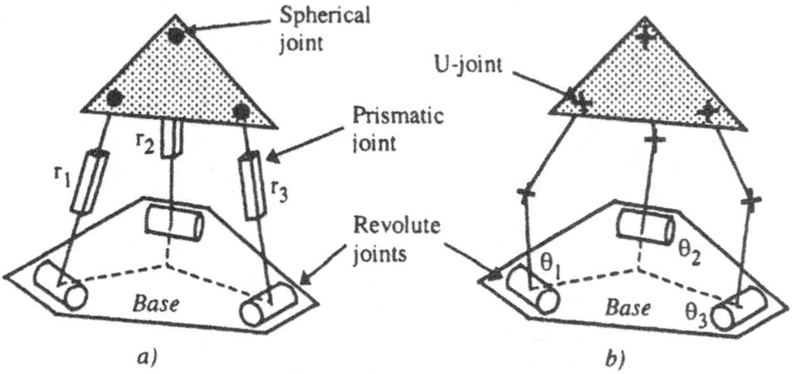

Figure 8.5. *Three degree-of-freedom spatial robots*

8.5.2.2. *Six degree-of-freedom spatial robots*

We consider the Gough-Stewart structure as representative of the six-degree-of-freedom spatial robots. Merlet [Merlet 00] describes three concepts of six degree-of-freedom architectures where the base and the mobile platform are connected by six legs driven by prismatic actuated joints (Figure 8.6):

- *SSM robot (Simplified Symmetric Manipulator)* in which the base and the mobile platform are hexagons. The legs are connected to the vertices of the hexagons;

- *TSSM robot (Triangular Simplified Symmetric Manipulator)* in which the mobile platform is triangular whereas the base is hexagonal. Two legs are connected on the same vertex of the triangle;
- *MSSM robot (Minimal Simplified Symmetric Manipulator)* in which the base and the mobile platform are triangular. Legs are mounted by pairs at both ends. The architecture forms an octahedron.

With six legs in parallel, there are five loops, and the application of equation [8.2] (N = 6d – 6x5) results in six degrees of freedom per leg (five passive and one active). The passive degrees of freedom can be distributed according to the combinations (0,5), (1,4) or (2,3). The (2,3) combination is the most popular one. It is composed of a universal joint (RR) and a spherical joint (RRR). The six degrees of freedom of the mobile platform are obtained by actuating each leg by either a prismatic joint (Figure 8.7) or a revolute joint (Figure 8.8). In the first case, the leg architecture is composed of (RR)-P-(RRR) joints, and in the second case, of R-(RR)-(RRR) joints.

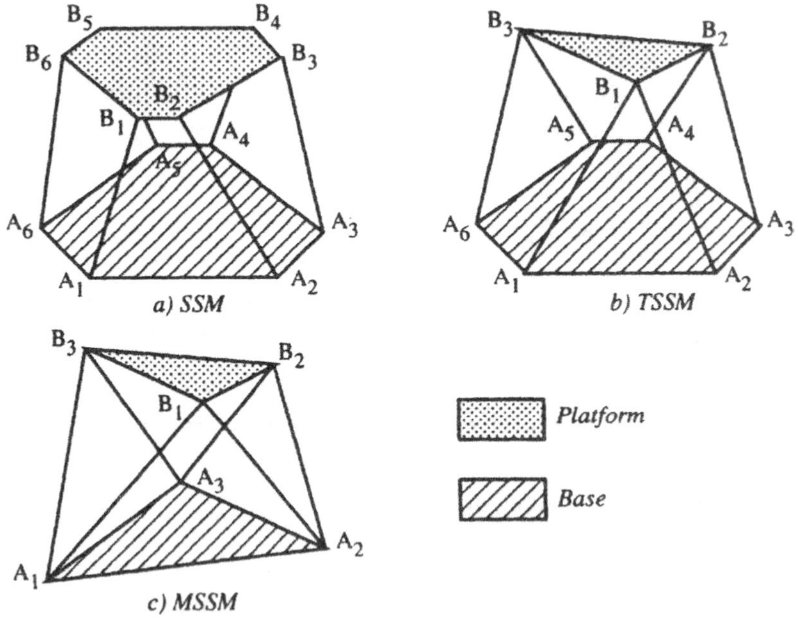

Figure 8.6. *SSM, TSSM and MSSM parallel robots*

Figure 8.9 presents a robot with a (0,5) combination for the passive joints. In this case, the leg orientation is fixed with respect to the base, whereas the mobile platform is connected to each leg through a five degree-of-freedom passive joint, a so-called C5 joint [Dafaoui 94]. The C5 joint is realized by a spherical joint fixed to two perpendicular prismatic joints. The leg lengths are actuated by prismatic joints, giving for each leg a P-(RRRPP) architecture.

Note that we can find a six degree-of-freedom spatial robot with only three legs with R-R-P-(RRR) architecture as shown in Figure 8.7a. In this case, each leg has two active joints, namely the revolute joint, which is close to the base, and the prismatic joint. Such structure has been achieved on the so-called *Space robot* [Beji 97]. The application of equation [8.2] gives N = 6 for this structure.

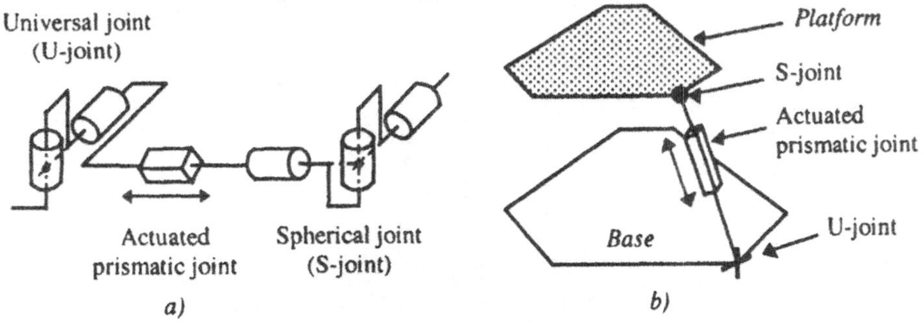

Figure 8.7. ***(RR)-P-(RRR) architecture***

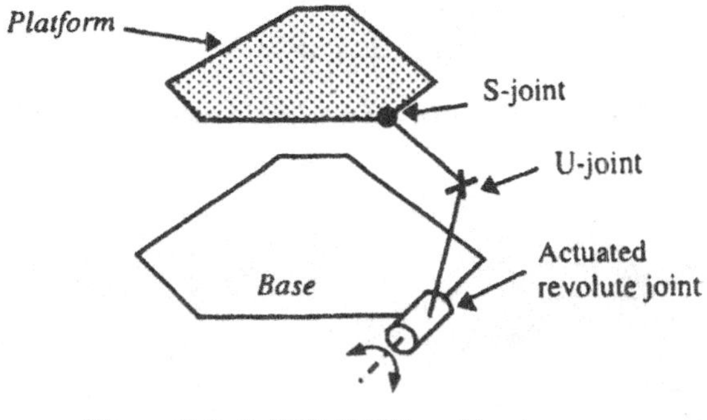

Figure 8.8. ***R-(RR)-(RRR) architectures***

8.5.3. *The Delta robot and its family*

An interesting realization actually being implemented in several industrial applications is the Delta robot designed by Clavel [Clavel 89] (Figure 8.10a). This robot has four degrees of freedom, the fourth being fixed on the mobile platform and allowing the end-effector to rotate around the vertical axis. The moving platform always remains parallel to the base. It is connected to the base by three identical kinematic chains having a R-(RR)-(RR) architecture. The parallel chains are actuated by the revolute joints, which are close to the base, using DC motors fixed to the base. With this architecture, the Delta has a very low inertia and can manipulate light pieces within a very short cycle time (typically, two pieces of 10g per second). This robot also presents the advantage of having a relatively large workspace. One can find in [Hervé 91], [Goudali 96] examples of robots derived from this architecture.

Pierrot [Pierrot 91b] has extended the Delta robot concept into a six degree-of-freedom robot, the *Hexa robot* (Figure 8.10b). The six actuators are fixed to the base and provide a speed of 8 m/s and an acceleration of 22 g to the mobile platform.

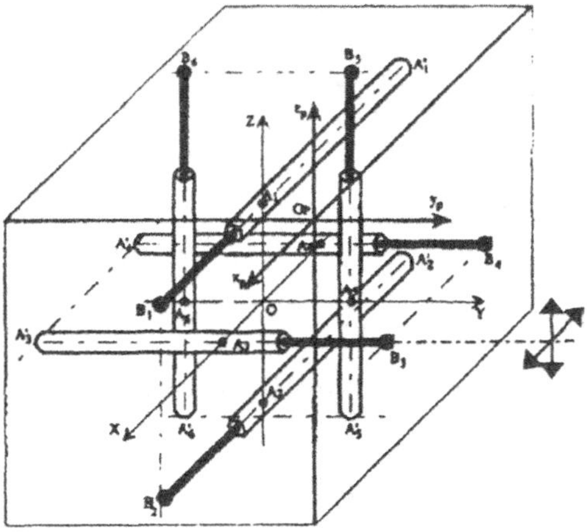

Figure 8.9. *Parallel robot with C5 joints [Dafaoui 94]*

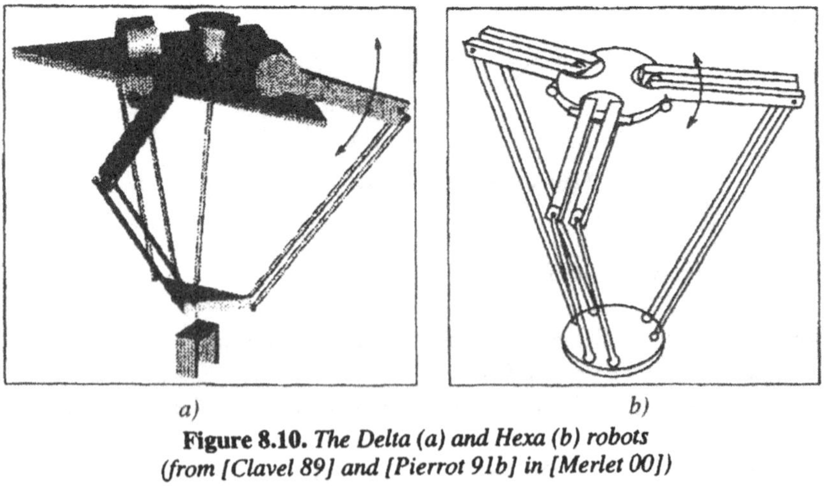

a) *b)*

Figure 8.10. *The Delta (a) and Hexa (b) robots (from [Clavel 89] and [Pierrot 91b] in [Merlet 00])*

8.6. Modeling the six degree-of-freedom parallel robots

In this section, we present the geometric and kinematic models of the Gough-Stewart parallel robots. The techniques developed in Chapter 7 for closed loop structures, namely the geometric description notations and the geometric and kinematic modeling methods can also be used for parallel robots. However, specific methods are usually more efficient. The proposed approach has direct relevance to the entire class of parallel robots such as the Delta robot [Clavel 89] and Hexa robot [Pierrot 91b].

8.6.1. *Geometric description*

We assume that the universal joints (U-joint) and the spherical joints (S-joint) are perfect, and that the prismatic joints are perfectly assembled. The centers of the U-joints and S-joints are denoted by A_i and B_i, for i = 1, ..., 6, respectively (Figure 8.11). Two coordinate systems need to be set up for the geometric description of a parallel structure: frame R_0 is attached to the base and frame R_m is attached to the mobile platform. They are defined as follows:

- A_1 is the origin of frame R_0, the x_0 axis is along A_1A_2, and the x_0y_0 plane is determined by A_1, A_2 and A_6;
- similarly, B_1 is the origin of frame R_m, the $\mathbf{x}_m$ axis is along B_1B_2, and B_1, B_2, B_6 are in the $\mathbf{x}_m\mathbf{y}_m$ plane.

The geometry of such a robot is described by:

- the (6x1) joint variable vector $\mathbf{q}$ representing the leg lengths A_iB_i for i = 1, ..., 6;
- the coordinates of the connection points A_i and B_i in frames R_0 and R_m respectively (${}^0\mathbf{P}_{Ai}$ and ${}^m\mathbf{P}_{Bi}$ for i = 1, ..., 6). Note that the points A_i may not be necessarily in the same plane, nor the points B_i.

According to the definition of frames R_0 and R_m, we obtain:

$$
\begin{aligned}
{}^0P_{XA1} &= {}^0P_{YA1} = {}^0P_{ZA1} = {}^0P_{YA2} = {}^0P_{ZA2} = {}^0P_{ZA6} = 0 \\
{}^mP_{XB1} &= {}^mP_{YB1} = {}^mP_{ZB1} = {}^mP_{YB2} = {}^mP_{ZB2} = {}^mP_{ZB6} = 0
\end{aligned}
$$

where ${}^j\mathbf{P}_{Pi}$ denotes the position of a point P_i with respect to frame R_j:

$$
{}^j\mathbf{P}_{Pi} = [\ {}^jP_{XPi} \quad {}^jP_{YPi} \quad {}^jP_{ZPi}\ ]^T
$$

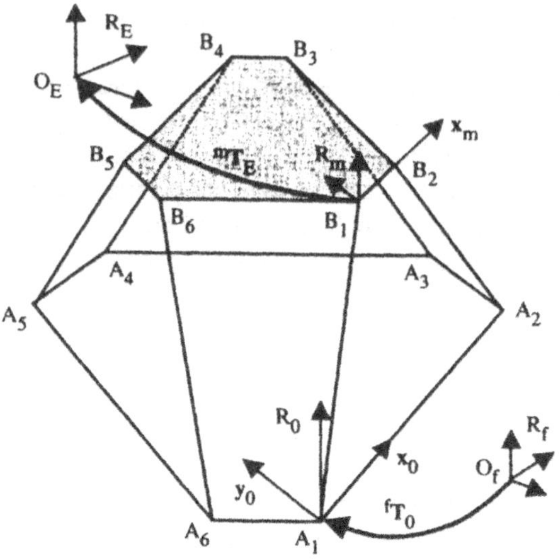

Figure 8.11. *Geometric notations*

Thus, the robot is described by 24 constant parameters that may be not zero.

In order to describe the location of the robot base frame R_0 with respect to the environment world reference frame R_f, we use the matrix $\mathbf{Z} = {}^f\mathbf{T}_0$. Similarly, to define a general end-effector frame R_E with respect to frame R_m, we use the matrix $\mathbf{E} = {}^m\mathbf{T}_E$. Consequently, the location of the end-effector frame relative to the world reference frame is:

$$ {}^f\mathbf{T}_E = \mathbf{Z}\,{}^0\mathbf{T}_m(\mathbf{q})\,\mathbf{E} \quad [8.5] $$

The coordinates of a connection point A_i relative to frame R_f are given by:

$$ \begin{bmatrix} {}^f\mathbf{P}_{Ai} \\ 1 \end{bmatrix} = {}^f\mathbf{T}_0 \begin{bmatrix} {}^0\mathbf{P}_{Ai} \\ 1 \end{bmatrix} = \mathbf{Z} \begin{bmatrix} {}^0\mathbf{P}_{Ai} \\ 1 \end{bmatrix} \quad [8.6] $$

The coordinates of the connection point B_i relative to frame R_E are:

$$ \begin{bmatrix} {}^E\mathbf{P}_{Bi} \\ 1 \end{bmatrix} = {}^E\mathbf{T}_m \begin{bmatrix} {}^m\mathbf{P}_{Bi} \\ 1 \end{bmatrix} = \mathbf{E} \begin{bmatrix} {}^m\mathbf{P}_{Bi} \\ 1 \end{bmatrix} \quad [8.7] $$

The matrices **Z** and **E** can be defined arbitrarily; therefore, six independent parameters are needed to define each of them.

To conclude, we can define the end-effector frame with respect to the reference frame using 42 geometric parameters (36 constant geometric parameters and 6 joint variables representing the leg lengths). The 36 constant parameters can be defined either by ${}^{f}\mathbf{P}_{A_i}$ and ${}^{E}\mathbf{P}_{B_i}$, or by the 24 coordinates of ${}^{0}\mathbf{P}_{A_i}$ and ${}^{m}\mathbf{P}_{B_i}$, which may be not zero, and the matrices **E** and **Z**. These parameters allow us to calculate the kinematic and geometric models. The second set of parameters is interesting when calibrating the geometric parameters using autonomous methods and when developing symbolic geometric or kinematic models.

8.6.2. *Inverse geometric model*

The inverse geometric model (IGM) provides the joint variables **q** corresponding to a given location ${}^{f}\mathbf{T}_{E}$ of the end-effector. It is represented by:

$$\mathbf{q} = \mathrm{IGM}({}^{f}\mathbf{T}_{E}) \qquad [8.8]$$

with:

$$\mathbf{q} = [q_1 \; \dots \; q_6]^T \qquad [8.9]$$

Since ${}^{E}\mathbf{P}_{B_i}$ and ${}^{f}\mathbf{T}_{E}$ are known, we can compute the coordinates of the connection points B_i with respect to the reference frame by the following relation:

$$\begin{bmatrix} {}^{f}\mathbf{P}_{B_i} \\ 1 \end{bmatrix} = {}^{f}\mathbf{T}_{E} \begin{bmatrix} {}^{E}\mathbf{P}_{B_i} \\ 1 \end{bmatrix} \qquad [8.10]$$

The prismatic variable q_i is equal to the distance between the connection points A_i and B_i:

$$q_i^2 = ({}^{f}\mathbf{P}_{B_i} - {}^{f}\mathbf{P}_{A_i})^T ({}^{f}\mathbf{P}_{B_i} - {}^{f}\mathbf{P}_{A_i}) = ({}^{f}\mathbf{A}_i\mathbf{B}_i)^T \, {}^{f}\mathbf{A}_i\mathbf{B}_i \qquad [8.11]$$

This equation shows that the IGM of the Gough-Stewart parallel robot is unique and can be easily determined. The equations giving the leg lengths are independent and can be computed in parallel [Merlet 00]. This result is not general for all parallel robots. For example, the IGM of a three degree-of-freedom spatial robot (Figure 8.5) is not unique: for a given position of the endpoint, there are four possible solutions for the leg lengths [Gosselin 88].

8.6.3. *Inverse kinematic model*

The inverse kinematic model (IKM) provides the actuated joint velocity $\dot{\mathbf{q}}$ corresponding to a given kinematic screw of the end-effector. It is denoted by:

$$\dot{\mathbf{q}} = {}^{f}\mathbf{J}_E^{-1} \begin{bmatrix} {}^{f}\mathbf{V}_E \\ {}^{f}\omega_E \end{bmatrix} \qquad [8.12]$$

where ${}^{f}\mathbf{V}_E$ is the linear velocity of O_E, origin of frame R_E, and ${}^{f}\omega_E$ is the angular velocity of the end-effector.

The inverse differential model can be defined by:

$$\mathbf{dq} = {}^{f}\mathbf{J}_E^{-1} \begin{bmatrix} {}^{f}\mathbf{dP}_E \\ {}^{f}\delta_E \end{bmatrix} \qquad [8.13]$$

The computation of ${}^{f}\mathbf{J}_E^{-1}$ is obtained by projecting the velocity of the connection points of the mobile platform onto the leg directions. Let ${}^{f}\mathbf{V}_{B_i}$ be the velocity of the point B_i with respect to frame R_E; hence we can write:

$${}^{f}\mathbf{V}_{B_i} = {}^{f}\mathbf{V}_E + {}^{f}\mathbf{B}_i\mathbf{O}_E \times {}^{f}\omega_E \qquad [8.14]$$

Thus, the joint velocity $\dot{q}_i$, for i = 1, …, 6, can be computed by:

$$\dot{q}_i = {}^{f}\mathbf{u}_i^T \, {}^{f}\mathbf{V}_{B_i} \qquad [8.15]$$

where $\mathbf{u}_i$ is the unit vector along the ith leg:

$$\mathbf{u}_i = \frac{\mathbf{A}_i\mathbf{B}_i}{\|\mathbf{A}_i\mathbf{B}_i\|} = \frac{\mathbf{A}_i\mathbf{B}_i}{q_i} \qquad [8.16]$$

Combining equations [8.14] and [8.15] leads to:

$$\dot{q}_i = {}^{f}\mathbf{u}_i^T \, {}^{f}\mathbf{V}_E + {}^{f}\mathbf{u}_i^T \, ({}^{f}\mathbf{B}_i\mathbf{O}_E \times {}^{f}\omega_E) \qquad [8.17]$$

which can be rewritten as:

$$\dot{q}_i = {}^{f}\mathbf{u}_i^T \, {}^{f}\mathbf{V}_E + ({}^{f}\mathbf{u}_i \times {}^{f}\mathbf{B}_i\mathbf{O}_E)^T \, {}^{f}\omega_E \qquad [8.18]$$

Consequently, the ith row of the inverse Jacobian matrix is given by:

$$\mathbf{L}_i = [\ {}^f\mathbf{u}_i^T \quad ({}^f\mathbf{u}_i \times {}^f\mathbf{B}_i\mathbf{O}_E)^T\ ] \qquad [8.19]$$

and finally:

$${}^f\mathbf{J}_E^{-1} = \begin{bmatrix} {}^f\mathbf{u}_1^T & ({}^f\mathbf{u}_1 \times {}^f\mathbf{B}_1\mathbf{O}_E)^T \\ {}^f\mathbf{u}_2^T & ({}^f\mathbf{u}_2 \times {}^f\mathbf{B}_2\mathbf{O}_E)^T \\ \dots & \dots \\ {}^f\mathbf{u}_6^T & ({}^f\mathbf{u}_6 \times {}^f\mathbf{B}_6\mathbf{O}_E)^T \end{bmatrix} \qquad [8.20]$$

The direct kinematic model is obtained by inverting equation [8.20].

8.6.4. *Direct geometric model*

The direct geometric model (DGM) provides the location of the end-effector corresponding to a given joint configuration $\mathbf{q}$. It is written as:

$${}^f\mathbf{T}_E = \mathrm{DGM}(\mathbf{q}) \qquad [8.21]$$

8.6.4.1. *Closed-form solution*

The solution of the DGM is relatively complicated to derive. In general, for a given set of joint variables, the mobile platform can take several different locations.

In order to find an analytical solution to the general six degree-of-freedom Gough-Stewart parallel robot, some authors propose to incorporate additional position sensors on some selected passive joints [Cheok 93], [Merlet 93], [Baron 94], [Han 95], [Tancredi 95].

Recently, Husty [Husty 96] presented a method giving all the DGM solutions of a general Gough-Stewart parallel robot. It is based on the resolution of a 40th degree polynomial in a single variable. The other variables are then uniquely determined. Some of these solutions may be complex numbers. However, this method does not give the maximum number of real solutions. It does not indicate if the solutions can be reached from a given configuration without crossing a singularity or after having crossed a singularity. The Husty algorithm confirms Raghavan's work [Raghavan 91] and Lazard's work [Lazard 92] that found the same number of solutions by numerical methods. In [Dietmaier 98], we find an example of a robot with 40 real solutions.

Closed-form solutions of the DGM have been proposed for the special (non-exhaustive) following architectures:

1) TSSM architectures (Figure 8.6b): the notion of equivalent mechanism allows us to derive the direct geometric model of spatial robots with a triangular platform. This approach has been used by Hunt [Hunt 83] and [Charentus 90] to compute the DGM of TSSM architectures. The problem is reduced to solving three equations in three unknowns, which are the rotation angles of these triangles about the base.

Indeed, in the TSSM architecture, each of the triangular faces $A_1A_2B_1$, $A_3A_4B_2$ and $A_5A_6B_3$ can only rotate around the axes A_1A_2, A_3A_4 and A_5A_6 respectively. This results in the equivalent mechanism shown in Figure 8.12.

The three equivalent segments r_1, r_2 and r_3 sweep three circles in the space and it can be shown that the solution of the DGM is given by an 8th degree polynomial. Thus, the number of possible configurations for a given vector **q** is 16 while taking into account the symmetry of solutions with respect to the base.

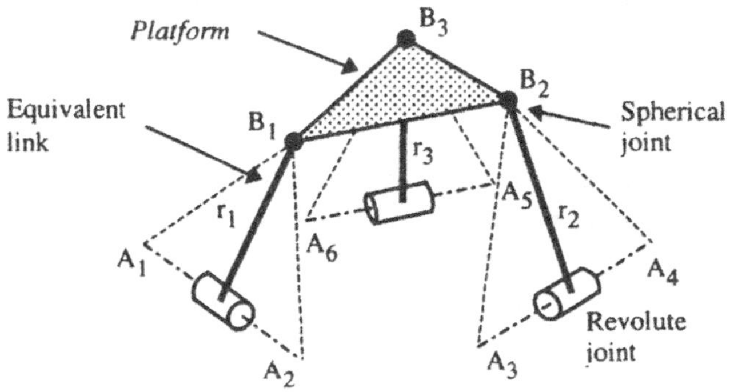

Figure 8.12. *Equivalent TSSM architecture with three rigid triangular faces*

2) Gough-Stewart robots with five collinear connection points: Zhang and Song [Zhang 92] showed that the DGM of such a parallel robot has an analytical solution (4th degree polynomial at most) if the connection points of five legs on either the base or the platform are collinear. Such structure decouples a rotational degree of freedom of the platform from the other five (Figure 8.13). The number of solutions is eight, four with P_{zm} positive, the other four with P_{zm} negative. A study of the duality of this architecture with a six degree-of-freedom serial robot having a spherical wrist can be found in [Khalil 96c].

3) 4-4 Gough-Stewart robots: [Lin 92] gives the polynomial solution of the DGM of a Gough-Stewart robot in which two pairs of connection points of the mobile platform are coincident, as well as two pairs of connection points of the base plate. The connection points of the base lie on one plane and those of the mobile platform lie on another one. The solution is given by a 12th degree polynomial at most.

4) 5-4 Parallel robots: Innocenti and Parenti-Castelli [Innocenti 93] give the solution of a Gough-Stewart robot in which two connection points of the base are coincident as well as two pairs of connection points of the mobile platform (Figure 8.14). The solution is given by a 24th degree polynomial in one variable, the other variables being uniquely determined. A similar result has been obtained for the 5-5 robot [Innocenti 95].

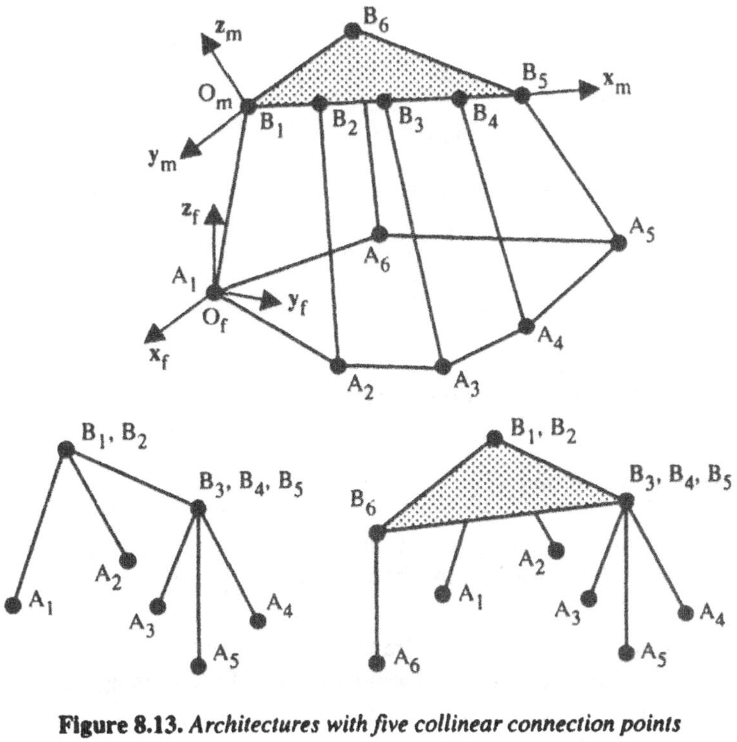

Figure 8.13. *Architectures with five collinear connection points*

5) Gough-Stewart robot with similar base and platform: it has been shown that all the joint variables can be obtained from linear or quadratic equations when the connection points of the base lie on one plane and those of the mobile platform lie on another one, and the form of the base and the mobile platform are similar (same form but different sizes). The number of solutions is 16 [Lee 93].

6) Gough-Stewart robot with three coincident connection points on the base as well as on the mobile platform: this so-called $(3\text{-}1\text{-}1\text{-}1)^2$ robot has eight solutions that can be analytically computed [Bruyninckx 98].

7) Six degree-of-freedom parallel robot with a C5 joint: it can be shown that such an architecture (Figure 8.9), which has a very small workspace, has a unique solution for both the DGM and the IGM [Dafaoui 94].

8) Delta family: the DGM may be obtained by solving a 2nd degree polynomial [Pierrot 91a].

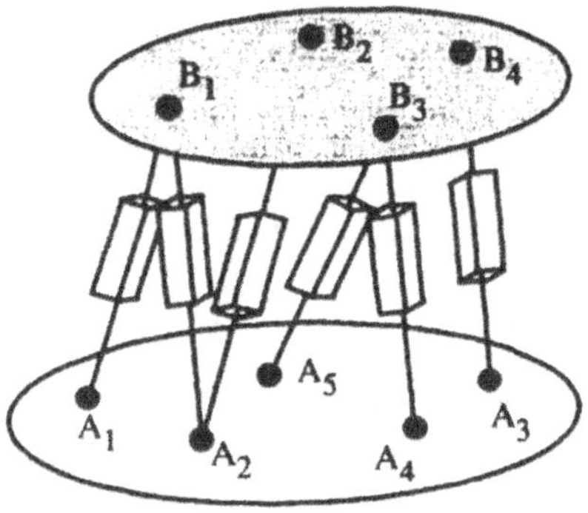

Figure 8.14. *Six degree-of-freedom (5-4) parallel robot*

8.6.4.2. *Numerical solution*

Practically, we can use the inverse differential model to compute a numerical solution of the DGM in an iterative manner. For a given $\mathbf{q}^d$, the algorithm is as follows:

- from an initial location ${}^f\mathbf{T}_E^c$ (random or current location) of the mobile platform, compute the corresponding joint variables $\mathbf{q}^c$ using the IGM;
- compute the difference between $\mathbf{q}^d$ and the current $\mathbf{q}^c$: $\mathbf{dq} = \mathbf{q}^d - \mathbf{q}^c$. If $\mathbf{dq}$ is small enough, ${}^f\mathbf{T}_E^d = {}^f\mathbf{T}_E^c$, then stop the computation;
- using equation [8.20], compute the inverse Jacobian matrix ${}^f\mathbf{J}_E^{-1}$;
- compute numerically the direct Jacobian matrix ${}^f\mathbf{J}_E = ({}^f\mathbf{J}_E^{-1})^+$;
- compute the position error ${}^f\mathbf{dP}_E$ and the orientation error ${}^f\delta_E = \theta\,\mathbf{u}$ (where $\mathbf{u}$ is a unit vector) corresponding to $\mathbf{dq}$ by using the relation:

$$\begin{bmatrix} {}^f\mathbf{dP}_E \\ {}^f\delta_E \end{bmatrix} = {}^f\mathbf{J}_E\,\mathbf{dq} \qquad [8.22]$$

– update the current position and orientation of the mobile platform:

$$ {}^{f}\mathbf{T}_{E}^{c} = \begin{bmatrix} {}^{f}\mathbf{A}_{E}^{c} & {}^{f}\mathbf{p}_{E}^{c} \\ 0\ \ 0\ \ 0 & 1 \end{bmatrix} \qquad [8.23] $$

with:

$$ {}^{f}\mathbf{P}_{E}^{c} = {}^{f}\mathbf{P}_{E}^{c} + {}^{f}\mathbf{dP}_{E} \qquad [8.24] $$

$$ {}^{f}\mathbf{A}_{E}^{c} = \mathbf{rot}(\mathbf{u}, \theta)\, {}^{f}\mathbf{A}_{E}^{c} \qquad [8.25] $$

– return to the first step.

This algorithm is efficient and can be computed in real time.

8.7. Singular configurations

Singular configurations of parallel robots are particular configurations at which the robot loses its natural rigidity. At these configurations, one or more degrees of freedom of the platform become uncontrollable. From such an initial configuration, the mobile platform moves toward an equilibrium location under the effect of the wrench applied to the mobile platform, for example, under the effect of the gravity loading of the platform and the manipulated load. Such a motion is due to the passive joints, while the actuated joints are fixed (§ 7.11).

Mathematically, the singular configurations can be determined by analyzing the static equilibrium of the robot. Let $\mathbf{\Gamma}$ be the vector of the joint torques and let $\mathbf{f}$ be the static wrench applied to the mobile platform. The static equilibrium of the robot is defined as (§ 5.9.2):

$$ \mathbf{f} = (\mathbf{J}^{-1})^{T}\, \mathbf{\Gamma} \qquad [8.26] $$

In order to maintain the equilibrium of the robot, the inverse Jacobian matrix has to be regular, such that for a given wrench $\mathbf{f}$, there is a corresponding finite joint torque vector $\mathbf{\Gamma}$.

The singular configurations can be determined by analyzing the rank of the matrix $\mathbf{J}^{-1}$. From equation [8.12], we can deduce that at a singular configuration, a motion in the null space of $\mathbf{J}^{-1}$ is possible even though $\dot{\mathbf{q}} = \mathbf{0}$ (§ 7.11). From equation [8.26], we see that at a singular configuration the motor torques $\mathbf{\Gamma}$ can be infinity, which may damage the robot. Thus, parallel robots should be designed without singularities in the reachable space. This can be attained by good selection of joint geometric parameters and joint limits, and even by providing the robot with redundant actuators on some passive joints [Merlet 00].

Merlet [Merlet 89] proposed to study the singularity using another interesting geometric method based on Grassmann manifolds.

The analysis of singularities has led some authors to propose design-related rules in order to avoid singular structures. Ma and Angeles [Ma 91] demonstrated that the inverse Jacobian matrix is singular throughout the whole workspace of the robot if the mobile platform and the base are made of regular and similar polygons with six connecting points. Such singularity is called *architecture singularity* and the corresponding architecture is called *singular architecture*. Figure 8.15 depicts this architecture, the mobile platform and the base being homothetical.

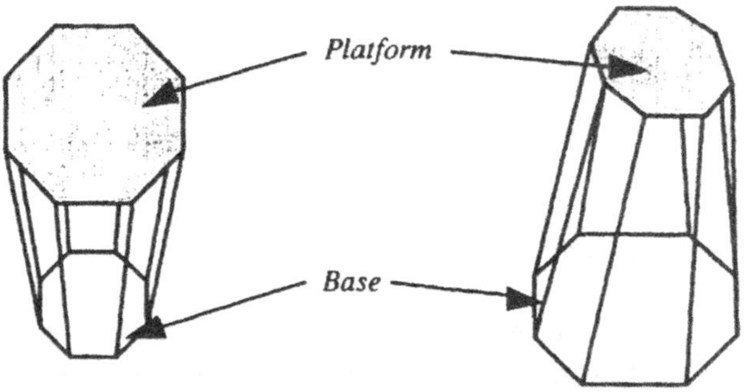

Figure 8.15. *Singular architectures*

8.8. Conclusion

In this chapter, we have presented the geometric and kinematic models of Gough-Stewart structures, which are considered to be representative of parallel robots. We have shown that the techniques developed for serial robots are often not appropriate and special approaches have to be used. The IGM and IKM are simple and straightforward to derive. On the contrary, the analytical DGM is not easy to compute in the general case since 40 solutions are possible. It was observed that merging some of the connection points on the platform or the base or both, by groups of two or three, simplifies the closed-form solution of the problem and also reduces the maximum number of possible solutions.

In addition, the numerical solution of the DGM can be used in most practical applications where only one real solution is required provided that a good initial location is available.

Chapter 9

Dynamic modeling of serial robots

9.1. Introduction

The *inverse dynamic model* provides the joint torques and forces in terms of the joint positions, velocities and accelerations. It is described by:

$$\mathbf{\Gamma} = \mathbf{f}(\mathbf{q}, \dot{\mathbf{q}}, \ddot{\mathbf{q}}, \mathbf{f}_e) \quad [9.1]$$

with:

- $\mathbf{\Gamma}$: vector of joint torques or forces, depending on whether the joint is revolute or prismatic respectively. In the sequel, we will only write *joint torques*;
- $\mathbf{q}$: vector of joint positions;
- $\dot{\mathbf{q}}$: vector of joint velocities;
- $\ddot{\mathbf{q}}$: vector of joint accelerations;
- $\mathbf{f}_e$: vector of forces and moments exerted by the robot on the environment.

Equation [9.1] is an inverse dynamic model because it defines the system input as a function of the output variables. It is often called the *dynamic model*.

The *direct dynamic model* describes the joint accelerations in terms of the joint positions, velocities and torques. It is represented by the relation:

$$\ddot{\mathbf{q}} = \mathbf{g}(\mathbf{q}, \dot{\mathbf{q}}, \mathbf{\Gamma}, \mathbf{f}_e) \quad [9.2]$$

The dynamic model of robots plays an important role in their design and operation. For robot design, the inverse dynamic model can be used to select the actuators [Chedmail 90b], [Potkonjak 86], while the direct dynamic model is employed to carry out simulations (§ 9.7) for the purpose of testing the performance of the robot and to study the relative merits of possible control schemes. Regarding robot operations, the inverse dynamic model is used to compute the actuator torques, which are needed to achieve a desired motion (Chapter 14). It is also used to identify the dynamic parameters that are necessary for both control and simulation applications (Chapter 12).

Several approaches have been proposed to model the dynamics of robots [Renaud 75], [Coiffet 81], [Vukobratovic 82]. The most frequently employed in robotics are the Lagrange formulation [Uicker 69], [Khalil 76], [Renaud 80a], [Hollerbach 80], [Paul 81], [Megahed 84], [Renaud 85] and the Newton-Euler formulation [Hooker 65], [Armstrong 79], [Luh 80b], [Orin 79], [Khalil 85a], [Khosla 86], [Khalil 87b], [Renaud 87].

In this chapter, we present the dynamic modeling of serial robots using these two formulations. The problem of calculating a minimum set of inertial parameters will be covered in detail. We will focus our study on the minimization of the number of operations of the dynamic model in view of its real time computation for control purposes. Lastly, the computation of the direct dynamic model will be addressed. In Chapter 10, these results will be generalized for tree structured and closed chain robots.

9.2. Notations

The main notations used in this chapter are compiled below:

- $\mathbf{a}_j$ unit vector along axis $\mathbf{z}_j$;
- $\mathbf{F}_j$ external forces on link j;
- $\mathbf{f}_j$ force exerted on link j by link j − 1;
- $\mathbf{f}_{ej}$ force exerted by link j on the environment;
- F_{cj} parameter of Coulomb friction acting at joint j;
- F_{vj} parameter of viscous friction acting at joint j;
- $\mathbf{g}$ gravitational acceleration;
- G_j center-of-mass of link j;
- $\mathbf{I}_{Gj}$ inertia tensor of link j about G_j and with respect to a frame parallel to frame R_j;
- Ia_j moment of inertia of the rotor and the transmission system of actuator j referred to the joint side;
- ${}^j\mathbf{J}_j$ inertia tensor of link j with respect to frame R_j. It is described by:

$$
{}^{j}\mathbf{J}_j = \begin{bmatrix} \int(y^2+z^2)dm & -\int xydm & -\int xzdm \\ -\int xydm & \int(x^2+z^2)dm & -\int yzdm \\ -\int xzdm & -\int yzdm & \int(x^2+y^2)dm \end{bmatrix} = \begin{bmatrix} XX_j & XY_j & XZ_j \\ XY_j & YY_j & YZ_j \\ XZ_j & YZ_j & ZZ_j \end{bmatrix} \quad [9.3]
$$

$\mathbb{J}_j$ (6x6) spatial inertia matrix of link j (relation [9.21]);
$\mathbf{L}_j$ position vector between O_{j-1} and O_j;
M_j mass of link j;
$\mathbf{MS}_j$ first moments of link j with respect to frame R_j, equal to $M_j\ \mathbf{S}_j$. The components of ${}^{j}\mathbf{MS}_j$ are denoted by $[\ MX_j\ \ MY_j\ \ MZ_j\ ]^T$;
$\mathbf{M}_{Gj}$ moment of external forces on link j about G_j;
$\mathbf{M}_j$ moment of external forces on link j about O_j;
$\mathbf{m}_j$ moment about O_j exerted on link j by link j – 1;
$\mathbf{m}_{ej}$ moment about O_j exerted by link j on the environment;
$\mathbf{S}_j$ vector of the center-of-mass coordinates of link j. It is equal to $\mathbf{O}_j\mathbf{G}_j$;
$\mathbf{V}_j$ linear velocity of O_j;
$\mathbb{V}_j$ (6x1) kinematic screw vector of link j, formed by the components of $\mathbf{V}_j$ and $\boldsymbol{\omega}_j$;
$\dot{\mathbf{V}}_j$ linear acceleration of O_j;
$\mathbf{V}_{Gj}$ linear velocity of the center-of-mass of link j;
$\dot{\mathbf{V}}_{Gj}$ linear acceleration of the center-of-mass of link j;
$\boldsymbol{\omega}_j$ angular velocity of link j;
$\dot{\boldsymbol{\omega}}_j$ angular acceleration of link j.

9.3. Lagrange formulation

9.3.1. *Introduction*

The dynamic model of a robot with several degrees of freedom represents a complicated system. The Newton-Euler method developed in § 9.5 presents an efficient and systematic approach to solving this problem. In this section, we develop a simple Lagrange method to present the general form of the dynamic model of robots and to get an insight into its properties. Firstly, we consider an ideal system without friction or elasticity, exerting neither forces nor moments on the environment. These phenomena will be covered in § 9.3.4 through 9.3.8.

The Lagrange formulation describes the behavior of a dynamic system in terms of work and energy stored in the system. The Lagrange equations are commonly written in the form:

$$\Gamma_i = \frac{d}{dt}\frac{\partial L}{\partial \dot{q}_i} - \frac{\partial L}{\partial q_i} \qquad \text{for } i = 1, \ldots, n \qquad [9.4a]$$

where L is the Lagrangian of the robot defined as the difference between the kinetic energy E and the potential energy U of the system:

$$L = E - U \qquad [9.4b]$$

9.3.2. *General form of the dynamic equations*

The kinetic energy of the system is a quadratic function in the joint velocities such that:

$$E = \frac{1}{2}\dot{\mathbf{q}}^T \mathbf{A}\,\dot{\mathbf{q}} \qquad [9.5]$$

where **A** is the (nxn) symmetric and positive definite *inertia matrix* of the robot. Its elements are functions of the joint positions. The (i, j) element of **A** is denoted by A_{ij}.

Since the potential energy is a function of the joint positions, equations [9.4] and [9.5] lead to:

$$\boldsymbol{\Gamma} = \mathbf{A}(\mathbf{q})\,\ddot{\mathbf{q}} + \mathbf{C}(\mathbf{q}, \dot{\mathbf{q}})\,\dot{\mathbf{q}} + \mathbf{Q}(\mathbf{q}) \qquad [9.6]$$

where:

- $\mathbf{C}(\mathbf{q}, \dot{\mathbf{q}})\,\dot{\mathbf{q}}$ is the (nx1) vector of Coriolis and centrifugal torques, such that:

$$\mathbf{C}\,\dot{\mathbf{q}} = \dot{\mathbf{A}}\,\dot{\mathbf{q}} - \frac{\partial E}{\partial \mathbf{q}}$$

- $\mathbf{Q} = [Q_1 \;\ldots\; Q_n]^T$ is the vector of gravity torques.

Consequently, the dynamic model of a robot is described by n coupled and nonlinear second order differential equations.

There exist several forms for the vector $\mathbf{C}(\mathbf{q}, \dot{\mathbf{q}})\,\dot{\mathbf{q}}$. Using the *Christoffell symbols* $c_{i,jk}$, the (i, j) element of the matrix **C** can be written as:

$$\begin{cases} C_{ij} = \sum_{k=1}^{n} c_{i,jk}\, \dot{q}_k \\ c_{i,jk} = \frac{1}{2}\left[\frac{\partial A_{ij}}{\partial q_k} + \frac{\partial A_{ik}}{\partial q_j} - \frac{\partial A_{jk}}{\partial q_i}\right] \end{cases} \qquad [9.7]$$

The Q_i element of the vector **Q** is calculated according to:

$$Q_i = \frac{\partial U}{\partial q_i} \qquad [9.8]$$

The elements of **A**, **C** and **Q** are functions of the geometric and inertial parameters of the robot.

9.3.3. *Computation of the elements of A, C and Q*

To compute the elements of **A**, **C** and **Q**, we begin by symbolically computing the expressions of the kinetic and potential energies of all the links of the robot. Then, we proceed as follows:

- the element A_{ii} is equal to the coefficient of ($\dot{q}_i^2/2$) in the expression of the kinetic energy, while A_{ij}, for $i \neq j$, is equal to the coefficient of $\dot{q}_i\, \dot{q}_j$;
- the elements of **C** are obtained from equation [9.7];
- the elements of **Q** are obtained from equation [9.8].

9.3.3.1. *Computation of the kinetic energy*

The kinetic energy of the robot is given as:

$$E = \sum_{j=1}^{n} E_j \qquad [9.9]$$

where E_j denotes the kinetic energy of link j, which can be computed by:

$$E_j = \frac{1}{2}(\omega_j^T\, I_{Gj}\, \omega_j + M_j\, V_{Gj}^T\, V_{Gj}) \qquad [9.10]$$

Since the velocity of the center-of-mass can be expressed as (Figure 9.1):

$$V_{Gj} = V_j + \omega_j \times S_j \qquad [9.11]$$

and since:

$$\mathbf{J}_j = \mathbf{I}_{Gj} - M_j \hat{\mathbf{S}}_j \hat{\mathbf{S}}_j \tag{9.12}$$

equation [9.10] becomes:

$$E_j = \frac{1}{2}[\boldsymbol{\omega}_j^T \mathbf{J}_j \boldsymbol{\omega}_j + M_j \mathbf{V}_j^T \mathbf{V}_j + 2\,\mathbf{MS}_j^T (\mathbf{V}_j \times \boldsymbol{\omega}_j)] \tag{9.13}$$

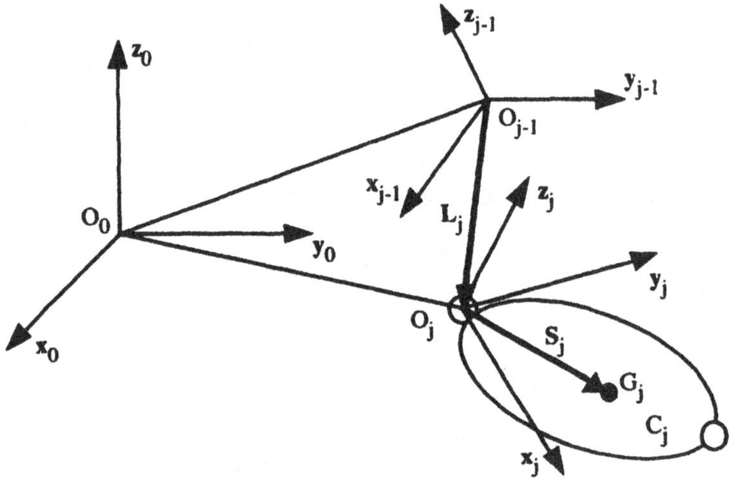

Figure 9.1. *Composition of velocities*

Equation [9.10] is not linear in the coordinates of the vector $\mathbf{S}_j$. On the contrary, equation [9.13] is linear in the elements of M_j, $\mathbf{MS}_j$ and $\mathbf{J}_j$, which we call the *standard inertial parameters*. The linear and angular velocities $\mathbf{V}_j$ and $\boldsymbol{\omega}_j$ are computed using the following recursive equations (Figure 9.1):

$$\boldsymbol{\omega}_j = \boldsymbol{\omega}_{j-1} + \bar{\sigma}_j \dot{q}_j \mathbf{a}_j \tag{9.14}$$

$$\mathbf{V}_j = \mathbf{V}_{j-1} + \boldsymbol{\omega}_{j-1} \times \mathbf{L}_j + \sigma_j \dot{q}_j \mathbf{a}_j \tag{9.15}$$

If the base of the robot is fixed, the previous equations are initialized by $\mathbf{V}_0 = \mathbf{0}$ and $\boldsymbol{\omega}_0 = \mathbf{0}$.

All the elements appearing in equation [9.13] must be expressed in the same frame. The most efficient way is to express them relative to frame R_j. Therefore, equations [9.13], [9.14] and [9.15] are rewritten as:

$$E_j = \frac{1}{2}\left[{}^j\omega_j^T\,{}^j\mathbf{J}_j\,{}^j\omega_j + M_j\,{}^j\mathbf{V}_j^T\,{}^j\mathbf{V}_j + 2\,{}^j\mathbf{MS}_j^T\,({}^j\mathbf{V}_j \times {}^j\omega_j)\right] \qquad [9.16]$$

$${}^j\omega_j = {}^j\mathbf{A}_{j-1}\,{}^{j-1}\omega_{j-1} + \bar{\sigma}_j\,\dot{q}_j\,{}^j\mathbf{a}_j = {}^j\omega_{j-1} + \bar{\sigma}_j\,\dot{q}_j\,{}^j\mathbf{a}_j \qquad [9.17]$$

$${}^j\mathbf{V}_j = {}^j\mathbf{A}_{j-1}\,({}^{j-1}\mathbf{V}_{j-1} + {}^{j-1}\omega_{j-1} \times {}^{j-1}\mathbf{P}_j) + \sigma_j\,\dot{q}_j\,{}^j\mathbf{a}_j \qquad [9.18]$$

The parameters ${}^j\mathbf{J}_j$ and ${}^j\mathbf{MS}_j$ are constants.

Using the spatial notation, the kinetic energy can be written in the following compact form:

$$E_j = \frac{1}{2}\,{}^j\mathbb{V}_j^T\,{}^j\mathbb{J}_j\,{}^j\mathbb{V}_j \qquad [9.19]$$

where:

$${}^j\mathbb{V}_j = \begin{bmatrix} {}^j\mathbf{V}_j \\ {}^j\omega_j \end{bmatrix} \qquad [9.20]$$

$${}^j\mathbb{J}_j = \begin{bmatrix} M_j\mathbf{I}_3 & -{}^j\hat{\mathbf{MS}}_j \\ {}^j\hat{\mathbf{MS}}_j & {}^j\mathbf{J}_j \end{bmatrix} \qquad [9.21]$$

The recursive velocity relations [9.17] and [9.18] can be combined as follows:

$${}^j\mathbb{V}_j = {}^j\mathbb{T}_{j-1}\,{}^{j-1}\mathbb{V}_{j-1} + \dot{q}_j\,{}^j\mathbb{a}_j \qquad [9.22]$$

where ${}^j\mathbb{T}_{j-1}$ is the (6x6) screw transformation matrix defined in [2.46] as:

$${}^j\mathbb{T}_{j-1} = \begin{bmatrix} {}^j\mathbf{A}_{j-1} & -{}^j\mathbf{A}_{j-1}\,{}^{j-1}\hat{\mathbf{P}}_j \\ \mathbf{0}_3 & {}^j\mathbf{A}_{j-1} \end{bmatrix} = \begin{bmatrix} {}^j\mathbf{A}_{j-1} & {}^j\hat{\mathbf{P}}_{j-1}\,{}^j\mathbf{A}_{j-1} \\ \mathbf{0}_3 & {}^j\mathbf{A}_{j-1} \end{bmatrix} \qquad [9.23a]$$

and where ${}^j\mathbb{a}_j$ is the (6x1) column matrix:

$${}^j\mathbb{a}_j = \begin{bmatrix} \sigma_j\,{}^j\mathbf{a}_j \\ \bar{\sigma}_j\,{}^j\mathbf{a}_j \end{bmatrix} \qquad [9.23b]$$

9.3.3.2. *Computation of the potential energy*

The potential energy is given by:

$$U = \sum_{j=1}^{n} U_j = \sum_{j=1}^{n} -M_j \, \mathbf{g}^T (\mathbf{L}_{0,j} + \mathbf{S}_j) \qquad [9.24]$$

where $\mathbf{L}_{0,j}$ is the position vector from the origin O_0 to O_j. Projecting the vectors appearing in [9.24] into frame R_0, we obtain:

$$U_j = -M_j \, {}^0\mathbf{g}^T ({}^0\mathbf{P}_j + {}^0\mathbf{A}_j \, {}^j\mathbf{S}_j) \qquad [9.25a]$$

an expression that can be rewritten linearly in M_j and the elements of ${}^j\mathbf{MS}_j$ as:

$$U_j = -{}^0\mathbf{g}^T (M_j \, {}^0\mathbf{P}_j + {}^0\mathbf{A}_j \, {}^j\mathbf{MS}_j) = -[\, {}^0\mathbf{g}^T \quad 0 \,] \, {}^0\mathbf{T}_j \begin{bmatrix} {}^j\mathbf{MS}_j \\ M_j \end{bmatrix} \qquad [9.25b]$$

Since the kinetic and potential energies are linear in the elements of ${}^j\mathbf{J}_j$, ${}^j\mathbf{MS}_j$, M_j, we deduce that the dynamic model is also linear in these parameters.

9.3.3.3. *Dynamic model properties*

In this section, we summarize some important properties of the dynamic model of robots:

a) the inertia matrix $\mathbf{A}$ is symmetric and positive definite;

b) the energy of link j is a function of $(q_1, \ldots, q_j)$ and $(\dot{q}_1, \ldots, \dot{q}_j)$;

c) the element A_{ij} is a function of $q_{k+1}, \ldots, q_n$, with $k = \min(i, j)$, and of the inertial parameters of links r, ..., n, with $r = \max(i, j)$;

d) from property b and equation [9.4], we deduce that Γ_i is a function of the inertial parameters of links i, ..., n;

e) the matrix $[\frac{d}{dt}\mathbf{A} - 2\mathbf{C}(\mathbf{q}, \dot{\mathbf{q}})]$ is skew-symmetric for the choice of the matrix $\mathbf{C}$ given by equation [9.7] [Koditschek 84], [Arimoto 84]. This property is used in Chapter 14 for the stability analysis of certain control schemes;

f) the inverse dynamic model is linear in the elements of the standard inertial parameters M_j, ${}^j\mathbf{MS}_j$ and ${}^j\mathbf{J}_j$. This property is exploited to identify the dynamic parameters (inertial and friction parameters), to reduce the computation burden of the dynamic model, and to develop adaptive control schemes;

g) there exist some positive real numbers $a_1, ..., a_7$ such that for any values of $\mathbf{q}$ and $\dot{\mathbf{q}}$ we have [Samson 87]:

$\| \mathbf{A}(\mathbf{q}) \| \leq a_1 + a_2 \| \mathbf{q} \| + a_3 \| \mathbf{q} \|^2$

$\| \mathbf{C}(\mathbf{q}, \dot{\mathbf{q}}) \| \leq \| \dot{\mathbf{q}} \| (a_4 + a_5 \| \mathbf{q} \|)$

$\| \mathbf{Q} \| \leq a_6 + a_7 \| \mathbf{q} \|$

where $\|*\|$ indicates a matrix or vector norm. If the robot has only revolute joints, these relations become:

$\| \mathbf{A}(\mathbf{q}) \| \leq a_1$

$\| \mathbf{C}(\mathbf{q}, \dot{\mathbf{q}}) \| \leq a_4 \| \dot{\mathbf{q}} \|$

$\| \mathbf{Q} \| \leq a_6$

h) a robot is a passive system which dissipates energy. This property is related to property e).

9.3.4. *Considering friction*

Friction plays a dominant role in limiting the quality of robot performance. Non-compensated friction produces static error, delay, and limit cycle behavior [Canudas de Wit 90]. Many works have been devoted to studying friction torque in the joint and transmission systems. Various friction models have been proposed in the literature [Dahl 77], [Canudas de Wit 89], [Armstrong 88], [Armstrong 91], [Armstrong 94]. In general, three kinds of frictions are noted: Coulomb friction, static friction, and viscous friction.

The model based on Coulomb friction assumes a constant friction component that is independent of the magnitude of the velocity. The static friction is the torque necessary to initiate motion from rest. It is often greater than the Coulomb friction (Figure 9.2a). The viscous friction is generally represented as being proportional to the velocity, but experimental studies [Armstrong 88] have pointed out the Stribeck phenomenon that arises from the use of fluid lubrication. It results in decreasing friction with increasing velocity at low velocity, then the friction becomes proportional to velocity (Figure 9.2b). A general friction model describing these components is given by:

$$\Gamma_{fi} = F_{ci}\,\mathrm{sign}(\dot{q}_i) + F_{vi}\,\dot{q}_i + (F_{sti} - F_{ci})\,\mathrm{sign}(\dot{q}_i)\,e^{-|\dot{q}_i|B_i} \qquad [9.26]$$

In this expression, Γ_{fi} denotes the friction torque of joint i, F_{ci} and F_{vi} indicate the Coulomb and viscous friction parameters respectively. The static torque is equal to $F_{sti}\,\mathrm{sign}(\dot{q}_i)$.

The most often employed model is composed of Coulomb friction together with viscous friction (Figure 9.2c). Therefore, the friction torque at joint i is written as:

$$\Gamma_{fi} = F_{ci}\,\mathrm{sign}(\dot{q}_i) + F_{vi}\,\dot{q}_i \qquad [9.27]$$

To take into account the friction in the dynamic model of a robot, we add the vector $\mathbf{\Gamma}_f$ on the right side of equation [9.6] such that:

$$\Gamma_f = \mathbf{diag}(\dot{\mathbf{q}})\mathbf{F}_v + \mathbf{diag}[\mathrm{sign}(\dot{\mathbf{q}})]\mathbf{F}_c \qquad [9.28]$$

where:

- $\mathbf{F}_v = [\, F_{v1} \ \cdots \ F_{vn} \,]^T$;
- $\mathbf{F}_c = [\, F_{c1} \ \cdots \ F_{cn} \,]^T$;
- $\mathbf{diag}(\dot{\mathbf{q}})$ is the diagonal matrix whose elements are the components of $\dot{\mathbf{q}}$.

This friction model can be approximated by a piecewise linear model as shown in Figure 9.2d.

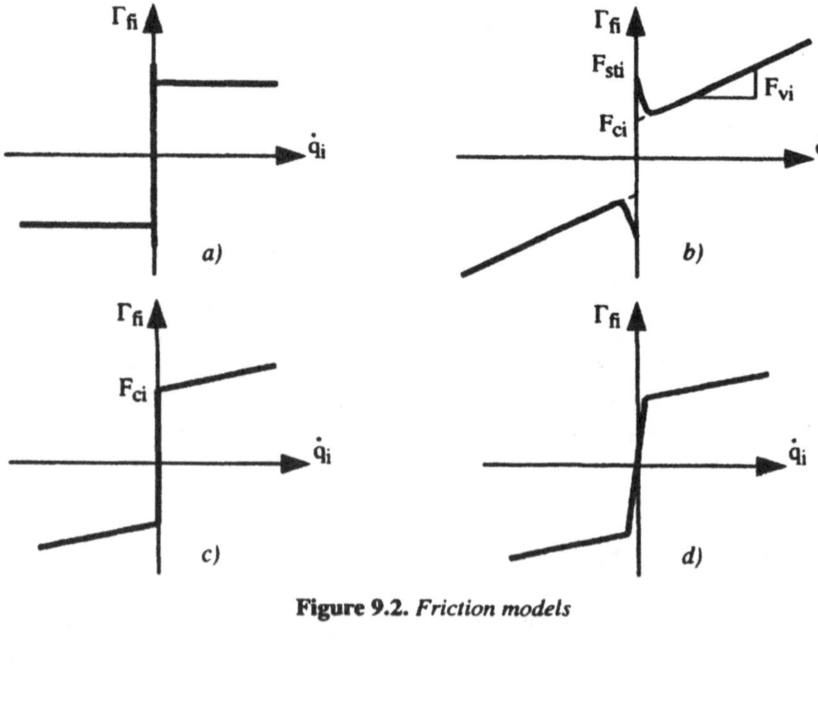

Figure 9.2. *Friction models*

9.3.5. *Considering the rotor inertia of actuators*

The kinetic energy of the rotor (and transmission system) of actuator j, is given by the expression $\frac{1}{2}\, Ia_j\, \dot{q}_j^2$. The inertial parameter Ia_j denotes the equivalent inertia referred to the joint velocity. It is given by:

$$Ia_j = N_j^2\, J_{mj} \qquad [9.29]$$

where J_{mj} is the moment of inertia of the rotor and transmissions of actuator j, N_j is the transmission ratio of joint j axis, equal to $\dot{q}_{mj} / \dot{q}_j$ where $\dot{q}_{mj}$ denotes the rotor velocity of actuator j. In the case of a prismatic joint, Ia_j is an equivalent mass.

In order to consider the rotor inertia in the dynamic model of the robot, we add the inertia (or mass) Ia_j to the A_{jj} element of the matrix **A**.

Note that such modeling neglects the gyroscopic effects of the rotors, which take place when the actuator is fixed on a moving link. However, this approximation is justified for high gear transmission ratios. For more accurate modeling of the rotors the reader is referred to [Llibre 83], [Chedmail 86], [Murphy 93], [Sciavicco 94].

9.3.6. *Considering the forces and moments exerted by the end-effector on the environment*

We have seen in § 5.9 that the joint torque vector Γ_e necessary to exert a given wrench f_{en} on the environment is obtained using the basic static equation:

$$\Gamma_e = J_n^T\, f_{en} \qquad [9.30]$$

Thus, we have to add the vector Γ_e on the right side of equation [9.6].

9.3.7. *Relation between joint torques and actuator torques*

In general, the joint variables are not equal to the motor variables because of the existence of transmission systems or because of couplings between the actuator variables. The relation between joint torques and actuator torques can be obtained using the principle of virtual work. Let the relationship between the infinitesimal joint displacement **dq** and the infinitesimal actuator variable $\mathbf{dq}_m$ be given by:

$$\mathbf{dq} = \mathbf{J}_{qm}\, \mathbf{dq}_m \qquad [9.31]$$

where $\mathbf{J}_{qm}$ is the Jacobian of $\mathbf{q}$ with respect to $\mathbf{q}_m$, equal to $\frac{\partial \mathbf{q}}{\partial \mathbf{q}_m}$.

The virtual work can be written as:

$$\boldsymbol{\Gamma}^T \mathbf{dq}^* = \boldsymbol{\tau}_m^T \mathbf{dq}_m^* \qquad [9.32]$$

where $\boldsymbol{\tau}_m$ is the actuator torque vector and the superscript (*) indicates virtual displacements.

Combining equations [9.31] and [9.32] yields:

$$\boldsymbol{\tau}_m = \mathbf{J}_{q_m}^T \boldsymbol{\Gamma} \qquad [9.33]$$

9.3.8. *Modeling of robots with elastic joints*

The presence of joint flexibility is a common feature of many current industrial robots. The joint elasticity may arise from several sources, such as elasticity in the gears, transmission belts, harmonic drives, etc. It follows that a time-varying displacement is introduced between the position of the driving actuator and the joint position. The joint elasticity is modeled as a linear torsional spring for revolute joints and a linear spring for prismatic joints [Khalil 78], [Spong 87]. Thus, the dynamic model requires twice the number of generalized coordinates to completely characterize the configuration of the links and the rotors of the actuators. Let $\mathbf{q}_M$ denote the (nx1) vector of rotor positions as reflected through the gear ratios (Figure 9.3). Consequently, the vector of joint deformations is given by $(\mathbf{q} - \mathbf{q}_M)$. Note that the direct geometric model is only a function of the joint variables $\mathbf{q}$.

The potential energy of the springs is given by:

$$U_k = \frac{1}{2} (\mathbf{q} - \mathbf{q}_M)^T \mathbf{k} (\mathbf{q} - \mathbf{q}_M) \qquad [9.34]$$

where $\mathbf{k}$ is the (nxn) definite positive joint stiffness matrix.

The dynamic equations are obtained using the Lagrange equation, i.e.:

$$\mathbf{A}\ddot{\mathbf{q}} + \mathbf{C}\dot{\mathbf{q}} + \mathbf{Q} + \mathbf{k}(\mathbf{q} - \mathbf{q}_M) = \mathbf{0} \qquad [9.35a]$$

$$\mathbf{I}_a \ddot{\mathbf{q}}_M + \mathbf{diag}(\dot{\mathbf{q}}_M)\, \mathbf{F}_{vm} + \mathbf{diag}(\mathbf{sign}(\dot{\mathbf{q}}_M))\, \mathbf{F}_{cm} - \mathbf{k}(\mathbf{q} - \mathbf{q}_M) = \boldsymbol{\Gamma} \qquad [9.35b]$$

where $\mathbf{I}_a$ is the (nxn) diagonal matrix containing the inertia of the rotors, $\mathbf{F}_{vm}$ and $\mathbf{F}_{cm}$ are the (nx1) vectors containing the viscous and Coulomb parameters of the actuators and transmissions referred to the joint side. The joint friction terms can easily be included in equation [9.35a].

A general and systematic method to model systems with lumped elasticity is presented in [Khalil 00a].

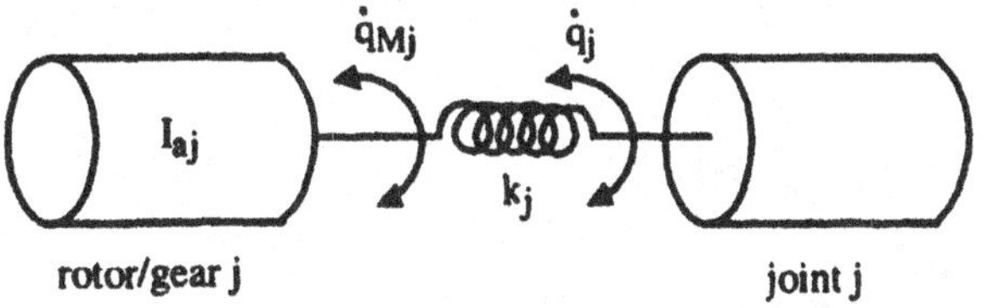

Figure 9.3. *Modeling of joint flexibility*

• **Example 9.1.** Computation of the elements of the matrices **A** and **Q** for the first three links of the Stäubli RX-90 robot whose geometric parameters are given in Example 3.1.

i) computation of the angular velocities (equation [9.17])

$${}^0\omega_0 = \mathbf{0}$$

$${}^1\omega_1 = [0 \quad 0 \quad \dot{q}_1]^T$$

$${}^2\omega_2 = {}^2\mathbf{A}_1\,{}^1\omega_1 + \dot{q}_2\,{}^2\mathbf{a}_2$$

$$= \begin{bmatrix} C2 & 0 & S2 \\ -S2 & 0 & C2 \\ 0 & -1 & 0 \end{bmatrix} \begin{bmatrix} 0 \\ 0 \\ \dot{q}_1 \end{bmatrix} + \begin{bmatrix} 0 \\ 0 \\ \dot{q}_2 \end{bmatrix} = [S2\,\dot{q}_1 \quad C2\,\dot{q}_1 \quad \dot{q}_2]^T$$

$${}^3\omega_3 = {}^3\mathbf{A}_2\,{}^2\omega_2 + \dot{q}_3\,{}^3\mathbf{a}_3$$

$$= \begin{bmatrix} C3 & S3 & 0 \\ -S3 & C3 & 0 \\ 0 & 0 & 1 \end{bmatrix} \begin{bmatrix} S2\,\dot{q}_1 \\ C2\,\dot{q}_1 \\ \dot{q}_2 \end{bmatrix} + \begin{bmatrix} 0 \\ 0 \\ \dot{q}_3 \end{bmatrix} = [S23\,\dot{q}_1 \quad C23\,\dot{q}_1 \quad \dot{q}_1 + \dot{q}_2]^T$$

ii) computation of the linear velocities (equation [9.18])

$${}^0\mathbf{V}_0 = \mathbf{0}$$

$$
\begin{aligned}
{}^{1}\mathbf{V}_1 &= \mathbf{0} \\
{}^{1}\mathbf{V}_2 &= {}^{1}\mathbf{V}_1 + {}^{1}\boldsymbol{\omega}_1 \times {}^{1}\mathbf{P}_2 = \mathbf{0} \\
{}^{2}\mathbf{V}_2 &= \mathbf{0} \\
{}^{2}\mathbf{V}_3 &= {}^{2}\boldsymbol{\omega}_2 \times {}^{2}\mathbf{P}_3 = [0 \quad D3\,\dot{q}_2 \quad -C2D3\,\dot{q}_1]^T \\
{}^{3}\mathbf{V}_3 &= {}^{3}\mathbf{A}_2\,{}^{2}\mathbf{V}_3 = [S3\,D3\,\dot{q}_2 \quad C3\,D3\,\dot{q}_2 \quad -C2D3\,\dot{q}_1]^T
\end{aligned}
$$

iii) computation of the inertia matrix **A**. With the following general inertial parameters:

$$
{}^{j}\mathbf{MS}_j = [MX_j \quad MY_j \quad MZ_j]^T
$$

$$
{}^{j}\mathbf{J}_j = \begin{bmatrix} XX_j & XY_j & XZ_j \\ XY_j & YY_j & YZ_j \\ XZ_j & YZ_j & ZZ_j \end{bmatrix}, \quad \mathbf{I}_a = \begin{bmatrix} I_{a1} & 0 & 0 \\ 0 & I_{a2} & 0 \\ 0 & 0 & I_{a3} \end{bmatrix}
$$

we obtain the elements of the robot inertia matrix as:

$$
\begin{aligned}
A_{11} &= Ia_1 + ZZ_1 + SS2\,XX_2 + 2CS2\,XY_2 + CC2\,YY_2 + SS23\,XX_3 + 2CS23 \\
&\quad XY_3 + CC23\,YY_3 + 2C2\,C23\,D3\,MX_3 - 2C2\,S23\,D3\,MY_3 + CC2\,D3^2 \\
&\quad M_3 \\
A_{12} &= S2\,XZ_2 + C2\,YZ_2 + S23\,XZ_3 + C23\,YZ_3 - S2\,D3\,MZ_3 \\
A_{13} &= S23\,XZ_3 + C23\,YZ_3 \\
A_{22} &= Ia_2 + ZZ_2 + ZZ_3 + 2C3\,D3\,MX_3 - 2S3\,D3\,MY_3 + D3^2\,M_3 \\
A_{23} &= ZZ_3 + C3\,D3\,MX_3 - S3\,D3\,MY_3 \\
A_{33} &= Ia_3 + ZZ_3
\end{aligned}
$$

where $SSj = (\sin\theta_j)^2$, $CCj = (\cos\theta_j)^2$ and $CSj = \cos\theta_j \sin\theta_j$. The elements of the matrix **C** can be computed by equation [9.7];

iv) computation of the gravity forces. Assuming that gravitational acceleration is given as ${}^{0}\mathbf{g} = [0 \quad 0 \quad G3]^T$, and using equation [9.25], the potential energy is obtained as:

$$
U = -G3\,(MZ_1 + S2MX_2 + C2MY_2 + S23MX_3 + C23MY_3 + D3S2M_3)
$$

Using equation [9.8] gives:

$$
\begin{aligned}
Q_1 &= 0 \\
Q_2 &= -G3\,(C2\,MX_2 - S2\,MY_2 + C23\,MX_3 - S23\,MY_3 + D3\,C2\,M_3)
\end{aligned}
$$

$$Q_3 = -G3\,(C23\,MX_3 - S23\,MY_3)$$

9.4. Determination of the base inertial parameters

In this section, we study the concept of *base inertial parameters* or *identifiable parameters*. We develop a straightforward closed-form method to determine them. These parameters constitute the minimum set of inertial parameters that are needed to compute the dynamic model of a robot [Mayeda 90]. The use of the base inertial parameters in a customized Newton-Euler algorithm reduces its computational cost [Khalil 86c], [Khalil 87b]. The determination of the base parameters is also essential for the identification of the dynamic parameters (Chapter 12), since they constitute the only identifiable parameters. The base inertial parameters can be deduced from the standard parameters by eliminating those that have no effect on the dynamic model and by grouping some others.

9.4.1. *Computation of the base parameters using the dynamic model*

Since the kinetic energy and the potential energy are linear in the standard inertial parameters, we deduce that the dynamic model is also linear in these parameters. It can be written as:

$$\boldsymbol{\Gamma} = \sum_{j=1}^{N_p} \mathbf{D}^j K_j = \mathbf{D}\,\mathbf{K} \qquad [9.36]$$

where:

- $\mathbf{D}$: (nxN_p) matrix, which is a function of $\mathbf{q}$, $\dot{\mathbf{q}}$, $\ddot{\mathbf{q}}$ and the geometric parameters;
- $\mathbf{K}$: (N_px1) vector of the standard inertial parameters of the links representing for each link a mass, three elements for the first moments, six elements for the inertia tensor, and one element for the rotor inertia ($N_p = 11n$);
- $\mathbf{D}^j$: j^{th} column of $\mathbf{D}$.

From equation [9.36], we deduce that:

a) a parameter K_j has no effect on the dynamic model if:

$$\mathbf{D}^j = \mathbf{0} \qquad [9.37]$$

Consequently, we can set $K_j = 0$ in equation [9.36] without changing the value of $\boldsymbol{\Gamma}$, which means that K_j can be eliminated;

b) a parameter K_j can be grouped with some other parameters $K_{j1}, \ldots, K_{jr}$, if the $\mathbf{D}^j$ column is linearly dependent of $\mathbf{D}^{j1}, \ldots, \mathbf{D}^{jr}$ such that:

$$\mathbf{D}^j = t_{j1}\, \mathbf{D}^{j1} + \ldots + t_{jr}\, \mathbf{D}^{jr} \qquad [9.38]$$

where all t_{jk} are constants.

In that case, the column $\mathbf{D}^j$ and the parameter K_j can be eliminated while the parameters $K_{j1}, \ldots, K_{jr}$, will be replaced by $KR_{j1}, \ldots, KR_{jr}$, where $KR_{jp} = K_{jp} + t_{jp} K_j$, for $p = 1, \ldots, r$. This operation will be repeated until the elimination of all the parameters with dependent columns. At the end, we obtain the minimum inertial parameter vector $\mathbf{K_B}$.

The selection of the parameters to be eliminated is not unique. We choose to eliminate those with the highest subscript in **K**. The search for dependent columns of **D** starts with the last column and moves backwards toward the first one. The link parameters are arranged in **K** such that we first place the parameters of link 1, and lastly, those of link n. The parameters of link j, defined by the vector $\mathbf{K}_j$, are given in the following order: XX_j, XY_j, XZ_j, YY_j, YZ_j, ZZ_j, MX_j, MY_j, MZ_j, M_j, Ia_j.

In summary, the calculation of $\mathbf{K_B}$ is based on the study of the linear dependence of the columns of the matrix **D**. Assuming b to be the rank of the matrix **D**, the determination of the base parameters can be illustrated in a compact and global form by writing equation [9.36] as:

$$\Gamma = [\,\mathbf{D1}\;\;\mathbf{D2}\,]\begin{bmatrix}\mathbf{K1}\\ \mathbf{K2}\end{bmatrix} \qquad [9.39]$$

where:

- **D1** represents the first b independent columns of **D**;
- **D2** represents the dependent columns of **D** such that $\mathbf{D2} = \mathbf{D1}\beta$, where β is a constant matrix.

We deduce that the parameters **K2** can be grouped with **K1** as follows:

$$\Gamma = \mathbf{D1}\,[\mathbf{K1} + \beta\, \mathbf{K2}] = \mathbf{D1}\, \mathbf{K_B} \qquad [9.40]$$

In Appendix 5, we present a general numerical method for determining the base inertial parameters [Gautier 91]. This numerical approach is based on the use of the QR decomposition. It can be used to determine the base parameters of closed chain robots and the identifiable geometric parameters of robots (Chapter 11).

9.4.2. *Determination of the base parameters using the energy model*

The use of the dynamic model to compute the base inertial parameters is tedious and error prone, owing to the complicated expressions of $\mathbf{D}^j$. In this section, we present a straightforward closed-form method for determining the base parameters. The demonstration of this method is based on the energy formulation, but the algorithm itself consists of simple rules, which do not need to calculate the energy expressions [Gautier 90b], [Khalil 94a].

Since the total energy of link j is linear in the inertial parameters, it can be written as:

$$H_j = E_j + U_j = \mathbf{h}_j \mathbf{K}_j = (\mathbf{e}_j + \mathbf{u}_j)\,{}^j\mathbf{K}_j \qquad [9.41]$$

$${}^j\mathbf{K}_j = [\; XX_j \;\; XY_j \;\; XZ_j \;\; YY_j \;\; YZ_j \;\; ZZ_j \;\; MX_j \;\; MY_j \;\; MZ_j \;\; M_j \;]^T \qquad [9.42]$$

$$\mathbf{h}_j = [\; h_{XXj} \;\; h_{XYj} \;\; h_{XZj} \;\; h_{YYj} \;\; h_{YZj} \;\; h_{ZZj} \;\; h_{MXj} \;\; h_{MYj} \;\; h_{MZj} \;\; h_{Mj} \;] \qquad [9.43]$$

with:

- $\mathbf{K}_j$: (10x1) vector of the standard inertial parameters of link j (the parameter I_{aj} will be considered in § 9.4.2.5);
- $\mathbf{h}_j$: (1x10) row vector of the total energy functions of link j;
- $\mathbf{e}_j$: (1x10) row vector of the kinetic energy functions of link j;
- $\mathbf{u}_j$: (1x10) row vector of the potential energy functions of link j.

The elements of $\mathbf{h}_j$ are obtained from equations [9.13] and [9.25] as:

$$\begin{cases} h_{XXj} = \frac{1}{2}\omega_{1,j}\,\omega_{1,j} \\ h_{XYj} = \omega_{1,j}\,\omega_{2,j} \\ h_{XZj} = \omega_{1,j}\,\omega_{3,j} \\ h_{YYj} = \frac{1}{2}\omega_{2,j}\,\omega_{2,j} \\ h_{YZj} = \omega_{2,j}\,\omega_{3,j} \\ h_{ZZj} = \frac{1}{2}\omega_{3,j}\,\omega_{3,j} \\ h_{MXj} = \omega_{3,j}\,V_{2,j} - \omega_{2,j}\,V_{3,j} - {}^0\mathbf{g}^T\,{}^0\mathbf{s}_j \\ h_{MYj} = \omega_{1,j}\,V_{3,j} - \omega_{3,j}\,V_{1,j} - {}^0\mathbf{g}^T\,{}^0\mathbf{n}_j \\ h_{MZj} = \omega_{2,j}\,V_{1,j} - \omega_{1,j}\,V_{2,j} - {}^0\mathbf{g}^T\,{}^0\mathbf{a}_j \\ h_{Mj} = \frac{1}{2}\,{}^j\mathbf{V}_j^T\,{}^j\mathbf{V}_j - {}^0\mathbf{g}^T\,{}^0\mathbf{P}_j \end{cases} \qquad [9.44]$$

where ${}^j\omega_j = [\omega_{1,j} \quad \omega_{2,j} \quad \omega_{3,j}]^T$ and ${}^j\mathbf{V}_j = [V_{1,j} \quad V_{2,j} \quad V_{3,j}]^T$.

From equations [9.13] and [9.25], we deduce the following recursive relationship between the energy functions of link j and link j – 1 (Appendix 6):

$$\mathbf{h}_j = \mathbf{h}_{j-1}\,{}^{j-1}\boldsymbol{\lambda}_j + \dot{q}_j\,\boldsymbol{\eta}_j \qquad [9.45]$$

where ${}^{j-1}\boldsymbol{\lambda}_j$ is a (10x10) matrix whose elements are given in Table 9.1. It is a function of the geometric parameters of frame R_j [Gautier 90a]. The matrix ${}^{j-1}\boldsymbol{\lambda}_j$ represents the transformation matrix of the inertial parameters of a link from frame R_j into frame R_{j-1} such that:

$${}^{j-1}\mathbf{K}_j = {}^{j-1}\boldsymbol{\lambda}_j\,{}^j\mathbf{K}_j \qquad [9.46]$$

The row vector $\boldsymbol{\eta}_j$ is written as:

$$\begin{aligned}\boldsymbol{\eta}_j = {} & \bar{\sigma}_j\,[0 \quad 0 \quad \omega_{1,j} \quad 0 \quad \omega_{2,j} \quad (\omega_{3,j} - \tfrac{1}{2}\dot{q}_j) \quad V_{2,j} \quad -V_{1,j} \quad 0 \quad 0] \\ & + \sigma_j\,[0 \quad 0 \quad 0 \quad 0 \quad 0 \quad 0 \quad -\omega_{2,j} \quad \omega_{1,j} \quad 0 \quad (V_{3,j} - \tfrac{1}{2}\dot{q}_j)]\end{aligned} \qquad [9.47]$$

9.4.2.1. *Determination of the parameters having no effect on the dynamic model*

From equation [9.37], we deduce that an inertial parameter K_j has no effect on the dynamic model if the corresponding energy function h_j is constant;

$$h_j \text{ constant} \rightarrow K_j \text{ has no effect on the dynamic model} \qquad [9.48]$$

Referring to the velocity equations [9.17] and [9.18] and to the $\mathbf{h}_j$ functions [9.44], we can obtain general rules to determine the parameters that have no effect on the dynamic model [Khalil 94a]. Let us assume that r_1 is the first revolute joint and r_2 is the subsequent revolute joint whose axis is not parallel to the axis of joint r_1 (Figure 9.4). The parameters that have no effect on the dynamic model belong to the links 1, ..., r_2, owing to the restricted motion of these links. The rules allowing us to determine them are given in the general algorithm presented in § 9.4.2.4.

Table 9.1. *Expression of the elements of matrix* ${}^{j-1}\lambda_j$

CCθ	−2CSθ	0	SSθ	0	0	0	0	2r	r^2
CSθCα	(CCθ−SSθ)Cα	−CθSα	−CSθCα	SθSα	0	−dSθCα+rCθSα	−dCθCα−rSθSα	dSα	drSα
CSθSα	(CCθ−SSθ)Sα	CθCα	−CSθSα	−SθCα	0	−dSθSα−rCθCα	−dCθSα+rSθCα	−dCα	−drCα
SSθCCα	2CSθCCα	−2SθCSα	CCθCCα	−2CθCSα	SSα	2(dCθ+rSθCSα)	2(−dSθ+rCθCSα)	2rCCα	d^2+r^2CCα
SSθCSα	2CSθCSα	Sθ(CCα−SSα)	CCθCSα	Cθ(CCα−SSα)	−CSα	rSθ(SSα−CCα)	rCθ(SSα−CCα)	2rCSα	r^2CSα
SSθSSα	2CSθSSα	2SθCSα	CCθSSα	2CθCSα	CCα	2(dCθ−rSθCSα)	2(−dSθ−rCθCSα)	2rSSα	d^2+r^2SSα
0	0	0	0	0	0	Cθ	−Sθ	0	d
0	0	0	0	0	0	SθCα	CθCα	−Sα	−rSα
0	0	0	0	0	0	SθSα	CθSα	Cα	rCα
0	0	0	0	0	0	0	0	0	1

The subscript j of the geometric parameters has been omitted.

The following notations are used: $CS^* = \cos(^*)\sin(^*)$, $CC^* = \cos^2(^*)$, $SS^* = \sin^2(^*)$

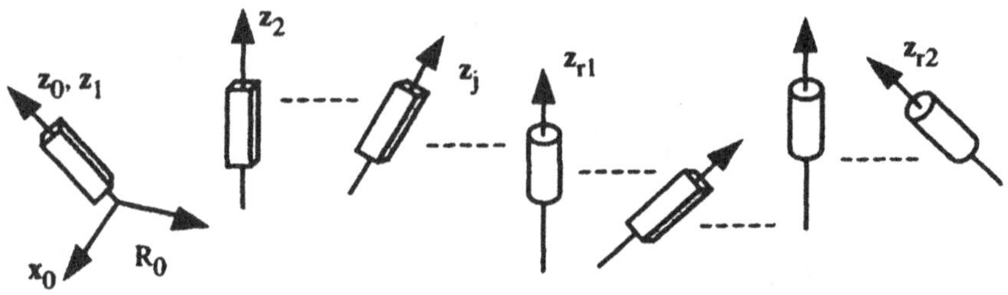

Figure 9.4. *Definition of joints r_1 and r_2*

9.4.2.2. *General grouping relations*

From condition [9.38], a parameter K_j is grouped with the parameters K_{jp} for $p = 1, \ldots, r$ if the corresponding energy function h_j can be written as:

$$h_j = \sum_{p=1}^{r} t_{jp}\, h_{jp} + \text{constant} \qquad [9.49]$$

The grouping relations are the same as in § 9.4.1, that is to say if equation [9.49] holds, then the parameter K_j can be grouped with the parameters K_{jp} for $p = 1, \ldots, r$ using the relation $KR_{jp} = K_{jp} + t_{jp} K_j$.

Using the recursive relation [9.45] between the energy functions $\mathbf{h}_j$ and $\mathbf{h}_{j-1}$, we can find general energy functions that satisfy equation [9.49]. These functions depend on the type of joint.

Let us first consider the case where joint j is revolute. The following three relations always hold:

$$h_{XXj} + h_{YYj} = \mathbf{h}_{j-1}\,({}^{j-1}\boldsymbol{\lambda}_j^1 + {}^{j-1}\boldsymbol{\lambda}_j^4) \qquad [9.50a]$$

$$h_{MZj} = \mathbf{h}_{j-1}\,{}^{j-1}\boldsymbol{\lambda}_j^9 \qquad [9.50b]$$

$$h_{Mj} = \mathbf{h}_{j-1}\,{}^{j-1}\boldsymbol{\lambda}_j^{10} \qquad [9.50c]$$

where ${}^{j-1}\boldsymbol{\lambda}_j^k$ denotes the kth column of the matrix ${}^{j-1}\boldsymbol{\lambda}_j$.

Consequently, three inertial parameters can be grouped with the others. The choice of these parameters is not unique. By choosing to group the parameters YY_j, MZ_j and M_j we obtain:

$$XXR_j = XX_j - YY_j \qquad [9.51]$$

$$\mathbf{KR}_{j-1} = \mathbf{K}_{j-1} + YY_j\,({}^{j-1}\lambda_j^1 + {}^{j-1}\lambda_j^4) + MZ_j\,{}^{j-1}\lambda_j^9 + M_j\,{}^{j-1}\lambda_j^{10} \qquad [9.52]$$

Substituting for ${}^{j-1}\lambda_j^1$, ${}^{j-1}\lambda_j^4$, ${}^{1}\lambda_j^9$, ${}^{j-1}\lambda_j^{10}$ from Table 9.1 into equations [9.51] and [9.52] gives the following theorem:

Theorem 9.1. If joint j is revolute, the parameters YY_j, MZ_j and M_j can be grouped with the parameters of link j and link j – 1. The resulting grouped parameters are:

$$\begin{aligned}
XXR_j &= XX_j - YY_j\\
XXR_{j-1} &= XX_{j-1} + YY_j + 2\,r_j\,MZ_j + r_j^2\,M_j\\
XYR_{j-1} &= XY_{j-1} + d_j\,S\alpha_j\,MZ_j + d_j\,r_j\,S\alpha_j\,M_j\\
XZR_{j-1} &= XZ_{j-1} - d_j\,C\alpha_j\,MZ_j - d_j\,r_j\,C\alpha_j\,M_j\\
YYR_{j-1} &= YY_{j-1} + CC\alpha_j\,YY_j + 2\,r_j\,CC\alpha_j\,MZ_j + (d_j^2 + r_j^2\,CC\alpha_j)\,M_j\\
YZR_{j-1} &= YZ_{j-1} + CS\alpha_j\,YY_j + 2\,r_j\,CS\alpha_j\,MZ_j + r_j^2\,CS\alpha_j\,M_j\\
ZZR_{j-1} &= ZZ_{j-1} + SS\alpha_j\,YY_j + 2\,r_jSS\alpha_j\,MZ_j + (d_j^2 + r_j^2\,SS\alpha_j)\,M_j\\
MXR_{j-1} &= MX_{j-1} + d_j\,M_j\\
MYR_{j-1} &= MY_{j-1} - S\alpha_j\,MZ_j - r_j\,S\alpha_j\,M_j\\
MZR_{j-1} &= MZ_{j-1} + C\alpha_j\,MZ_j + r_j\,C\alpha_j\,M_j\\
MR_{j-1} &= M_{j-1} + M_j
\end{aligned} \qquad [9.53]$$

where SS(*) = S(*) S(*), CC(*) = C(*) C(*) and CS(*) = C(*) S(*).

On the other hand, if joint j is prismatic, the following six relations always hold:

$$h_{XXj} = \mathbf{h}_{j-1}\,{}^{j-1}\lambda_j^1,\ h_{XYj} = \mathbf{h}_{j-1}\,{}^{j-1}\lambda_j^2,\ \ldots,\ h_{ZZj} = \mathbf{h}_{j-1}\,{}^{j-1}\lambda_j^6 \qquad [9.54]$$

Therefore, six parameters can be grouped. Choosing to group the parameters of link j with those of link j – 1 yields:

$$\mathbf{KR}_{j-1} = \mathbf{K}_{j-1} + {}^{j-1}\lambda_j^1\,XX_j + {}^{j-1}\lambda_j^2\,XY_j + \ldots + {}^{j-1}\lambda_j^6\,ZZ_j \qquad [9.55]$$

Expanding equation [9.55] gives the following theorem:

Theorem 9.2. If joint j is prismatic, the parameters of the inertia tensor of link j can be grouped with those of link j – 1. The grouping relations are:

$$\begin{aligned}
XXR_{j-1} &= XX_{j-1} + CC\theta_j\,XX_j - 2\,CS\theta_j\,XY_j + SS\theta_j\,YY_j\\
XYR_{j-1} &= XY_{j-1} + CS\theta_j\,C\alpha_j\,XX_j + (CC\theta_j - SS\theta_j)\,C\alpha_j\,XY_j - C\theta_j\,S\alpha_j\,XZ_j\\
&\quad - CS\theta_j\,C\alpha_j\,YY_j + S\theta_j\,S\alpha_j\,YZ_j
\end{aligned}$$

$$
\begin{aligned}
XZR_{j-1} &= XZ_{j-1} + CS\theta_j\, S\alpha_j\, XX_j + (CC\theta_j - SS\theta_j)\, S\alpha_j\, XY_j + C\theta_j\, C\alpha_j\, XZ_j \\
&\quad - CS\theta_j\, S\alpha_j\, YY_j - S\theta_j\, C\alpha_j\, YZ_j \\
YYR_{j-1} &= YY_{j-1} + SS\theta_j\, CC\alpha_j\, XX_j + 2CS\theta_j\, CC\alpha_j\, XY_j - 2S\theta_j\, CS\alpha_j\, XZ_j \\
&\quad + CC\theta_j\, CC\alpha_j\, YY_j - 2C\theta_j\, CS\alpha_j\, YZ_j + SS\alpha_j\, ZZ_j \\
YZR_{j-1} &= YZ_{j-1} + SS\theta_j\, CS\alpha_j\, XX_j + 2CS\theta_j\, CS\alpha_j\, XY_j + S\theta_j\, (CC\alpha_j - SS\alpha_j)\, XZ_j \\
&\quad + CC\theta_j\, CS\alpha_j\, YY_j + C\theta_j\, (CC\alpha_j - SS\alpha_j)\, YZ_j - CS\alpha_j\, ZZ_j \\
ZZR_{j-1} &= ZZ_{j-1} + SS\theta_j\, SS\alpha_j\, XX_j + 2CS\theta_j\, SS\alpha_j\, XY_j + 2S\theta_j\, CS\alpha_j\, XZ_j \\
&\quad + CC\theta_j\, SS\alpha_j\, YY_j + 2C\theta_j\, CS\alpha_j\, YZ_j + CC\alpha_j\, ZZ_j
\end{aligned}
\qquad [9.56]
$$

Equation [9.56] can be rewritten under the following matrix form:

$$
{}^{j-1}\mathbf{JR}_{j-1} = {}^{j-1}\mathbf{J}_{j-1} + {}^{j-1}\mathbf{A}_j\, {}^{j}\mathbf{J}_j\, {}^{j}\mathbf{A}_{j-1} \qquad [9.57]
$$

where ${}^{j}\mathbf{J}_j$ is the inertia tensor of link j, and ${}^{j-1}\mathbf{A}_j$ is the (3x3) orientation matrix of frame R_j relative to frame R_{j-1}.

Relation [9.57] can be also obtained by calculating the sum of the rotational kinetic energy of link j – 1 and link j in terms of the angular velocity of link j – 1, and by noting that when joint j is prismatic the angular velocity of link j is equal to the angular velocity of link j – 1.

9.4.2.3. *Particular grouped parameters*

Equations [9.53] and [9.56] allow us to compute most of the grouped inertial parameters of any serial robot. Additional grouping of inertial parameters concerns the parameters of the prismatic links between link r_1 and link r_2 (joints r_1 and r_2 are defined in § 9.4.2.1). The following two cases are considered [Khalil 94a]:

i) if the axis of the prismatic joint j, $r_1 < j < r_2$, is not parallel to the r_1 axis, then the functions h_{MXj}, h_{MYj} and h_{MZj} satisfy the relation:

$$
{}^{j}a_{xr1}\, h_{MXj} + {}^{j}a_{yr1}\, h_{MYj} + {}^{j}a_{zr1}\, h_{MZj} = \text{constant} \qquad [9.58]
$$

where ${}^{j}\mathbf{a}_{r1} = [{}^{j}a_{xr1} \quad {}^{j}a_{yr1} \quad {}^{j}a_{zr1}]^T$ is the unit vector along the $\mathbf{z}_{r1}$ axis referred to frame R_j.

Therefore, depending on the particular values of the components of ${}^{j}\mathbf{a}_{r1}$, one parameter can be eliminated or grouped as shown in Table 9.2.

Table 9.2. *Grouped parameters if $r_1 < j < r_2$, $\sigma_j = 1$ and j is not parallel to the r_1 axis*

Condition	Grouping or elimination
${}^{j}a_{zr1} \neq 0$	$MXR_j = MX_j - \frac{{}^{j}a_{xr1}}{{}^{j}a_{zr1}} MZ_j$ $MYR_j = MY_j - \frac{{}^{j}a_{yr1}}{{}^{j}a_{zr1}} MZ_j$
${}^{j}a_{zr1} = 0, {}^{j}a_{xr1}\, {}^{j}a_{yr1} \neq 0$	$MXR_j = MX_j - \frac{{}^{j}a_{xr1}}{{}^{j}a_{yr1}} MY_j$
${}^{j}a_{zr1} = 0, {}^{j}a_{xr1} = 0$	$MY_j = 0$ (has no effect)
${}^{j}a_{zr1} = 0, {}^{j}a_{yr1} = 0$	$MX_j = 0$ (has no effect)

ii) if the axis of the prismatic joint j, $r_1 < j < r_2$, is parallel to the r_1 axis, then the following relation is obtained:

$$\mathbf{h}_{MSj}^{T} = {}^{j}\mathbf{A}_{j-1}\, \mathbf{h}_{MSj-1}^{T} + \begin{bmatrix} 2\, d_j C\theta_j\, h_{ZZi} & -2 d_j S\theta_j\, h_{ZZi} & 0 \end{bmatrix}^{T} \qquad [9.59]$$

where i denotes the nearest revolute joint from j back to the base, $i \geq r_1$, and:

$$\mathbf{h}_{MSj} = \begin{bmatrix} h_{MXj} & h_{MYj} & h_{MZj} \end{bmatrix}$$

Therefore, we deduce that the parameter MZ_j has no effect on the dynamic model and that the parameters MX_j and MY_j can be grouped using the relations:

$$\begin{aligned} MXR_{j-1} &= MX_{j-1} + C\theta_j\, MX_j - S\theta_j\, MY_j \\ MYR_{j-1} &= MY_{j-1} + S\theta_j\, C\alpha_j\, MX_j + C\theta_j\, C\alpha_j\, MY_j \\ MZR_{j-1} &= MZ_{j-1} + S\theta_j\, S\alpha_j\, MX_j + C\theta_j\, S\alpha_j\, MY_j \\ ZZR_i &= ZZ_i + 2\, d_j\, C\theta_j\, MX_j - 2\, d_j\, S\theta_j\, MY_j \end{aligned} \qquad [9.60]$$

9.4.2.4. *Practical determination of the base parameters*

The following algorithm can be used to determine all the parameters that can be eliminated or grouped. The remaining parameters constitute the set of base inertial parameters of the links. The grouped relations make use of closed-form symbolic expressions.

For $j = n, ..., 1$:

1) use the general grouping relations [9.53] or [9.56] to group:
   a) YY_j, MZ_j and M_j if joint j is revolute ($\sigma_j = 0$);
   b) XX_j, XY_j, XZ_j, YY_j, YZ_j and ZZ_j if joint j is prismatic ($\sigma_j = 1$);
2) if joint j is prismatic and $\mathbf{a}_j$ is parallel to $\mathbf{a}_{r1}$, for $r_1 < j < r_2$, then eliminate MZ_j and group MX_j and MY_j using relation [9.60];
3) if joint j is prismatic and $\mathbf{a}_j$ is not parallel to $\mathbf{a}_{r1}$, for $r_1 < j < r_2$, then group or eliminate one of the parameters MX_j, MY_j, MZ_j using Table 9.2;
4) if joint j is revolute, for $r_1 \leq j < r_2$, then eliminate XX_j, XY_j, XZ_j and YZ_j. Note that the axis of this joint is parallel to the axis of joint r_1, and that the parameter YY_j has been eliminated by rule 1;
5) if joint j is revolute, for $r_1 \leq j < r_2$, and the $\mathbf{a}_j$ axis is along $\mathbf{a}_{r1}$, and if $\mathbf{a}_{r1}$ is parallel to $\mathbf{a}_i$ and to gravity $\mathbf{g}$, for all $i < j$, then eliminate the parameters MX_j, MY_j. Note that MZ_j is eliminated by rule 1;
6) if joint j is prismatic and $j < r_1$, then eliminate the parameters MX_j, MY_j, MZ_j.

From this algorithm, we deduce that the number of minimum inertial parameters of the links for a general serial robot (without considering the inertia of the rotors) is given by:

$$b_m \leq 7\, n_r + 4\, n_p - 3 - \bar{\sigma}_1 - 2\, n_{g0} \qquad [9.61]$$

with:

- n_r: number of revolute joints $= \sum \bar{\sigma}_j$;
- n_p: number of prismatic joints $= \sum \sigma_j$;
- $n_{g0} = 1$ if the first joint is revolute and parallel to gravity, $n_{g0} = 0$ otherwise.

This equation gives in most cases the exact number of base inertial parameters.

9.4.2.5. *Considering the inertia of rotors*

Considering the rotor inertial parameter Ia_j, the number of standard inertial parameters per link becomes equal to 11. The corresponding components in the matrices $\mathbf{h}_j$ and $\mathbf{K}_j$ are given as:

$$h_{11,j} = \frac{1}{2}\dot{q}_j^2 \qquad [9.62]$$

$$K_{11,j} = Ia_j \qquad [9.63]$$

To verify if a parameter Ia_j can be grouped with other parameters, let us look for the existence of a linear relationship between $h_{11,j}$ and the other energy functions $\{h_k\}$. Such a relationship always holds for $j = r_1$, and exists under some conditions for $j = r_2$, and for the first prismatic joint denoted as p_1:

i) for joint r_1, we deduce that:

$$h_{6,r1} = \frac{1}{2}\dot{q}_{r1}^2 \qquad [9.64]$$

Consequently, the parameter Ia_{r1} can be grouped with ZZ_{r1} using the relation:

$$ZZR_{r1} = ZZ_{r1} + Ia_{r1} \qquad [9.65]$$

ii) if the axis of joint r_2 is orthogonal to that of r_1, we obtain:

$$h_{6,r2} = \frac{1}{2}\dot{q}_{r2}^2 \qquad [9.66]$$

Thus, the parameter Ia_{r2} can be grouped with ZZ_{r2} using the relation:

$$ZZR_{r2} = ZZ_{r2} + Ia_{r2} \qquad [9.67]$$

iii) if the axis of the first prismatic joint p_1 is orthogonal to gravity, and if $p_1 = 1$ or its axis is aligned with the revolute axes preceding it (such that M_1 has no effect on the gravity force of joints 1, …, p_1), then we can group Ia_{p1} with M_{p1}:

$$MR_{p1} = M_{p1} + Ia_{p1} \qquad [9.68]$$

• **Example 9.2.** Find the base inertial parameters of the Stäubli RX-90 robot. For this robot, $r_1 = 1$ and $r_2 = 2$. Since all the joints are revolute, the use of equation [9.53] for j = n, …, 1 gives all the grouped parameters:

link 6:

$$XXR_6 = XX_6 - YY_6$$
$$XXR_5 = XX_5 + YY_6$$
$$ZZR_5 = ZZ_5 + YY_6$$
$$MYR_5 = MY_5 + MZ_6$$
$$MR_5 = M_5 + M_6$$

The minimum inertial parameters of link 6 are: XXR_6, XY_6, XZ_6, YZ_6, ZZ_6, MX_6 and MY_6.

link 5:

$$XXR_5 = XX_5 + YY_6 - YY_5$$
$$XXR_4 = XX_4 + YY_5$$
$$ZZR_4 = ZZ_4 + YY_5$$
$$MYR_4 = MY_4 - MZ_5$$
$$MR_4 = M_4 + MR_5 = M_4 + M_5 + M_6$$

The minimum parameters of link 5 are: XXR_5, XY_5, XZ_5, YZ_5, ZZR_5, MX_5 and MYR_5.

link4:

$$XXR_4 = XX_4 + YY_5 - YY_4$$
$$XXR_3 = XX_3 + YY_4 + 2\,RL4\,MZ_4 + RL4^2\,(M_4 + M_5 + M_6)$$
$$ZZR_3 = ZZ_3 + YY_4 + 2\,RL4\,MZ_4 + RL4^2\,(M_4 + M_5 + M_6)$$
$$MYR_3 = MY_3 + MZ_4 + RL4\,(M_4 + M_5 + M_6)$$
$$MR_3 = M_3 + M_4 + M_5 + M_6$$

The minimum parameters of link 4 are: XXR_4, XY_4, XZ_4, YZ_4, ZZR_4, MX_4 and MYR_4.

link 3:

$$XXR_3 = XX_3 + YY_4 + 2\,RL4\,MZ_4 + RL4^2\,(M_4 + M_5 + M_6) - YY_3$$
$$XXR_2 = XX_2 + YY_3$$
$$XZR_2 = XZ_2 - D3\,MZ_3$$
$$YYR_2 = YY_2 + D3^2\,(M_3 + M_4 + M_5 + M_6) + YY_3$$
$$ZZR_2 = ZZ_2 + D3^2\,(M_3 + M_4 + M_5 + M_6)$$
$$MXR_2 = MX_2 + D3\,(M_3 + M_4 + M_5 + M_6)$$
$$MZR_2 = MZ_2 + MZ_3$$
$$MR_2 = M_2 + M_3 + M_4 + M_5 + M_6$$

The minimum parameters of link 3 are: XXR_3, XY_3, XZ_3, YZ_3, ZZR_3, MX_3 and MYR_3.

link 2:

$$XXR_2 = XX_2 - YY_2 - D3^2\,(M_3 + M_4 + M_5 + M_6)$$
$$ZZR_1 = ZZ_1 + YY_2 + D3^2\,(M_3 + M_4 + M_5 + M_6) + YY_3$$

The minimum parameters of link 2 are: XXR_2, XY_2, XZR_2, YZ_2, ZZR_2, MXR_2 and MY_2.

link 1: from § 9.4.2.4 (rules 4 and 5), we deduce that the parameters XX_1, XY_1, XZ_1, YY_1, YZ_1, MX_1, MY_1, MZ_1 and M_1 have no effect on the dynamic model. Note that MZ_2 and M_2 have no effect because they are grouped with parameters having no effect. The only parameter of link 1 is ZZR_1.

Rotor inertia. From § 9.4.2.5, we can group the parameter Ia_1 with ZZ_1 and the parameter Ia_2 with ZZ_2:

$$ZZR_1 = ZZ_1 + Ia_1 + YY_2 + D3^2 (M_3 + M_4 + M_5 + M_6) + YY_3$$
$$ZZR_2 = ZZ_2 + Ia_2 + D3^2 (M_3 + M_4 + M_5 + M_6)$$

The final result can be summarized as follows:

- the parameters that have no effect on the dynamic model are: XX_1, XY_1, XZ_1, YY_1, YZ_1, MX_1, MY_1, MZ_1, M_1, MZ_2 and M_2;
- the parameters that are grouped are: Ia_1, YY_2, Ia_2, YY_3, MZ_3, M_3, YY_4, MZ_4, M_4,YY_5, MZ_5, M_5, YY_6, MZ_6 and M_6;
- the grouping equations are:

$$ZZR_1 = ZZ_1 + Ia_1 + YY_2 + D3^2 (M_3 + M_4 + M_5 + M_6) + YY_3$$
$$XXR_2 = XX_2 - YY_2 - D3^2 (M_3 + M_4 + M_5 + M_6)$$
$$XZR_2 = XZ_2 - D3\, MZ_3$$
$$ZZR_2 = ZZ_2 + Ia_2 + D3^2 (M_3 + M_4 + M_5 + M_6)$$
$$MXR_2 = MX_2 + D3 (M_3 + M_4 + M_5 + M_6)$$
$$XXR_3 = XX_3 - YY_3 + YY_4 + 2\, RL4\, MZ_4 + RL4^2 (M_4 + M_5 + M_6)$$
$$ZZR_3 = ZZ_3 + YY_4 + 2RL4\, MZ_4 + RL4^2 (M_4 + M_5 + M_6)$$
$$MYR_3 = MY_3 + MZ_4 + RL4 (M_4 + M_5 + M_6)$$
$$XXR_4 = XX_4 + YY_5 - YY_4$$
$$ZZR_4 = ZZ_4 + YY_5$$
$$MYR_4 = MY_4 - MZ_5$$
$$XXR_5 = XX_5 + YY_6 - YY_5$$
$$ZZR_5 = ZZ_5 + YY_6$$
$$MYR_5 = MY_5 + MZ_6$$
$$XXR_6 = XX_6 - YY_6$$

Table 9.3 gives the 40 base inertial parameters of the Stäubli RX-90 robot.

• **Example 9.3.** Let us assume that the inertia tensors ${}^{j}J_j$, for j = 1, ..., 6, are diagonal and that the first moments of the links are given by:

$${}^{1}MS_1 = [\,0 \quad 0 \quad MZ1\,]^T$$
$${}^{2}MS_2 = [\,MX2 \quad MY2 \quad 0\,]^T$$

$^3\mathbf{MS}_3 = [\,0 \;\; MY3 \;\; 0\,]^T$
$^4\mathbf{MS}_4 = [\,0 \;\; 0 \;\; MZ4\,]^T$
$^5\mathbf{MS}_5 = [\,0 \;\; MY5 \;\; 0\,]^T$
$^6\mathbf{MS}_6 = [\,0 \;\; 0 \;\; MZ6\,]^T$

The corresponding 19 base inertial parameters are given in Table 9.4. They are derived using the general grouping relations after eliminating the parameters whose values are zero.

Table 9.3. *Base inertial parameters of the Stäubli RX-90 robot*

j	XX_j	XY_j	XZ_j	YY_j	YZ_j	ZZ_j	MX_j	MY_j	MZ_j	M_j	Ia_j
1	0	0	0	0	0	ZZR_1	0	0	0	0	0
2	XXR_2	XY_2	XZR_2	0	YZ_2	ZZR_2	MXR_2	MY_2	0	0	0
3	XXR_3	XY_3	XZ_3	0	YZ_3	ZZR_3	MX_3	MYR_3	0	0	Ia_3
4	XXR_4	XY_4	XZ_4	0	YZ_4	ZZR_4	MX_4	MYR_4	0	0	Ia_4
5	XXR_5	XY_5	XZ_5	0	YZ_5	ZZR_5	MX_5	MYR_5	0	0	Ia_5
6	XXR_6	XY_6	XZ_6	0	YZ_6	ZZ_6	MX_6	MY_6	0	0	Ia_6

Table 9.4. *Simplified base inertial parameters for the Stäubli RX-90 robot*

j	XX_j	XY_j	XZ_j	YY_j	YZ_j	ZZ_j	MX_j	MY_j	MZ_j	M_j	Ia_j
1	0	0	0	0	0	ZZR_1	0	0	0	0	0
2	XXR_2	0	0	0	0	ZZR_2	MXR_2	MY_2	0	0	0
3	XXR_3	0	0	0	0	ZZR_3	0	MYR_3	0	0	Ia_3
4	XXR_4	0	0	0	0	ZZR_4	0	0	0	0	Ia_4
5	XXR_5	0	0	0	0	ZZR_5	0	MYR_5	0	0	Ia_5
6	XXR_6	0	0	0	0	ZZ_6	0	0	0	0	Ia_6

9.5. Newton-Euler formulation

9.5.1. *Introduction*

The Newton-Euler equations describing the forces and moments (wrench) acting on the center-of-mass of link j are given as:

$$\mathbf{F}_j = M_j \dot{\mathbf{V}}_{Gj} \qquad [9.69]$$

$$\mathbf{M}_{Gj} = \mathbf{I}_{Gj} \dot{\omega}_j + \omega_j \times (\mathbf{I}_{Gj} \omega_j) \qquad [9.70]$$

The Newton-Euler algorithm of Luh, Walker and Paul [Luh 80b], which is considered as one of the most efficient algorithms for real time computation of the inverse dynamic model, consists of two recursive computations: forward recursion and backward recursion. The forward recursion, from the base to the terminal link, computes the link velocities and accelerations and consequently the dynamic wrench on each link. The backward recursion, from the terminal link to the base, provides the reaction wrenches on the links and consequently the joint torques.

This method gives the joint torques as defined by equation [9.1] without explicitly computing the matrices **A**, **C** and **Q**. The model obtained is not linear in the inertial parameters because it makes use of M_j, S_j and $\mathbf{I}_{Gj}$.

9.5.2. *Newton-Euler inverse dynamics linear in the inertial parameters*

In this section, we develop a Newton-Euler algorithm based on the double recursive computations of Luh et al. [Luh 80b], but which uses as inertial parameters the elements of $\mathbf{J}_j$, $\mathbf{MS}_j$ and M_j [Khalil 87b], [Khosla 86]. The dynamic wrench on link j is calculated on O_j and not on the center of gravity G_j. Therefore, the resulting model is linear in the dynamic parameters. This reformulation allows us to compute the dynamic model in terms of the base inertial parameters and to use it for the identification of the dynamic parameters.

The Newton-Euler equations giving the forces and moments of link j at the origin of frame R_j are given as:

$$\mathbf{F}_j = M_j \dot{\mathbf{V}}_j + \dot{\omega}_j \times \mathbf{MS}_j + \omega_j \times (\omega_j \times \mathbf{MS}_j) \qquad [9.71]$$

$$\mathbf{M}_j = \mathbf{J}_j \dot{\omega}_j + \omega_j \times (\mathbf{J}_j \omega_j) + \mathbf{MS}_j \times \dot{\mathbf{V}}_j \qquad [9.72]$$

Using the spatial notation, we can write these equations as:

$$\mathbb{F}_j = \mathbb{J}_j \dot{\mathbb{V}}_j + \begin{bmatrix} \omega_j \times (\omega_j \times \mathbf{MS}_j) \\ \omega_j \times (\mathbf{J}_j \, \omega_j) \end{bmatrix} \quad [9.73]$$

where $\mathbb{F}_j = \begin{bmatrix} \mathbf{F}_j \\ \mathbf{M}_j \end{bmatrix}$, $\dot{\mathbb{V}} = \begin{bmatrix} \dot{\mathbf{V}} \\ \dot{\omega} \end{bmatrix}$, $\mathbb{J}_j$ is the spatial inertia matrix (equation [9.21]).

i) forward recursive computation: to compute $\mathbf{F}_j$ and $\mathbf{M}_j$, for j = 1, ..., n, using equations [9.71] and [9.72], we need ω_j, $\dot{\omega}_j$ and $\dot{\mathbf{V}}_j$. The velocities are given by the recursive equations [9.14] and [9.15] rewritten hereafter as:

$$\omega_j = \omega_{j-1} + \bar{\sigma}_j \, \dot{q}_j \, \mathbf{a}_j \quad [9.74]$$

$$\mathbf{V}_j = \mathbf{V}_{j-1} + \omega_{j-1} \times \mathbf{L}_j + \sigma_j \, \dot{q}_j \, \mathbf{a}_j \quad [9.75]$$

Differentiating equations [9.74] and [9.75] with respect to time gives:

$$\dot{\omega}_j = \dot{\omega}_{j-1} + \bar{\sigma}_j \, (\ddot{q}_j \, \mathbf{a}_j + \omega_{j-1} \times \dot{q}_j \, \mathbf{a}_j) \quad [9.76]$$

$$\dot{\mathbf{V}}_j = \dot{\mathbf{V}}_{j-1} + \dot{\omega}_{j-1} \times \mathbf{L}_j + \omega_{j-1} \times (\omega_{j-1} \times \mathbf{L}_j) + \sigma_j \, (\ddot{q}_j \, \mathbf{a}_j + 2 \, \omega_{j-1} \times \dot{q}_j \, \mathbf{a}_j) \quad [9.77]$$

The initial conditions for a robot with a fixed base are $\omega_0 = \mathbf{0}$, $\dot{\omega}_0 = \mathbf{0}$ and $\dot{\mathbf{V}}_0 = \mathbf{0}$;

ii) backward recursive computation: this is based on writing for each link j, for j = n,...,1, the Newton-Euler equations at the origin O_j, as follows (Figure 9.5):

$$\mathbf{F}_j = \mathbf{f}_j - \mathbf{f}_{j+1} + M_j \, \mathbf{g} - \mathbf{f}_{ej} \quad [9.78]$$

$$\mathbf{M}_j = \mathbf{m}_j - \mathbf{m}_{j+1} - \mathbf{L}_{j+1} \times \mathbf{f}_{j+1} + \mathbf{S}_j \times M_j \, \mathbf{g} - \mathbf{m}_{ej} \quad [9.79]$$

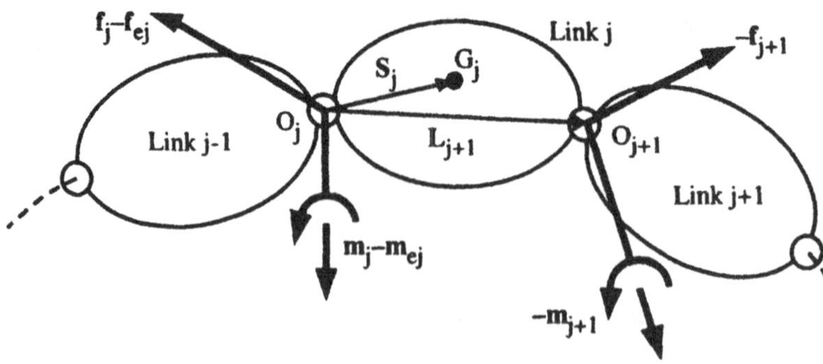

Figure 9.5. *Forces and moments on link j*

We note that $\mathbf{f}_{ej}$ and $\mathbf{m}_{ej}$, which represent the force and moment exerted by link j on the environment, may include contributions from springs, dampers, contact with the environment, etc. Their values are assumed to be known, or at least to be calculated from known quantities.

We can cancel the gravity terms from equations [9.78] and [9.79] and take into account their effects by setting up the initial linear acceleration such that:

$$\dot{\mathbf{V}}_0 = -\mathbf{g} \qquad [9.80]$$

Thus, using equations [9.78] and [9.79], we obtain:

$$\mathbf{f}_j = \mathbf{F}_j + \mathbf{f}_{j+1} + \mathbf{f}_{ej} \qquad [9.81]$$

$$\mathbf{m}_j = \mathbf{M}_j + \mathbf{m}_{j+1} + \mathbf{L}_{j+1} \times \mathbf{f}_{j+1} + \mathbf{m}_{ej} \qquad [9.82]$$

This backward recursive algorithm is initialized by $\mathbf{f}_{n+1} = \mathbf{0}$ and $\mathbf{m}_{n+1} = \mathbf{0}$.

Finally, the joint torque Γ_j can be obtained by projecting $\mathbf{f}_j$ or $\mathbf{m}_j$ on the joint axis, depending whether the joint is prismatic or revolute respectively. We can also consider the friction forces and the rotor inertia as shown in the Lagrange method:

$$\Gamma_j = (\sigma_j\, \mathbf{f}_j + \bar{\sigma}_j\, \mathbf{m}_j)^T \mathbf{a}_j + F_{cj}\, \mathrm{sign}(\dot{q}_j) + F_{vj}\, \dot{q}_j + Ia_j\, \ddot{q}_j \qquad [9.83]$$

From equations [9.81], [9.82] and [9.83], we deduce that Γ_j is a function of the inertial parameters of links j, j+1, ..., n. This property has been outlined in § 9.3.3.3.

9.5.3. *Practical form of the Newton-Euler algorithm*

Since $\mathbf{J}_j$ and $\mathbf{MS}_j$ are constants when referred to their own link coordinates, the Newton-Euler algorithm can be efficiently computed by referring the velocities, accelerations, forces and moments to the local link coordinate system [Luh 80b]. The forward recursive equations become, for j = 1, ..., n:

$$^j\boldsymbol{\omega}_{j-1} = {}^j\mathbf{A}_{j-1}\, {}^{j-1}\boldsymbol{\omega}_{j-1} \qquad [9.84]$$

$$^j\boldsymbol{\omega}_j = {}^j\boldsymbol{\omega}_{j-1} + \bar{\sigma}_j\, \dot{q}_j\, {}^j\mathbf{a}_j \qquad [9.85]$$

$$^j\dot{\boldsymbol{\omega}}_j = {}^j\mathbf{A}_{j-1}\, {}^{j-1}\dot{\boldsymbol{\omega}}_{j-1} + \bar{\sigma}_j\, (\ddot{q}_j\, {}^j\mathbf{a}_j + {}^j\boldsymbol{\omega}_{j-1} \times \dot{q}_j\, {}^j\mathbf{a}_j) \qquad [9.86]$$

$$^j\dot{\mathbf{V}}_j = {}^j\mathbf{A}_{j-1}\, ({}^{j-1}\dot{\mathbf{V}}_{j-1} + {}^{j-1}\mathbf{U}_{j-1}\, {}^{j-1}\mathbf{P}_j) + \sigma_j\, (\ddot{q}_j\, {}^j\mathbf{a}_j + 2\, {}^j\boldsymbol{\omega}_{j-1} \times \dot{q}_j\, {}^j\mathbf{a}_j) \qquad [9.87]$$

$$^j\mathbf{F}_j = M_j\, {}^j\dot{\mathbf{V}}_j + {}^j\mathbf{U}_j\, {}^j\mathbf{MS}_j \qquad [9.88]$$

$$^j\mathbf{M}_j = {}^j\mathbf{J}_j\, {}^j\dot{\boldsymbol{\omega}}_j + {}^j\boldsymbol{\omega}_j \times ({}^j\mathbf{J}_j\, {}^j\boldsymbol{\omega}_j) + {}^j\mathbf{MS}_j \times {}^j\dot{\mathbf{V}}_j \qquad [9.89]$$

$${}^{j}U_j = {}^{j}\hat{\dot{\omega}}_j + {}^{j}\hat{\omega}_j\,{}^{j}\hat{\omega}_j \qquad [9.90]$$

For a stationary base, the initial conditions are $\boldsymbol{\omega}_0 = \mathbf{0}$, $\dot{\boldsymbol{\omega}}_0 = \mathbf{0}$ and $\dot{\mathbf{V}}_0 = -\mathbf{g}$.

The use of ${}^{j}\mathbf{U}_j$ saves 21n multiplications and 6n additions in the computation of the inverse dynamic model of a general robot [Kleinfinger 86a].
It is to be noted that ${}^{j}\dot{\mathbf{V}}_j$ means ${}^{j}\mathbf{A}_0{}^{0}\dot{\mathbf{V}}_j$, and not the time derivative of ${}^{j}\mathbf{V}_j$, since ${}^{j}\dot{\mathbf{V}}_j = \frac{d}{dt}{}^{j}\mathbf{V}_j + {}^{j}\boldsymbol{\omega}_j \times {}^{j}\mathbf{V}_j$. On the contrary, ${}^{j}\dot{\boldsymbol{\omega}}_j$ is equal to the time derivative of ${}^{j}\boldsymbol{\omega}_j$.

The backward recursive equations, for j = n, ..., 1, are:

$${}^{j}\mathbf{f}_j = {}^{j}\mathbf{F}_j + {}^{j}\mathbf{f}_{j+1} + {}^{j}\mathbf{f}_{ej} \qquad [9.91]$$

$${}^{j-1}\mathbf{f}_j = {}^{j-1}\mathbf{A}_j\,{}^{j}\mathbf{f}_j \qquad [9.92]$$

$${}^{j}\mathbf{m}_j = {}^{j}\mathbf{M}_j + {}^{j}\mathbf{A}_{j+1}\,{}^{j+1}\mathbf{m}_{j+1} + {}^{j}\mathbf{P}_{j+1} \times {}^{j}\mathbf{f}_{j+1} + {}^{j}\mathbf{m}_{ej} \qquad [9.93]$$

$$\Gamma_j = (\sigma_j\,{}^{j}\mathbf{f}_j + \bar{\sigma}_j\,{}^{j}\mathbf{m}_j)^T\,{}^{j}\mathbf{a}_j + F_{sj}\,\mathrm{sign}(\dot{q}_j) + F_{vj}\,\dot{q}_j + Ia_j\,\ddot{q}_j \qquad [9.94]$$

The previous algorithm can be numerically programmed for a general serial robot. Its computational complexity is $\mathcal{O}(n)$, which means that the number of operations is linear in the number of degrees of freedom. However, as we will see in the next section, the use of the base inertial parameters in a customized symbolic algorithm considerably reduces the number of operations of the dynamic model.

9.6. Real time computation of the inverse dynamic model

9.6.1. *Introduction*

Much work has been focused on the problem of computational efficiency of the inverse dynamic model of robots in order to realize real time dynamic control. This objective is now recognized as being attained (Table 9.5).

Before describing our method, which is based on a customized symbolic Newton-Euler formulation linear in the inertial parameters [Khalil 87b], [Kleinfinger 86a], we review the main approaches presented in the literature.

The Lagrangian formulation of Uicker and Kahn [Uicker 69], [Kahn 69] served as a standard robot dynamics formulation for over a decade. In this form, the complexity of the equations precluded the real time computation of the inverse dynamic model. For example, for a six degree-of-freedom robot, this formulation requires 66271 multiplications and 51548 additions [Hollerbach 80]. This led researchers to investigate either simplification or tabulation approaches to achieve real time implementation.

The first approach towards simplification is to neglect the Coriolis and centrifugal terms with the assumption that they are not significant except at high speeds [Paul 72], [Bejczy 74]. Unfortunately, this belief is not valid: firstly, Luh et al. [Luh 80b] have shown that such simplifications leads to large errors not only in the value of the computed joint torques but also in its sign; later, Hollerbach [Hollerbach 84a] has shown that the velocity terms have the same significance relative to the acceleration terms whatever the velocity.

An alternative simplification approach has been proposed by Bejczy [Bejczy 79]. This approach is based on an exhaustive term-by-term analysis of the elements A_{ij}, $C_{i,jk}$ and Q_i so that the most significant terms are only retained. A similar procedure has been utilized by Armstrong et al. [Armstrong 86] who also proposed computing the elements A_{ij}, $C_{i,jk}$ and Q_i with a low frequency rate with respect to that of the servo rate. Using such an analysis for a six degree-of-freedom robot becomes very laborious and tedious.

In the tabulation approach, some terms of the dynamic equations are precomputed and tabulated. The combination of a look-up table with reduced analytical computations renders them feasible in real time. Two methods based on a trade-off between memory space and computation time have been investigated by Raibert [Raibert 77]. In the first method *SSM (State Space Model)*, the table was carried out as a function of the joint positions and velocities ($\mathbf{q}$ and $\dot{\mathbf{q}}$), but the required memory was too big to consider in real applications at that time. In the *Configuration Space Method (CSM)*, the table is computed as a function of the joint positions. Another technique, proposed by Aldon [Aldon 82] consists of tabulating the elements A_{ij} and Q_i and of calculating the elements $C_{i,jk}$ on-line in terms of the A_{ij} elements. This method considerably reduces the required memory but increases the number of on-line operations, which becomes proportional to n^3. We note that the tabulated elements are functions of the load inertial parameters, which means making a table for each load.

Luh et al. [Luh 80a] proposed to determine the inverse dynamic model using a Newton-Euler formulation (§ 9.5). The complexity of this method is $\mathcal{O}(n)$. This method has proved the importance of the recursive computations and the organization of the different steps of the dynamic algorithm. Therefore, other researchers tried to improve the existing Lagrange formulations by introducing recursive computations. For example, Hollerbach [Hollerbach 80] proposed a new recursive Lagrange formulation with complexity $\mathcal{O}(n)$, whereas Megahed [Megahed 84] developed a new recursive computational procedure for the Lagrange method of Uicker and Kahn. However, these methods are less efficient than the Newton-Euler formulation of Luh et al. [Luh 80a].

More recently, researchers investigated alternative formulations [Kane 83], [Vukobratovic 85], [Renaud 85], [Kazerounian 86], but the recursive Newton-Euler proved to be computationally more efficient.

The most efficient models proposed until now are based on a customized symbolic Newton-Euler formulation that takes into account the particularity of the geometric and inertial parameters of each robot [Kanade 84], [Khalil 85a], [Khalil 87b], [Renaud 87]. Moreover, the use of the base inertial parameters in this algorithm reduces the computational cost by about 25%. We note that the number of operations for this method is even less than that of the tabulated CSM method.

Table 9.5. *Computational cost of the inverse dynamic modeling methods*

Method	Robot	General case		2RP(3R) Stanford		R(2P)(3R) TH8		6R Stäubli RX-90	
	Operations	n ddl	n=6	General	Simplified*	General	Simplified*	General	Simplified*
Raibert 78**	Mult.	n^3+2n^2	288	288	288	288	288	288	288
	Add.	3^n+n^2	252	252	252	252	252	252	252
Luh 80b	Mult.	137n-22	800	800	800	800	800	800	800
	Add.	101n-11	595	595	595	595	595	595	595
Hollerbach 80**	Mult.	412n-277	2195	2195	2195	2195	2195	2195	2195
	Add.	320n-201	1719	1719	1719	1719	1719	1719	1719
Kane 83**	Mult.	?	?	646	?	?	?	?	?
	Add.	?	?	394	?	?	?	?	?
Vukobratovic 85**	Mult.	?	?	?	>372	?	>193	?	?
	Add.	?	?	?	>167	?	>80	?	?
Renaud 85**	Mult.	?	?	?	?	?	?	?	368
	Add.	?	?	?	?	?	?	?	271
Presented method Khalil [87b]	Mult.	92n-127	425	240	142	175	104	253	159
	Add.	81n-117	369	227	99	162	72	238	113

? the number of operations is not given.

* the matrix $\mathbf{J}_j$ is diagonal and two components of the first moments are zero.

** forces and moments exerted by the end-effector on the environment are not considered.

> the given number of operations corresponds to the computation of the elements of **A**, $C_{i,jk}$ and **Q**.

Before closing this section, it is worth noting the formidable technological progress in the field of computer processors, to the point that the dynamic model can be calculated at control rate with standard personal computers (Chapter 14).

9.6.2. *Customization of the Newton-Euler formulation*

The recursive Newton-Euler formulation of robot dynamics has become a standard algorithm for real time control and simulation (§ 9.5.3). To increase the efficiency of the Newton-Euler algorithm, we generate a customized symbolic model for each specific robot and utilize the base dynamic parameters. To obtain this model, we expand the recursive equations to transform them into scalar equations without incorporating loop computations. The elements of a vector or a matrix containing at least one mathematical operation are replaced by an intermediate variable. This variable is written in the output file, which contains the customized model [Kanade 84], [Khalil 85a]. The elements that do not contain any operation are not modified. We propagate the obtained vectors and matrices in the subsequent equations. Consequently, customizing eliminates multiplications by one (and minus one) and zero, and additions with zero. A good choice of the intermediate variables allows us to avoid redundant computations. In the end, the dynamic model is obtained as a set of intermediate variables. Those that have no effect on the desired output, the joint torques in the case of inverse dynamics, can be eliminated by scanning the intermediate variables from the end to the beginning.

The customization technique allows us to reduce the computational load for a general robot, but this reduction is larger when carried out for a specific robot [Kleinfinger 86a]. The computational efficiency in customization is obtained at the cost of a software symbolic iterative structure [Khalil 97] and a relatively increased program memory requirement.

• **Example 9.4.** To illustrate how to generate a customized symbolic model, we develop in this example the computation of the link angular velocities ${}^{j}\omega_j$ for the Stäubli RX-90 robot. The computation of the orientation matrices ${}^{j-1}\mathbf{A}_j$ (Example 3.3) generates the 12 sinus and cosinus intermediate variables:

$$\mathrm{S}j = \sin(q_j)$$
$$\mathrm{C}j = \cos(q_j) \qquad \text{for } j = 1, \ldots, 6$$

The computation of the angular velocities for j = 1, ..., 6 is given as:

$${}^{1}\omega_1 = \begin{bmatrix} 0 \\ 0 \\ \mathrm{QP1} \end{bmatrix}$$

Computation of ${}^{1}\omega_1$ does not generate any intermediate variable.

$$ {}^{2}\omega_1 = \begin{bmatrix} S2*QP1 \\ C2*QP1 \\ 0 \end{bmatrix} = \begin{bmatrix} WI12 \\ WI22 \\ 0 \end{bmatrix} $$

Computation of ${}^{2}\omega_2$ generates the following intermediate variables:

WI12 = S2*QP1
WI22 = C2*QP1

In the following, the vector ${}^{2}\omega_2$ is set as:

$$ {}^{2}\omega_2 = \begin{bmatrix} WI12 \\ WI22 \\ QP2 \end{bmatrix} $$

Continuing the recursive computation leads to:

$$ {}^{3}\omega_2 = \begin{bmatrix} C3*WI12 + S3*WI22 \\ -S3*WI12 + C3*WI22 \\ QP2 \end{bmatrix} = \begin{bmatrix} WI13 \\ WI23 \\ QP2 \end{bmatrix} $$

$$ {}^{3}\omega_3 = \begin{bmatrix} WI13 \\ WI23 \\ QP2 + QP3 \end{bmatrix} = \begin{bmatrix} WI13 \\ WI23 \\ W33 \end{bmatrix} $$

$$ {}^{4}\omega_3 = \begin{bmatrix} C4*WI13 - S4*W33 \\ -S4*WI12 - C4*W33 \\ WI23 \end{bmatrix} = \begin{bmatrix} WI14 \\ WI24 \\ WI23 \end{bmatrix} $$

$$ {}^{4}\omega_4 = \begin{bmatrix} WI14 \\ WI24 \\ WI23 + QP4 \end{bmatrix} = \begin{bmatrix} WI14 \\ WI24 \\ W34 \end{bmatrix} $$

$$ {}^{5}\omega_4 = \begin{bmatrix} C5*WI14 + S5*W34 \\ -S5*WI14 + C5*W34 \\ -WI24 \end{bmatrix} = \begin{bmatrix} WI15 \\ WI25 \\ -WI24 \end{bmatrix} $$

$$ {}^{5}\omega_5 = \begin{bmatrix} WI15 \\ WI25 \\ -WI24 + QP5 \end{bmatrix} = \begin{bmatrix} WI15 \\ WI25 \\ W35 \end{bmatrix} $$

$$^6\omega_5 = \begin{bmatrix} C6*WI15 - S6*W35 \\ -S6*WI15 - C6*W35 \\ WI25 \end{bmatrix} = \begin{bmatrix} WI16 \\ WI26 \\ WI25 \end{bmatrix}$$

$$^6\omega_6 = \begin{bmatrix} WI16 \\ WI26 \\ WI25 + QP6 \end{bmatrix} = \begin{bmatrix} WI16 \\ WI26 \\ W36 \end{bmatrix}$$

Finally, the computation of $^j\omega_j$, for j = 1, ..., 6, requires the following intermediate variables: WI12, WI22, WI13, WI23, W33, WI14, WI24, W34, WI15, WI25, W35, WI16, WI26 and W36, in addition to the variables Sj and Cj for j = 2, ..., 6. The variables S1 and C1 can be eliminated because they have no effect on the angular velocities.

9.6.3. *Utilization of the base inertial parameters*

It is obvious that the use of the base inertial parameters in a customized Newton-Euler formulation that is linear in the inertial parameters will reduce the number of operations because the parameters that have no effect on the model or have been grouped are set equal to zero. Practically, the number of operations of the inverse dynamic model when using the base inertial parameters for a general n revolute degree-of-freedom robot is $92n - 127$ multiplications and $81n - 117$ additions ($n > 2$), which gives 425 multiplications and 369 additions for $n = 6$. By general robot, we mean:

- the geometric parameters r_1, d_1, α_1 and r_n are zero (this assumption holds for any robot);
- the other geometric parameters, all the inertial parameters, and the forces and moments exerted by the terminal link on the environment can have an arbitrary real value;
- the friction forces are assumed to be negligible, otherwise, with a Coulomb and viscous friction model, we add n multiplications, 2n additions, and n sign functions.

Table 9.6. shows the computational complexity of the inverse dynamic model for the Stäubli RX-90 robot using the customized Newton-Euler formulation. In Appendix 7, we give the dynamic model of the Stäubli RX-90 robot when using the base inertial parameters of Table 9.4, which takes into account the symmetry of the links. The corresponding computational cost is 160 multiplications and 113 additions.

Table 9.6. *Computational complexity of the inverse dynamic model for the Stäubli RX-90 robot*

	Inertial parameters	Multiplications	Additions
General inertial parameters	Standard parameters	294	283
	Base parameters	253	238
Simplified inertial parameters	Standard parameters	202	153
	Base parameters	160	113

9.7. Direct dynamic model

The computation of the direct dynamic model is employed to carry out simulations for the purpose of testing the robot performances and studying the synthesis of the control laws. During simulation, the dynamic equations are solved for the joint accelerations given the input torques and the current state of the robot (joint positions and velocities). Through integration of the joint accelerations, the robot trajectory is then determined. Although the simulation may be carried out off-line, it is interesting to have an efficient direct dynamic model to reduce the simulation time. In this section, we consider two methods: the first is based on using a specialized Newton-Euler inverse dynamic model and needs to compute the inverse of the inertia matrix **A** of the robot; the second method is based on a recursive Newton-Euler algorithm that does not explicitly use the matrix **A** and has a computational cost that varies linearly with the number of degrees of freedom of the robot.

9.7.1. *Using the inverse dynamic model to solve the direct dynamic problem*

From equation [9.6], we can express the direct dynamic problem as the solution of the joint accelerations from the following equation:

$$\mathbf{A}\,\ddot{\mathbf{q}} = [\mathbf{\Gamma} - \mathbf{H}(\mathbf{q}, \dot{\mathbf{q}})] \qquad [9.95]$$

where **H** contains the Coriolis, centrifugal, gravity, friction and external torques:

$$\mathbf{H}(\mathbf{q}, \dot{\mathbf{q}}) = \mathbf{C}(\mathbf{q}, \dot{\mathbf{q}})\,\dot{\mathbf{q}} + \mathbf{Q} + \mathbf{diag}(\dot{\mathbf{q}})\mathbf{F}_v + \mathbf{diag}[\mathrm{sign}(\dot{\mathbf{q}})]\mathbf{F}_c + \mathbf{J}^T\,\mathbf{f}_{en}$$

Although in practice we do not explicitly calculate the inverse of the matrix **A**, the solution of equation [9.95] is generally denoted by:

$$\ddot{\mathbf{q}} = \mathbf{A}^{-1} [\mathbf{\Gamma} - \mathbf{H}(\mathbf{q}, \dot{\mathbf{q}})] \qquad [9.96]$$

The computation of the direct dynamics can be broken down into three steps: the calculation of $\mathbf{H}(\mathbf{q}, \dot{\mathbf{q}})$, the calculation of $\mathbf{A}$, and the solution of the linear equation [9.95] for $\ddot{\mathbf{q}}$.

The computational complexity of the first step is minimized by the use of a specialized version of the inverse dynamics algorithm in which the desired joint accelerations are zero [Walker 82]. By comparing equations [9.1] and [9.95], we deduce that $\mathbf{H}(\mathbf{q}, \dot{\mathbf{q}})$ is equal to $\mathbf{\Gamma}$ if $\ddot{\mathbf{q}} = \mathbf{0}$.

The inertia matrix can also be calculated one column at a time, using Newton-Euler inverse dynamic model [Walker 82]. From relation [9.95], we deduce that the ith column of $\mathbf{A}$ is equal to $\mathbf{\Gamma}$ if:

$$\ddot{\mathbf{q}} = \mathbf{u}_i, \ \dot{\mathbf{q}} = \mathbf{0}, \ \mathbf{g} = \mathbf{0}, \ \mathbf{F}_c = \mathbf{0} \ (\mathbf{f}_{ej} = \mathbf{0}, \ \mathbf{m}_{ej} = \mathbf{0} \quad \text{for } j = 1, \ldots, n) \qquad [9.97]$$

where $\mathbf{u}_i$ is an (nx1) unit vector with 1 in the ith row and zeros elsewhere. Iterating the procedure for i = 1,..., n leads to the construction of the entire inertia matrix.

To reduce the computational complexity of this algorithm, we can make use of the base inertial parameters and the customized symbolic techniques. Moreover, we can take advantage of the fact that the inertia matrix $\mathbf{A}$ is symmetric. A more efficient procedure for computing the inertia matrix using the concept of composite links is given in Appendix 8. Alternative efficient approaches for computing the inertia matrix based on the Lagrange formulation are proposed in [Megahed 82], [Renaud 85].

NOTE.– The nonlinear state equation of a robot follows from relation [9.95] as:

$$\begin{bmatrix} \dot{\mathbf{q}} \\ \ddot{\mathbf{q}} \end{bmatrix} = \begin{bmatrix} \dot{\mathbf{q}} \\ \mathbf{A}^{-1} [\mathbf{\Gamma} - \mathbf{H}(\mathbf{q}, \dot{\mathbf{q}})] \end{bmatrix} \qquad [9.98]$$

and the output equation is written as:

$$\mathbf{y} = \mathbf{q} \ \text{ or } \ \mathbf{y} = \mathbf{X}(\mathbf{q}) \qquad [9.99]$$

In this formulation, the state variables are given by $[\mathbf{q}^T \ \ \dot{\mathbf{q}}^T]^T$, the equation $\mathbf{y} = \mathbf{q}$ gives the output vector in the joint space, and $\mathbf{y} = \mathbf{X}(\mathbf{q})$ denotes the coordinates of the end-effector frame in the task space.

9.7.2. *Recursive computation of the direct dynamic model*

This method is based on the recursive Newton-Euler equations and does not use explicitly the inertia matrix of the robot [Armstrong 79], [Featherstone 83b], [Brandl 86]. In this section, we utilize the compact spatial notation, which is also called *screw notation*. Consequently, by combining equations [9.87] and [9.86], which give ${}^j\dot{\mathbf{V}}_j$ and ${}^j\dot{\omega}_j$, we obtain:

$$ {}^j\dot{\mathbb{V}}_j = {}^j\mathbb{T}_{j-1}\,{}^{j-1}\dot{\mathbb{V}}_{j-1} + \ddot{q}_j\,{}^j\mathbb{a}_j + {}^j\gamma_j \qquad [9.100] $$

where ${}^j\mathbb{a}_j$ is defined by equation [9.23b], and:

$$ {}^j\gamma_j = \begin{bmatrix} {}^j\mathbf{A}_{j-1}[{}^{j-1}\omega_{j-1} \times ({}^{j-1}\omega_{j-1} \times {}^{j-1}\mathbf{P}_j)] + 2\sigma_j\,({}^j\omega_{j-1} \times \dot{q}_j\,{}^j\mathbf{a}_j) \\ \bar{\sigma}_j\,{}^j\omega_{j-1} \times \dot{q}_j\,{}^j\mathbf{a}_j \end{bmatrix} \qquad [9.101] $$

Equations [9.88], [9.89], [9.91] and [9.93], which represent the equilibrium equations of link j, can be combined as:

$$ {}^j\mathbb{J}_j\,{}^j\dot{\mathbb{V}}_j = {}^j\mathbb{f}_j - {}^{j+1}\mathbb{T}_j^T\,{}^{j+1}\mathbb{f}_{j+1} + {}^j\beta_j \qquad [9.102] $$

where:

$$ {}^j\beta_j = -{}^j\mathbb{f}_{ej} - \begin{bmatrix} {}^j\omega_j \times ({}^j\omega_j \times {}^j\mathbf{MS}_j) \\ {}^j\omega_j \times ({}^j\mathbf{J}_j\,{}^j\omega_j) \end{bmatrix} \qquad [9.103] $$

In equation [9.102], we use equation [2.63] to transform the dynamic wrench from frame R_{j+1} to frame R_j.

The joint accelerations are obtained as a result of three recursive computations:

i) first forward recursive computations for j = 1, ..., n: in this step, we compute the screw transformation matrices ${}^j\mathbb{T}_{j-1}$, the link angular velocities ${}^j\omega_j$ as well as ${}^j\gamma_j$ and ${}^j\beta_j$ vectors, which represent the link accelerations and the link wrenches respectively when $\ddot{\mathbf{q}} = \mathbf{0}$;

ii) backward recursive computations for j = n, ..., 1: we compute the vectors and matrices needed to express the joint acceleration $\ddot{q}_j$ and the wrench ${}^j\mathbb{f}_j$ in terms of ${}^{j-1}\dot{\mathbb{V}}_{j-1}$. To illustrate the equations required, we detail the case when j = n and j = n − 1. By combining equations [9.100] and [9.102] for j = n, and since ${}^{n+1}\mathbb{f}_{n+1} = \mathbf{0}$, we obtain:

$$ {}^{n}\mathbb{J}_n ({}^{n}\mathbb{T}_{n-1}\, {}^{n-1}\dot{\mathbb{V}}_{n-1} + \ddot{q}_n\, {}^{n}\mathbf{a}_n + {}^{n}\gamma_n) = {}^{n}\mathbb{f}_n + {}^{n}\beta_n \quad [9.104] $$

Since:

$$ {}^{j}\mathbf{a}_j^T\, {}^{j}\mathbb{f}_j = \tau_j - Ia_j\, \ddot{q}_j \quad [9.105] $$

$$ \tau_j = \Gamma_j - F_{sj}\, \mathrm{sign}(\dot{q}_j) - F_{vj}\, \dot{q}_j \quad [9.106] $$

multiplying equation [9.104] by ${}^{n}\mathbf{a}_n^T$ and using equation [9.105], we deduce the joint acceleration of joint n:

$$ \ddot{q}_n = H_n^{-1} (-\, {}^{n}\mathbf{a}_n^T\, {}^{n}\mathbb{J}_n ({}^{n}\mathbb{T}_{n-1}\, {}^{n-1}\dot{\mathbb{V}}_{n-1} + {}^{n}\gamma_n) + \tau_n + {}^{n}\mathbf{a}_n^T\, {}^{n}\beta_n) \quad [9.107] $$

where H_n is a scalar given as:

$$ H_n = ({}^{n}\mathbf{a}_n^T\, {}^{n}\mathbb{J}_n\, {}^{n}\mathbf{a}_n + Ia_n) \quad [9.108] $$

Substituting for $\ddot{q}_n$ from equation [9.107] into equation [9.104], we obtain the dynamic wrench ${}^{n}\mathbb{f}_n$ as:

$$ {}^{n}\mathbb{f}_n = \begin{bmatrix} {}^{n}\mathbf{f}_n \\ {}^{n}\mathbf{m}_n \end{bmatrix} = {}^{n}\mathbb{K}_n\, {}^{n}\mathbb{T}_{n-1}\, {}^{n-1}\dot{\mathbb{V}}_{n-1} + {}^{n}\alpha_n \quad [9.109] $$

where:

$$ {}^{n}\mathbb{K}_n = {}^{n}\mathbb{J}_n - {}^{n}\mathbb{J}_n\, {}^{n}\mathbf{a}_n\, H_n^{-1}\, {}^{n}\mathbf{a}_n^T\, {}^{n}\mathbb{J}_n \quad [9.110] $$

$$ {}^{n}\alpha_n = {}^{n}\mathbb{K}_n\, {}^{n}\gamma_n + {}^{n}\mathbb{J}_n\, {}^{n}\mathbf{a}_n\, H_n^{-1} (\tau_n + {}^{n}\mathbf{a}_n^T\, {}^{n}\beta_n) - {}^{n}\beta_n \quad [9.111] $$

We now have $\ddot{q}_n$ and ${}^{n}\mathbb{f}_n$ in terms of ${}^{n-1}\dot{\mathbb{V}}_{n-1}$. Iterating the procedure for $j = n-1$, we obtain, from equation [9.102]:

$$ {}^{n-1}\mathbb{J}_{n-1}\, {}^{n-1}\dot{\mathbb{V}}_{n-1} = {}^{n-1}\mathbb{f}_{n-1} + {}^{n}\mathbb{T}_{n-1}^T\, {}^{n}\mathbb{f}_n + {}^{n-1}\beta_{n-1} \quad [9.112] $$

which can be rewritten using equation [9.100] as:

$$ {}^{n-1}\mathbb{J}^*_{n-1} ({}^{n-1}\mathbb{T}_{n-2}\, {}^{n-2}\dot{\mathbb{V}}_{n-2} + \ddot{q}_{n-1}\, {}^{n-1}\mathbf{a}_{n-1} + {}^{n-1}\gamma_{n-1}) = {}^{n-1}\mathbb{f}_{n-1} + {}^{n-1}\beta^*_{n-1} \quad [9.113] $$

where:

$$^{n-1}\mathbb{J}^*_{n-1} = {}^{n-1}\mathbb{J}_{n-1} + {}^{n}\mathbb{T}^T_{n-1}\, {}^{n}\mathbb{K}_n\, {}^{n}\mathbb{T}_{n-1} \quad [9.114]$$

$$^{n-1}\beta^*_{n-1} = {}^{n-1}\beta_{n-1} - {}^{n}\mathbb{T}^T_{n-1}\, {}^{n}\alpha_n \quad [9.115]$$

Equation [9.113] has the same form as equation [9.104]. Consequently, we can express $\ddot{q}_{n-1}$ and ${}^{n-1}\mathbb{f}_{n-1}$ in terms of ${}^{n-2}\dot{\mathbb{V}}_{n-2}$. Iterating this procedure for j = n−2, ..., 1, we obtain $\ddot{q}_j$ and ${}^{j}\mathbb{f}_j$ in terms of ${}^{j-1}\dot{\mathbb{V}}_{j-1}$ for j = n−1, ..., 1. Since ${}^{0}\dot{\mathbb{V}}_0$ is composed of the linear and angular accelerations of the base that are assumed to be known ($\dot{\mathbf{V}}_0 = -\mathbf{g}$, $\dot{\omega}_0 = \mathbf{0}$), the third recursive computation allows us to compute $\ddot{q}_j$ and ${}^{j}\mathbb{f}_j$ for j = 1, ..., n. These backward recursive equations are summarized as follows:

For j = n, ..., 1, compute:

$$H_j = ({}^{j}\mathbb{a}_j^T\, {}^{j}\mathbb{J}^*_j\, {}^{j}\mathbb{a}_j + Ia_j) \quad [9.116]$$

$$^{j}\mathbb{K}_j = {}^{j}\mathbb{J}^*_j - {}^{j}\mathbb{J}^*_j\, {}^{j}\mathbb{a}_j\, H_j^{-1}\, {}^{j}\mathbb{a}_j^T\, {}^{j}\mathbb{J}^*_j \quad [9.117]$$

$$^{j}\alpha_j = {}^{j}\mathbb{K}_j\, {}^{j}\gamma_j + {}^{j}\mathbb{J}^*_j\, {}^{j}\mathbb{a}_j\, H_j^{-1}\, (\tau_j + {}^{j}\mathbb{a}_j^T\, {}^{j}\beta^*_j) - {}^{j}\beta^*_j \quad [9.118]$$

$$^{j-1}\beta^*_{j-1} = {}^{j-1}\beta_{j-1} - {}^{j}\mathbb{T}^T_{j-1}\, {}^{j}\alpha_j \quad [9.119]$$

$$^{j-1}\mathbb{J}^*_{j-1} = {}^{j-1}\mathbb{J}_{j-1} + {}^{j}\mathbb{T}^T_{j-1}\, {}^{j}\mathbb{K}_j\, {}^{j}\mathbb{T}_{j-1} \quad [9.120]$$

Note that these equations are initialized by ${}^{j}\mathbb{J}^*_j = {}^{j}\mathbb{J}_j$ and that equations [9.119] and [9.120] are not calculated for j = 1;

iii) second forward recursive computations for j = 1, ..., n. The joint acceleration $\ddot{q}_j$ and the dynamic wrench ${}^{j}\mathbb{f}_j$ (if needed) are then obtained from the following equations (see equations [9.107] and [9.109]):

$$^{j}\dot{\mathbb{V}}_{j-1} = {}^{j}\mathbb{T}_{j-1}\, {}^{j-1}\dot{\mathbb{V}}_{j-1} \quad [9.121]$$

$$\ddot{q}_j = H_j^{-1}\, [-{}^{j}\mathbb{a}_j^T\, {}^{j}\mathbb{J}^*_j\, ({}^{j}\dot{\mathbb{V}}_{j-1} + {}^{j}\gamma_j) + \tau_j + {}^{j}\mathbb{a}_j^T\, {}^{j}\beta^*_j] \quad [9.122]$$

$$^{j}\mathbb{f}_j = \begin{bmatrix} {}^{j}\mathbf{f}_j \\ {}^{j}\mathbf{m}_j \end{bmatrix} = {}^{j}\mathbb{K}_j\, {}^{j}\dot{\mathbb{V}}_{j-1} + {}^{j}\alpha_j \quad [9.123]$$

$$^{j}\dot{\mathbb{V}}_j = {}^{j}\dot{\mathbb{V}}_{j-1} + {}^{j}\mathbb{a}_j\, \ddot{q}_j + {}^{j}\gamma_j \quad [9.124]$$

NOTES.–

- to reduce the number of operations of this algorithm, we can make use of the base inertial parameters and the customized symbolic technique. Thereby, the number of operations of the direct dynamic model for the Stäubli RX-90 robot is 889 multiplications and 653 additions [Khalil 97]. In the case of the use of simplified inertial parameters (Table 9.4), the computational cost becomes 637 multiplications and 423 additions;
- the computational complexity of this method is $\mathcal{O}(n)$, while the method requiring the inverse of the robot inertia matrix is of complexity $\mathcal{O}(n^3)$;
- from the numerical point of view, this method is more stable than the method requiring the inverse of the robot inertia matrix [Cloutier 95].

9.8. Conclusion

In this chapter, we have presented the dynamics of serial robots using Lagrange and Newton-Euler formulations that are linear in the inertial parameters. The Lagrange formulation allowed us to study the characteristics and properties of the dynamic model of robots, while the Newton-Euler was shown to be the most efficient for real time implementation. We have illustrated that the base inertial parameters can be determined using simple closed-form rules without calculating neither the dynamic model nor the energy functions. In order to increase the efficiency of the Newton-Euler algorithms, we have proposed the use of the base inertial parameters in a customized symbolic programming algorithm. The problem of computing the direct dynamic model for simulating the dynamics of robots has been treated using two methods; the first is based on the Newton-Euler inverse dynamic algorithm, while the second is based on another Newton-Euler algorithm that does not require the computation of the robot inertia matrix.

In the next chapter, we extend these results to tree structured and closed loop robots. Note that the inverse and direct dynamic algorithms using recursive Newton-Euler equations have been generalized to flexible robots [Boyer 98] and to systems with lumped elasticities [Khalil 00a].

Chapter 10

Dynamics of robots with complex structure

10.1. Introduction

In this chapter, we present the dynamic modeling of tree structured robots and of closed chain mechanisms. We also derive the base inertial parameters for these structures. The algorithms given constitute a generalization of the results developed for serial robots in Chapter 9. We make use of the notations of § 9.2 and we assume that the reader is familiar with the geometric description of complex structures exposed in Chapter 7.

10.2. Dynamic modeling of tree structured robots

10.2.1. *Lagrange equations*

Since the joint variables are independent in a tree structure, we can make use of the Lagrange formulation in a similar way as for a serial structure. Thus, the kinetic energy and the potential energy will be computed by equations [9.16] and [9.25]. The recursive equations for computing the angular and linear velocities of link j must take into consideration that the antecedent of link j is link i, denoted as $i = a(j)$, and not $j - 1$ as in the case of simple open chain structures. The vector L_j denotes the position of frame R_j with respect to its antecedent frame R_i. Consequently, the linear and angular velocities of link j can be obtained from equations [9.17] and [9.18] by replacing $j - 1$ by i:

$$ {}^{j}\omega_{j} = {}^{j}\mathbf{A}_{i}\, {}^{i}\omega_{i} + \bar{\sigma}_{j}\, \dot{q}_{j}\, {}^{j}\mathbf{a}_{j} \qquad [10.1] $$

$$ {}^{j}\mathbf{V}_{j} = {}^{j}\mathbf{A}_{i}\, ({}^{i}\mathbf{V}_{i} + {}^{i}\omega_{i} \times {}^{i}\mathbf{P}_{j}) + \sigma_{j}\, \dot{q}_{j}\, {}^{j}\mathbf{a}_{j} \qquad [10.2] $$

10.2.2. *Newton-Euler formulation*

The forward recursive equations of the Newton-Euler inverse dynamic model (§ 9.5) can be generalized for tree structured robots by replacing j – 1 by i, with i = a(j):

$$^{j}\omega_{i} = {}^{j}A_{i}\, {}^{i}\omega_{i} \qquad [10.3]$$

$$^{j}\omega_{j} = {}^{j}\omega_{i} + \bar{\sigma}_{j}\, \dot{q}_{j}\, {}^{j}a_{j} \qquad [10.4]$$

$$^{j}\dot{\omega}_{j} = {}^{j}A_{i}\, {}^{i}\dot{\omega}_{i} + \bar{\sigma}_{j}\, (\ddot{q}_{j}\, {}^{j}a_{j} + {}^{j}\omega_{i} \times \dot{q}_{j}\, {}^{j}a_{j}) \qquad [10.5]$$

$$^{j}\dot{V}_{j} = {}^{j}A_{i}\, ({}^{i}\dot{V}_{i} + {}^{i}U_{i}\, {}^{i}P_{j}) + \sigma_{j}\, (\ddot{q}_{j}\, {}^{j}a_{j} + 2\, {}^{j}\omega_{i} \times \dot{q}_{j}\, {}^{j}a_{j}) \qquad [10.6]$$

$$^{j}F_{j} = M_{j}\, {}^{j}\dot{V}_{j} + {}^{j}U_{j}\, {}^{j}MS_{j} \qquad [10.7]$$

$$^{j}M_{j} = {}^{j}J_{j}\, {}^{j}\dot{\omega}_{j} + {}^{j}\omega_{j} \times ({}^{j}J_{j}\, {}^{j}\omega_{j}) + {}^{j}MS_{j} \times {}^{j}\dot{V}_{j} \qquad [10.8]$$

with $\omega_0 = 0$, $\dot{\omega}_0 = 0$, $\dot{V}_0 = -g$ and $^{j}U_{j} = {}^{j}\hat{\dot{\omega}}_{j} + {}^{j}\hat{\omega}_{j}\, {}^{j}\hat{\omega}_{j}$.

These equations will be computed recursively for j = 1, ..., n.

Let us suppose that k denotes all the links such that a(k) = j (Figure 10.1). The backward recursive equations for j = n, ..., 1 are written as follows:

$$^{j}f_{j} = {}^{j}F_{j} + {}^{j}f_{ej} + \sum_{k/a(k)=j} {}^{j}f_{k} \qquad [10.9]$$

$$^{i}f_{j} = {}^{i}A_{j}\, {}^{j}f_{j} \qquad [10.10]$$

$$^{j}m_{j} = {}^{j}M_{j} + {}^{j}m_{ej} + \sum_{k/a(k)=j} ({}^{j}A_{k}\, {}^{k}m_{k} + {}^{j}P_{k} \times {}^{j}f_{k}) \qquad [10.11]$$

$$\Gamma_{j} = (\sigma_{j}\, {}^{j}f_{j} + \bar{\sigma}_{j}\, {}^{j}m_{j})^{T}\, {}^{j}a_{j} + F_{sj}\, \mathrm{sign}\,(\dot{q}_{j}) + F_{vj}\, \dot{q}_{j} + Ia_{j}\, \ddot{q}_{j} \qquad [10.12]$$

For a terminal link, $^{j}m_{k}$ and $^{j}f_{k}$ are zero.

10.2.3. *Direct dynamic model of tree structured robots*

Similarly, the computation of the direct dynamic model of tree structured robots can be obtained using the two methods presented in § 9.7 without any particular difficulty.

10.2.4. *Determination of the base inertial parameters*

All the results concerning the base inertial parameters of serial robots can be generalized for tree structured robots [Khalil 89b], [Khalil 95a]. Thereby, equations [9.37] and [9.38], or [9.48] and [9.49], giving the conditions of elimination or grouping of the inertial parameters, are valid. To expose the computation of the base parameters of tree structured robots, we recall that a main branch is composed of the set of links of a path connecting the base to a terminal link. Thus, there are as many main branches as the number of terminal links.

The parameters having no effect on the dynamic model for a tree structure can be obtained by applying the rules derived for serial robots to each main branch.

As in the case of serial robots (§ 9.4.2.2), the general grouped parameters for tree structures will concern the parameters YY_j, MZ_j and M_j if joint j is revolute, and the elements of the inertia tensor J_j if joint j is prismatic. The general grouping equations are different than those of Chapter 9, because certain frames may be defined by six geometric parameters.

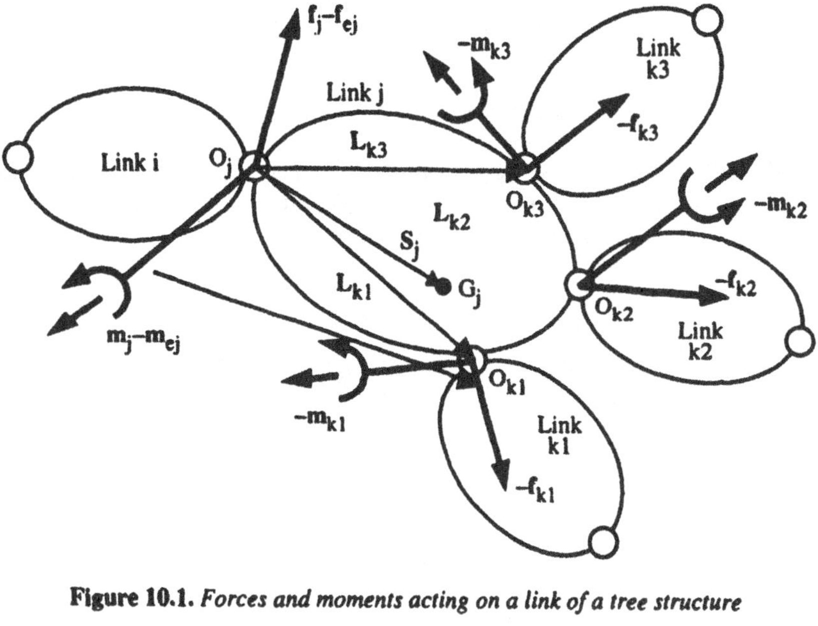

Figure 10.1. *Forces and moments acting on a link of a tree structure*

10.2.4.1. *General grouping equations*

The recursive equation between the energy functions of two successive links is given by:

$$\mathbf{h}_j = \mathbf{h}_i \, {}^i\lambda_j + \dot{q}_j \, \eta_j \qquad [10.13]$$

where $\mathbf{h}_j$ is the (1x10) row matrix containing the energy functions of link j denoted by $[\, h_{XXj} \;\; h_{XYj} \;\; \cdots \;\; h_{Mj} \,]$, while ${}^i\lambda_j$ is the (10x10) matrix expressing the transformation of the inertial parameters of a link from frame R_j to frame R_i. The general form of η_j and ${}^i\lambda_j$, with i = a(j), is developed in Appendix 6. We deduce that equations [9.50] and [9.54] are valid for tree structures after replacing j − 1 by i, which leads to the following theorem:

Theorem 10.1. If joint j is revolute, then the parameters YY_j, MZ_j and M_j can be grouped with the parameters of link i and link j, with i = a(j). The general grouping equations are the following:

$$XXR_i = XX_i - YY_j \qquad [10.14a]$$

$$\mathbf{KR}_i = \mathbf{K}_i + YY_j\,({}^i\lambda_j^1 + {}^i\lambda_j^4) + MZ_j\,{}^i\lambda_j^9 + M_j\,{}^i\lambda_j^{10} \qquad [10.14b]$$

where ${}^i\lambda_j^k$ is the kth column of the matrix ${}^i\lambda_j$, which is a function of the geometric parameters defining frame R_j. In the following formulas, the corresponding subscript j has been dropped for simplicity. From Appendix 6, we obtain:

$${}^i\lambda_j^1 + {}^i\lambda_j^4 = \begin{bmatrix} 1-SS\gamma SS\alpha \\ CS\gamma SS\alpha \\ -S\gamma CS\alpha \\ 1-CC\gamma SS\alpha \\ C\gamma CS\alpha \\ SS\alpha \\ 0 \\ 0 \\ 0 \\ 0 \end{bmatrix} \qquad [10.15]$$

$$
{}^{i}\lambda_j^9 = \begin{bmatrix} 2P_zC\alpha - 2P_yC\gamma S\alpha \\ P_xC\gamma S\alpha - P_yS\gamma S\alpha \\ -P_xC\alpha - P_zS\gamma S\alpha \\ 2P_xS\gamma S\alpha + 2P_zC\alpha \\ -P_yC\alpha + P_zC\gamma S\alpha \\ 2P_xS\gamma S\alpha - 2P_yC\gamma S\alpha \\ S\gamma S\alpha \\ -C\gamma S\alpha \\ C\alpha \\ 0 \end{bmatrix} \quad [10.16], \qquad {}^{i}\lambda_j^{10} = \begin{bmatrix} P_y^2 + P_z^2 \\ -P_xP_y \\ -P_xP_z \\ P_x^2 + P_z^2 \\ -P_yP_z \\ P_x^2 + P_y^2 \\ P_x \\ P_y \\ P_z \\ 1 \end{bmatrix} \quad [10.17]
$$

where P_x, P_y and P_z denote the coordinates of the vector ${}^{i}\mathbf{P}_j$, which can be obtained from equation [7.3] giving the general transformation matrix ${}^{i}\mathbf{T}_j$ such that:

$$
{}^{i}\mathbf{P}_j = \begin{bmatrix} P_x \\ P_y \\ P_z \end{bmatrix} = \begin{bmatrix} d_j\,C\gamma_j + r_jS\gamma_jS\alpha_j \\ d_j\,S\gamma_j - r_jC\gamma_jS\alpha_j \\ r_jC\alpha_j + b_j \end{bmatrix} \qquad [10.18]
$$

After expanding equations [10.14], we obtain:

$$
\begin{aligned}
XXR_j &= XX_j - YY_j \\
XXR_i &= XX_i + YY_j\,(1 - SS\gamma_jSS\alpha_j) + 2\,MZ_j\,(P_zC\alpha_j - P_yC\gamma_jS\alpha_j) + M_j\,(P_y^2 + P_z^2) \\
XYR_i &= XY_i + YY_j\,(CS\gamma_jSS\alpha_j) + MZ_j\,(P_xC\gamma_jS\alpha_j - P_yS\gamma_jS\alpha_j) + M_j\,(-P_xP_y) \\
XZR_i &= XZ_i - YY_j\,(S\gamma_jCS\alpha_j) + MZ_j\,(-P_xC\alpha_j - P_zS\gamma_jS\alpha_j) + M_j\,(-P_xP_z) \\
YYR_i &= YY_i + YY_j\,(1 - CC\gamma_jSS\alpha_j) + 2\,MZ_j\,(P_xS\gamma_jS\alpha_j + P_zC\alpha_j) + M_j\,(P_x^2 + P_z^2) \\
YZR_i &= YZi + YY_j\,(C\gamma_jCS\alpha_j) + MZ_j\,(-P_y\,C\alpha_j + P_zC\gamma_jS\alpha_j) + M_j\,(-P_yP_z) \qquad [10.19] \\
ZZR_i &= ZZ_i + YY_jSS\alpha_j + 2MZ_j\,(P_x\,S\gamma_jS\alpha_j - P_yC\gamma_jS\alpha_j) + M_j(P_x^2 + P_y^2) \\
MXR_i &= MX_i + MZ_j\,(S\gamma_jS\alpha_j) + M_j\,P_x \\
MYR_i &= MY_i - MZ_j\,(C\gamma_jS\alpha_j) + M_j\,P_y \\
MZR_i &= MZ_i + MZ_j\,C\alpha_j + M_j\,P_z \\
MR_i &= M_i + M_j
\end{aligned}
$$

with SS(*) = sin(*) sin(*), CC(*) = cos(*) cos(*) and CS(*) = cos(*) sin(*).

Theorem 10.2. If joint j is prismatic, then the elements of the inertia tensor ${}^{j}\mathbf{J}_j$ can be grouped with those of ${}^{i}\mathbf{J}_i$ using the following equation:

$$
\mathbf{KR}_i = \mathbf{K}_i + {}^{i}\lambda_j^1\,XX_j + {}^{i}\lambda_j^2\,XY_j + \ldots + {}^{i}\lambda_j^6\,ZZ_j \qquad [10.20a]
$$

which is equivalent to:

$$ {}^{i}\mathbf{JR}_i = {}^{i}\mathbf{J}_i + {}^{i}\mathbf{A}_j\, {}^{j}\mathbf{J}_j\, {}^{j}\mathbf{A}_i \qquad [10.20b] $$

The expanded expressions of these equations are too complicated to be developed here.

10.2.4.2. *Particular grouped parameters*

Particular grouped parameters can be obtained by applying the results of serial robots to each main branch b. Let r_{1b} be the first revolute joint of branch b, and r_{2b} be the subsequent revolute joint whose axis is not parallel to the r_{1b} axis. Additional grouping and/or elimination of certain elements among MX_j, MY_j and MZ_j takes place if j is prismatic and lies between r_{1b} and r_{2b}. For simplicity, the subscript b is dropped in the remainder of this section. Two cases are considered:

i) the axis of the prismatic joint j is not parallel to the r_1 axis. In this case, the coefficients h_{MXj}, h_{MYj} and h_{MZj} satisfy the equation:

$$ {}^{j}a_{xr1}\, h_{MXj} + {}^{j}a_{yr1}\, h_{MYj} + {}^{j}a_{zr1}\, h_{MZj} = \text{constant} \qquad [10.21] $$

where ${}^{j}\mathbf{a}_{r1} = [\, {}^{j}a_{xr1} \;\; {}^{j}a_{yr1} \;\; {}^{j}a_{zr1} \,]^T$ is the unit vector of the $\mathbf{z}_{r1}$ axis referred to frame R_j. The corresponding grouping equations are given in Table 10.1;

ii) the axis of the prismatic joint j is parallel to the r_1 axis. The following equation is satisfied:

$$ [\mathbf{h}_{MSj}]^T = {}^{j}\mathbf{A}_i\, [\mathbf{h}_{MSi}]^T - [\, 2P_x h_{ZZk} \;\; 2P_y h_{ZZk} \;\; 0 \,]^T \qquad [10.22] $$

where k denotes the nearest revolute joint from j back to the base, $k \geq r_1$; $\mathbf{h}_{MSj} = [\, h_{MXj} \;\; h_{MYj} \;\; h_{MZj} \,]$; P_x and P_y are the first and second coordinates of ${}^{j}\mathbf{P}_i$ respectively, with $i = a(j)$. Using equation [7.4], we obtain:

$$ {}^{j}\mathbf{P}_i = [\, P_x \;\; P_y \;\; P_z \,]^T = [\, -b_j S\theta_j S\alpha_j - d_j C\theta_j \;\; -b_j C\theta_j S\alpha_j + d_j S\theta_j \;\; -b_j C\alpha_j - r_j \,]^T $$

Therefore, we deduce that the parameter MZ_j has no effect on the dynamic model and the parameters MX_j and MY_j can be grouped with the first moments of link i, and with the parameter ZZ_k of link k using the following equations:

$$ \begin{aligned} MXR_i &= MX_i + (C\gamma_j C\theta_j - S\gamma_j C\alpha_j S\theta_j)\, MX_j - (C\gamma_j S\theta_j + S\gamma_j C\alpha_j C\theta_j)\, MY_j \\ MYR_i &= MY_i + (S\gamma_j C\theta_j + C\gamma_j C\alpha_j S\theta_j)\, MX_j + (-S\gamma_j S\theta_j + C\gamma_j C\alpha_j C\theta_j)\, MY_j \\ MZR_i &= MZ_i + S\theta_j\, S\alpha_j\, MX_j + C\theta_j\, S\alpha_j\, MY_j \\ ZZR_k &= ZZ_k + 2\,(d_j\, C\theta_j + b_j\, S\theta_j\, S\alpha_j)\, MX_j - 2\,(d_j\, S\theta_j + b_j\, C\theta_j\, S\alpha_j)\, MY_j \end{aligned} \qquad [10.23] $$

Table 10.1. *Grouped parameters if $r_1 < j < r_2$, $\sigma_j = 1$, and joint j axis is not parallel to the r_1 axis*

Conditions	Grouping or elimination
${}^{j}a_{zr1} \neq 0$	$MXR_j = MX_j - \frac{{}^{j}a_{xr1}}{{}^{j}a_{zr1}} MZ_j$ $MYR_j = MY_j - \frac{{}^{j}a_{yr1}}{{}^{j}a_{zr1}} MZ_j$
${}^{j}a_{zr1} = 0, {}^{j}a_{xr1}{}^{j}a_{yr1} \neq 0$	$MXR_j = MX_j - \frac{{}^{j}a_{xr1}}{{}^{j}a_{yr1}} MY_j$
${}^{j}a_{zr1} = 0, {}^{j}a_{xr1} = 0$	$MY_j \equiv 0$
${}^{j}a_{zr1} = 0, {}^{j}a_{yr1} = 0$	$MX_j \equiv 0$

Therefore, the practical rules for computing the base inertial parameters given in § 9.4.2.4 can be applied for the tree structure case. The only difference is that the joints r_1 and r_2 should be defined for each main branch b as r_{1b} and r_{2b} respectively. Thus, a rule like "if j is such that $r_1 < j < r_2$" means in the tree structure case "if j is lying between r_{1b} and r_{2b}". From this algorithm, it occurs that the number of minimum inertial parameters for the links (without considering the inertia of rotors) of a general robot is less than b_m, such that:

$$b_m \leq 7\,n_r + 4\,n_p - 4\,n_{r0} - 3\,n_{p0} - 2\,n_{g0} \qquad [10.24]$$

with:

- n_r: number of revolute joints $= \sum \bar{\sigma}_j$;
- n_p: number of prismatic joints $= \sum \sigma_j$;
- n_{r0}: number of revolute joints connected directly to the base;
- n_{p0}: number of prismatic joints connected directly to the base;
- n_{g0}: number of revolute joints connected directly to the base and whose axes are parallel to gravity.

The grouped parameters of the rotor inertias concern those of the actuators of joints (p_{1b}, r_{1b} and r_{2b}), where p_{1b} denotes the first prismatic joint of the main branch b. They can be obtained by applying the results of serial robots (§ 9.4.2.5) for each branch of the tree structure.

10.3. Dynamic model of robots with closed kinematic chains

Many methods have been proposed in the literature to compute the dynamic models of robots containing closed kinematic chains. Among them, let us mention the works of [Chace 67], [Uicker 69], [Chace 71], [Baumgarte 72], [Wittenburg 77], [Megahed 84], [Touron 84], [Luh 85b], [Wittenburg 85], [Kleinfinger 86b], [Giordano 86]. The dynamic model developed in this section is based on firstly computing the dynamic model of an equivalent tree structure, then by multiplying it by the Jacobian matrix representing the derivative of the tree structure variables with respect to the actuated variables [Kleinfinger 86b].

10.3.1. *Description of the system*

The geometry of the robot is described using the method presented in Chapter 7. The system is composed of L joints and n + 1 links, where link 0 is the base. N joints are actuated (active) and the other L – N joints are unactuated (passive). The number of independent closed loops B is equal to L – n. We assume that the structure is controllable and has the minimum number of actuators, thus the number of actuated joints that represent the independent variables is equal to the number of degrees of freedom of the mechanism.

We construct an equivalent tree structure by virtually cutting each loop at one of its passive joints as has already been explained in § 7.3. Since a closed loop contains several unactuated joints, we select the cut joint in such a way that the difference between the number of links of the branches from the root of the loop to the cut joint is as small as possible. This choice reduces the computational complexity of the dynamic model [Kleinfinger 86a].

We represent the tree structure variables by the (nx1) vector $\mathbf{q}_{tr}$, and the cut joints by the (Bx1) vector $\mathbf{q}_c$. The total joint variables are given by equation [7.7]:

$$\mathbf{q} = \begin{bmatrix} \mathbf{q}_{tr} \\ \mathbf{q}_c \end{bmatrix} \qquad [10.25]$$

The vector $\mathbf{q}_{tr}$ is partitioned into the (Nx1) vector of active joints $\mathbf{q}_a$ and the (px1) vector of passive joints $\mathbf{q}_p$:

$$\mathbf{q}_{tr} = \begin{bmatrix} \mathbf{q}_a \\ \mathbf{q}_p \end{bmatrix} \qquad [10.26]$$

The relation between $\mathbf{q}_a$ and $\mathbf{q}_p$ is obtained by solving the loop closure equations (§ 7.3). The constraint kinematic equations of first and second order have already been derived in § 7.8 and are rewritten here as:

$$[\,\mathbf{W_a} \;\; \mathbf{W_p}\,]\begin{bmatrix}\dot{\mathbf{q}}_a \\ \dot{\mathbf{q}}_p\end{bmatrix} = \mathbf{0} \qquad [10.27]$$

$$[\,\mathbf{W_a} \;\; \mathbf{W_p}\,]\begin{bmatrix}\ddot{\mathbf{q}}_a \\ \ddot{\mathbf{q}}_p\end{bmatrix} + \mathbf{\Psi} = \mathbf{0} \qquad [10.28]$$

where $\mathbf{W_a}$ and $\mathbf{W_p}$ are (pxN) and (pxp) matrices respectively. In regular configurations, the rank of $\mathbf{W_p}$ is equal to p. Thus, from equation [10.27], we obtain:

$$\dot{\mathbf{q}}_p = \mathbf{W}\,\dot{\mathbf{q}}_a \qquad [10.29]$$

where:

$$\mathbf{W} = -\mathbf{W}_p^{-1}\,\mathbf{W_a} \qquad [10.30]$$

10.3.2. *Computation of the inverse dynamic model*

If the joint positions and velocities can be expressed in terms of the independent actuated variables, we can use the standard Lagrange equation [9.4] to get the dynamic model of the closed chain structure [Desbats 90]. Otherwise, we have to use the Lagrange equation with constraints such that:

$$\Gamma_i = \frac{d}{dt}\frac{\partial L}{\partial \dot{q}_i} - \frac{\partial L}{\partial q_i} + \left[\frac{\partial \boldsymbol{\phi}(\mathbf{q}_{tr})}{\partial q_i}\right]^T \boldsymbol{\lambda} \qquad i = 1, \ldots, n \qquad [10.31a]$$

where $L(\mathbf{q}_{tr}, \dot{\mathbf{q}}_{tr}, \ddot{\mathbf{q}}_{tr})$ is the Lagrangian of the equivalent tree structure; $\boldsymbol{\phi}(\mathbf{q}_{tr}) = \mathbf{0}$ is the vector containing the p independent constraint functions of the loop closure equations; $\boldsymbol{\lambda} = [\,\lambda_1 \;\cdots\; \lambda_p\,]^T$ is the Lagrange multiplier vector.

This equation can be rewritten as:

$$\boldsymbol{\Gamma} = \boldsymbol{\Gamma}_{tr}(\mathbf{q}_{tr}, \dot{\mathbf{q}}_{tr}, \ddot{\mathbf{q}}_{tr}) + \left[\frac{\partial \boldsymbol{\phi}(\mathbf{q}_{tr})}{\partial \mathbf{q}_{tr}}\right]^T \boldsymbol{\lambda} \qquad [10.31b]$$

where $\boldsymbol{\Gamma}_{tr}$ represents the inverse dynamic model of the equivalent tree structure. It is a function of $\mathbf{q}_{tr}$, $\dot{\mathbf{q}}_{tr}$ and $\ddot{\mathbf{q}}_{tr}$.

The general form of the dynamic model of the tree structure is given by:

$$\Gamma_{tr} = A_{tr}\begin{bmatrix} \ddot{q}_a \\ \ddot{q}_p \end{bmatrix} + H_{tr} \qquad [10.32]$$

where A_{tr} is the inertia matrix of the equivalent tree structure, and H_{tr} is the vector of centrifugal, Coriolis and gravity torques of the equivalent tree structure.

Using equation [10.27], we deduce that:

$$\frac{\partial \phi(q_{tr})}{\partial q_{tr}} = [\ W_a \ \ W_p\ ] \qquad [10.33]$$

The term containing the Lagrange multipliers represents the reaction forces transmitted by the cut joints to ensure that the loops remain closed. Let us decompose Γ_{tr} in a similar way to equation [10.26]:

$$\Gamma_{tr} = \begin{bmatrix} \Gamma_a \\ \Gamma_p \end{bmatrix} \qquad [10.34]$$

where Γ_a and Γ_p denote the torques of the actuated and unactuated joints of the equivalent tree structure respectively.

The joint torques of the closed chain robot are given as:

$$\Gamma = \begin{bmatrix} \Gamma_{cl} \\ 0_{px1} \end{bmatrix} \qquad [10.35]$$

where Γ_{cl} denotes the torques of the N actuated joints, and the zero vector corresponds to the torques of the passive joints:

$$\Gamma = \begin{bmatrix} \Gamma_{cl} \\ 0_{px1} \end{bmatrix} = \begin{bmatrix} \Gamma_a \\ \Gamma_p \end{bmatrix} + \begin{bmatrix} W_a^T \lambda \\ W_p^T \lambda \end{bmatrix} \qquad [10.36]$$

We have thus a system of n equations where the unknowns are Γ_{cl} and λ. Computing the Lagrange multipliers from the lower part of equation [10.36] leads to:

$$\boldsymbol{\lambda} = -[\mathbf{W}_p^T]^{-1}\,\boldsymbol{\Gamma}_p \qquad [10.37]$$

Substituting [10.37] into the upper part of [10.36] yields:

$$\boldsymbol{\Gamma}_{cl} = \boldsymbol{\Gamma}_a - \mathbf{W}_a^T\,[\mathbf{W}_p^T]^{-1}\,\boldsymbol{\Gamma}_p \qquad [10.38]$$

Using equation [10.30], the actuator torque vector of the closed chain robot is written as:

$$\boldsymbol{\Gamma}_{cl} = \boldsymbol{\Gamma}_a + \mathbf{W}^T\boldsymbol{\Gamma}_p = [\;\mathbf{I}_N \;\; \mathbf{W}^T\;]\begin{bmatrix}\boldsymbol{\Gamma}_a \\ \boldsymbol{\Gamma}_p\end{bmatrix} \qquad [10.39]$$

which can be rewritten as:

$$\boldsymbol{\Gamma}_{cl} = [\;\mathbf{I}_N \;\; \mathbf{W}^T\;]\,\boldsymbol{\Gamma}_{tr} = \begin{bmatrix} [\frac{\partial \mathbf{q}_a}{\partial \mathbf{q}_a}]^T & [\frac{\partial \mathbf{q}_p}{\partial \mathbf{q}_a}]^T \end{bmatrix}\boldsymbol{\Gamma}_{tr} = \mathbf{G}^T\boldsymbol{\Gamma}_{tr} \qquad [10.40]$$

where $\mathbf{I}_N$ is the (NxN) identity matrix; $\mathbf{W}$ is the Jacobian matrix representing the derivative of the passive joint positions with respect to the actuated ones; $\mathbf{G}$ is the Jacobian matrix representing the derivative of $\mathbf{q}_{tr}$ with respect to $\mathbf{q}_a$, equal to $\partial\mathbf{q}_{tr}/\partial\mathbf{q}_a$.

Equation [10.40] constitutes the inverse dynamic model of the closed chain structure. The vector $\boldsymbol{\Gamma}_{tr}$ can be computed using the efficient recursive Newton-Euler algorithm described in § 10.2.2.

10.3.3. *Computation of the direct dynamic model*

To simulate the dynamics of a closed chain robot with a given input torque for the active joints $\boldsymbol{\Gamma}_{cl}$ and a given state ($\mathbf{q}_a$, $\dot{\mathbf{q}}_a$), the dynamic equation [10.40] is formulated and solved for the independent accelerations $\ddot{\mathbf{q}}_a$. The accelerations are then numerically integrated to obtain the velocities and positions at the next sampling time. This process is repeated until integration through the time interval of interest is completed. The direct dynamic model can be obtained by formulating the Lagrange dynamic model as follows:

$$\boldsymbol{\Gamma}_{cl} = \mathbf{A}_{cl}\,\ddot{\mathbf{q}}_a + \mathbf{H}_{cl} \qquad [10.41]$$

where $\mathbf{A}_{cl}$ is the inertia matrix and $\mathbf{H}_{cl}$ is the vector of centrifugal, Coriolis and gravity torques of the closed chain structure.

To derive A_{cl} and H_{cl}, we express $\ddot{q}_p$ as a function of $\ddot{q}_a$ in equation [10.40], then we identify the result with equation [10.41]. Using equations [10.28] and [10.30], we deduce that:

$$\ddot{q}_p = W\,\ddot{q}_a - W_p^{-1}\,\Psi \qquad [10.42]$$

Using equation [10.32], we rewrite equation [10.40] as:

$$\Gamma_{cl} = [\,I_N \quad W^T\,]\,A_{tr}\begin{bmatrix}\ddot{q}_a \\ \ddot{q}_p\end{bmatrix} + [\,I_N \quad W^T\,]\,H_{tr} \qquad [10.43]$$

Partitioning the matrix A_{tr} and the vector H_{tr} to explicit the terms corresponding to the active and passive joints gives:

$$A_{tr} = \begin{bmatrix}A_{aa} & A_{ap} \\ A_{pa} & A_{pp}\end{bmatrix},\quad H_{tr} = \begin{bmatrix}H_a \\ H_p\end{bmatrix} \qquad [10.44]$$

where $A_{pa} = [A_{ap}]^T$.

By combining equations [10.43] and [10.44], we obtain:

$$\Gamma_{cl} = A_{aa}\,\ddot{q}_a + A_{ap}\,[W\,\ddot{q}_a - W_p^{-1}\,\Psi] + W^T A_{pa}\,\ddot{q}_a + W^T A_{pp}\,[W\,\ddot{q}_a - W_p^{-1}\,\Psi] + H_a + W^T H_p \qquad [10.45]$$

Identifying equations [10.41] and [10.45] leads to:

$$A_{cl} = A_{aa} + A_{ap}\,W + W^T A_{pa} + W^T A_{pp}\,W \qquad [10.46]$$

$$H_{cl} = H_a + W^T H_p - (A_{ap} + W^T A_{pp})\,W_p^{-1}\,\Psi \qquad [10.47]$$

The solution of [10.41] gives the active joint accelerations, then the passive joint accelerations can be computed from equation [10.42]. Although the active joint accelerations are obtained by solving the linear system [10.41] without inverting the inertia matrix, we generally denote the direct dynamic model by:

$$\ddot{q}_a = A_{cl}^{-1}\,(\Gamma_{cl} - H_{cl}) \qquad [10.48]$$

• **Example 10.1.** Dynamic model of the Acma SR400 robot. The geometry and the constraint equations of the loop of this robot are treated in Example 7.1. The inverse dynamic model of the tree structured robot is computed using the recursive Newton-Euler algorithm quoted in § 10.2.2. To compute the dynamic model of the closed chain, we have to calculate the Jacobian matrix **G** representing the derivative of the variables $\mathbf{q}_{tr}$ with respect to the variables $\mathbf{q}_a$. We recall that:

$$\mathbf{q}_a = [\theta_1 \ \ \theta_2 \ \ \theta_4 \ \ \theta_5 \ \ \theta_6 \ \ \theta_7]^T$$
$$\mathbf{q}_p = [\theta_3 \ \ \theta_8]^T$$
$$\mathbf{q}_{tr} = [\theta_1 \ \ \theta_2 \ \ \theta_4 \ \ \theta_5 \ \ \theta_6 \ \ \theta_7 \ \ \theta_3 \ \ \theta_8]^T$$

The constraint equations are obtained in Example 7.1 as:

$$\theta_3 = \theta_7 + \pi/2 - \theta_2$$
$$\theta_8 = -\theta_7 + \theta_2$$

From these equations, we obtain:

$$\mathbf{G}^T = \begin{bmatrix} 1 & 0 & 0 & 0 & 0 & 0 & 0 & 0 \\ 0 & 1 & 0 & 0 & 0 & 0 & -1 & 1 \\ 0 & 0 & 1 & 0 & 0 & 0 & 0 & 0 \\ 0 & 0 & 0 & 1 & 0 & 0 & 0 & 0 \\ 0 & 0 & 0 & 0 & 1 & 0 & 0 & 0 \\ 0 & 0 & 0 & 0 & 0 & 1 & 1 & -1 \end{bmatrix}$$

The actuated torques of the closed chain robot is computed in terms of the joint torques of the tree structure as:

$$\Gamma_{cl1} = \Gamma_{tr1}$$
$$\Gamma_{cl2} = \Gamma_{tr2} - \Gamma_{tr3} + \Gamma_{tr8}$$
$$\Gamma_{cl4} = \Gamma_{tr4}$$
$$\Gamma_{cl5} = \Gamma_{tr5}$$
$$\Gamma_{cl6} = \Gamma_{tr6}$$
$$\Gamma_{cl7} = \Gamma_{tr7} + \Gamma_{tr3} - \Gamma_{tr8}$$

where Γ_{clj} and Γ_{trj} denote the torque of joint j in the closed structure and in the tree structure respectively.

10.3.4. *Base inertial parameters of closed chain robots*

Since the matrix G is a function of the geometric parameters, we deduce from equation [10.40] that the minimum inertial parameters of the tree structure are valid to compute the dynamic model of the closed chain robot. The constraint equations of the loops may lead to additional elimination or grouping of certain inertial parameters. To compute the grouped parameters for the closed chain robot, we have to find the linear relations between the energy functions of the inertial parameters. We note that the expressions of the energy functions of the closed chain robot are obtained from those of the tree structure after expressing them as a function of the positions and velocities of the active joints. There is no complete symbolic solution for a general closed chain robot. Therefore, the numerical method developed in Appendix 5 [Gautier 91] can be used for this purpose. However, certain general grouped parameters can be obtained without solving the closure equations of the loops. These parameters belong to the links connected to the cut joints [Khalil 95a]. Furthermore, in § 10.3.5, we will show that the grouped parameters of a parallelogram closed loop can be computed explicitly.

Referring to the notations of the closed chain robots described in Chapter 7, we assume that frame R_k and frame R_{k+B} denote the frames placed on the cut joint k connecting link i to link j, with i = a(k) and j = a(k + B). Since frames R_k and R_{k+B} are aligned, then the kinematic screws of these frames are the same. Consequently, we deduce from equation [9.45] that the energy functions $\mathbf{h}_k$ are equal to $\mathbf{h}_{k+B}$:

$$\mathbf{h}_k = \mathbf{h}_{k+B} \qquad [10.49]$$

Using the recursive equation of the energy functions [10.13] and by noting that i = a(k), j = a(k + B) and $\dot{q}_{k+B} = 0$, we obtain the following equation:

$$\mathbf{h}_i \, {}^i\boldsymbol{\lambda}_k + \dot{q}_k \, \boldsymbol{\eta}_k = \mathbf{h}_j \, {}^j\boldsymbol{\lambda}_{k+B} \qquad [10.50]$$

Since the elements of the matrix ${}^j\boldsymbol{\lambda}_{k+B}$ are constants, two cases are considered to identify the linear combinations between the terms of $\mathbf{h}_i$ and $\mathbf{h}_j$:

i) for a revolute cut joint, we obtain the following three linear equations between the energy functions of links i and j:

$$\mathbf{h}_i \, ({}^i\lambda_k^1 + {}^i\lambda_k^4) = \mathbf{h}_j \, ({}^j\lambda_{k+B}^1 + {}^j\lambda_{k+B}^4) \qquad [10.51a]$$

$$\mathbf{h}_i \, {}^i\lambda_k^9 = \mathbf{h}_j \, {}^j\lambda_{k+B}^9 \qquad [10.51b]$$

$$\mathbf{h}_i \, {}^i\lambda_k^{10} = \mathbf{h}_j \, {}^j\lambda_{k+B}^{10} \qquad [10.51c]$$

The left side terms of equations [10.51] are functions of the geometric parameters of frame R_k, while those of the right side are functions of the geometric parameters of frame R_{k+B}. The expressions of λ are given by equations [10.15], [10.16] and [10.17] after considering the appropriate subscript.

We deduce that for any closed loop, if the cut joint is revolute, then we have three linear relations between the elements of $\mathbf{h}_i$ and $\mathbf{h}_j$. There is no general systematic choice for the parameters to be grouped. They must be studied on a case-by-case basis. Furthermore, in some cases, these relations may not lead to three additional grouping parameters with respect to those obtained for the equivalent tree structure;

ii) for a prismatic cut joint, we obtain the following six linear equations between the energy functions of links i and j:

$$\mathbf{h}_i\, {}^i\lambda_k^r = \mathbf{h}_j\, {}^j\lambda_{k+B}^r \quad \text{for } r = 1, \ldots, 6 \qquad [10.52]$$

In this case, we can group the parameters of the inertia tensor of link j with those of link i, with j > i, using the following equation:

$$^i\mathbf{JR}_i = {}^i\mathbf{J}_i + {}^i\mathbf{A}_{k+B}\, {}^k\mathbf{A}_j\, {}^j\mathbf{J}_j\, {}^j\mathbf{A}_k\, {}^{k+B}\mathbf{A}_i \qquad [10.53]$$

10.3.5. *Base inertial parameters of parallelogram loops*

For parallelogram loops and planar loops, additional general linear relations between the energy functions can be deduced [Khalil 95a]. We develop in this section the grouping relations for parallelogram loops, where all the grouped parameters can be obtained systematically without computing explicitly the energy functions [Bennis 91b]. In fact, we can prove that one parameter among the first moments (MX or MY) of one link of the parallelogram can be grouped using equation [10.51c], and that the inertia tensors of two links can be grouped.

Let us consider a parallelogram loop composed of links k1, k2, k3, k4. We assume that the loop is cut between links k3 and k4 and that link k1 is parallel to link k3, and link k2 is parallel to link k4 (Figure 10.2). Thus:

$$\begin{cases} \omega_{k1} = \omega_{k3} \\ \omega_{k2} = \omega_{k4} \end{cases} \qquad [10.54]$$

Consequently, we can group the inertia tensor $^{k3}\mathbf{J}_{k3}$ with $^{k1}\mathbf{J}_{k1}$ using the equation:

$$^{k1}\mathbf{JR}_{k1} = {}^{k1}\mathbf{J}_{k1} + {}^{k1}\mathbf{A}_{k3}\, {}^{k3}\mathbf{J}_{k3}\, {}^{k3}\mathbf{A}_{k1} \qquad [10.55]$$

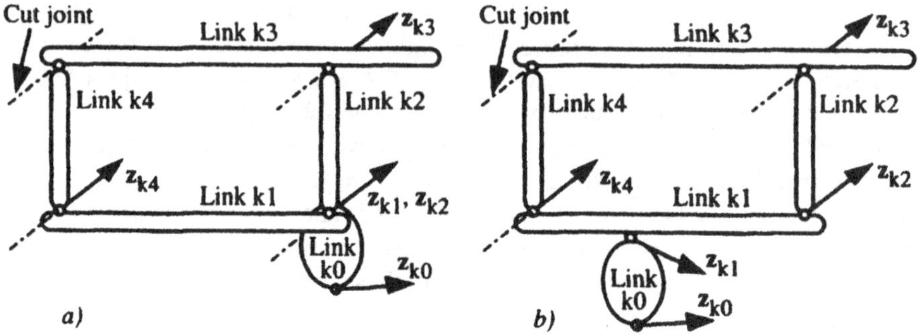

Figure 10.2. *Examples of parallelograms*

Similarly, ${}^{k4}\mathbf{J}_{k4}$ can be grouped with ${}^{k2}\mathbf{J}_{k2}$ using the following equation:

$$ {}^{k2}\mathbf{JR}_{k2} = {}^{k2}\mathbf{J}_{k2} + {}^{k2}\mathbf{A}_{k4}\, {}^{k4}\mathbf{J}_{k4}\, {}^{k4}\mathbf{A}_{k2} \qquad [10.56] $$

10.3.6. *Practical computation of the base inertial parameters*

Most of the parameters to be eliminated or grouped can be computed by applying the following rules. Firstly, the joints r_1 and r_2 for each main branch of the equivalent tree structure must be determined. Then, apply the following rules for $j = n, \ldots, 1$:

1) if link j constitutes a link that is connected to a cut joint, that is to say $j = a(k)$ with $k > n$, then apply either the general grouping equations or the grouping equations of the parallelogram depending on the type of the corresponding loop;

2) group:
   a) YY_j, MZ_j and M_j if $\sigma_j = 0$, using Theorem 10.1;
   b) XX_j, XY_j, XZ_j, YY_j, YZ_j and ZZ_j if $\sigma_j = 1$, using Theorem 10.2;

3) if joint j is prismatic and $\mathbf{a}_j$ is parallel to $\mathbf{a}_{r1}$ for $r_1 < j < r_2$, then eliminate MZ_j and group MX_j and MY_j using equation [10.23];

4) if joint j is prismatic and $\mathbf{a}_j$ is not parallel to $\mathbf{a}_{r1}$ for $r_1 < j < r_2$, then group or eliminate one of the parameters MX_j, MY_j, MZ_j using Table 10.1;

5) if joint j is revolute and $r_1 \le j < r_2$, then eliminate XX_j, XY_j, XZ_j and YZ_j. Notice that the axis of this joint is parallel to the axis of joint r_1, and that the parameter YY_j has been eliminated by rule 2;

6) if j is revolute and $r_1 \le j < r_2$, and $\mathbf{a}_j$ is along $\mathbf{a}_{r1}$, and if $\mathbf{a}_{r1}$ is parallel to both $\mathbf{a}_i$ and gravity $\mathbf{g}$ for all $i < j$, then eliminate the parameters MX_j, MY_j. Notice that MZ_j has been eliminated by rule 2;

7) if j is prismatic and $j < r_1$, then eliminate the parameters MX_j, MY_j, MZ_j.

From this algorithm, we deduce that the number of minimum inertial parameters for the links (without considering the inertia of rotors) of a general robot is less than b_m, such that:

$$b_m \leq 7\, n_r + 4\, n_p - 4\, n_{r0} - 3\, n_{p0} - 11\, n_{par} - 2\, n_{g0} \qquad [10.57]$$

with:

- n_r: number of revolute joints of the equivalent tree structure;
- n_p: number of prismatic joints of the equivalent tree structure;
- n_{g0}: number of revolute joints that are directly connected to the base and whose axes are parallel to gravity;
- n_{r0}: number of revolute joints directly connected to the base;
- n_{p0}: number of prismatic joints directly connected to the base;
- n_{par}: number of parallelogram loops in the mechanism.

• **Example 10.2.** Computation of the base inertial parameters of the Acma SR400 robot. This structure has two main branches: the first contains links 1, ..., 6, while the second is composed of links 1, 7 and 8. For the first branch, we obtain $r_1 = 1$ and $r_2 = 2$; for the second branch, we find $r_1 = 1$ and $r_2 = 7$. Applying the general algorithm for j = 8, ..., 1, we obtain:

Link 8. This link constitutes a terminal link in a parallelogram loop. Equation [10.51c] gives:

$$h_{XX3}\, d_8^2 + h_{ZZ3}\, d_8^2 - h_{MY3}\, d_8 + h_{M3} = h_{YY8}\, d_3^2 + h_{ZZ8}\, d_3^2 + h_{MX8}\, d_3 + h_{M8}$$

We choose to group MX_8 as follows:

$$YYR_8 = YY_8 - d_3\, MX_8$$
$$ZZR_8 = ZZ_8 - d_3\, MX_8$$
$$MR_8 = M_8 - \frac{MX_8}{d_3}$$
$$XXR_3 = XX_3 + MX_8 \frac{d_8^2}{d_3}$$
$$ZZR_3 = ZZ_3 + MX_8 \frac{d_8^2}{d_3}$$
$$MYR_3 = MY_3 - MX_8 \frac{d_8}{d_3}$$

$$MR_3 = M_3 + \frac{MX_8}{d_3}$$

Equation [10.55] allows us to group $\mathbf{J}_8$ with $\mathbf{J}_2$:

$$^2\mathbf{JR}_2 = {}^2\mathbf{J}_2 + {}^2\mathbf{A}_8\, {}^8\mathbf{J}_8\, {}^8\mathbf{A}_2$$

Since the orientation matrix $^2\mathbf{A}_8 = \mathbf{I}_3$ (equation [7.39]), we obtain:

$$\begin{aligned}
XXR_2 &= XX_2 + XX_8 \\
XYR_2 &= XY_2 + XY_8 \\
XZR_2 &= XZ_2 + XZ_8 \\
YYR_2 &= YY_2 + (YY_8 - d_3\, MX_8) \\
YZR_2 &= YZ_2 + YZ_8 \\
ZZR_2 &= ZZ_2 + (ZZ_8 - d_3\, MX_8)
\end{aligned}$$

Finally, since joint 8 is revolute, we group the parameters MZ_8 and MR_8 with the parameters of link 7 using equations [10.19]:

$$\begin{aligned}
XZR_7 &= XZ_7 - MZ_8\, d_8 \\
YYR_7 &= YY_7 + M_8\, d_8 - MX_8 \frac{d_8{}^2}{d_3} \\
ZZR_7 &= ZZ_7 + d_8{}^2\, M_8 - MX_8 \frac{d_8{}^2}{d_3} \\
MXR_7 &= MX_7 - MX_8 \frac{d8}{d3} + M_8\, d_8 \\
MR_7 &= M_7 + M_8 - \frac{MX_8}{d_3}
\end{aligned}$$

Thus, concerning link 8, only the parameter MY_8 belongs to the base inertial parameters of the robot.

Link 7. This link is a terminal link of a parallelogram loop. We group $\mathbf{J}_7$ with $\mathbf{J}_3$ using equation [10.55]:

$$^3\mathbf{JR}_3 = {}^3\mathbf{J}_3 + {}^3\mathbf{A}_7\, {}^7\mathbf{J}_7\, {}^7\mathbf{A}_3$$

Since $^7\mathbf{A}_3 = \mathbf{rot}(\mathbf{z}, \frac{\pi}{2})$, (equation [7.38]), we obtain:

$$\begin{aligned}
XXR_3 &= XX_3 + YY_7 + M_8\, d_8{}^2 \\
XYR_3 &= XY_3 - XY_7 \\
XZR_3 &= XZ_3 + YZ_7 \\
YYR_3 &= YY_3 + XXR7 = YY_3 + XX_7
\end{aligned}$$

$$YZR_3 = YZ_3 - XZR_7 = YZ_3 - XZ_7 + MZ_8\, d_8$$
$$ZZR_3 = ZZ_3 + ZZ_7 + d_8^2\, M_8$$

We group the parameters MR_7 and MZ_7 with the parameters of link 1 using the following equation:

$$ZZR_1 = ZZ_1 + (M_7 + M_8)\, d_2^2 - MX_8 \frac{d_2^2}{d_3}$$

Note that MZ_7 does not appear in this expression. Thus, it has no effect on the dynamic model. The minimum parameters of link 7 are MXR_7 and MY_7.

Link 6. From Theorem 10.1 and since $a(6) = 5$, we group the parameters YY_6, MZ_6 and MR_6 as follows:

$$XXR_6 = XX_6 - YY_6$$
$$XXR_5 = XX_5 + YY_6$$
$$ZZR_5 = ZZ_5 + YY_6$$
$$MYR_5 = MY_5 + MZ_6$$
$$MR_5 = M_5 + M_6$$

The minimum parameters of link 6 are: XXR_6, XY_6, XZ_6, YZ_6, ZZ_6, MX_6 and MY_6.

Link 5. We group the parameters YY_5, MZ_5 and MR_5 with those of link 4:

$$XXR_5 = XX_5 + YY_6 - YY_5$$
$$XXR_4 = XX_4 + YY_5$$
$$ZZR_4 = ZZ_4 + YY_5$$
$$MYR_4 = MY_4 - MZ_5$$
$$MR_4 = M_4 + M_5 + M_6$$

The minimum parameters of link 5 are: XXR_5, XY_5, XZ_5, YZ_5, ZZR_5, MX_5 and MYR_5.

Link 4. We group the parameters YY_4, MZ_4 and MR_4 with those of link 3:

$$XXR_4 = XX_4 + YY_5 - YY_4$$
$$XXR_3 = XX3 + YY7 + M_8\, d_8^2 + YY_4 + 2\, RL_4\, MZ_4 + RL_4^2\, MR_4$$
$$XYR_3 = XY_3 - XY_7 - d_4\, MZ_4 - d_4\, RL_4\, MR_4$$
$$XZR_3 = XZ_3 + YZ_7$$
$$YYR_3 = YY3 + XX7 + d_4^2\, MR_4$$
$$YZR_3 = YZ_3 - XZ_7 + MZ_8\, d_8$$

$$\begin{aligned}
ZZR_3 &= ZZ_3 + ZZ_7 + d_8{}^2 M_8 + YY_4 + 2\, RL_4\, MZ_4 + (d_4{}^2 + RL_4{}^2)\, MR_4 \\
MXR_3 &= MX_3 + d_4\, MR_4 \\
MYR_3 &= MY_3 - d_8 \frac{MX_8}{d_3} + MZ_4 + RL_4\, MR_4 \\
MR_3 &= M_3 + \frac{MX_8}{d_3} + M_4 + M_5 + M_6
\end{aligned}$$

The minimum parameters of link 4 are: XXR_4, XY_4, XZ_4, YZ_4, ZZR_4, MX_4 and MYR_4.

Link 3. We group the parameters YYR_3, MZ_3 and MR_3 with those of link 2:

$$\begin{aligned}
XXR_3 &= XX_3 + YY_7 + M8\, d_8{}^2 + YY_4 + 2RL_4MZ_4 + RL_4{}^2MR_4 - YY_3 - XX_7 - d_4{}^2MR_4 \\
XXR_2 &= XX_2 + XX_8 + YY_3 + XX_7 + d_4{}^2\, MR_4 \\
XZR_2 &= XZ_2 + XZ_8 - d_3\, MZ_3 \\
YYR_2 &= YY_2 + YY_8 + YY_3 - d_3\, MX_8 + XX_7 + d_4{}^2\, MR_4 + d3^2\, MR_3 \\
ZZR_2 &= ZZ_2 + ZZ_8 - d_3\, MX_8 + d_3{}^2\, MR_3 \\
MXR_2 &= MX_2 + d_3\, MR_3 \\
MR_2 &= M_2 + MR_3
\end{aligned}$$

The minimum parameters of link 3 are: XXR_3, XYR_3, XZR_3, YZR_3, ZZR_3, MXR_3 and MYR_3.

Link 2. We group the parameters YYR_2, MZ_2 and MR_2 with those of link 1:

$$\begin{aligned}
XXR_2 &= XX_2 + XX_8 - YY_2 - YY_8 - d_3{}^2\, MR_3 + d_3\, MX_8 \\
ZZR_1 &= ZZ_1 + (M_7 + M_8)d_2{}^2 - d_2{}^2 \frac{MX_8}{d_3} + YY_2 + YY_8 + YY_3 + XX_7 + d_4{}^2MR_4 + \\
&\quad d_3{}^2MR_3 - d_3\, MX_8 + d_2{}^2MR_2
\end{aligned}$$

The minimum parameters of link 2 are: XXR_2, XYR_2, XZR_2, YZR_2, ZZR_2, MXR_2 and MY_2. Note that MZ_2 does not appear in this expression. Thus, it has no effect on the dynamic model.

Link 1. Only the parameter ZZR_1 belongs to the base inertial parameters.

Finally, the rotor inertias are treated as shown in § 9.4.2.5, leading to group I_{a1}, I_{a2} and I_{a7} with ZZR_1, ZZR_2 and ZZR_3 respectively.

The final result is summarized as follows:

- the following 11 parameters have no effect on the dynamic model: XX_1, XY_1, XZ_1, YY_1, YZ_1, MX_1, MY_1, MZ_1, M_1, MZ_2 and MZ_7;
- the following 33 parameters have been grouped: I_{a1}, YY_2, M_2, I_{a2}, YY_3, MZ_3, M_3, YY_4, MZ_4, M_4, YY_5, MZ_5, M_5, YY_6, MZ_6, M_6, XX_7, XY_7, XZ_7, YY_7, YZ_7, ZZ_7, M_7, I_{a7}, XX_8, XY_8, XZ_8, YY_8, YZ_8, ZZ_8, MX_8, MZ_8 and M_8;
- the SR400 robot has 42 base parameters (Table 10.2);
- the grouping equations are:

$$\begin{aligned}
ZZR_1 &= Ia_1 + ZZ_1 + YY_2 + YY_3 + XX_7 + YY_8 + d_4^2(M_4 + M_5 + M_6) + \\
&\quad d_2^2(M_3 + M_4 + M_5 + M_6) + d_3^2(M_2 + M_3 + M_4 + M_5 + M_6) + d_2^2(M_7 + M_8) \\
XXR_2 &= XX_2 - YY_2 + XX_8 - YY_8 - d3^2(M_3 + M_4 + M_5 + M_6) \\
XYR_2 &= XY_2 + XY_8 \\
XZR_2 &= XZ_2 + XZ_8 - d_3MZ_3 \\
YZR_2 &= YZ_2 + YZ_8 \\
ZZR_2 &= Ia_2 + ZZ_2 + ZZ_8 + d_3^2(M_3 + M_4 + M_5 + M_6) \\
MXR_2 &= MX_2 + MX_8 + d_3(M_3 + M_4 + M_5 + M_6) \\
XXR_3 &= XX_3 - YY_3 + YY_4 - XX_7 + YY_7 - d4^2(M_4 + M_5 + M_6) + 2MZ_4RL_4 + \\
&\quad (M_4 + M_5 + M_6)RL_4^2 + d_8^2M_8 \\
XYR_3 &= XY_3 - XY_7 - d_4MZ_4 - d_4RL_4(M_4 + M_5 + M_6) \\
XZR_3 &= XZ_3 + YZ_7 \\
YZR_3 &= YZ_3 - XZ_7 + d_8MZ_8 \\
ZZR_3 &= Ia_7 + ZZ_3 + YY4 + ZZ_7 + d_8^2M_8 + 2MZ_4RL_4 + (M4 + M5 + M6)*(d4^2 + RL4^2) \\
MXR_3 &= MX_3 + d_4(M_4 + M_5 + M_6) \\
MYR_3 &= MY_3 + MZ_4 + (M_4 + M_5 + M_6)RL_4 \\
XXR_4 &= XX_4 - YY_4 + YY_5 \\
ZZR_4 &= YY_5 + ZZ_4 \\
MYR_4 &= MY_4 - MZ_5 \\
XXR_5 &= XX_5 - YY_5 + YY_6 \\
ZZR5 &= YY6 + ZZ_5 \\
MYR_5 &= MY_5 + MZ_6 \\
XXR_6 &= XX_6 - YY_6 \\
MXR_7 &= MX_7 - \frac{d_8}{d_3}MX_8 + d_8M8
\end{aligned}$$

Table 10.3 illustrates the computational complexity of the inverse dynamic model for the Acma SR400 robot. Two cases are considered: general inertial parameters where all the parameters are assumed to have real values different from zero; and the simplified case where the links are assumed to be symmetric. For each case, the dynamic model is computed twice: firstly with the standard inertial parameters, and secondly with the base inertial parameters. We note that the real time computation of the inverse dynamic model for this robot can be realized using classical personal computers.

Table 10.2. *Base inertial parameters of the Acma SR400 robot*

j	XXj	XYj	XZj	YYj	YZj	ZZj	MXj	MYj	MZj	Mj	Iaj
1	0	0	0	0	0	ZZR1	0	0	0	0	0
2	XXR2	XYR2	XZR2	0	YZR2	ZZR2	MXR2	MY2	0	0	0
3	XXR3	XYR3	XZR3	0	YZR3	ZZR3	MXR3	MYR3	0	0	0
4	XXR4	XY4	XZ4	0	YZ4	ZZR4	MX4	MYR4	0	0	Ia4
5	XXR5	XY5	XZ5	0	YZ5	ZZR5	MX5	MYR5	0	0	Ia5
6	XXR6	XY6	XZ6	0	YZ6	ZZ6	MX6	MY6	0	0	Ia6
7	0	0	0	0	0	0	MXR7	MY7	0	0	0
8	0	0	0	0	0	0	0	MY8	0	0	0

Table 10.3. *Computational complexity of the inverse dynamic model of the Acma SR400 robot*

Set of inertial parameters	Complete		Simplified	
	Multiplicat.	Additions	Multiplicat.	Additions
Standard parameters	430	420	295	245
Base parameters	304	326	243	118

10.4. Conclusion

In this chapter, we have developed the dynamics of robots with tree structure or containing closed chains. This treatment constitutes a generalization of the results presented in Chapter 9 for serial robots. We can use the efficient Newton-Euler method for computing the inverse and direct dynamic models of tree structured systems. The corresponding base inertial parameters can be determined using the symbolic algorithm, which is composed of simple rules and makes use of closed form grouping equations. Concerning the systems with closed chain, the inverse dynamic model is computed from the inverse dynamic model of the equivalent tree structure and the Jacobian matrix representing the derivative of the joint positions of the equivalent tree structure with respect to the actuated joint positions. The base parameters of general closed chain robots can be completely determined using the numerical method presented in Appendix 5. However, most of them and even all of them in many cases can be computed using the rules of the symbolic algorithm.

From this study, we can conclude that the computation of the dynamic model in real time is now possible using classical personal computers. In Chapters 11 and 12, we direct our attention toward the identification of the geometric and dynamic parameters appearing in the different models of robots.

Chapter 11

Geometric calibration of robots

11.1. Introduction

A high level of positioning accuracy is an essential requirement in a wide range of applications involving industrial robots. This accuracy is affected by geometric factors, such as geometric parameter errors, as well as non-geometric factors, such as flexibility of links and gear trains, gear backlashes, encoder resolution errors, wear, and thermal effects. Positioning accuracy of an industrial robot can be improved to approach its repeatability by a calibration procedure that determines current values of the geometrical dimensions and mechanical characteristics of the structure. Practical techniques to compensate for all geometric and non-geometric effects are not yet developed. Based on investigation of the error contribution from various sources, Judd and Knasinski concluded that the error due to geometric factors accounted for 95% of the total error [Judd 90]. Hence, a reasonable approach would be to calibrate the current geometric parameters and treat the non-geometric factors as a randomly-distributed error. This calibration procedure is also important for robot programming using CAD systems where the simulated robot must reflect accurately the real robot [Craig 93], [Dombre 94], [Chedmail 98]. In recent years, considerable attention has been paid to the problem of geometric calibration. A partial list of these works is given in references [Schefer 82], [Wu 84], [Khalil 85b], [Payannet 85], [Sugimoto 85], [Aldon 86], [Veitschegger 86], [Whitney 86], [Roth 87], [Hollerbach 89], [Mooring 91], [Lavallée 92], [Caenen 93], [Guyot 95], [Damak 96], [Maurine 96], [Besnard 00a].

The problem of geometric calibration can be divided into four distinct steps. The first step is concerned with a particular mathematical formulation that results in a model, which is a function of the geometric parameters $\boldsymbol{\xi}$, the joint variables $\mathbf{q}$, and eventually some external measurements $\mathbf{x}$. The second step is devoted to the collection of experimental data for a sufficient number of configurations. The third step is concerned with the identification of the geometric parameters and validation of the results obtained. The last step is concerned with compensating the geometric parameter errors in the direct and inverse geometric models.

This chapter addresses all the four steps outlined above for serial robots. The case of the parallel robot is covered briefly at the end.

11.2. Geometric parameters

The geometric parameters concerned with geometric calibration are the parameters required to compute the direct and inverse geometric models. The direct geometric model, which gives the location of the end-effector frame R_{n+1} relative to a fixed world frame R_{-1}, is given by equation [3.13] and is rewritten as:

$$^{-1}\mathbf{T}_{n+1} = \mathbf{Z}\,{}^{0}\mathbf{T}_n(\mathbf{q})\,\mathbf{E} = {}^{-1}\mathbf{T}_0\,{}^{0}\mathbf{T}_1\,{}^{1}\mathbf{T}_2 \ldots {}^{n-1}\mathbf{T}_n\,{}^{n}\mathbf{T}_{n+1} \qquad [11.1]$$

where:

- $\mathbf{Z} = {}^{-1}\mathbf{T}_0$ denotes the transformation matrix defining the robot base frame R_0 relative to the world reference frame R_{-1};
- $\mathbf{E} = {}^{n}\mathbf{T}_{n+1}$ is the transformation matrix defining the end-effector frame with respect to the terminal link frame R_n;
- ${}^{0}\mathbf{T}_n$ is the transformation matrix of the robot defining frame R_n relative to frame R_0.

Equation [11.1] contains three kinds of parameters: the robot geometric parameters appearing in ${}^{0}\mathbf{T}_n$, the base frame parameters defining the matrix $\mathbf{Z}$ and the end-effector parameters defining the matrix $\mathbf{E}$. We add to these parameters the joint gear transmission ratios that can be calibrated in the same manner as the geometric parameters.

For convenience, in the remainder of the chapter, we denote the origin O_{n+1} of the end-effector frame R_{n+1} as the *endpoint* of the robot.

11.2.1. *Robot parameters*

The robot parameters are deduced from the notations developed in Chapter 3. According to these notations, frame R_j is fixed with link j. It is located relative to frame R_{j-1} by the homogeneous transformation matrix ${}^{j-1}\mathbf{T}_j$, which is a function of the four geometric parameters (α_j, d_j, θ_j, r_j) (Figure 3.2), such that:

$$^{j-1}\mathbf{T}_j = \mathbf{Rot}(\mathbf{x}, \alpha_j)\,\mathbf{Trans}(\mathbf{x}, d_j)\,\mathbf{Rot}(\mathbf{z}, \theta_j)\,\mathbf{Trans}(\mathbf{z}, r_j) \qquad [11.2]$$

Note that the parameters α_1 and d_1 can be taken to be equal to zero by assigning the base frame R_0 aligned with frame R_1 when $q_1 = 0$.

If two consecutive joint axes $j-1$ and j are parallel, the $\mathbf{x}_{j-1}$ axis is taken arbitrarily along one of the common normals between them. When $\mathbf{z}_{j-1}$ or $\mathbf{z}_j$ becomes slightly misaligned, the common normal is uniquely defined and the corresponding variation in the parameter r_{j-1} can be very large. To ensure that small variation in axis alignment produces proportionally small variations in the parameters, we make use of a fifth parameter β_j [Hayati 83] representing a rotation around the $\mathbf{y}_{j-1}$ axis. The general transformation matrix ${}^{j-1}\mathbf{T}_j$ becomes:

$$ {}^{j-1}\mathbf{T}_j = \mathbf{Rot}(\mathbf{y}, \beta_j)\, \mathbf{Rot}(\mathbf{x}, \alpha_j)\, \mathbf{Trans}(\mathbf{x}, d_j)\, \mathbf{Rot}(\mathbf{z}, \theta_j)\, \mathbf{Trans}(\mathbf{z}, r_j) \qquad [11.3] $$

The nominal value of β_j is zero. If $\mathbf{z}_{j-1}$ and $\mathbf{z}_j$ are not parallel, β_j is not identifiable. We note that when $\mathbf{z}_{j-1}$ and $\mathbf{z}_j$ are parallel, we can identify either r_{j-1} or r_j (§ 11.4.2), thus the number of identifiable parameters for each frame is at most four.

11.2.2. *Parameters of the base frame*

Since the reference frame can be chosen arbitrarily by the user, six parameters are needed to locate the robot base relative to the world frame. As developed in § 7.2, these parameters can be taken as $(\gamma_z, b_z, \alpha_z, d_z, \theta_z, r_z)$ (Figure 11.1):

$$ \mathbf{Z} = {}^{-1}\mathbf{T}_0 = \mathbf{Rot}(\mathbf{z}, \gamma_z)\, \mathbf{Trans}(\mathbf{z}, b_z)\, \mathbf{Rot}(\mathbf{x}, \alpha_z)\, \mathbf{Trans}(\mathbf{x}, d_z)\, \mathbf{Rot}(\mathbf{z}, \theta_z)\, \mathbf{Trans}(\mathbf{z}, r_z) \qquad [11.4] $$

The transformation matrix ${}^{-1}\mathbf{T}_1$ is given by:

$$ {}^{-1}\mathbf{T}_1 = {}^{-1}\mathbf{T}_0\, {}^{0}\mathbf{T}_1 = \mathbf{Rot}(\mathbf{z}, \gamma_z)\, \mathbf{Trans}(\mathbf{z}, b_z)\, \mathbf{Rot}(\mathbf{x}, \alpha_z)\, \mathbf{Trans}(\mathbf{x}, d_z)\, \mathbf{Rot}(\mathbf{z}, \theta_z)\, \mathbf{Trans}(\mathbf{z}, r_z)\, \mathbf{Rot}(\mathbf{x}, \alpha_1)\, \mathbf{Trans}(\mathbf{x}, d_1)\, \mathbf{Rot}(\mathbf{z}, \theta_1)\, \mathbf{Trans}(\mathbf{z}, r_1) \qquad [11.5] $$

Since $\alpha_1 = 0$ and $d_1 = 0$, we can write that:

$$ {}^{-1}\mathbf{T}_0\, {}^{0}\mathbf{T}_1 = \mathbf{Rot}(\mathbf{x}, \alpha_0)\, \mathbf{Trans}(\mathbf{x}, d_0)\, \mathbf{Rot}(\mathbf{z}, \theta_0)\, \mathbf{Trans}(\mathbf{z}, r_0)\, \mathbf{Rot}(\mathbf{x}, \alpha'_1)\, \mathbf{Trans}(\mathbf{x}, d'_1)\, \mathbf{Rot}(\mathbf{z}, \theta'_1)\, \mathbf{Trans}(\mathbf{z}, r'_1) \qquad [11.6] $$

with $\alpha_0 = 0$, $d_0 = 0$, $\theta_0 = \gamma_z$, $r_0 = b_z$, $\alpha'_1 = \alpha_z$, $d'_1 = d_z$, $\theta'_1 = \theta_1 + \theta_z$, $r'_1 = r_1 + r_z$

Equation [11.6] represents two transformations having the same kind of parameters as those of equation [11.2]. We note that the parameters θ_z and r_z are grouped with θ_1 and r_1 respectively. This proves that consecutive frames are represented at most by four independent parameters.

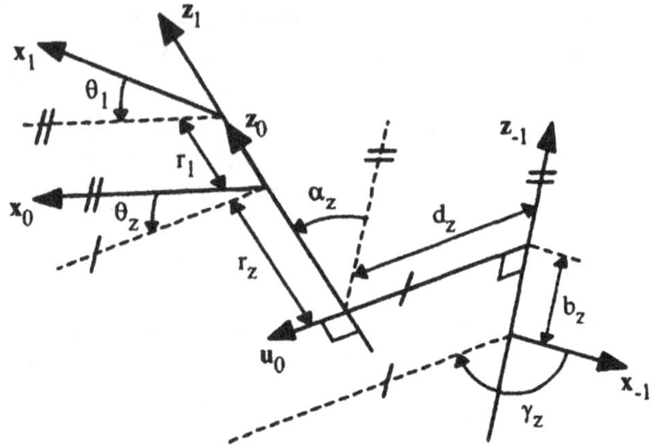

Figure 11.1. *Description of frame R_0 relative to frame R_{-1}*

11.2.3. *End-effector parameters*

Since the end-effector frame R_{n+1} can be defined arbitrarily with respect to the terminal link frame R_n, six parameters (γ_e, b_e, α_e, d_e, θ_e, r_e) are needed to define the matrix **E**. As previously, we can extend the robot notations to the definition of the end-effector frame:

$$^{n}\mathbf{T}_{n+1} = \mathbf{Rot}(\mathbf{z}, \gamma_e)\, \mathbf{Trans}(\mathbf{z}, b_e)\, \mathbf{Rot}(\mathbf{x}, \alpha_e)\, \mathbf{Trans}(\mathbf{x}, d_e)\, \mathbf{Rot}(\mathbf{z}, \theta_e)\, \mathbf{Trans}(\mathbf{z}, r_e) \qquad [11.7]$$

The transformation matrix $^{n-1}\mathbf{T}_{n+1}$ can be written as:

$$\begin{aligned} ^{n-1}\mathbf{T}_{n+1} &= {}^{n-1}\mathbf{T}_n\, {}^{n}\mathbf{T}_{n+1} \\ &= \mathbf{Rot}(\mathbf{x}, \alpha_n)\, \mathbf{Trans}(\mathbf{x}, d_n)\, \mathbf{Rot}(\mathbf{z}, \theta_n)\, \mathbf{Trans}(\mathbf{z}, r_n)\, \mathbf{Rot}(\mathbf{z}, \gamma_e) \\ &\quad \mathbf{Trans}(\mathbf{z}, b_e)\, \mathbf{Rot}(\mathbf{x}, \alpha_e)\, \mathbf{Trans}(\mathbf{x}, d_e)\, \mathbf{Rot}(\mathbf{z}, \theta_e)\, \mathbf{Trans}(\mathbf{z}, r_e) \end{aligned} \qquad [11.8]$$

which gives:

$$^{n-1}\mathbf{T}_{n+1} = \mathbf{Rot}(\mathbf{x}, \alpha_n)\, \mathbf{Trans}(\mathbf{x}, d_n)\, \mathbf{Rot}(\mathbf{z}, \theta'_n)\, \mathbf{Trans}(\mathbf{z}, r'_n)\, \mathbf{Rot}(\mathbf{x}, \alpha_{n+1})\, \mathbf{Trans}(\mathbf{x}, d_{n+1})\, \mathbf{Rot}(\mathbf{z}, \theta_{n+1})\, \mathbf{Trans}(\mathbf{z}, r_{n+1}) \qquad [11.9]$$

with $\theta'_n = \theta_n + \gamma_e$, $r'_n = r_n + b_e$, $\alpha_{n+1} = \alpha_e$, $d_{n+1} = d_e$, $\theta_{n+1} = \theta_e$, $r_{n+1} = r_e$.

Thus, the end-effector frame introduces four independent parameters α_e, d_e, θ_e, r_e, whereas the parameters γ_e and b_e are grouped with θ_n and r_n respectively.

Finally, the description of the location of the end-effector frame R_{n+1} in the reference frame R_{-1} of an n degree-of-freedom robot, needs at most $(4n + 6)$ independent parameters. More precisely, since in the case of a prismatic joint only two parameters can be identified, the maximum number of independent parameters reduces to $(4n_r + 2n_p + 6)$, where n_r and n_p are the numbers of revolute and prismatic joints of the robot respectively [Everett 88].

11.3. Generalized differential model of a robot

The generalized differential model provides the differential variation of the location of the end-effector as a function of the differential variation of the geometric parameters. It is represented by:

$$\Delta \mathbf{X} = \begin{bmatrix} \mathbf{dP}_{n+1} \\ \boldsymbol{\delta}_{n+1} \end{bmatrix} = \boldsymbol{\Psi}\, \Delta\boldsymbol{\xi} \qquad [11.10]$$

with:

- $\mathbf{dP}_{n+1}$: (3x1) differential translation vector of the origin O_{n+1};
- $\boldsymbol{\delta}_{n+1}$: (3x1) differential rotation vector of frame R_{n+1};
- $\boldsymbol{\Psi}$: $(6xN_{par})$ generalized Jacobian matrix;
- $\Delta\boldsymbol{\xi}$: $(N_{par}x1)$ vector of the differential variation of the geometric parameters.

The columns of the generalized Jacobian matrix $\boldsymbol{\Psi}$ can be computed using simple vector relationships as we did in Chapter 5 for the computation of the base Jacobian matrix. According to the kind of parameter, we obtain [Khalil 89c]:

i) column corresponding to the parameter $\Delta\beta_i$: the parameter β_i represents a rotation about the y_{i-1} axis. A differential variation on β_i generates a differential position on frame R_{n+1} equal to $(\mathbf{n}_{i-1} \times \mathbf{L}_{i-1,n+1})\, \Delta\beta_i$ and a differential orientation equal to $\mathbf{n}_{i-1}\, \Delta\beta_i$. The column of $\boldsymbol{\Psi}$ corresponding to the parameter $\Delta\beta_i$ is given by:

$$\boldsymbol{\Psi}\beta_i = \begin{bmatrix} \mathbf{n}_{i-1} \times \mathbf{L}_{i-1,n+1} \\ \mathbf{n}_{i-1} \end{bmatrix} \qquad [11.11]$$

where $\mathbf{L}_{i-1,n+1}$ is the vector connecting O_{i-1} to O_{n+1}, and $\mathbf{n}_{i-1}$ is the unit vector along the y_{i-1} axis;

ii) column corresponding to the parameter $\Delta\alpha_i$: the parameter α_i represents a rotation about the $\mathbf{x}_{i-1}$ axis. A differential variation on α_i generates a differential position on frame R_{n+1} equal to $(\mathbf{s}_{i-1} \times \mathbf{L}_{i-1,n+1})\ \Delta\alpha_i$ and a differential orientation equal to $\mathbf{s}_{i-1}\ \Delta\alpha_i$. The column of $\boldsymbol{\Psi}$ corresponding to the parameter $\Delta\alpha_i$ is given by:

$$\boldsymbol{\Psi}\alpha_i = \begin{bmatrix} \mathbf{s}_{i-1} \times \mathbf{L}_{i-1,n+1} \\ \mathbf{s}_{i-1} \end{bmatrix} \qquad [11.12]$$

where $\mathbf{s}_{i-1}$ is the unit vector along the $\mathbf{x}_{i-1}$ axis;

iii) column corresponding to the parameter Δd_i: the parameter d_i represents a translation along the $\mathbf{x}_{i-1}$ axis. A differential variation on d_i generates a differential position on frame R_{n+1} equal to $\mathbf{s}_{i-1}\ \Delta d_i$, but produces no differential orientation. Thus, the corresponding column is expressed as:

$$\boldsymbol{\Psi}d_i = \begin{bmatrix} \mathbf{s}_{i-1} \\ \mathbf{0}_{3x1} \end{bmatrix} \qquad [11.13]$$

iv) column corresponding to the parameter $\Delta\theta_i$: this case has been handled in Chapter 5 while calculating the base Jacobian matrix. The corresponding column is given by:

$$\boldsymbol{\Psi}\theta_i = \begin{bmatrix} \mathbf{a}_i \times \mathbf{L}_{i,n+1} \\ \mathbf{a}_i \end{bmatrix} \qquad [11.14]$$

where $\mathbf{a}_i$ is the unit vector along the $\mathbf{z}_i$ axis;

v) column corresponding to the parameter Δr_i: this case has also been developed in Chapter 5. The corresponding column is given by:

$$\boldsymbol{\Psi}r_i = \begin{bmatrix} \mathbf{a}_i \\ \mathbf{0}_{3x1} \end{bmatrix} \qquad [11.15]$$

vi) column corresponding to a differential variation in the gear transmission ratio: Let us denote $K_i = 1/N_i$, where N_i is the gear transmission ratio. In general, the joint variables are given by:

$$\mathbf{q} = \mathbf{C}_a\ \mathbf{diag}(K_1, ..., K_n)\ \mathbf{C}_m\ \mathbf{q}_m + \mathbf{q}_0 \qquad [11.16]$$

with:

- $\mathbf{C}_m$: (nxn) matrix representing the coupling between the motor variables;

- $\mathbf{C_a}$: (nxn) matrix representing the coupling between the variables after the gear transmission;
- $\mathbf{q}_m$: vector of motor variables;
- $\mathbf{q}_0$: constant vector representing the offset values on the joint variables.

Thus, the column corresponding to K_i is given by:

$$\Psi_{ki} = \mathbf{J}_{n+1} \frac{\partial \mathbf{q}}{\partial K_i} \qquad [11.17]$$

where $\mathbf{J}_{n+1}$ is the base Jacobian matrix of the robot (§ 5.3) whose i^{th} column is either Ψr_i if joint i is prismatic or $\Psi \theta_i$ if joint i is revolute.

If the joints are actuated independently, we have:

$$\begin{cases} \Psi k_i = q_{mi}\, \Psi r_i & \text{if joint i is prismatic} \\ \Psi k_i = q_{mi}\, \Psi \theta_i & \text{if joint i is revolute} \end{cases} \qquad [11.18]$$

NOTE.– All the vectors used in the computation of Ψ are derived from the matrices $^{-1}\mathbf{T}_i$, i = 0, ..., n + 1. The vectors $^{-1}\mathbf{s}_i$,$^{-1}\mathbf{n}_i$ and $^{-1}\mathbf{a}_i$ are obtained directly from these matrices, whereas the vector $\mathbf{L}_{i,n+1}$ is computed using the following equation:

$$^{-1}\mathbf{L}_{i,n+1} = {}^{-1}\mathbf{P}_{n+1} - {}^{-1}\mathbf{P}_i \qquad [11.19]$$

11.4. Principle of geometric calibration

The calibration of the geometric parameters is based on estimating the parameters minimizing the difference between a function of the real robot variables and its mathematical model. The methods proposed in the literature differ according to the variables used to define this function. In § 11.5, we will present some of these methods. In the following section, they are formulated using a unified approach.

11.4.1. *General calibration model*

The calibration model can be represented by the general nonlinear equation [11.20] and by the general linearized equation [11.21]:

$$\mathbf{0} = \mathbf{f}(\mathbf{q}, \mathbf{x}, \boldsymbol{\xi}) \qquad [11.20]$$

$$\Delta\mathbf{y}(\mathbf{q}, \mathbf{x}, \boldsymbol{\xi}) = \boldsymbol{\phi}(\mathbf{q}, \boldsymbol{\xi})\, \Delta\boldsymbol{\xi} \qquad [11.21]$$

where:

- $\mathbf{x}$ represents the external measured variables, such as the Cartesian variables giving the position and orientation of the end-effector frame, or the distance traveled by the endpoint between two configurations;
- $\mathbf{q}$ is the (nx1) vector of the joint variables;
- $\boldsymbol{\xi}$ is the (N_{par}x1) vector of the geometric parameters;
- $\boldsymbol{\phi}$ is the (pxN_{par}) calibration Jacobian matrix, whose elements are computed as functions of the generalized Jacobian matrix $\boldsymbol{\Psi}$;
- $\Delta\mathbf{y}$ is the (px1) prediction error vector.

To estimate $\Delta\boldsymbol{\xi}$, we apply equation [11.20] and/or equation [11.21] for a sufficient number of configurations. Combining all the equations results in the following nonlinear and linear systems of (pxe) equations, where e is the number of configurations:

$$\mathbf{0} = \mathbf{F}(\mathbf{Q}_t, \mathbf{X}_t, \boldsymbol{\xi}) + \boldsymbol{\rho}' \qquad [11.22]$$

$$\Delta\mathbf{Y} = \mathbf{W}(\mathbf{Q}_t, \boldsymbol{\xi})\, \Delta\boldsymbol{\xi} + \boldsymbol{\rho} \qquad [11.23]$$

with:

$$\mathbf{F} = \begin{bmatrix} \mathbf{f}(\mathbf{q}^1, \mathbf{x}^1, \boldsymbol{\xi}) \\ \cdots \\ \mathbf{f}(\mathbf{q}^e, \mathbf{x}^e, \boldsymbol{\xi}) \end{bmatrix} \qquad [11.24]$$

$$\Delta\mathbf{Y} = \begin{bmatrix} \Delta\mathbf{y}^1(\mathbf{q}^1, \mathbf{x}^1, \boldsymbol{\xi}) \\ \cdots \\ \Delta\mathbf{y}^e(\mathbf{q}^e, \mathbf{x}^e, \boldsymbol{\xi}) \end{bmatrix} \qquad [11.25]$$

where $\mathbf{Q}_t = [\mathbf{q}^{1T} \ldots \mathbf{q}^{eT}]^T$, $\mathbf{X}_t = [\mathbf{x}^{1T} \ldots \mathbf{x}^{eT}]^T$ and $\mathbf{W}$ is the (rxN_{par}) observation matrix. $\boldsymbol{\rho}$ and $\boldsymbol{\rho}'$ are the modeling error vector for the nonlinear and linear models respectively, including the effects of unmodeled non-geometric parameters:

$$\mathbf{W} = \begin{bmatrix} \boldsymbol{\phi}^1(\mathbf{q}^1, \boldsymbol{\xi}) \\ \cdots \\ \boldsymbol{\phi}^e(\mathbf{q}^e, \boldsymbol{\xi}) \end{bmatrix} \qquad [11.26]$$

The number of configurations e must be chosen such that the number of equations, $r = p \times e$, is greater than N_{par}. In practice, good results can be obtained by taking $r \geq 5N_{par}$ and by choosing configurations optimizing the observability measure (§ 11.4.2.2).

Equation [11.22] and/or equation [11.23] can be used to estimate the geometric parameters. However, before solving these equations, we have to rewrite them such that the unknown vector is only composed of the identifiable parameters. Of course, if a parameter is exactly known, it will not be included in the unknown vector $\boldsymbol{\xi}$.

NOTES.–

- the errors corresponding to the joint variables represent the joint offset errors;
- the number of equations of the function **f** is also called the *calibration index* [Hollerbach 96];
- in autonomous calibration methods, both **f** and $\Delta\mathbf{y}$ are computed in terms of the joint variables and the robot parameters. No external measuring device is needed.

11.4.2. *Identifiability of the geometric parameters*

It may happen that some parameters are not uniquely determined by the identification equation. All sources of parameter ambiguity can be linked to the rank of the matrix **W**. If some columns of **W** are linearly dependent, then the corresponding parameters may vary arbitrarily such that these variations only satisfy the linear dependence. For example, in the conventional calibration method, which uses the measurements of the end-effector location or position (§ 11.5.1), if the joint axes $i-1$ and i are parallel, then the columns corresponding to the parameters r_{i-1} and r_i are equal in the calibration Jacobian matrix ($\Psi r_{i-1} = \Psi r_i$). Thus, an infinite set of solutions can be obtained for the errors in r_{i-1} and r_i. The basic solution consists of identifying the error in one of these parameters while assuming that the error in the other parameter is zero.

The loss of identifiability of some parameters may be caused by two kinds of problems, namely structural identifiability and selection of calibration configurations. Unidentifiability of some parameters constitutes a structural problem when some columns of the observation matrix are zero or are linearly dependent whatever the number and the values of the configurations used to construct the observation matrix. The structurally unidentifiable parameters depend on the calibration model and on the structure of the robot. In § 11.4.2.1, we will give an algorithm to determine the structurally identifiable parameters.

The problem of the identifiability as a function of the calibration configurations is known as the *excitation problem*. It will be addressed in § 11.4.2.2 by selecting the calibration configurations that optimize an observability measure. For example, it is obvious that if one joint has a constant value in all the calibration configurations, some parameters concerning this joint will not be identified.

The determination of the identifiable geometric parameters is based on the determination of the independent columns of the calibration Jacobian matrix $\boldsymbol{\phi}$. A symbolic method is presented in [Khalil 91b] to determine these parameters for the conventional calibration methods. However, in the following, we develop a general method based on the QR decomposition to determine the identifiable parameters. This method is similar to that which has been presented in Appendix 5 for the determination of the dynamic base parameters.

11.4.2.1. *Determination of the identifiable parameters*

Numerically, the study of the identifiable parameters, also termed *base geometric parameters*, is equivalent to the study of the space spanned by the columns of an (rxNpar) matrix **W** similar to that defined in equation [11.23] but obtained using random configurations satisfying the constraints of the calibration method. The identifiable parameters are determined through the following three steps:

- if a column of **W** is zero, then the corresponding parameter has no effect on the geometric calibration model. Eliminating such parameters and the corresponding columns reduces **W** to an (rxc) matrix; for convenience, we continue to indicate this new matrix by **W**;
- the rank of **W**, denoted by b, gives the number of identifiable parameters;
- a set of identifiable parameters can be chosen as those corresponding to b independent columns of **W**. The other parameters are not identifiable.

To carry out the last two steps, we make use of the QR decomposition of the (rxc) matrix **W**. The matrix **W** can be written as [Dongarra 79], [Lawson 74], [Golub 83]:

$$\mathbf{W} = \mathbf{Q}\begin{bmatrix} \mathbf{R} \\ \mathbf{0}_{(r-c)xc} \end{bmatrix} \qquad [11.27]$$

where **Q** is an (rxr) orthogonal matrix, **R** is a (cxc) upper triangular matrix, and $\mathbf{0}_{ixj}$ is the (ixj) null matrix.

Theoretically, the non-identifiable parameters are those whose corresponding elements on the diagonal of the matrix **R** are zero. In practice, they are defined using a numerical tolerance $\tau \neq 0$ [Forsythe 77]. Thus, if $|R_{ii}|$, which represents the absolute value of the (i, i) element of **R**, is less than τ, the corresponding parameter is not identifiable. The numerical zero τ can be taken as [Dongarra 79]:

$$\tau = r.\varepsilon.\max |R_{ii}| \qquad [11.28]$$

where ε is the computer precision, and r is the number of rows.

It is obvious that the identifiable parameters are not uniquely defined. The QR method will provide as base parameters those corresponding to the first b independent columns of the matrix **W**. It is convenient to identify, if possible, the parameters that can be updated in the control system without changing the closed-form geometric models of the robot. Therefore, we permute the columns of **W** and the elements of ξ to first place the following parameters:

- the joint offsets and the gear transmission ratios;
- the parameters r_j and d_j whose nominal values are not zero;
- the angles α_j and θ_j whose nominal values are not $k\,\pi/2$, where k is an integer;
- the parameters defining the matrices **Z** and **E**.

Having determined the identifiable parameters, the linearized identification equation [11.23] is rewritten as:

$$\Delta \mathbf{Y} = \mathbf{W}_b \Delta \xi_b + \rho \qquad [11.29]$$

$\mathbf{W}_b$ is composed of the columns of **W** corresponding to the identifiable parameters. The nonlinear model will be solved in ξ_b. In the remainder of the chapter, the subscript b will be dropped for simplicity. Hence, **W** and $\Delta\xi$ will stand for $\mathbf{W}_b$ and $\Delta\xi_b$ respectively.

11.4.2.2. *Optimum calibration configurations*

The goal is to select a set of robot configurations that yield maximum observability of the model parameters and minimize the effect of noise on the parameter estimation. The condition number of the observation matrix **W** (Appendix 4) gives a good estimate of the observability of the parameters [Driels 90], [Khalil 91b]. Thus, optimum calibration configurations provide a condition number of **W** close to one. We have either to determine the calibration configurations by solving a nonlinear optimization problem that minimizes the condition number, or to verify that the randomly collected data give a good condition number.

The optimization problem has (exn) unknowns (n is the number of joints, e is the number of configurations). It can be formulated as follows [Khalil 91b]:

> <u>Find the configurations</u> $\mathbf{Q}_t = [\mathbf{q}^{1T} \dots \mathbf{q}^{eT}]^T$
> <u>minimizing the criterion</u> $C(\mathbf{Q}_t) = \text{cond}[\mathbf{W}(\mathbf{Q}_t)] = \|\mathbf{W}\|.\|\mathbf{W}^+\|$,
> <u>under the calibration method constraints and the following joint limit constraints:</u>
> $q_{i,min} \leq \mathbf{Q}_t\,[i+(j-1)n] \leq q_{i,max}$ where i = 1, ..., n and j = 1, ..., e

where $\mathbf{q}^j$ is the (nx1) joint position vector corresponding to configuration j, $\mathbf{W}^+$ is the pseudoinverse of **W**, $\|\mathbf{W}\|$ is a norm of **W**, and $q_{i,min}$, $q_{i,max}$ give the minimum and maximum values of the position of joint i respectively.

We recall that, when using the 2-norm, the condition number is given by (Appendix 4):

$$\text{cond}_2(\mathbf{W}) = \frac{\sigma_{max}}{\sigma_{min}} \qquad [11.30]$$

where σ_{max} and σ_{min} are the largest and smallest singular values of **W**.

This algorithm has been applied in [Khalil 91b] for the conventional calibration methods that use the Cartesian end-effector coordinates (§ 11.5.1). The optimization algorithm was based on the gradient conjugate method proposed by Powell [Powell 64].

Other observability measures have been proposed in the literature, namely the smallest singular value [Nahvi 94], and the product of the singular values of **W** [Borm 91], but the condition number is shown to be more efficient [Hollerbach 95].

It is worth noting that most of the geometric calibration methods give an acceptable condition number using random configurations [Khalil 00b].

11.4.3. *Solution of the identification equation*

The calibrated geometric parameters are obtained by solving the nonlinear algebraic equation [11.22] in order to minimize the least-squares error:

$$\hat{\xi} = \min_{\Delta\xi} \|\mathbf{F}\|^2$$

This optimization problem can be performed using the Levenberg-Marquardt algorithm, which is implemented in Matlab.

For rigid robot calibration, the linearized model can be used to solve iteratively this nonlinear optimization problem. Equation [11.23] is solved to get the least-squares error solution to the current parameter estimate. This procedure is iterated until the variation $\Delta\xi$ approaches zero and the parameters have converged to some stable value. At each iteration, the geometric parameters are updated by adding $\Delta\xi$ to the current value of ξ. The observation matrix **W** and the prediction error $\Delta\mathbf{Y}$ are updated as well.

The least-squares solution $\hat{\Delta\xi}$ of equation [11.23] is written as:

$$\hat{\Delta\xi} = \min_{\Delta\xi} \|\rho\|^2 = \min_{\Delta\xi} \|\Delta\mathbf{Y} - \mathbf{W}\,\hat{\Delta\xi}\|^2 \qquad [11.31]$$

The solution can be obtained using the pseudoinverse matrix (Appendix 4):

$$\hat{\Delta\xi} = \mathbf{W}^+ \,\Delta\mathbf{Y} \qquad [11.32]$$

where $\mathbf{W}^+$ denotes the pseudoinverse matrix of **W**. If **W** is of full rank, the explicit computation of $\mathbf{W}^+$ is given by $(\mathbf{W}^T\,\mathbf{W})^{-1}\,\mathbf{W}^T$.

In general, for rigid robots, the iterative least-squares method converges much faster than the Levenberg-Marquardt algorithm.

Standard deviation of the parameter estimation errors is calculated using the matrix **W** as a function of the estimated geometric parameters. Assuming that **W** is deterministic, and ρ is a zero mean additive independent noise with standard deviation σ_ρ, the variance-covariance matrix $\mathbf{C}_\rho$ is given by:

$$\mathbf{C}_\rho = E(\rho\,\rho^T) = \sigma_\rho^2\,\mathbf{I}_r \qquad [11.33]$$

where E is the expectation operator and $\mathbf{I}_r$ is the (rxr) identity matrix.

An unbiased estimation of σ_ρ can be computed using the following equation:

$$\sigma_\rho^2 = \frac{\|\Delta\mathbf{Y} - \mathbf{W}\,\hat{\Delta\xi}\|^2}{(r-c)} \qquad [11.34]$$

The variance-covariance matrix of the estimation error is given by [de Larminat 77]:

$$\mathbf{C}_{\hat{\xi}} = E[(\xi - \hat{\xi})\,(\xi - \hat{\xi})^T] = \mathbf{W}^+\,\mathbf{C}_\rho\,(\mathbf{W}^+)^T = \sigma_\rho^2\,(\mathbf{W}^T\,\mathbf{W})^{-1} \qquad [11.35]$$

The standard deviation of the estimation error on the jth parameter is obtained from the (j, j) element of $\mathbf{C}_{\hat{\xi}}$:

$$\sigma_{\hat{\xi}_j} = \sqrt{C_{\hat{\xi}}(j,j)} \qquad [11.36]$$

Equations [11.34] and [11.35] are valid when using the Levenberg-Marquardt method, but σ_ρ^2 is rather evaluated using the residual of $\|\mathbf{F}\|^2$:

$$\sigma_\rho^2 = \frac{\|\mathbf{F}(\mathbf{Q}_t, \mathbf{X}_t, \hat{\xi})\|^2}{(r - c)} \qquad [11.37]$$

In order to validate the success of the parameter estimation process, we can evaluate the residual error on some configurations that have not been used in the identification. We can also compare the values of the estimated parameters using different calibration methods.

11.5. Calibration methods

In this section, we present the most common calibration methods. The first method requires an external sensor, which provides either the location or the position coordinates of the end-effector; the second requires an external sensor to provide the distance traveled by the endpoint when moving from one configuration to another. The other methods are termed as autonomous because they only make use of the joint variables. They are based on realizing geometric constraints between the robot configurations or between the robot and the environment.

11.5.1. *Calibration using the end-effector coordinates*

This method is the most popular one and can be considered as the conventional approach. The function to be minimized is the difference between the measured and calculated end-effector locations. This method needs an external sensor to measure the location of the end-effector frame with respect to the world reference frame. In § 11.8, we describe some measurement systems that can be used for that. The nonlinear calibration model is given by:

$$ {}^{-1}\mathbf{T}_{n+1}(\mathbf{x}) - {}^{-1}\mathbf{T}_{n+1}(\mathbf{q}, \boldsymbol{\xi}) = \mathbf{0} \qquad [11.38] $$

where ${}^{-1}\mathbf{T}_{n+1}(\mathbf{x})$ is the measured location of the end-effector with respect to frame R_{-1}.

Equation [11.38] contains twelve elements that may be different from zero, but it has only six independent degrees of freedom. To obtain six independent elements, we rewrite this equation as:

$$ \Delta\mathbf{X} = \begin{bmatrix} \Delta\mathbf{X}_p(\mathbf{x}, \mathbf{q}, \boldsymbol{\xi}) \\ \Delta\mathbf{X}_r(\mathbf{x}, \mathbf{q}, \boldsymbol{\xi}) \end{bmatrix} = 0 \qquad [11.39] $$

with:

- $\Delta\mathbf{X}_p$: (3x1) vector of the position error, equal to:

$$ \Delta\mathbf{X}_p = {}^{-1}\mathbf{P}_{n+1}(\mathbf{x}) - {}^{-1}\mathbf{P}_{n+1}(\mathbf{q}, \boldsymbol{\xi}) \qquad [11.40] $$

- $\Delta\mathbf{X}_r$: (3x1) vector of the orientation error (representing the difference between the measured and computed ${}^{-1}\mathbf{A}_{n+1}$), given by:

$$ \Delta\mathbf{X}_r = \mathbf{u}\,\alpha \qquad [11.41] $$

where $\mathbf{u}$ and α are obtained by solving the following equation (§ 2.3.7):

$$ {}^{-1}\mathbf{A}^{r}_{n+1} = \mathbf{rot}(\mathbf{u}, \alpha)\, {}^{-1}\mathbf{A}^{m}_{n+1} \qquad [11.42] $$

where ${}^{-1}\mathbf{A}^{r}_{n+1}(\mathbf{x}) = [\, \mathbf{s}_r \;\; \mathbf{n}_r \;\; \mathbf{a}_r \,]$ is the measured (3x3) orientation matrix of frame R_{n+1}, and ${}^{-1}\mathbf{A}^{m}_{n+1}(\mathbf{q}, \boldsymbol{\xi}) = [\, \mathbf{s}_m \;\; \mathbf{n}_m \;\; \mathbf{a}_m \,]$ is the computed orientation matrix using the direct geometric model.

If the orientation error is small, the following equation can be used (§ 2.3.8 and equation [2.35]):

$$ \Delta\mathbf{X}_r = \mathbf{u}\sin(\alpha) = \frac{1}{2}\begin{bmatrix} n_z - a_y \\ a_x - s_z \\ s_y - n_x \end{bmatrix} \qquad [11.43] $$

where the **s**, **n**, **a** components are obtained from the equation:

$$[\,\mathbf{s}\ \ \mathbf{n}\ \ \mathbf{a}\,] = {}^{-1}\mathbf{A}^{r}_{n+1}\,({}^{-1}\mathbf{A}^{m}_{n+1})^{-1} \qquad [11.44]$$

The linear differential model defining the deviation of the end-effector location due to the differential errors in the geometric parameters can be obtained as:

$$\Delta\mathbf{X} = \begin{bmatrix} \Delta\mathbf{X}_p(\mathbf{x}, \mathbf{q}, \boldsymbol{\xi}) \\ \Delta\mathbf{X}_r(\mathbf{x}, \mathbf{q}, \boldsymbol{\xi}) \end{bmatrix} = \boldsymbol{\Psi}(\mathbf{q}, \boldsymbol{\xi})\,\Delta\boldsymbol{\xi} \qquad [11.45]$$

where:

- $\Delta\mathbf{X}$ represents the (6x1) vector of position and orientation errors (representing the difference between the measured and computed ${}^{-1}\mathbf{T}_{n+1}$);
- $\boldsymbol{\Psi}$ is the (6xN_{par}) generalized Jacobian matrix developed in § 11.3.

The calibration index of this method is 6. If we only measure the position of the endpoint, the first three equations of the nonlinear or linear calibration models only should be used. The calibration index reduces to 3.

It is worth noting that the geometric parameters have different units: meters (for distances), radians (for angles) or even no unit (for gear transmission ratios). The effect of this heterogeneity can be handled by introducing an appropriate weighting matrix. However, for industrial robots of about the size of a human arm, one obtains good results by using meters for the distances, radians for the angles, and by normalizing the encoder readings such that the elements K_i are of the order of one. Of course, if the links are much smaller (like fingers) or much larger (like excavators), the situation is different.

11.5.2. *Calibration using distance measurement*

In this method, we make use of the distance traveled by the endpoint when moving from one configuration to another [Goswami 93]. Thus, an external sensor measuring the distance such as an extendable ball bar system or a linear variable differential transformer (LVDT) is required. The calibration index of this method is 1. Let $D^{r}_{i,j}$ be the measured distance traveled by the endpoint between configurations $\mathbf{q}^i$ and $\mathbf{q}^j$. Thus, the nonlinear calibration equation is given by:

$$[P_x(\mathbf{q}^j,\boldsymbol{\xi}) - P_x(\mathbf{q}^i,\boldsymbol{\xi})]^2 + [P_y(\mathbf{q}^j,\boldsymbol{\xi}) - P_y(\mathbf{q}^i,\boldsymbol{\xi})]^2 + [P_z(\mathbf{q}^j,\boldsymbol{\xi}) - P_z(\mathbf{q}^i,\boldsymbol{\xi})]^2 = (D^{r}_{i,j})^2 \qquad [11.46]$$

The differential calibration model is given by:

$$2\{[P_x(q^j)-P_x(q^i)]\,[\Psi_x(q^j)-\Psi_x(q^i)] + [P_y(q^j)-P_y(q^i)]\,[\Psi_y(q^j)-\Psi_y(q^i)] + [P_z(q^j)-P_z(q^i)]\,[\Psi_z(q^j)-\Psi_z(q^i)]\}\,\Delta\xi = (D^r_{i,j})^2-(D_{i,j})^2 \quad [11.47]$$

where:

- $D_{i,j}$ is the computed distance traveled by the endpoint between configurations q^i and q^j, using the nominal parameters;
- Ψ_x, Ψ_y and Ψ_z denote the first, second and third rows of the generalized Jacobian matrix respectively.

11.5.3. *Calibration using location constraint and position constraint*

The main limitation of the previous approaches is that they require an accurate, fast and inexpensive external sensor to measure the Cartesian variables. Location constraint and position constraint methods are autonomous methods that do not require an external sensor. These methods can be used when the specified end-effector locations (or positions) can be realized by multiple configurations. It is worth noting that a robot with more than three degrees of freedom ($n > 3$) can be calibrated by the position constraint method, whereas for the location constraint method we must have $n \geq 6$ [Khalil 95b].

Let q^i and q^j represent two configurations giving the same location of the end-effector. Then, the nonlinear calibration model is given by:

$$^{-1}T_{n+1}(q^j, \xi) - {}^{-1}T_{n+1}(q^i, \xi) = 0 \quad [11.48]$$

This equation can be transformed into a (6x1) vectorial equation as illustrated in § 11.5.1. The resulting equation is as follows:

$$\Delta X(q^i, q^j, \xi) = \begin{bmatrix} \Delta X_p(q^i, q^j, \xi) \\ \Delta X_r(q^i, q^j, \xi) \end{bmatrix} = 0 \quad [11.49]$$

The differential calibration model is given by:

$$\Delta X(q^i, q^j, \xi) = \left[\Psi(q^j, \xi) - \Psi(q^i, \xi)\right] \Delta\xi \quad [11.50]$$

where Ψ is the generalized Jacobian matrix developed in § 11.3, ΔX is the (6x1) vector representing the position and orientation differences between the locations $^{-1}T_{n+1}(q^j, \xi)$ and $^{-1}T_{n+1}(q^i, \xi)$.

The calibration index for this method is 6. In the position constraint method, the endpoint at configuration $\mathbf{q}^i$ should coincide with the endpoint at configuration $\mathbf{q}^j$. Thus, the first three equations of [11.49] and [11.50] only are considered. Hence, the calibration index is 3.

We note that the calibration Jacobian matrices of these methods are obtained by subtracting the generalized Jacobian matrices of two configurations. Thus, the number of identifiable parameters is less than that of the conventional methods (Example 11.1). For example:

- if a column of the generalized Jacobian matrix is constant, then the corresponding parameter will not be identified. This is the case of the parameters $\Delta\gamma_z$, Δb_z, $\Delta\alpha_z$, Δd_z, $\Delta\theta_{1'}$, $\Delta r_{1'}$, $\Delta\beta_0$ for both location constraint and position constraint methods;
- if a column of the generalized Jacobian matrix is identical for any two configurations $\mathbf{q}^i$ and $\mathbf{q}^j$ satisfying the calibration constraint, then the corresponding parameter cannot be identified. For example, since in the location constraint method ${}^{-1}\mathbf{T}_{n+1}(\mathbf{q}^i) = {}^{-1}\mathbf{T}_{n+1}(\mathbf{q}^j)$, then ${}^{-1}\mathbf{T}_n(\mathbf{q}^i) = {}^{-1}\mathbf{T}_n(\mathbf{q}^j)$. Thus, the parameters $\Delta r_{n'}$, $\Delta\theta_{n'}$, $\Delta\beta_{n'}$, Δd_e, $\Delta\alpha_e$, $\Delta\theta_e$, Δr_e are not identifiable.

11.5.4. *Calibration methods using plane constraint*

In this approach, the calibration is carried out using the values of the joint variables of a set of configurations for which the endpoint of the robot is constrained to lie in the same plane. Several methods based on this technique have been proposed [Tang 94], [Zhong 95], [Khalil 96b], [Ikits 97]. The main advantage of this autonomous method is the possibility to collect the calibration points automatically using touch or tactile sensor (such as LVDT, a trigger probe, or a laser telemeter). Two methods are developed in this section: the first makes use of the plane equation while the second uses the coordinates of the normal to the plane. The calibration index of these methods is 1. Special care has to be taken to locate the constraint plane and to select the calibration points in order to obtain a good condition number for the observation matrix [Ikits 97], [Khalil 00b].

11.5.4.1. *Calibration using plane equation*

Since the endpoints are in the same plane, and assuming that the plane does not intersect the origin, the nonlinear calibration model is:

$$aP_x(\mathbf{q}, \xi) + bP_y(\mathbf{q}, \xi) + cP_z(\mathbf{q}, \xi) + 1 = 0 \qquad [11.51]$$

where:

- a, b, c represent the plane coefficients referred to the reference world frame;
- P_x, P_y, P_z represent the Cartesian coordinates of the endpoint relative to the reference frame.

Applying equation [11.51] to a sufficient number of configurations, the resulting system of nonlinear equations can be solved to estimate the plane coefficients and the identifiable geometric parameters.

Using a first order development for equation [11.51] leads to the following linearized calibration model:

$$\begin{bmatrix} P_x(\mathbf{q}) & P_y(\mathbf{q}) & P_z(\mathbf{q}) & a\Psi_x(\mathbf{q})+b\Psi_y(\mathbf{q})+c\Psi_z(\mathbf{q}) \end{bmatrix} \begin{bmatrix} \Delta a \\ \Delta b \\ \Delta c \\ \Delta\xi \end{bmatrix} = -aP_x(\mathbf{q}) - bP_y(\mathbf{q}) - cP_z(\mathbf{q}) - 1 \qquad [11.52]$$

where:

- Ψ_x, Ψ_y and Ψ_z are the first, second and third rows of the generalized Jacobian matrix respectively;
- $P_x(\mathbf{q})$, $P_y(\mathbf{q})$, $P_z(\mathbf{q})$ represent the computed Cartesian coordinates of the endpoint in the reference frame.

The coefficients of the plane are initialized by solving the equation of the plane for the collected configurations:

$$\begin{bmatrix} -1 \\ \cdots \\ -1 \end{bmatrix} = \begin{bmatrix} P_x^1 & P_y^1 & P_z^1 \\ \cdots & \cdots & \cdots \\ P_x^e & P_y^e & P_z^e \end{bmatrix} \begin{bmatrix} a \\ b \\ c \end{bmatrix} + \rho \qquad [11.53]$$

where P_x^j, P_y^j and P_z^j are the coordinates of the endpoint as given by the DGM with the nominal values of the geometric parameters for configuration $\mathbf{q}^j$.

If the coefficients a, b, and c of the plane are known, the corresponding columns and unknowns in equation [11.52] are eliminated. The resulting linear calibration model is given by:

$$[a\Psi_x(\mathbf{q}) + b\Psi_y(\mathbf{q}) + c\Psi_z(\mathbf{q})]\, \Delta\xi = -aP_x(\mathbf{q}) - bP_y(\mathbf{q}) - cP_z(\mathbf{q}) - 1 \qquad [11.54]$$

11.5.4.2. *Calibration using normal coordinates to the plane*

In this method, the calibration model is established by using the fact that the scalar product of the vector normal to the plane and the vector between two points in the plane is zero [Zhong 95], [Maurine 96]. The coordinates of the normal can be obtained with inclinometers. Since this method is independent of the plane position, it may give poor results if the initial values of the robot parameters are not close to the real values.

The nonlinear calibration model for the configurations $\mathbf{q}^i$ and $\mathbf{q}^j$ is such that:

$$a[P_x(\mathbf{q}^j, \xi) - P_x(\mathbf{q}^i, \xi)] + b[P_y(\mathbf{q}^j, \xi) - P_y(\mathbf{q}^i, \xi)] + c[P_z(\mathbf{q}^j, \xi) - P_z(\mathbf{q}^i, \xi)] = 0 \quad [11.55]$$

Assuming that the normal coordinates are known, we obtain the following linearized equation:

$$\left\{a[\Psi_x(\mathbf{q}^j) - \Psi_x(\mathbf{q}^i)] + b[\Psi_y(\mathbf{q}^j) - \Psi_y(\mathbf{q}^i)] + c[\Psi_z(\mathbf{q}^j) - \Psi_z(\mathbf{q}^i)]\right\}\Delta\xi = -a[P_x(\mathbf{q}^j) - P_x(\mathbf{q}^i)] - b[P_y(\mathbf{q}^j) - P_y(\mathbf{q}^i)] - c[P_z(\mathbf{q}^j) - P_z(\mathbf{q}^i)] \quad [11.56]$$

where Ψ_u, $\mathbf{P}_u$, for u = x, y, z, are computed in terms of $\mathbf{q}$ and ξ.

NOTE.– The concatenation of the equations of several planes in a unique system of equations increases the number of identifiable parameters [Zhuang 99]. The use of three planes gives the same identifiable parameters as in the position measurement method if the robot has at least one prismatic joint, while four planes are needed if the robot has only revolute joints [Besnard 00b].

• **Example 11.1.** Determination of the identifiable parameters with the previous methods for the Stäubli RX-90 robot (Figure 3.3b) and the Stanford robot (Figure 11.2). The geometric parameters of these robots are given in Tables 11.1 and 11.2 respectively. For the plane constraint methods, we assume that the plane coefficients are known.

Tables 11.3 and 11.4 present the identifiable geometric parameters of the two robots as provided by the software package GECARO "GEometric CAlibration of Robots" [Khalil 99a], [Khalil 00b]. The parameters indicated by "0" are not identifiable, because they have no effect on the identification model, while the parameters indicated by "n" are not identified because they have been grouped with some other parameters.

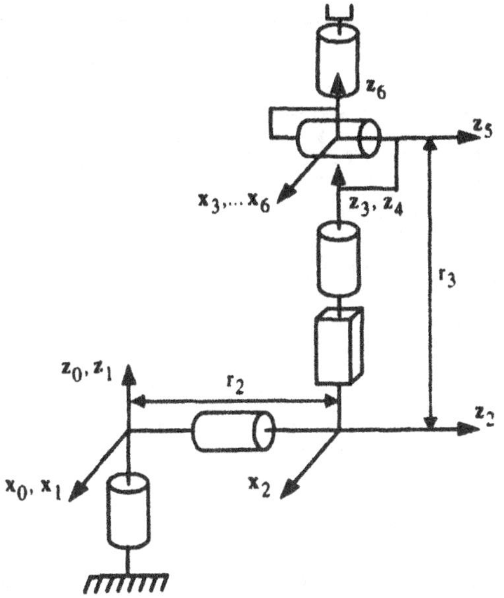

Figure 11.2. *Stanford robot*

Table 11.1. *Geometric parameters of the RX-90 Stäubli robot*[1]

j	σ_j	α_j	d_j	θ_j	r_j	β_j	K_j
0	2	0	0	$\pi/2$	0.5	0	0
1	0	0.1	0	θ_1	0	0	1
2	0	$-\pi/2$	0	θ_2	0	0	1
3	0	0	0.5	θ_3	0	0	1
4	0	$\pi/2$	0	θ_4	0.5	0	1
5	0	$-\pi/2$	0	θ_5	0	0	1
6	0	$\pi/2$	0	θ_6	0.3[2]	0	1
7	2	1.3	0.2	$\pi/2$	0.1	0	0

[1] Distances are in meters and angles are in radians.

[2] r_6 is equal to the end-effector frame parameter b_e.

Table 11.2. *Geometric parameters of the Stanford robot*[3]

j	σ_j	α_j	d_j	θ_j	r_j	β_j	K_j
0	2	0	0	$\pi/2$	0.5	0	0
1	0	0.1	0	θ_1	0	0	1
2	0	$-\pi/2$	0	θ_2	0.2	0	1
3	1	$\pi/2$	0	0	r_3	0	1
4	0	0	0	θ_4	0	0	1
5	0	$-\pi/2$	0	θ_5	0	0	1
6	0	$\pi/2$	0	θ_6	0.3[1]	0	1
7	2	1.3	0.2	$\pi/2$	0.1	0	0

From Tables 11.3 and 11.4, the following general results are deduced:

1) the location measurement method allows the identification of the maximum number of parameters (36 parameters for the Stäubli RX-90 robot and 34 for the Stanford robot), which corresponds to the following general equation [Everett 88]:

$$b = 4(n_r + 1) + 2 + 2n_p + n$$

including:

- $4(n_r + 1)$ parameters for the n_r revolute joints and for frame R_{n+1};
- 2 parameters for R_0;
- 2 n_p parameters for the prismatic joints;
- n parameters for the joint gear transmission ratios;

2) the parameters of frames R_0 and R_1 have no effect on the model and cannot be identified when using the following methods: distance measurement, position constraint and location constraint;

3) most of the parameters of frames R_n and R_{n+1}, (R_6 and R_7), are not identifiable with the location constraint method;

4) most of the parameters of frames R_0, R_1 and R_7 are not identifiable with the planar methods. Some of them are grouped with other parameters;

5) the parameter β_j is not identifiable when $\alpha_j \neq 0$;

6) the offsets of joint variables 2, …, n – 1 are identifiable with all the methods;

---

[3] Distances are in meters and angles are in radians.

7) the offset of joint 1 is not identifiable with the following methods: distance measurement, position constraint, location constraint; whereas the offset of joint n is not identifiable with the location constraint method;

8) all the gear transmission ratios K_j are identifiable;

9) the parameter r_6 is not identifiable with the position constraint and the plane constraint methods for both robots. It represents the scale factor of these methods. Note that the constraint equations are also verified when all the distances are zero.

10) in the location constraint method, the parameter r_6 has no effect. The scale factor is represented by r_4 for the Stäubli robot and r_2 for the Stanford robot. It is worth noting that in the case of the Stanford robot, the scale factor could be the prismatic variable r_3 (instead of r_6 or r_2) if we assume that the gear transmission ratio K_3 is known and has not to be identified;

11) the parameters α_7, θ_7 and r_7 are not identifiable with the position measurement, position constraint, and plane constraint methods. This is because the end-effector reduces to a point that is defined by the three parameters r_6, θ_6 and d_7.

11.6. Correction and compensation of errors

Once the identification is performed, the estimated parameters must be integrated into the robot controller. Computing the DGM with the general identified parameters is more time consuming than the analytical solution, but it can be performed on-line. If the IGM were computed using a general numerical iterative algorithm, we could compensate all the calibrated parameters. But, a major problem stems for industrial robots whose IGM is analytically implemented. The calibration may result in a non-analytically solvable robot. For example, a spherical wrist could be found not to be spherical. In this case, the following geometric parameters can be updated in the controller straightaway: end-effector parameters, robot base parameters, joint offsets, r_j and d_j whose nominal values are not zero, and the angles α_j, θ_j whose nominal values are not $k\,\pi/2$, with k being an integer. For the other parameters, an iterative approach must be implemented. A possible approach is to use the closed-form solution in order to compute a good first guess. Then, an accurate solution is obtained using an iterative algorithm. Such a process converges in a small number of iterations, or even in a single iteration if the end-effector location error due to the geometric parameters to be compensated is not too high. The iterative tuning process is summarized as follows (Figure 11.3):

Table 11.3. *Identifiable parameters of the RX-90 Stäubli robot*

Parameter	Position measurement	Location measurement	Distance measurement	Position constraint	Location constraint	Plane equation	Plane normal
θ0			0	0	0	n	n
r0			0	0	0		0
α1			0	0	0		
d1			0	0	0	n	0
θ1			0	0	0		
r1			0	0	0	n	0
β1	n	n	0	0	0	n	n
α2							
d2							
θ2							
r2							
β2	n	n	n	n	n	n	n
α3							
d3							
θ3							
r3	n	n	n	n	n	n	n
β3							
α4							
d4							
θ4							
r4					n		
β4	n	n	n	n	n	n	n
α5							
d5							
θ5							
r5							
β5	n	n	n	n	n	n	n
α6							
d6							
θ6					0		
r6				n	0	n	n
β6	n	n	n	n	0	n	n
α7	n		n	n	0	n	n
d7					0		
θ7	0		0	0	0	0	0
r7	n		n	n	0	n	n
β7	n	n	n	n	0	n	n
K1,...K6							
Total	**33**	**36**	**27**	**26**	**23**	**29**	**28**

(n: non-identifiable parameter. Its effect is grouped with some other parameters.
0: non-identifiable parameter having no effect on the model)

Table 11.4. *Identifiable parameters of the Stanford robot*

Parameter	Position measurement	Location measurement	Distance measurement	Position constraint	Location constraint	Plane equation	Plane normal
θ0			0	0	0	n	n
r0			0	0	0		0
α1			0	0	0		
d1			0	0	0	n	0
θ1			0	0	0		
r1			0	0	0	n	0
β1	n	n	0	0	0	n	n
α2							
d2							
θ2							
r2					n		
β2	n	n	n	n	n	n	n
α3							
d3							
θ3	n	n	n	n	n	n	n
r3							
β3	n	n	n	n	n	n	n
α4							
d4	n	n	n	n	n	n	n
θ4							
r4	n	n	n	n	n	n	n
β4							
α5							
d5							
θ5							
r5							
β5	n	n	n	n	n	n	n
α6							
d6							
θ6					0		
r6				n	0	n	n
β6	n	n	n	n	0	n	n
α7	n		n	n	0	n	n
d7					0		
θ7	0		0	0	0	0	0
r7	n		n	n	0	n	n
β7	n	n	n	n	0	n	n
K1, ...K6							
Total	**31**	**34**	**25**	**24**	**21**	**27**	**26**

(n: non-identifiable parameter. Its effect is grouped with some other parameters.
0: non-identifiable parameter having no effect on the model)

1) use the analytical IGM to compute the joint variables $\mathbf{q}$ corresponding to the desired end-effector location ${}^{-1}\mathbf{T}^{d}_{n+1}$;

2) compute the differential error $\Delta\mathbf{X}$ between ${}^{-1}\mathbf{T}^{d}_{n+1}$ and ${}^{-1}\mathbf{T}^{c}_{n+1}(\hat{\xi}, \mathbf{q})$, where ${}^{-1}\mathbf{T}^{c}_{n+1}(\hat{\xi}, \mathbf{q})$ indicates the direct geometric model using the estimated parameters. Note that $\Delta\mathbf{X}$ can also be computed using the generalized differential model (§ 11.3);

3) if $\Delta\mathbf{X}$ is sufficiently small, stop the tuning process. Otherwise, compute $\Delta\mathbf{q}$ corresponding to the error $\Delta\mathbf{X}$ using the classical inverse differential model $\Delta\mathbf{q} = \mathbf{J}^{+}\Delta\mathbf{X}$;

4) update the joint variables such that $\mathbf{q} = \mathbf{q} + \Delta\mathbf{q}$;

5) return to step 2.

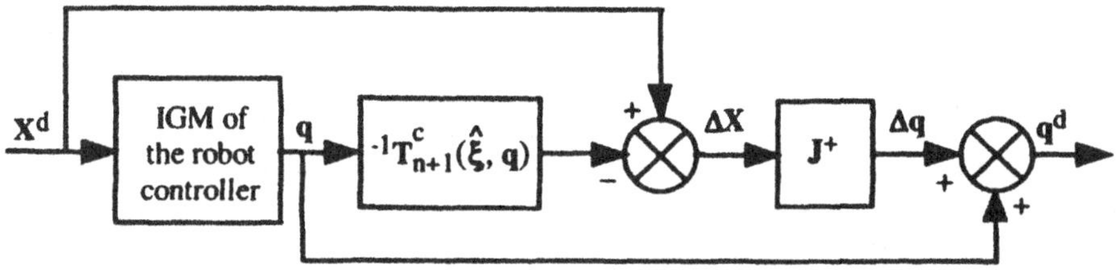

Figure 11.3. *Principle of compensation*

In the context of off-line programming systems, once a calibration has been performed, the compensated joint values can be downloaded directly to the controller.

11.7. Calibration of parallel robots

One of the attractive feature of parallel robots is their potential for higher accuracy as compared to serial robots, mainly due to the higher stiffness of their closed-loop structure. However, this stiffness does not result directly into better accuracy, but rather into higher repeatability. Therefore, good calibration of the geometric parameters is also necessary for a parallel robot to improve its accuracy.

In this section, we consider the case of a six degree-of-freedom parallel robot of the Gough-Stewart family (Figures 8.1 and 8.11). It is composed of six legs of (RR)-P-(RRR) architecture, a fixed base, and a mobile platform to which the tool is attached. The prismatic joints are actuated, while the universal joints (U-joints) and the spherical joints (S-joints) are passive. The reference frame R_f is assumed to be attached to the base and the end-effector frame R_E is attached to the platform. We assume that the universal and spherical joints are perfect, and that the prismatic joints are perfectly assembled. Consequently, 42 geometric parameters are needed to compute the geometric model, namely: six leg lengths denoted by the joint variable vector $\mathbf{q}$, 3x6 coordinates of the centers of the U-joints with respect to the reference frame (${}^{f}\mathbf{P}_{Ai}$, for i = 1, ..., 6), and 3x6 coordinates of the centers of the S-joints with respect to the end-effector frame (${}^{E}\mathbf{P}_{Bi}$, for i = 1, ..., 6). The geometric calibration consists of estimating these parameters accurately. As in serial robots, the calibration procedure consists of four steps: construction of a calibration model, collection of a sufficient number of configurations, identification of the geometric parameters from the calibration equation, and implementation of error compensation. It is worth noting that for parallel robots the parameters estimated can be introduced into the controller straightaway.

Various calibration models are proposed in the literature. We will only consider the conventional techniques where the calibration model is a function of the Cartesian coordinates of the location of the mobile platform frame, which must be provided by an external sensor. In this case, the calibration model can be derived from the IGM [Zhuang 95] or the DGM. Recall that the IGM is easy to compute using closed-form expressions and that it gives a unique solution $\mathbf{q}$ for a desired location ${}^{f}\mathbf{T}_{E}$, whereas the DGM has multiple solutions and is computed iteratively.

11.7.1. *IGM calibration model*

The error function in the IGM calibration model is the difference between the measured and computed joint variables. It is represented by the following equation:

$$\mathbf{q} - \mathrm{IGM}({}^{f}\mathbf{T}_{E}(\mathbf{x}), \boldsymbol{\xi}) = \mathbf{0} \qquad [11.57]$$

where $\boldsymbol{\xi}$ denotes the current geometric parameters, and ${}^{f}\mathbf{T}_{E}(\mathbf{x})$ is the measured location of the end-effector frame relative to the fixed reference frame.

The IGM providing the joint position q_i in terms of the desired transformation matrix ${}^{f}\mathbf{T}_{E}$ and of the fixed geometric parameters ${}^{f}\mathbf{P}_{Ai}$ and ${}^{E}\mathbf{P}_{Bi}$ is given by equation [8.11]. It is rewritten as:

$$q_i^2 = ({}^{f}\mathbf{P}_{Bi} - {}^{f}\mathbf{P}_{Ai})^T ({}^{f}\mathbf{P}_{Bi} - {}^{f}\mathbf{P}_{Ai}) \qquad [11.58]$$

The joint variable q_i is given by:

$$q_i = D_i + q_{ci} \qquad [11.59]$$

where q_{ci} represents the joint i sensor value and D_i is a constant offset.

The nonlinear calibration model is obtained from equation [11.58] as:

$$q_i^2 - ({}^f\mathbf{A}_E\,{}^E\mathbf{P}_{Bi} + {}^f\mathbf{P}_E - {}^f\mathbf{P}_{Ai})^T ({}^f\mathbf{A}_E\,{}^E\mathbf{P}_{Bi} + {}^f\mathbf{P}_E - {}^f\mathbf{P}_{Ai}) = 0 \qquad [11.60]$$

Equation [11.60] shows that the parameters of each leg can be identified separately. Thus, we can split the identification problem into six independent systems of equations where each system is a function of the seven parameters of one leg. Applying equation [11.60] for a sufficient number of configurations e and concatenating all the equations together leads to the following nonlinear equation:

$$\mathbf{0} = \mathbf{F}_i(\mathbf{Q}_{ti}, \mathbf{X}_t, \boldsymbol{\xi}_i) + \boldsymbol{\rho}_i \qquad [11.61]$$

where $\mathbf{Q}_{ti} = q_i^1, \ldots, q_i^e$ represents the calibration configurations of leg i, $\mathbf{X}_t = \mathbf{x}^1, \ldots, \mathbf{x}^e$ indicates the corresponding locations of the mobile platform, $\boldsymbol{\xi}_i$ indicates the geometric parameters of leg i, and $\boldsymbol{\rho}_i$ is the vector of modeling error for leg i.

This nonlinear optimization problem can be solved by the Levenberg-Marquardt algorithm as described in § 11.4.3.

The Jacobian matrix of the IGM calibration method is formulated for leg i by:

$$\Delta q_i = \boldsymbol{\phi}_i(\mathbf{x}, \boldsymbol{\xi}_i)\,\Delta\boldsymbol{\xi}_i \qquad [11.62]$$

The closed-form expressions of the columns of the calibration Jacobian matrix $\boldsymbol{\phi}_i$ can be obtained by differentiating equations [11.58] and [11.59] with respect to the elements of ${}^E\mathbf{P}_{Bi}$, ${}^f\mathbf{P}_{Ai}$ and D_i:

$$\Delta q_i = \frac{1}{q_i}\left[\; (-{}^f\mathbf{P}_{Bi}+{}^f\mathbf{P}_{Ai})^T \quad ({}^f\mathbf{P}_{Bi}-{}^f\mathbf{P}_{Ai})^T\,{}^f\mathbf{A}_E \quad q_i \;\right] \begin{bmatrix} \Delta{}^f\mathbf{P}_{Ai} \\ \Delta{}^E\mathbf{P}_{Bi} \\ \Delta D_i \end{bmatrix} \qquad [11.63]$$

Since the Jacobian matrix can be obtained using closed-form expressions, we can use the iterative pseudoinverse solution to solve this problem as for serial robots. The prediction error function of all the joints at configuration $\mathbf{q}^j$ is calculated in terms of the measured location ${}^f\mathbf{T}_E{}^j$ and the current geometric parameters $\boldsymbol{\xi}$ by:

$$\Delta\mathbf{q}^j = \mathbf{q}^j - \mathrm{IGM}({}^f\mathbf{T}_E{}^j, \boldsymbol{\xi}) \qquad [11.64]$$

11.7.2. *DGM calibration model*

The second nonlinear calibration model, which makes use of the iterative DGM, is given by:

$$\Delta \mathbf{X}(\mathbf{q}, {}^{f}\mathbf{T}_{E}(\mathbf{x}), \xi) = \mathbf{0} \qquad [11.65]$$

where $\Delta\mathbf{X}$ indicates the (6x1) vector of the difference between the measured location ${}^{f}\mathbf{T}_{E}(\mathbf{x})$ and the computed location of the platform $DGM(\mathbf{q}^{j}, \xi)$.

The vector $\Delta\mathbf{X}$ can be determined as given in § 11.5.1 by computing $\Delta\mathbf{X}_{p}$ and $\Delta\mathbf{X}_{r}$. The identification equation can be solved using the Levenberg-Marquardt method.

The advantage of the DGM calibration approach is the possibility to use in the calibration equation partial elements of the location of the end-effector [Besnard 99]. For example, we can carry out the calibration using the position coordinates of the platform by considering the first three components of equation [11.65]. In the same manner, we can also make use of two inclinometers [Besnard 99]. By comparison, the IGM calibration approach needs complete measurement of the location. The main drawback of the DGM method is its computational complexity. Besides, the corresponding Jacobian matrix cannot be computed analytically.

Before closing this section, we have to mention the autonomous calibration methods that are based on measuring some of the passive joint variables [Zhuang 97] or by locking some of them [Murareci 97], [Besnard 00a], [Daney 00], [Khalil 99b]. Recall that providing some passive joints with sensors simplifies the direct geometric model solution. With these autonomous methods, the coordinates of points A_i and B_i, for $i = 1, ..., 6$, are estimated with respect to frames R_0 and R_m respectively and not with respect to frames R_f and R_E (see § 8.6.1 for the definition of these frames). The identifiable parameters of all these methods are presented in [Besnard 01].

11.8. Measurement techniques for robot calibration

Conventional calibration methods, as well as the evaluation of positioning accuracy and repeatability of robots, requires measurement of either the end-effector location or position with high accuracy. Most of the current measurement schemes are based on vision systems, measuring machines, laser interferometers, laser tracking systems, and theodolites [Mooring 91], [Bernhardt 93].

Ideally, the measurement system should be accurate, inexpensive and should be operated automatically. The goal is to minimize the calibration time and the robot unavailability. At this time, such devices are not yet available. Nevertheless, we present in this section some principles that have given place to industrial realization.

11.8.1. *Three-cable system*

Such a system is basically composed of three high resolution optical encoders P1, P2, P3. Low mass cables are fixed to one of their ends on the encoder shafts whereas the other ends are fixed on the endpoint M of the robot (Figure 11.4). The encoder readings give the cable lengths, which represent the radii ρ_1, ρ_2, ρ_3 of three spheres whose centers are on the encoder shafts. The intersection of the spheres determines the coordinates of M.

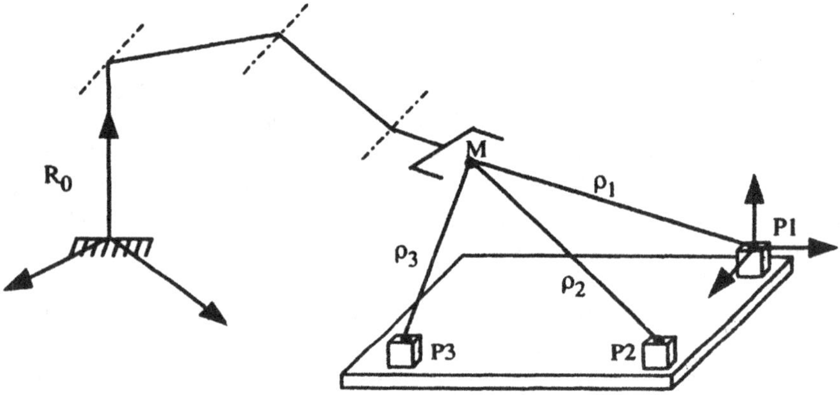

Figure 11.4. *A three-cable system*

This low cost device provides automatically the coordinates of the endpoint M. As a commercial example of such a system, we can mention the *3D CompuGauge* from Dynalog whose accuracy is about 0.1 mm for a cubic measuring space of 1.5 m of side.

11.8.2. *Theodolites*

A theodolite is a telescope where the two angles giving the orientation of the line of sight can be measured precisely. The Cartesian coordinates of a target ball M on the end-effector can be obtained in terms of the readings of two theodolites Th1 and Th2 pointing this ball and of the transformation matrix **T** between the two theodolites (Figure 11.5).

The accuracy of this system is excellent (of the order of 0.02 mm at about 1 m distance). The cost of a theodolite is rather high. The first systems had to be manually operated, but the data were read and stored by a computer (ECDS3 from Leica for example). Now, we can find motorized theodolites that can track automatically an illuminated target ball (TCA from Leica, Elta from Zeiss). Their cost is naturally greater.

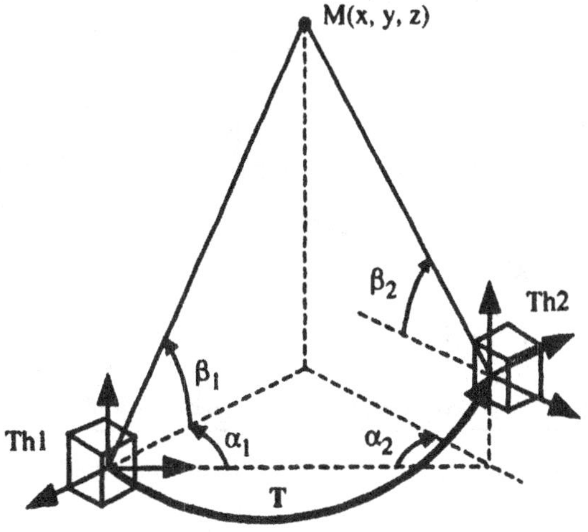

Figure 11.5. *Measurement system using two theodolites*

11.8.3. *Laser tracking system*

This device is composed of two 2-D scanner systems T1 and T2 external to the robot. Each of the two systems deflects a laser beam in a vertical plane and a horizontal plane, thanks to two motorized mirrors. The direction of the beam is controlled to track a retroreflector fixed on the terminal link of the robot (Figure 11.6). The position of the target point is calculated automatically using the angles of the laser beams. The precision is of the order of 0.1 mm for a target at 1 m distance. As examples of this system, we can mention the LASERTRACES system from ASL (UK) and OPTOTRAC from the University of Surrey (UK). The limitation of this system is the requirement of a dedicated end-effector.

11.8.4. *Camera-type devices*

The principle is to acquire at least simultaneously two images of the robot configuration using two cameras. The two images are processed in real time to estimate the 3D coordinates of target markers attached to the robot links.

Practically, the existing systems differ by the number and the type of the cameras used. We can mention as an example the RODYM6D from Krypton, which uses an OPTOTRAK sensor from Northern Digital. The sensor is composed of 3 CCD cameras and can handle up to 24 infrared light emitting diode markers. The precision is of the order of 0.2 mm for points at 2.5 m distance.

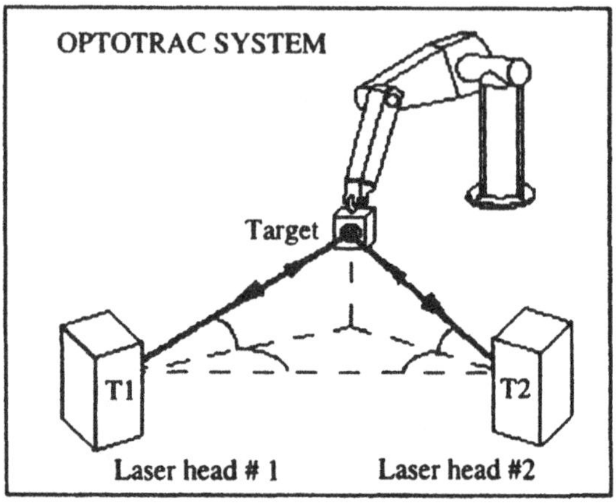

Figure 11.6. *Laser tracking system (University of Surrey)*

11.9. Conclusion

We have presented various approaches for the geometric calibration of serial robots. The geometric parameters of the robot, the base frame parameters and the end-effector frame parameters are defined using the Hayati modification of Khalil-Kleinfinger notations. All of the calibration methods are described by a unified nonlinear equation and a general linear equation. The Jacobian matrix of each calibration method is obtained as a function of the generalized Jacobian matrix relating the variation of the end-effector location with the geometric parameter variation. The generalized Jacobian matrix is computed using an efficient method making use of the elements of the transformation matrices of the link frames. The identifiable parameters are determined numerically by studying the QR decomposition of the observation matrix using random configurations satisfying the constraints of the calibration method. The nonlinear estimation problem is resolved using the Levenberg-Marquardt method or using an iterative pseudoinverse method. The optimum selection of the calibration configurations is treated by minimizing the condition number of the observation matrix. These methods can be extended to include the calibration of joint elasticity and link flexibility [Besnard 00a].

We have also presented the geometric calibration of parallel robots when the measurement of the end-effector location is available. The problem can be formulated either with inverse or with direct geometric models. References are given for autonomous calibration approaches for these structures.

The instrumentation for geometric calibration is quite varied. Good commercial systems measuring the end-effector coordinates are beginning to emerge, and some of them have been described. The autonomous calibration methods do not need such external sensors. They are only based on the joint sensor readings, which are available with good precision on all the robots.

In the next chapter, we address the problem of estimating the inertial parameters and friction parameters.

Chapter 12

Identification of the dynamic parameters

12.1. Introduction

Most advanced control schemes formulated in the recent literature for robots require dynamic models (Chapter 14). The precision, performance, stability, and robustness of these schemes depend, to a large extent, on the accuracy of the parameters that describe the dynamic model. Adaptive and robust schemes can tolerate some errors in the dynamic parameters, while other schemes aimed at achieving perfect feedback linearization, such as the computed torque technique, assume precise knowledge of the dynamic parameters. In view of this, *a priori* precise determination of the dynamic parameters is useful to most schemes and is crucial to some others. Furthermore, these values are necessary to simulate the dynamic equations.

Accurate values of the dynamic parameters are typically unknown, even to the robot manufacturers. In this chapter, we will exploit the fact that the dynamic model and the energy model are linear in these parameters in order to identify them. The problem will be reduced to the least-squares solution of an overdetermined linear system of equations.

We assume *a priori* knowledge of the geometric parameters (Chapter 11). The dynamic parameters of link j and actuator j are composed of the inertial parameters of the link, the actuator rotor inertia, and the friction parameters (Chapter 9). We combine these parameters in the vector χ^j:

$$\chi^j = [XXj \;\; Xyj \;\; XZj \;\; YYj \;\; YZj \;\; ZZj \;\; MXj \;\; Myj \;\; MZj \;\; M_j \;\; Iaj \;\; F_{cj} \;\; F_{vj}]^T \quad [12.1]$$

The dynamic parameters of a robot with n mobile links are represented by the vector χ such that:

$$\chi = [\chi^{1T}\ \chi^{2T} \cdots \chi^{n^T}]^T \qquad [12.2]$$

To simplify the notations, we will address the case of serial robots only. The extension to tree structure or closed-chain robots can easily be carried out using the results of Chapter10.

12.2. Estimation of inertial parameters

There are three main methods for estimating the inertial parameters of a robot:

i) physical experiments: if we could disassemble the robot to isolate each link, the following parameters could be obtained by physical experiment [Armstrong 86]:

- the mass could be weighed directly;
- the coordinates of the center-of-mass could be estimated by determining counterbalanced points of the link;
- the diagonal elements of the inertia tensor could be obtained by pendular motions.

This method is very tedious and should be realized by the manufacturer before assembling the robot;

ii) using CAD/CAM models: all robotics CAD/CAM packages provide tools to calculate the inertia parameters from 3D models. This method is prone to errors due to the fact that the geometry of the links is complicated to define precisely, and that certain parts such as bearings, bolts, nuts, and washers are generally neglected;

iii) identification: this approach is based on the analysis of the "input/output" behavior of the robot on some planned motion and on estimating the parameter values by minimizing the difference between a function of the real robot variables and its mathematical model. This method has been used extensively and was found to be the best in terms of ease of experimentation and precision of the obtained values. In this chapter, we consider off-line identification methods, for which we collect all the input-output data prior to analysis. The on-line identification will be treated in Chapter 14 when presenting the adaptive control techniques.

12.3. Principle of the identification procedure

Several schemes have been proposed in the literature to identify the dynamic parameters [Ferreira 84], [Mayeda 84], [An 85], [Khosla 85], [Atkeson 86], [Gautier 86], [Olsen 86], [Aldon 86], [Kawasaki 88], [Bouzouia 89], [Raucent 90],

[Aubin 91], [Prüfer 94], [Gautier 95], [Restrepo 96]. These methods present the following common features:

- the use of a linear model in the dynamic parameters (dynamic model, energy model, power model, model of the wrench exerted on the base of the robot);
- the construction of an overdetermined linear system of equations by applying the identification model at a sufficient number of points along some trajectories of the robot. In general, a constant sampling rate is used between the different points;
- the estimation of the parameter values using linear regression techniques (ordinary least-squares solution or any other alternative method).

12.3.1. *Resolution of the identification equations*

All the identification models can be written in the following general form:

$$\mathbf{y}(\Gamma, \dot{\mathbf{q}}) = \mathbf{w}(\mathbf{q}, \dot{\mathbf{q}}, \ddot{\mathbf{q}})\, \boldsymbol{\chi} \qquad [12.3]$$

Applying the identification model at a sufficient number of points on some trajectories, we construct the following overdetermined linear system of equations in $\boldsymbol{\chi}$:

$$\mathbf{Y}(\Gamma, \dot{\mathbf{q}}) = \mathbf{W}(\mathbf{q}, \dot{\mathbf{q}}, \ddot{\mathbf{q}})\, \boldsymbol{\chi} + \boldsymbol{\rho} \qquad [12.4]$$

where $\mathbf{W}$ is an (rxc) observation matrix, or regressor, r is the total number of equations, c is the number of parameters such that r >> c, and $\boldsymbol{\rho}$ is the residual error vector.

The identification handbooks provide a large variety of deterministic and stochastic methods to estimate $\boldsymbol{\chi}$ from the previous system of equations. The use of ordinary least-squares solution of linear overdetermined system of equations, such as those based on the SVD or QR decomposition (Appendix 4), gives good results if some care is taken in processing the data measured and the elements of the matrices $\mathbf{Y}$ and $\mathbf{W}$ as we will show in this chapter. Note that the use of scientific software packages such as Matlab or Mathematica facilitates the application of the proposed data processing. Consequently, the estimated values $\hat{\boldsymbol{\chi}}$ can be obtained from equation [12.4] such that:

$$\hat{\boldsymbol{\chi}} = \underset{\boldsymbol{\chi}}{\mathrm{Min}} \|\boldsymbol{\rho}\|^2$$

If $\mathbf{W}$ is of full rank, the explicit solution of this problem is given by (Appendix 4):

$$\hat{\chi} = (\mathbf{W}^T \mathbf{W})^{-1} \mathbf{W}^T \mathbf{Y} = \mathbf{W}^+ \mathbf{Y} \qquad [12.5]$$

where $\mathbf{W}^+$ is the pseudoinverse matrix of $\mathbf{W}$.

It should be noted that this least-squares estimation is biased because the observation matrix $\mathbf{W}$ is random, and because $\mathbf{W}$ and ρ are realizations of random and correlated variables [Mendel 73], [Eykhoff 74], [de Larminat 77], [Gautier 86]. Furthermore, the elements of the matrix $\mathbf{W}$ are nonlinear functions in $\mathbf{q}$ and $\dot{\mathbf{q}}$, which leads one to assume some statistical properties of the noise in order to calculate the quality of the estimation process (bias and standard deviation) [Armstrong 89], [Raucent 90]. Consequently, it is important to verify the accuracy of the values obtained using appropriate validation procedures (§ 12.7).

The standard deviations of the estimated values are calculated by assuming that $\mathbf{W}$ is deterministic, and ρ is a zero mean additive independent noise, with standard deviation σ_ρ. As stated in § 11.4.3, the variance-covariance matrix $\mathbf{C}_\rho$ is given by:

$$\mathbf{C}_\rho = E(\rho\, \rho^T) = \sigma_\rho^2\, \mathbf{I}_r \qquad [12.6]$$

where E is the expectation operator, and $\mathbf{I}_r$ is the (rxr) identity matrix.

An unbiased estimation of σ_ρ can be calculated using the following equation:

$$\sigma_\rho^2 = \frac{\|\mathbf{Y} - \mathbf{W}\,\hat{\chi}\|^2}{(r - c)} \qquad [12.7]$$

The variance-covariance matrix of the estimation error is given by [de Larminat 77]:

$$\mathbf{C}_{\hat{\chi}} = E[(\chi - \hat{\chi})\,(\chi - \hat{\chi})^T] = \mathbf{W}^+\, \mathbf{C}_\rho\, (\mathbf{W}^+)^T = \sigma_\rho^2\, (\mathbf{W}^T \mathbf{W})^{-1} \qquad [12.8]$$

The standard deviation on the jth parameter is obtained from the (j, j) element of $\mathbf{C}_{\hat{\chi}}$:

$$\sigma_{\hat{\chi}_j} = \sqrt{\mathbf{C}_{\hat{\chi}}(j,j)} \qquad [12.9]$$

This interpretation has been proposed by Raucent [Raucent 90], but we should be careful with the results obtained because the corresponding assumptions are not verified.

The relative standard deviation can be used as a criterion to measure the quality of the identification value for each parameter (§ 12.7). It is obtained as:

$$\sigma_{\hat{\chi}_{jr}}\% = 100\frac{\sigma_{\hat{\chi}_j}}{|\hat{\chi}_j|} \qquad [12.10]$$

For example, if the relative standard deviation of a parameter is greater than ten times the minimum relative standard deviation value, this parameter can be considered as poorly identified.

12.3.2. *Identifiability of the dynamic parameters*

The dynamic parameters can be classified into three groups: fully identifiable, identifiable in linear combinations, and completely unidentifiable. Consequently, the observation matrix **W** corresponding to the set of parameters χ is rank deficient (some columns of **W** are linearly dependent whatever the values of **q**, $\dot{\mathbf{q}}$ and $\ddot{\mathbf{q}}$). In order to obtain a unique solution, we have to determine a set of independent *identifiable parameters*, which are also called *base dynamic parameters* or *minimum dynamic parameters*.

We have shown how to calculate the base inertial parameters using symbolic methods (Chapters 9 and 10) or numerical methods (Appendix 5). It is easy to demonstrate that the columns corresponding to the Coulomb and viscous friction parameters are independent. The determination of the base parameters is a prerequisite for the identification algorithms. It should be noted that the grouping equations do not need to be computed since the identification will give directly the grouped values.

To simplify the notations, we assume in the following that χ represents the base inertial parameters and the friction parameters, and that **W** is composed of the corresponding columns.

12.3.3. *Estimation of the friction parameters*

Using the classical friction model at non-zero velocity, which is represented by viscous and Coulomb frictions, we can write the friction torque on joint j as (§ 9.3.4):

$$\Gamma_{fj} = F_{cj}\,\mathrm{sign}(\dot{q}_j) + F_{vj}\,\dot{q}_j \qquad [12.11]$$

Two approaches can be used to identify the joint friction parameters:

i) global identification of the inertial and friction parameters: in this approach, we consider the estimation of the inertial parameters together with the friction parameters;

ii) separate identification of the friction parameters: in this approach, we identify at first the friction parameters using constant velocity motion on an axis-by-axis basis [Spetch 88], [Held 88]. These parameters are then considered to be known for the identification of the inertial parameters. This simple method induces the risk of error accumulation between the two steps.

12.3.4. *Trajectory selection*

In order to improve the convergence rate and the noise immunity of the least-squares estimation, the trajectory used in the identification must be carefully selected. Such a trajectory is known as a *persistently exciting trajectory*. To obtain an exciting trajectory, two schemes are generally used:

- calculation of a trajectory satisfying some optimization criteria;
- use of sequential sets of special test motions, where each motion will excite some dynamic parameters. As the number of the parameters to be identified is reduced with respect to the global problem, it is easier in this case to find an exciting trajectory.

12.3.4.1. *Trajectory optimization*

The sensitivity of the least-squares solution with respect to the modeling errors and noise can be measured by the condition number of the observation matrix. Thus, the planification of an exciting trajectory can be formulated by calculating a trajectory whose points give a "good" conditioned observation matrix. This is a nonlinear optimization problem whose degrees of freedom are the starting point, the intermediate points, the maximum joint velocities and accelerations, etc. In the literature, the following criteria have been used to define the exciting condition [Armstrong 89], [Lu 93], [Gautier 92a], [Benhlima 93]:

i) the condition number of the matrix **W**, which is defined using the 2-norm, as:

$$\text{cond}(\mathbf{W}) = \frac{\sigma_{max}}{\sigma_{min}} \geq 1 \qquad [12.12]$$

where σ_{max} and σ_{min} denote the maximum and minimum singular values of **W** (Appendix 4). The optimization problem consists of determining the trajectory, which provides a condition number of **W** that is close to 1;

ii) the sum of the condition number of **W** *with a parameter equilibrating the values of the elements of* **W** such that they will be of the same order of magnitude [Gautier 90a]:

$$C = \text{cond}(\mathbf{W}) + k_1 \frac{\max |W_{i,j}|}{\min |W_{i,j}|} \quad \text{with } \min |W_{i,j}| \neq 0 \qquad [12.13]$$

where $|W_{i,j}|$ is the absolute value of the (i, j) element of **W**, and $k_1 > 0$ is a weighting scalar parameter;

iii) the sum of the condition number of **W** *and the inverse of the smallest singular value of* **W***:*

$$C = \text{cond}(\mathbf{W}) + k_2 \frac{1}{\sigma_{min}} \qquad [12.14]$$

This criterion prevents a trajectory of good condition number but with small singular values being obtained [Pressé 93]. It equilibrates the standard deviation $\sigma_{\hat{\chi}}$ on the different parameters, but the relative standard deviation of the parameters with small values will be too high;

iv) the condition number of a weighted observation matrix. If we have *a priori* information about the order of magnitude of the dynamic parameters, the following cost function will equilibrate the contribution of each parameter on the identification model. This will result in equilibrating the relative standard deviation of the different parameters [Pressé 93]:

$$C = \text{cond}(\mathbf{W}\,\mathbf{diag}(\mathbf{Z})) \qquad [12.15]$$

where **diag**(**Z**) is the diagonal matrix formed by the elements of the vector **Z** representing the (bx1) vector of the *a priori* absolute values of the dynamic parameters.

The generation of an exciting trajectory by an optimization procedure for the identification of the dynamic parameters presents the following difficulties:

- there is no analytical expression for the cost functions;
- the algorithm must take into account the joint limits, the maximum velocities, and the maximum accelerations;

– the number of degrees of freedom of the optimization problem is very large.

The generation of an exciting trajectory for the energy identification model (§ 12.6) is easier than that for the dynamic model because the energy model expression is simpler and does not need the joint accelerations [Gautier 92b], [Pressé 94]. In any case, if we cannot determine an exciting trajectory using an optimization procedure, we make use of a random trajectory and verify the corresponding cost criterion.

12.3.4.2. *Sequential identification*

The most widespread approaches propose to use a set of different trajectories where each trajectory excites some parameters. For instance, we can move some joints while locking some others [Mayeda 84], [Olsen 86], [Atkeson 86], [Ha 89], [Aubin 91], [Gaudin 92]. This technique simplifies the identification equations. However, an accumulation of errors may occur since the values of some estimated parameters will be assumed to be known in subsequent identification.

Vandanjon [Vandanjon 95] has proposed to avoid this drawback by generating four different trajectories to excite four different physical phenomena, which are: inertial effect, centrifugal coupling, inertial coupling and gravity effect. The trajectories are periodic between two points (except for gravity). During an experiment, a limited number of joints move while the others are locked. The experiments are designed in order to ensure optimal condition number of the observation matrix. These trajectories are then combined in a global identification system of equations.

12.3.5. *Calculation of the joint velocities and accelerations*

The observation matrix elements are functions of $\mathbf{q}$, $\dot{\mathbf{q}}$ and also $\ddot{\mathbf{q}}$ in the case of the dynamic identification model. Industrial robots are generally equipped with position sensors with good accuracy although they can be corrupted by high frequency noise due to quantization errors. On the contrary, velocity sensors provide noisy information, and the acceleration sensors are not used in industrial robots. Consequently, the joint velocities and accelerations have to be obtained by numerically differentiating the joint positions. However, the derivative, and especially the second derivative, of the joint positions amplifies the high frequency noise because the differentiation process behaves like a high-pass filter.

A solution to this problem is to filter the joint position readings using a low-pass filter prior to compute the derivatives [Khosla 86], [Bouzouia 89], [Benhlima 93], [Gautier 95]. Such a strategy has been successfully used while identifying the dynamic parameters of the Acma SR400 robot [Restrepo 96]: the position filtering is carried out with a non-causal zero-phase digital filter by processing the input data through an IIR low-pass Butterworth filter in both the forward and reverse direction, thanks to the "filtfilt" procedure of Matlab. The numerical derivation is carried out using a central difference algorithm to avoid phase shift. Thus, the velocity is obtained from:

$$\dot{q}(k) = [q(k+1) - q(k-1)] / 2T \qquad [12.16]$$

where $q(k)$ indicates the joint positions at the kth sample, and T is the sampling period.

Another solution to avoid the calculation of joint accelerations consists of using an identification model devoid of joint accelerations such as the filtered dynamic model (§ 12.5.2), the energy model (§ 12.6.1), the power model (§ 12.6.2), or by using a stochastic filter such as the extended Kalman filter [Guglielmi 87], [Gautier 93].

12.3.6. *Calculation of joint torques*

The dynamic identification methods are based on estimating the parameters minimizing a cost function of the difference between the real robot variables and its mathematical model. Most of the proposed cost functions require joint torques. Since torque sensors are not used in industrial robots, the actuator torque may be estimated from the reference current of the amplifier current loop. Owing to the high bandwidth of the current loop, the relation between the actuator torque and the reference current can be represented by a constant gain in the operating range of the robot. For joint j, this relation can be written as (Figure 12.1):

$$\Gamma_j = G_{Tj}\, u_j \qquad [12.17a]$$

where:

$$G_{Tj} = N_j\, K_{aj}\, K_{Tj} \qquad [12.17b]$$

and where G_{Tj} is the torque gain of the drive chain of joint j, u_j is the reference current, N_j is the gear transmission ratio, K_{aj} is the current amplifier gain, and K_{Tj} is the actuator torque constant.

The parameters of equations [12.17] can be obtained from the manufacturers data sheets. A global estimation of G_{Tj} can be obtained using specific experimentation [Restrepo 95].

Note that the dynamic identification methods can be reformulated to be independent of the drive chain constant G_{Tj} [Khalil 93], [Gautier 94]. In this case, a new set of dynamic parameters is defined.

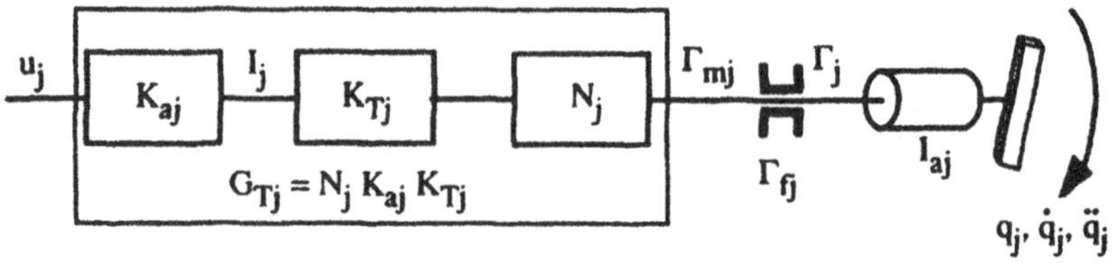

Figure 12.1. *Drive chain of joint j*

12.4. Dynamic identification model

The dynamic model is linear in the dynamic parameters. It is given by the following equation:

$$\Gamma = \Phi(\mathbf{q}, \dot{\mathbf{q}}, \ddot{\mathbf{q}})\, \chi \qquad [12.18]$$

where Φ is an (nxb) matrix, and b is the number of the base dynamic parameters.

From the above equation, we deduce that the ith column of Φ, denoted by Φ^i, is equal to:

$$\Phi^i = \Gamma(\mathbf{q}, \dot{\mathbf{q}}, \ddot{\mathbf{q}} \text{ with } \chi_i = 1, \chi_j = 0 \text{ for } j \neq i) \qquad [12.19]$$

Consequently, Φ^i can be computed using a specialized version of the inverse dynamic model in which the dynamic parameters are assumed to be $\chi_i = 1$, $\chi_j = 0$ for $j \neq i$. To increase the efficiency of this algorithm, we use the customized symbolic technique, taking into account that the forward recursive equations are the same for all the columns Φ^i. Moreover, we note that this customized symbolic technique is convenient for the computation of the observation matrix using an array multiply operator (.* of Matlab). Collecting $(\mathbf{q}, \dot{\mathbf{q}}, \ddot{\mathbf{q}}, \Gamma)_{(i)}$ for a sufficient large number of points i = 1, …, e on a given trajectory, and using equation [12.19], we can construct the following overdetermined system of equations:

$$\mathbf{Y}(\Gamma) = \mathbf{W}(\mathbf{q}, \dot{\mathbf{q}}, \ddot{\mathbf{q}})\, \chi + \rho \qquad [12.20]$$

with:

$$\mathbf{Y} = \begin{bmatrix} \Gamma(1) \\ \cdots \\ \Gamma(e) \end{bmatrix}, \ \mathbf{W} = \begin{bmatrix} \mathbf{\Phi}(1) \\ \cdots \\ \mathbf{\Phi}(e) \end{bmatrix} \qquad [12.21]$$

such that:

- $r = n \times e >> b$
- $\mathbf{\Phi}(i) = \mathbf{\Phi}[(\mathbf{q}, \dot{\mathbf{q}}, \ddot{\mathbf{q}})_{(i)}]$
- $(\mathbf{q}, \dot{\mathbf{q}}, \ddot{\mathbf{q}})_{(i)} = \mathbf{q}(t_i), \dot{\mathbf{q}}(t_i), \ddot{\mathbf{q}}(t_i)$
- $\Gamma(i) = \Gamma(t_i)$

In order to eliminate high frequency noise out of torque signals, we filter the vector **Y** and the columns of the observation matrix **W**. The values obtained are then decimated at a low rate. This procedure is known as *parallel filtering* [Richalet 98]. For the identification of the Acma SR400 robot [Restrepo 96], the authors used a low-pass Tchebychev filter of order 8 with a cut-off frequency of 40 Hz. The filtered signal was then decimated at order 10. The "decimate" procedure of Matlab (Signal Processing ToolBox) can be used to carry out these two steps.

12.5. Other approaches to the dynamic identification model

In this section, we present two different approaches to develop the identification model. The first is sequential and is based on a link-by-link identification starting from the terminal link. The second approach makes use of a filtered dynamic model, which is a function of $\mathbf{q}$ and $\dot{\mathbf{q}}$, and no more of $\ddot{\mathbf{q}}$. The two approaches can be combined to obtain a filtered sequential dynamic identification model.

12.5.1. *Sequential formulation of the dynamic model*

Since the torque of joint j is independent of the dynamic parameters of links 1, ..., j−1 (property d, § 9.3.3.3), we can write the dynamic model [12.18] such that the matrix $\mathbf{\Phi}$ is upper-block triangular:

$$\begin{bmatrix} \Gamma_1 \\ \Gamma_2 \\ \vdots \\ \Gamma_n \end{bmatrix} = \begin{bmatrix} \mathbf{\Phi}_{11}\mathbf{\Phi}_{12}\ldots\mathbf{\Phi}_{1n-1}\mathbf{\Phi}_{1n} \\ \mathbf{0} \ \ \mathbf{\Phi}_{22}\ldots\mathbf{\Phi}_{2n-1}\mathbf{\Phi}_{2n} \\ \vdots \ \ \mathbf{0} \ \ldots \ \vdots \ \ \vdots \\ \mathbf{0} \ \ \mathbf{0} \ \ldots \ \mathbf{0} \ \ \mathbf{\Phi}_{nn} \end{bmatrix} \begin{bmatrix} \chi^1 \\ \chi^2 \\ \vdots \\ \chi^n \end{bmatrix} \qquad [12.22]$$

where χ^j contains the identifiable parameters of both link and actuator j, and $\boldsymbol{\Phi}_{ij}$ is the row vector to be multiplied by χ^j in the equation for Γ_i.

This form of the dynamic equation can be exploited to estimate the dynamic parameters of each link individually, starting sequentially from link n and proceeding back to link 1. Thus, the dynamic parameters of link j are known when considering the torque equation of link j – 1. This sequential procedure reduces the number of the dynamic parameters that must be estimated at each step. The method can be summarized as follows:

- identify the base parameters of link n, χ^n, using the dynamic equation of joint n on a sufficient number of points:

$$\Gamma_n = \boldsymbol{\Phi}_{nn}(\mathbf{q}, \dot{\mathbf{q}}, \ddot{\mathbf{q}})\, \chi^n \qquad [12.23]$$

- then, identify χ^{n-1} using the torque equation of joint n – 1 assuming χ^n to be known:

$$\Gamma_{n-1} - \boldsymbol{\Phi}_{n-1,n}(\mathbf{q}, \dot{\mathbf{q}}, \ddot{\mathbf{q}})\, \hat{\chi}^n = \boldsymbol{\Phi}_{n-1,n-1}(\mathbf{q}, \dot{\mathbf{q}}, \ddot{\mathbf{q}})\, \chi^{n-1} \qquad [12.24]$$

- and so on until the estimation of the parameters of link 1.

Since the vector $\boldsymbol{\Phi}_{jj}$ is independent of the positions, velocities and accelerations of joints j + 1, ..., n, we can lock these joints while identifying the parameters of link j. Furthermore, the number of points needed for the identification is significantly reduced with respect to the global method, owing to the reduced number of parameters. Consequently, the problem of planning an exciting trajectory is greatly simplified.

The main drawback of this method is the possible accumulation of errors from one step to the next. To overcome this problem, a weighted least-squares solution can be used [Gautier 97].

12.5.2. *Filtered dynamic model (reduced order dynamic model)*

We can avoid calculating the joint accelerations by applying to the dynamic equations a low-pass filter of at least second order with unit gain at zero frequency [Aubin 91]. This approach has also been used for adaptive control schemes [Slotine 87], [Middleton 88]. To develop the new model, let us expand equation [12.18] to isolate the elements that are functions of the joint accelerations:

$$\Gamma = [\mathbf{A}^1(\mathbf{q})\,\ddot{\mathbf{q}} \;\; \ldots \;\; \mathbf{A}^m(\mathbf{q})\,\ddot{\mathbf{q}}]\,\mathbf{K} + \mathbf{H}_2(\mathbf{q},\dot{\mathbf{q}})\,\mathbf{K} + \mathbf{diag}(\mathrm{sign}(\dot{\mathbf{q}}))\,\mathbf{F}_c + \mathbf{diag}(\dot{\mathbf{q}})\,\mathbf{F}_v \qquad [12.25]$$

where **K** is the (mx1) vector of the base inertial parameters with $m = b - 2n$; $\mathbf{A}^j$ is the (nxn) matrix of the coefficients of the inertial parameter K_j in the robot inertia matrix **A**; $\mathbf{H}_2$ is the (nxm) matrix of the coefficients of the inertial parameters in the centrifugal, Coriolis and gravity terms; **diag**(**u**) is a diagonal matrix whose diagonal elements are composed of the elements of the vector **u**; $\mathbf{F}_c = [F_{c1} \ldots F_{cn}]^T$ and $\mathbf{F}_v = [F_{v1} \ldots F_{vn}]^T$ are the friction parameters.

The vector $\mathbf{A}^j\,\ddot{\mathbf{q}}$ can be written as:

$$\mathbf{A}^j\,\ddot{\mathbf{q}} = \frac{d}{dt}(\mathbf{A}^j\,\dot{\mathbf{q}}) - \dot{\mathbf{A}}^j\,\dot{\mathbf{q}} \qquad [12.26]$$

Using equations [12.25] and [12.26], we obtain:

$$\Gamma = [\frac{d}{dt}\mathbf{B1}(\mathbf{q},\dot{\mathbf{q}}) + \mathbf{B2}(\mathbf{q},\dot{\mathbf{q}})]\,\mathbf{K} + \mathbf{diag}(\mathrm{sign}(\dot{\mathbf{q}}))\,\mathbf{F}_c + \mathbf{diag}(\dot{\mathbf{q}})\,\mathbf{F}_v \qquad [12.27]$$

with:

$$\mathbf{B1}(\mathbf{q},\dot{\mathbf{q}}) = [\frac{d}{dt}(\mathbf{A}^1\,\dot{\mathbf{q}}) \;\; \frac{d}{dt}(\mathbf{A}^2\,\dot{\mathbf{q}}) \;\; \ldots \;\; \frac{d}{dt}(\mathbf{A}^m\,\dot{\mathbf{q}})]$$

$$\mathbf{B2}(\mathbf{q},\dot{\mathbf{q}}) = [-\dot{\mathbf{A}}^1\,\dot{\mathbf{q}} \;\; -\dot{\mathbf{A}}^2\,\dot{\mathbf{q}} \;\; \ldots \;\; -\dot{\mathbf{A}}^m\,\dot{\mathbf{q}}] + \mathbf{H}_2(\mathbf{q},\dot{\mathbf{q}})$$

Applying the Laplace transformation of the filter F(s) to equation [12.27] yields:

$$F(s)\,\Gamma = [sF(s)\,\mathbf{B1}(\mathbf{q},\dot{\mathbf{q}}) + F(s)\,\mathbf{B2}(\mathbf{q},\dot{\mathbf{q}})]\,\mathbf{K} + F(s)\,\mathbf{diag}(\mathrm{sign}(\dot{\mathbf{q}}))\,\mathbf{F}_c + F(s)\,\mathbf{diag}(\dot{\mathbf{q}})\,\mathbf{F}_v \qquad [12.28]$$

In [Aubin 91], the following second order filter was proposed:

$$F(s) = \frac{a^2}{(s+a)^2} \qquad [12.29]$$

The cut-off frequency value "a" should be chosen to be sufficiently large to maintain the dynamic behavior of the robot while rejecting the high frequency components ($10 \le a \le 40$).

Using equation [12.27] on a sufficient number of points leads to the overdetermined system of equations:

$$\mathbf{Y} = [(\frac{d}{dt}\mathbf{W1} + \mathbf{W2}) \quad \mathbf{W3} \quad \mathbf{W4}]\begin{bmatrix}\mathbf{K} \\ \mathbf{F_c} \\ \mathbf{F_v}\end{bmatrix} + \rho \qquad [12.30]$$

In practice, the application of F(s) is carried out in equation [12.30] as follows:

- numerically differentiating **W1** using a central difference algorithm;
- using the "decimate" function of Matlab on $\dot{\mathbf{W}}\mathbf{1}$, **Y**, **W2**, **W3**, and **W4**.

The least-squares solution is applied to the filtered model to estimate the values of $\hat{\mathbf{K}}$, $\hat{\mathbf{F}}_c$ and $\hat{\mathbf{F}}_v$.

The computation of equation [12.28] requires the symbolic expressions of $\mathbf{A}^j$ and $\dot{\mathbf{A}}^j$, for j = 1, ..., m. For a robot with more than three degrees of freedom, these expressions are dramatically complex. To overcome this difficulty, we propose to develop the filtered dynamic model starting from the Lagrange equation [9.4] [Khalil 96a]. In fact, taking into account the friction effects, the Lagrange equation can be written as:

$$\Gamma = \frac{d}{dt}\frac{\partial E}{\partial \dot{\mathbf{q}}} - \frac{\partial E}{\partial \mathbf{q}} + \frac{\partial U}{\partial \mathbf{q}} + \mathbf{diag}(\mathrm{sign}(\dot{\mathbf{q}}))\,\mathbf{F}_c + \mathbf{diag}(\dot{\mathbf{q}})\,\mathbf{F}_v \qquad [12.31]$$

Since the energy is linear in the inertial parameters (§ 9.4), we can write:

$$\begin{cases} E = \mathbf{e}(\mathbf{q}, \dot{\mathbf{q}})\,\mathbf{K} \\ U = \mathbf{u}(\mathbf{q})\,\mathbf{K} \end{cases} \qquad [12.32]$$

where E and U are the kinetic energy and potential energy of the robot respectively.

The elements of the row vector **e** are obtained using equations [9.44] and [9.25]. Consequently, we can write:

$$\begin{cases} e_{XXj} = \frac{1}{2}\omega_{1,j}\,\omega_{1,j} \\ e_{XYj} = \omega_{1,j}\,\omega_{2,j} \\ e_{XZj} = \omega_{1,j}\,\omega_{3,j} \\ e_{YYj} = \frac{1}{2}\omega_{2,j}\,\omega_{2,j} \\ e_{YZj} = \omega_{2,j}\,\omega_{3,j} \\ e_{ZZj} = \frac{1}{2}\omega_{3,j}\,\omega_{3,j} \\ e_{MXj} = \omega_{3,j}\,V_{2,j} - \omega_{2,j}\,V_{3,j} \\ e_{MYj} = \omega_{1,j}\,V_{3,j} - \omega_{3,j}\,V_{1,j} \\ e_{MZj} = \omega_{2,j}\,V_{1,j} - \omega_{1,j}\,V_{2,j} \\ e_{Mj} = \frac{1}{2}\,{}^{j}\mathbf{V}_j^{T}\,{}^{j}\mathbf{V}_j \\ e_{Iaj} = \frac{1}{2}\dot{q}_j^{2} \end{cases} \qquad [12.33]$$

where ${}^{j}\boldsymbol{\omega}_j = [\omega_{1,j} \;\; \omega_{2,j} \;\; \omega_{3,j}]^T$ and ${}^{j}\mathbf{V}_j = [V_{1,j} \;\; V_{2,j} \;\; V_{3,j}]^T$.

The elements of the row vector **u** are obtained from equation [9.25] as:

$$\begin{cases} u_{XXj} = u_{XYj} = \ldots = u_{ZZj} = 0 \\ u_{MXj} = -\,{}^{0}\mathbf{g}^{T}\,{}^{0}\mathbf{s}_j \\ u_{MYj} = -\,{}^{0}\mathbf{g}^{T}\,{}^{0}\mathbf{n}_j \\ u_{MZj} = -\,{}^{0}\mathbf{g}^{T}\,{}^{0}\mathbf{a}_j \\ u_{Mj} = -\,{}^{0}\mathbf{g}^{T}\,{}^{0}\mathbf{P}_j \\ u_{Iaj} = 0 \end{cases} \qquad [12.34]$$

where ${}^{0}\mathbf{s}_j$, ${}^{0}\mathbf{n}_j$, ${}^{0}\mathbf{a}_j$ and ${}^{0}\mathbf{P}_j$ are the (3x1) vectors appearing in the transformation matrix ${}^{0}\mathbf{T}_j$ and **g** is the acceleration of gravity.

Thus, equation [12.31] can be written as:

$$\Gamma = [\frac{d}{dt}\mathbf{B1}(\mathbf{q}, \dot{\mathbf{q}}) + \mathbf{B2}(\mathbf{q}, \dot{\mathbf{q}})]\,\mathbf{K} + \mathbf{diag}(\text{sign}(\dot{\mathbf{q}}))\,\mathbf{F}_c + \mathbf{diag}(\dot{\mathbf{q}})\,\mathbf{F}_v \qquad [12.35]$$

with:

$$\mathbf{B1}(\mathbf{q}, \dot{\mathbf{q}}) = \frac{\partial \mathbf{e}}{\partial \dot{\mathbf{q}}} \qquad [12.36]$$

$$\mathbf{B2}(\mathbf{q}, \dot{\mathbf{q}}) = \frac{\partial \mathbf{u}}{\partial \mathbf{q}} - \frac{\partial \mathbf{e}}{\partial \mathbf{q}} \qquad [12.37]$$

Since the expressions for the velocity vectors ${}^{j}\mathbf{V}_j$ and ${}^{j}\boldsymbol{\omega}_j$ are not so complicated, even for a six degree-of-freedom robot, we can symbolically compute **B1** and **B2**. We can also make use of efficient recursive algorithms to compute them using customized symbolic techniques [Stepanenko 93], [Khalil 96a]. However, the computational burden of the filtered dynamic model is still higher than that of the dynamic model.

12.6. Energy (or integral) identification model

In order to avoid the calculation of the joint accelerations, a model based on energy theorem has been proposed [Gautier 88]. In addition, this model has the following advantages:

- it is linear in the dynamic parameters, and the corresponding base dynamic parameters are the same as those of the dynamic model;
- the planning of an exciting trajectory is easier than for the dynamic model;
- the computation of the observation matrix is easier than that using the dynamic model.

12.6.1. *Principle of the energy model*

The energy (or integral) model is based on the energy theorem, which states that the total mechanical energy applied to the robot is equal to the sum of potential and kinetic energy contained in the system.

Let the total energy of the system, also termed *Hamiltonian*, be denoted by H = E + U. From the energy theorem, we can write (§ 14.5.2):

$$dH = \mathbf{T}^T \mathbf{dq} \qquad [12.38]$$

$$dH = \mathbf{T}^T \dot{\mathbf{q}}\, dt \qquad [12.39]$$

where:

$$\mathbf{T} = \mathbf{\Gamma} - \mathbf{diag}(\text{sign}(\dot{\mathbf{q}}))\, \mathbf{F}_c - \mathbf{diag}(\dot{\mathbf{q}})\, \mathbf{F}_v \qquad [12.40]$$

After integration:

$$y = \int_{t_a}^{t_b} \mathbf{T}^T \dot{\mathbf{q}}\, dt = H(t_b) - H(t_a) = \Delta H \qquad [12.41]$$

The total energy of the system can be written as equation [9.41]:

$$H = [\ \mathbf{h}_1 \ \ldots \ \mathbf{h}_n\ ]\ [\ \mathbf{K}_1^T \ \ldots \ \mathbf{K}_n^T\ ]^T = \mathbf{h}\,\mathbf{K} \qquad [12.42]$$

where $\mathbf{K}_j$ is the vector of the base inertial parameters of link j, and $\mathbf{h}_j$ is the row vector of the energy functions such that $\mathbf{h}_j = \mathbf{e}_j + \mathbf{u}_j$.

Hence, the energy equation is linear in the dynamic parameters of the robot. The expressions of the elements of $\mathbf{h}_j$ are given by equations [12.33] and [12.34]. Substituting equation [12.42] into equation [12.41] yields the scalar identification model:

$$y = \int_{t_a}^{t_b} \Gamma^T \dot{\mathbf{q}}\, dt = [\ \Delta\mathbf{h} \quad \Delta\mathbf{f}_c \quad \Delta\mathbf{f}_v\ ] \begin{bmatrix} \mathbf{K} \\ \mathbf{F}_c \\ \mathbf{F}_v \end{bmatrix} \qquad [12.43]$$

with:

$$\Delta\mathbf{h} = \mathbf{h}(t_b) - \mathbf{h}(t_a) \qquad [12.44]$$

$$\Delta f_{cj} = \int_{t_a}^{t_b} \dot{q}_j \operatorname{sign}(\dot{q}_j)\, dt \qquad [12.45]$$

$$\Delta f_{vj} = \int_{t_a}^{t_b} \dot{q}_j^2\, dt \qquad [12.46]$$

$$\Delta\mathbf{f}_c = [\ \Delta f_{c1} \ \ldots \ \Delta f_{cn}\ ] \qquad [12.47]$$

$$\Delta\mathbf{f}_v = [\ \Delta f_{v1} \ \ldots \ \Delta f_{vn}\ ] \qquad [12.48]$$

Applying equation [12.43] to a sufficient number of intervals ab(i) and collecting the corresponding $(\Gamma, \mathbf{q}, \dot{\mathbf{q}})_{ab(i)}$, we obtain a linear system of r equations in b unknowns such that r >> b, where b is the number of base dynamic parameters:

$$\mathbf{Y}(\Gamma, \dot{\mathbf{q}}) = \mathbf{W}(\mathbf{q}, \dot{\mathbf{q}}) \begin{bmatrix} \mathbf{K} \\ \mathbf{F}_c \\ \mathbf{F}_v \end{bmatrix} + \rho \qquad [12.49]$$

with:

- $\mathbf{Y} = \begin{bmatrix} y(1) \\ \cdots \\ y(r) \end{bmatrix}$, $\mathbf{W} = \begin{bmatrix} \Delta\mathbf{h}(1) & \Delta\mathbf{f}_c(1) & \Delta\mathbf{f}_v(1) \\ \cdots & \cdots & \cdots \\ \Delta\mathbf{h}(r) & \Delta\mathbf{f}_c(r) & \Delta\mathbf{f}_v(r) \end{bmatrix}$
- $\Delta\mathbf{h}(i) = \mathbf{h}[(\mathbf{q}, \dot{\mathbf{q}})_{b(i)}] - \mathbf{h}[(\mathbf{q}, \dot{\mathbf{q}})_{a(i)}]$
- $(\mathbf{q}, \dot{\mathbf{q}})_{a(i)} = [\mathbf{q}(t_{a(i)}), \dot{\mathbf{q}}(t_{a(i)})]$
- $(\mathbf{q}, \dot{\mathbf{q}})_{b(i)} = [\mathbf{q}(t_{b(i)}), \dot{\mathbf{q}}(t_{b(i)})]$
- $y(i) = \int_{t_{a(i)}}^{t_{b(i)}} \Gamma^T \dot{\mathbf{q}}\, dt$

The least-squares solution of equation [12.49] gives the estimated dynamic parameters $\hat{\mathbf{K}}$, $\hat{\mathbf{F}}_c$ and $\hat{\mathbf{F}}_v$.

12.6.2. *Power model*

The integrator appearing in the energy model is an infinite-gain filter at zero frequency. This means that small low-frequency errors such as offsets can produce large errors. To overcome this problem, we can make use of the differential equation [12.39] instead of the integral one [Gautier 96]. This leads to the power model, which is written as:

$$\Gamma^T \dot{\mathbf{q}} = \frac{d}{dt}(\mathbf{h}\,\mathbf{K}) + \dot{\mathbf{q}}^T\,[\mathbf{diag}(\mathrm{sign}(\dot{\mathbf{q}}))\,\mathbf{F}_c + \mathbf{diag}(\dot{\mathbf{q}})\,\mathbf{F}_v] \qquad [12.50]$$

or, in a linear form with respect to the dynamic parameters:

$$\Gamma^T \dot{\mathbf{q}} = \left[\frac{d}{dt}\mathbf{h} \quad \dot{\mathbf{q}}^T\mathbf{diag}(\mathrm{sign}(\dot{\mathbf{q}})) \quad \dot{\mathbf{q}}^T\mathbf{diag}(\dot{\mathbf{q}})\right] \begin{bmatrix} \mathbf{K} \\ \mathbf{F}_c \\ \mathbf{F}_v \end{bmatrix} \qquad [12.51]$$

Using a sufficient number of points, we obtain the system of linear equations:

$$\mathbf{Y} = \left[\frac{d}{dt}\mathbf{W1} \quad \mathbf{W2} \quad \mathbf{W3}\right] \begin{bmatrix} \mathbf{K} \\ \mathbf{F}_c \\ \mathbf{F}_v \end{bmatrix} + \rho \qquad [12.52]$$

As we did for the filtered dynamic model (§ 12.5.2), we process the columns **Y**, **W2** and **W3** using a low-pass filter F(s), while the columns of **W1** are filtered by sF(s). In practice, this process can be carried out using a central difference algorithm to obtain the time derivative of **W1**, then by using the "decimate" function of Matlab to filter all the model. The least-squares solution can then be used to estimate the dynamic parameters.

Another advantage of the power model with respect to the integral model comes from the fact that it is calculated for each point of the trajectory such that the problem of determining the integration period $[t_b - t_a]$ is avoided.

Before closing this section, we note that the energy and the power identification methods do not minimize the torque errors. Thus, minimum equation errors with these models do not guarantee that the torques will be correctly evaluated using the identified parameters. Therefore, when using these methods, the results must be validated by comparing the real torques and the prediction torques (inverse dynamic model) on a test trajectory.

12.7. Recommendations for experimental application

For experimental application of the identification algorithms, the following recommendations and remarks should be taken into account:

- the dynamic model consists of as many equations as the number of joints, while the energy (or power) model is composed of a single one. Thus, the dynamic model is basically more exciting and the energy model is more sensitive to the use of exciting trajectories;
- the energy identification method is robust with respect to high frequency perturbation, thus less sensitive to the filtering parameters (cut-off frequency, order of the filter). On the contrary, the dynamic identification model is more sensitive to the filtering parameters and to the quality of estimating the joint velocities and accelerations;
- after filtering the joint positions, the joint velocities and accelerations should be obtained with a central difference algorithm to avoid phase lag;
- when using the dynamic identification model, we must filter the vector of measured torques in order to reject the high frequency ripples. The same filter must be applied to the columns of the observation matrix. This process is called *parallel filtering*. We note that parallel filtering is automatically incorporated in the filtered dynamic model (reduced order dynamic model);
- it may appear that the filtered dynamic model has the advantages of the dynamic model (number of equations equal to the number of joints) and that of the energy model (no joint accelerations are needed). However, if the

dynamic model is correctly processed, it gives equivalent results to the filtered dynamic model [Restrepo 96];

- in the case of robots with several degrees of freedom, six for instance, it is recommended to carry out the identification sequentially in two steps: first, identify the parameters of the wrist links, then, identify the parameters of the shoulder links while locking the wrist joints and assuming that the wrist parameters are known. This procedure is especially efficient because the dimensions of the wrist links are generally not in the same order of magnitude as those of the shoulder;
- the number of equations must be at least 500 times the number of parameters to identify;
- the filtering parameters (cut-off frequency, order of the filter, ...) can be determined by simulating the identification method;
- the relative standard deviation given by equation [12.10] can be used as a criterion to measure the identification quality of a parameter. If the relative standard deviation of a parameter is greater than ten times the minimum relative standard deviation value, this parameter can be considered as poorly identified. Thus, either this parameter has not been sufficiently excited or its contribution to the model is negligible. If the same result is obtained with different trajectories, and if the value of this parameter is relatively small with respect to the other parameters, we can cancel this parameter and define a new set of essential parameters that can be better identified [Pham 91b];
- in order to validate the results obtained, the following tests can be carried out:
  - direct validation on the identification trajectory, by calculating the error vector;
  - cross validation on a new trajectory that has not been used for the identification;
  - verify that the inertia matrix of the robot computed with the estimated parameters is positive definite [Yoshida 00];
  - identification of the dynamic parameters twice: without load, and with a known load, to verify if the load parameter values can be correctly estimated;
  - carrying out the identification using different methods – dynamic model, filtered dynamic model and power model – to compare the results;
  - realizing a simulator of the robot using the identified parameters and comparing the response of the real robot with that of the simulator.

12.8. Conclusion

In this chapter, we have addressed the problem of identification of the dynamic parameters of robots. We have proposed several methods that are linear in these parameters and have the same set of base dynamic parameters. The filtered dynamic model, the energy model and the power model do not need the joint accelerations. Other identification methods, which have not been treated in this chapter, can be found in the literature, namely the method based on the minimization of the error between the measured and calculated reaction wrench (forces and moments) on the base of the robot [West 89], [Raucent 92], [Geffard 00] and the method based on the use of the extended Kalman filter [Guglielmi 87], [Poignet 00].

We have presented different criteria to measure the excitation of a given trajectory. We have pointed out that the dynamic and filtered dynamic models are basically more exciting than the energy and power models.

Having laid the foundation to identify the dynamic parameters, we can now proceed further with the generation of reference trajectories and the control schemes to track them.

Chapter 13

Trajectory generation

13.1. Introduction

A robotic motion task is specified by defining a path along which the robot must move. A *path* is a sequence of *points* defined either in task coordinates (end-effector coordinates) or in joint coordinates. The issue of trajectory generation is to compute for the control system the desired reference joint or end-effector variables as functions of time such that the robot tracks the desired path. Thus, a *trajectory* refers to a path and a *time history* along the path.

The trajectories of a robot can be classified as follows:

- trajectory between two points with free path between them;
- trajectory between two points via a sequence of desired intermediate points, also called *via points*, with free paths between via points;
- trajectory between two points with constrained path between the points (straight line segment for instance);
- trajectory between two points via intermediate points with constrained paths between the via points.

In the first two classes, the trajectory is generally generated in the joint space. In the last two classes, it is better to generate the trajectory in the task space.

In the next sections, we present trajectory generation techniques related to this classification, but we first analyze the reasons that motivate the choice of either the joint space or the task space for the generation.

13.2. Trajectory generation and control loops

The two approaches to trajectory generation – in the joint space and in the task space – are illustrated in Figures 13.1 and 13.2 (superscripts i and f designate the initial and final values respectively).

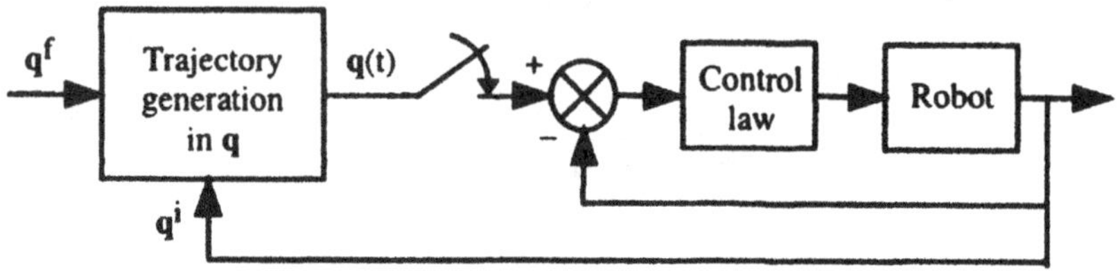

Figure 13.1. *Trajectory generation in the joint space*

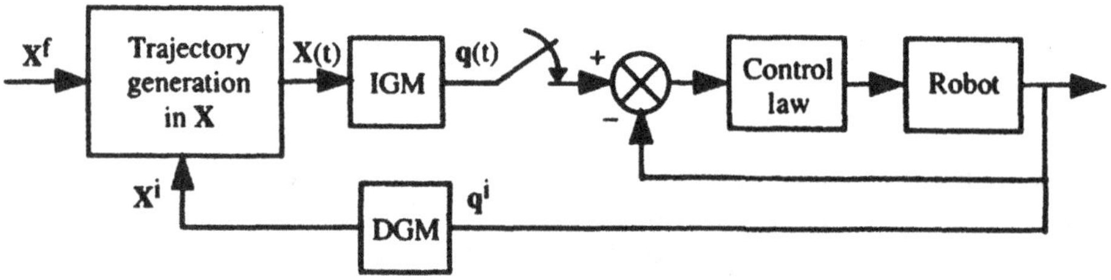

Figure 13.2. *Trajectory generation in the task space*

Trajectory generation in the joint space presents several advantages:

– it requires fewer on-line computations, since there is no need to compute the inverse geometric or kinematic models;

– the trajectory is not affected by crossing singular configurations;

– maximum velocities and torques are determined from actuator data sheets.

The drawback is that the corresponding end-effector path in the task space is not predictable, although it is repetitive, which increases risk of undesirable collisions when the robot works in a cluttered environment. In conclusion, the joint space scheme is appropriate to achieve fast motions in a free space.

Trajectory generation in the task space permits prediction of the geometry of the path. It has, however, a number of disadvantages:

– it may fail when the computed trajectory crosses a singular configuration;

– it fails when the generated points are out of the joint limits or when the robot is forced to change its current aspect (§ 5.7.4);

– velocities and accelerations of the task coordinates depend on the robot configuration. Lower bounds are generally used such that joint velocity and torque limits are satisfied. Consequently, the robot may work under its nominal performance.

The choice of a trajectory generation scheme depends on the application at hand. Each approach has its own limits, due to the fact that constraints are specified either in the joint space (velocities, torques, joint limits), or in the task space (accuracy, geometry of obstacles). Accounting for these remarks, the first two sections cover the problem of trajectory generation between two points in the joint space and the task space respectively. The last section extends the results to trajectory generation between several points.

13.3. Point-to-point trajectory in the joint space

We consider a robot with n degrees of freedom. Let $\mathbf{q}^i$ and $\mathbf{q}^f$ be the joint coordinate vectors corresponding to the initial and final configurations. Let $\mathbf{k}_v$ and $\mathbf{k}_a$ be the vectors of maximum joint velocities and maximum joint accelerations respectively. The value of k_{vj} can be exactly computed from the actuator specifications and transmission ratios, while the value of k_{aj} is approximated by the ratio of the maximum actuator torque to the maximum joint inertia (upper bound of the diagonal element A_{jj} of the inertia matrix defined in Chapter 9). Once the trajectory has been computed with these kinematic constraints, we can proceed to a time scaling in order to better match the maximum joint torques using the dynamic model [Hollerbach 84a].

The trajectory between $\mathbf{q}^i$ and $\mathbf{q}^f$ is determined by the following equation:

$$\mathbf{q}(t) = \mathbf{q}^i + r(t)\,\mathbf{D} \quad \text{for } 0 \leq t \leq t_f \qquad [13.1]$$

$$\dot{\mathbf{q}}(t) = \dot{r}(t)\,\mathbf{D} \qquad [13.2]$$

with $\mathbf{D} = \mathbf{q}^f - \mathbf{q}^i$.

The boundary conditions of the *interpolation function* r(t) are given by:

$$\begin{cases} r(0) = 0 \\ r(t_f) = 1 \end{cases}$$

Equation [13.1] can also be written as:

$$\mathbf{q}(t) = \mathbf{q}^f(t) - [1 - r(t)]\,\mathbf{D} \qquad [13.3]$$

which is more appropriate for tracking moving objects where $\mathbf{q}^f$ is time-varying [Taylor 79]. In this case, $\mathbf{D} = \mathbf{q}^f(0) - \mathbf{q}^i$.

Several interpolation functions can provide a trajectory such that $\mathbf{q}(0) = \mathbf{q}^i$ and $\mathbf{q}(t_f) = \mathbf{q}^f$. We will study successively the polynomial interpolation, the so-called *bang-bang acceleration profile*, and the bang-bang profile with a constant velocity phase termed *trapeze velocity profile*.

13.3.1. *Polynomial interpolation*

The most commonly used polynomials are the linear interpolation, the third degree polynomials (cubic) and the fifth degree polynomials (quintic).

13.3.1.1. *Linear interpolation*

The trajectory of each joint is described by a linear equation in time. The equation of the joint position is written as:

$$\mathbf{q}(t) = \mathbf{q}^i + \frac{t}{t_f}\mathbf{D} \qquad [13.4]$$

With this trajectory, the position is continuous but not the velocity. This induces undesirable vibrations on the robot and may cause early wear and tear of the mechanical parts.

13.3.1.2. *Cubic polynomial*

If the initial and final velocities are also set to zero, the minimum degree of the polynomial satisfying the constraints is at least three, and has the form:

$$\mathbf{q}(t) = \mathbf{a}_0 + \mathbf{a}_1 t + \mathbf{a}_2 t^2 + \mathbf{a}_3 t^3 \qquad [13.5]$$

The coefficients $\mathbf{a}_i$ are determined from the following boundary conditions:

$$\begin{cases} a_0 = q^i \\ a_1 = 0 \\ a_2 = \dfrac{3}{t_f^2} D \\ a_3 = -\dfrac{2}{t_f^3} D \end{cases} \qquad [13.6]$$

The expression [13.5] can also be written under the form [13.1] or [13.3] with the following interpolation function:

$$r(t) = 3\left(\frac{t}{t_f}\right)^2 - 2\left(\frac{t}{t_f}\right)^3 \qquad [13.7]$$

The cubic polynomial ensures the continuity of velocity but not of acceleration. Practically, the industrial robots are sufficiently rigid so that this discontinuity is filtered by the mechanical structure. Therefore, such a trajectory is generally satisfactory for most applications.

Figure 13.3 shows the position, velocity and acceleration profiles for joint j. The velocity is maximum at $t = t_f/2$ and its magnitude is given by:

$$|\dot{q}_{jmax}| = \frac{3|D_j|}{2t_f} \quad \text{with } |D_j| = |q_j^f - q_j^i| \qquad [13.8]$$

The maximum acceleration occurs at $t = 0$ and $t = t_f$ with the magnitude:

$$|\ddot{q}_{jmax}| = \frac{6|D_j|}{t_f^2} \qquad [13.9]$$

13.3.1.3. *Quintic polynomial*

For high speed robots or when a robot is handling heavy or delicate loads, it is worth ensuring the continuity of accelerations as well, in order to avoid exciting resonances in the mechanics. The trajectory is said to be of class C^2. Since six constraints have to be satisfied, the interpolation requires a polynomial of at least fifth degree [Binford 77]. The additional two constraints are written as:

$$\begin{cases} \ddot{q}(0) = 0 \\ \ddot{q}(t_f) = 0 \end{cases} \qquad [13.10]$$

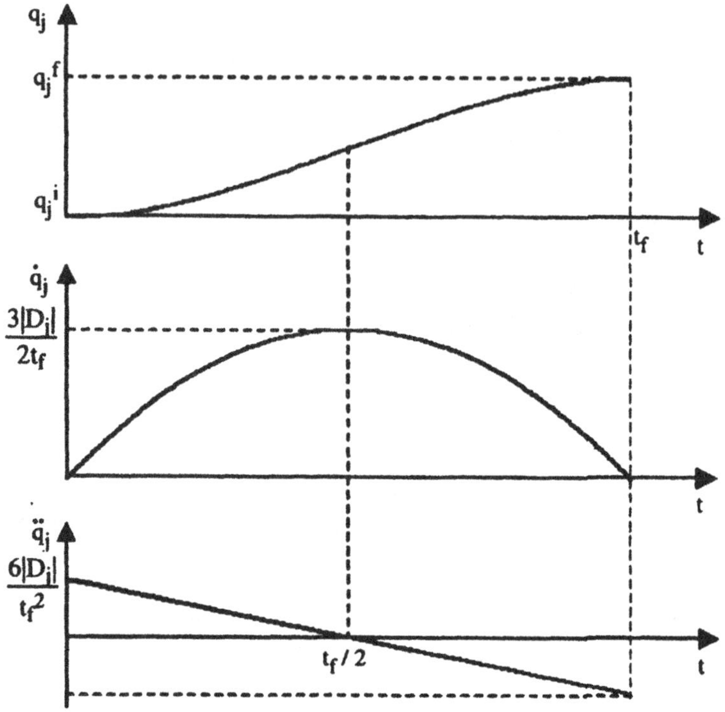

Figure 13.3. *Position, velocity and acceleration profiles for a cubic polynomial*

Solving for the six constraints yields the following interpolation function:

$$r(t) = 10\left(\frac{t}{t_f}\right)^3 - 15\left(\frac{t}{t_f}\right)^4 + 6\left(\frac{t}{t_f}\right)^5 \quad [13.11]$$

The position, velocity and acceleration with respect to time for joint j are plotted in Figure 13.4. Maximum velocity and acceleration are given by:

$$|\dot{q}_{jmax}| = \frac{15\,|D_j|}{8\,t_f} \quad [13.12]$$

$$|\ddot{q}_{jmax}| = \frac{10\,|D_j|}{\sqrt{3}\,t_f^2} \quad [13.13]$$

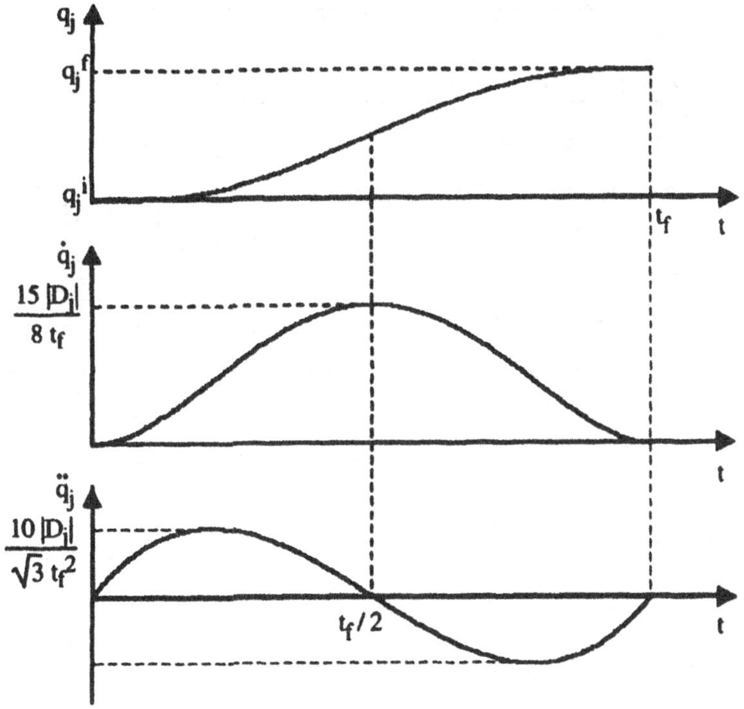

Figure 13.4. *Position, velocity and acceleration profiles for a quintic polynomial*

13.3.1.4. *Computation of the minimum traveling time*

Generally, the duration t_f of a trajectory is not specified. The goal is to minimize the time to travel from the initial configuration $\mathbf{q}^i$ to the final one $\mathbf{q}^f$ while satisfying velocity and acceleration constraints. The approach is to compute the minimum time separately for each joint, and then to synchronize all the joints at a common time.

The minimum traveling time t_{fj} for joint j occurs if either the velocity or the acceleration is saturated during the trajectory. This minimum time is computed from the maximum magnitudes of velocity and acceleration of the different polynomial interpolation functions (Table 13.1). The global minimum traveling time t_f is equal to the largest minimum time:

$$t_f = \max(t_{f1}, \ldots, t_{fn}) \quad [13.14]$$

Table 13.1. *Minimum traveling time for joint j*

Interpolation function	Minimum time
Linear interpolation	$t_{fj} = \frac{\|D_j\|}{k_{vj}}$
Cubic polynomial	$t_{fj} = \max\left[\frac{3\,\|D_j\|}{2\,k_{vj}}, \sqrt{\frac{6\,\|D_j\|}{k_{aj}}}\right]$
Quintic polynomial	$t_{fj} = \max\left[\frac{15\,\|D_j\|}{8\,k_{vj}}, \sqrt{\frac{10\,\|D_j\|}{\sqrt{3}\,k_{aj}}}\right]$
Bang-bang profile (§ 13.3.2)	$t_{fj} = \max\left[\frac{2\,\|D_j\|}{k_{vj}}, 2\sqrt{\frac{\|D_j\|}{k_{aj}}}\right]$

13.3.2. *Bang-bang acceleration profile*

A bang-bang acceleration profile consists of a constant acceleration phase until $t_f/2$ followed by a constant deceleration phase (Figure 13.5). The initial and final velocities are zero. Thus, the trajectory is continuous in position and velocity, but discontinuous in acceleration.

The position is given by:

$$\begin{cases} \mathbf{q}(t) = \mathbf{q}^i + 2\left(\frac{t}{t_f}\right)^2 \mathbf{D} & \text{for } 0 \le t \le \frac{t_f}{2} \\ \mathbf{q}(t) = \mathbf{q}^i + \left[-1 + 4\left(\frac{t}{t_f}\right) - 2\left(\frac{t}{t_f}\right)^2\right] \mathbf{D} & \text{for } \frac{t_f}{2} \le t \le t_f \end{cases} \qquad [13.15]$$

For joint j, the maximum velocity and acceleration are given by:

$$|\dot{q}_{jmax}| = \frac{2\,|D_j|}{t_f} \qquad [13.16]$$

$$|\ddot{q}_{jmax}| = \frac{4\,|D_j|}{t_f^2} \qquad [13.17]$$

The minimum traveling time is obtained as for a polynomial trajectory (Table 13.1).

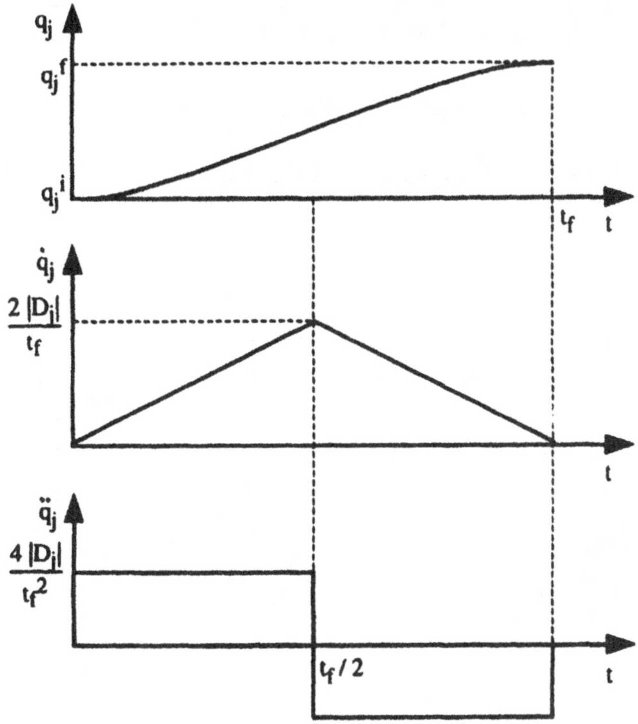

Figure 13.5. *Bang-bang velocity profile*

13.3.3. *Trapeze velocity profile*

With a bang-bang profile, when the velocity reaches its maximum, adding a constant velocity phase would allow us to saturate also the acceleration and to minimize the traveling time (Figure 13.6). According to equations [13.16] and [13.17], the condition to have a constant velocity phase on joint j is such that:

$$|D_j| > \frac{k_{vj}^2}{k_{aj}} \qquad [13.18]$$

The trapeze velocity trajectory results in the minimum traveling time among those providing the continuity of velocity. The joint j trajectory (Figure 13.7) is represented by the following equations:

$$\begin{cases} q_j(t) = q_j^i + \frac{1}{2} t^2 k_{aj} \,\text{sign}(D_j) & \text{for } 0 \le t \le \tau_j \\ q_j(t) = q_j^i + (t - \frac{\tau_j}{2}) k_{vj} \,\text{sign}(D_j) & \text{for } \tau_j \le t \le t_{fj} - \tau_j \\ q_j(t) = q_j^f - \frac{1}{2} (t_{fj} - t)^2 k_{aj} \,\text{sign}(D_j) & \text{for } t_{fj} - \tau_j \le t \le t_{fj} \end{cases} \quad [13.19]$$

with:

$$\tau_j = \frac{k_{vj}}{k_{aj}} \quad [13.20]$$

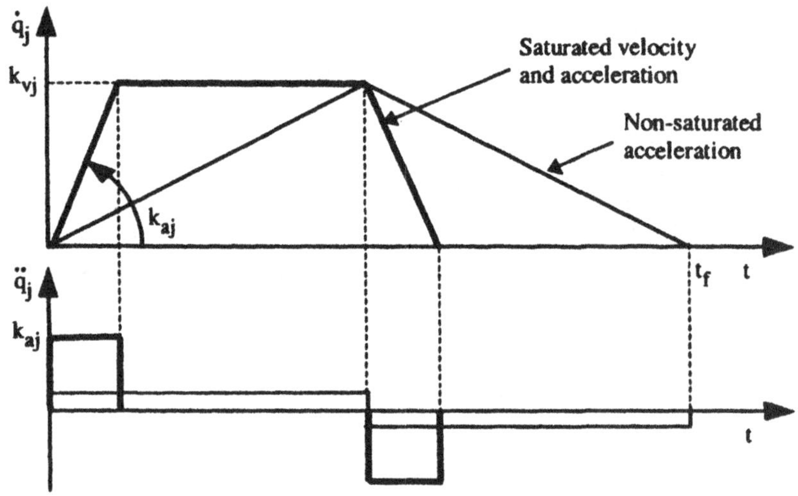

Figure 13.6. *Trapeze velocity profile versus bang-bang acceleration profile*

The area of the trapeze under the velocity curve is equal to the distance traveled in the interval $[0, t_{fj}]$, which can be written as:

$$|D_j| = |q_j^f - q_j^i| = 2\int_0^{\tau_j} k_{aj}\, t\, dt + \int_{\tau_j}^{t_{fj}-\tau_j} k_{vj} dt = k_{vj}\, t_{fj} - \frac{k_{vj}^2}{k_{aj}} \quad [13.21]$$

Hence, the minimum time for joint j is given by:

$$t_{fj} = \frac{k_{vj}}{k_{aj}} + \frac{|D_j|}{k_{vj}} = \tau_j + \frac{|D_j|}{k_{vj}} \quad [13.22]$$

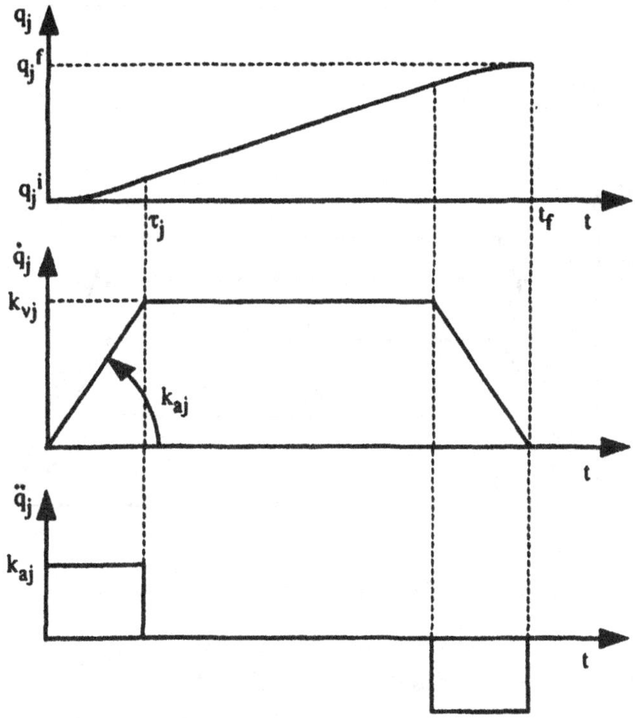

Figure 13.7. *Position, velocity and acceleration profiles for a trapeze trajectory*

In order to synchronize the joint trajectories, we present in the following a method giving homothetical trajectories with the same acceleration and deceleration duration for all joints. Such a method is the most common in use in industrial robot controllers. Let us designate by α_j the ratio between the velocity profile of joint j and an arbitrary joint k. We can write that (Figure 13.8):

$$\dot{q}_j(t) = \alpha_j \dot{q}_k(t) \quad \text{for } j = 1, \ldots, n \qquad [13.23]$$

Doing this, the duration τ of the acceleration phase of the synchronized trajectories is *a priori* not equal to any optimal τ_j computed for each joint separately (equation [13.20]).

Let $\lambda_j\, k_{vj}$ be the maximum velocity of the synchronized trajectory for joint j and let $\upsilon_j\, k_{aj}$ be the corresponding maximum acceleration. To calculate the optimal τ, resulting in a minimum time t_f, let us first solve the problem for two joints. According to equation [13.22], the minimum traveling time for each joint, if calculated separately, should be:

$$\begin{cases} t_{f1} = \tau_1 + \dfrac{|D_1|}{k_{v1}} \\ t_{f2} = \tau_2 + \dfrac{|D_2|}{k_{v2}} \end{cases} \qquad [13.24]$$

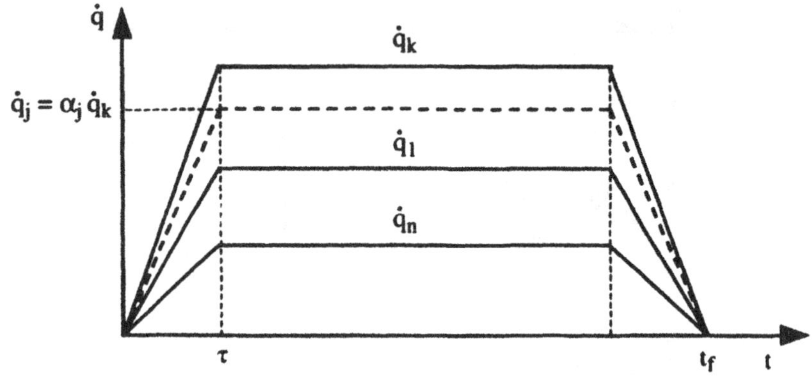

Figure 13.8. *Homothetical velocity profiles*

The synchronized trajectories should be such that:

$$t_f = \frac{\lambda_1 k_{v1}}{\upsilon_1 k_{a1}} + \frac{|D_1|}{\lambda_1 k_{v1}} = \frac{\lambda_2 k_{v2}}{\upsilon_2 k_{a2}} + \frac{|D_2|}{\lambda_2 k_{v2}} \qquad [13.25]$$

with $t_f \geq \max(t_{f1}, t_{f2})$.

From equation [13.25], it is straightforward to obtain:

$$\tau = \frac{\lambda_1 k_{v1}}{\upsilon_1 k_{a1}} = \frac{\lambda_2 k_{v2}}{\upsilon_2 k_{a2}} \qquad [13.26]$$

$$\lambda_2 = \lambda_1 \frac{k_{v1}|D_2|}{k_{v2}|D_1|} \qquad [13.27]$$

$$\upsilon_2 = \upsilon_1 \frac{k_{a1}|D_2|}{k_{a2}|D_1|} \qquad [13.28]$$

The velocity constraints imply that:

$$\begin{cases} 0 \le \lambda_1 \le 1 \\ 0 \le \lambda_2 \le 1 \end{cases} \quad [13.29]$$

Combining the last inequality with equation [13.27] yields:

$$0 \le \lambda_1 \le \frac{k_{v2}|D_1|}{k_{v1}|D_2|} \quad [13.30]$$

Likewise, from the acceleration constraints, we get:

$$\begin{cases} 0 \le \upsilon_1 \le 1 \\ 0 \le \upsilon_1 \le \dfrac{k_{a2}|D_1|}{k_{a1}|D_2|} \end{cases} \quad [13.31]$$

The minimum time t_f is obtained when the parameters λ_1 and υ_1 are the largest and satisfy simultaneously the above constraints, which results in:

$$\begin{cases} \lambda_1 = \min\left[1, \dfrac{k_{v2}|D_1|}{k_{v1}|D_2|}\right] \\ \upsilon_1 = \min\left[1, \dfrac{k_{a2}|D_1|}{k_{a1}|D_2|}\right] \end{cases} \quad [13.32]$$

and the corresponding duration of the acceleration phase is:

$$\tau = \frac{\lambda_1 k_{v1}}{\upsilon_1 k_{a1}} \quad [13.33]$$

These equations are easily generalized for n joints:

$$\begin{cases} \lambda_1 = \min\left[1, \dfrac{k_{vj}|D_1|}{k_{v1}|D_j|}\right] \\ \upsilon_1 = \min\left[1, \dfrac{k_{aj}|D_1|}{k_{a1}|D_j|}\right] \end{cases} \quad \text{for } j = 2, \ldots, n \quad [13.34]$$

assuming that $D_1 \neq 0$ and $D_j \neq 0$.

NOTE.– If, for a given joint j, the distance $|D_j|$ is such that the maximum velocity k_{vj} cannot be attained, we replace in the above formulas the term k_{vj} by the maximum achievable velocity. According to equation [13.18], this occurs when:

$$|D_j| < \frac{k_{vj}^2}{k_{aj}}$$

which implies that the maximum achievable velocity is:

$$k'_{vj} = \sqrt{|D_j|\, k_{aj}} \quad [13.35]$$

13.3.4. *Continuous acceleration profile with constant velocity phase*

We can modify the previous method to have a continuous trajectory in acceleration by replacing the acceleration and deceleration phases either by a second degree polynomial (Figure 13.9a) or by a trapeze acceleration profile (Figure 13.9b) [Castain 84]. In the following, we detail the first approach, which is simpler to implement. Let τ' be the new duration of the acceleration and let $\lambda_j\, k_{vj}$ be the maximum velocity of the trapeze profile. The boundary conditions for joint j are defined as:

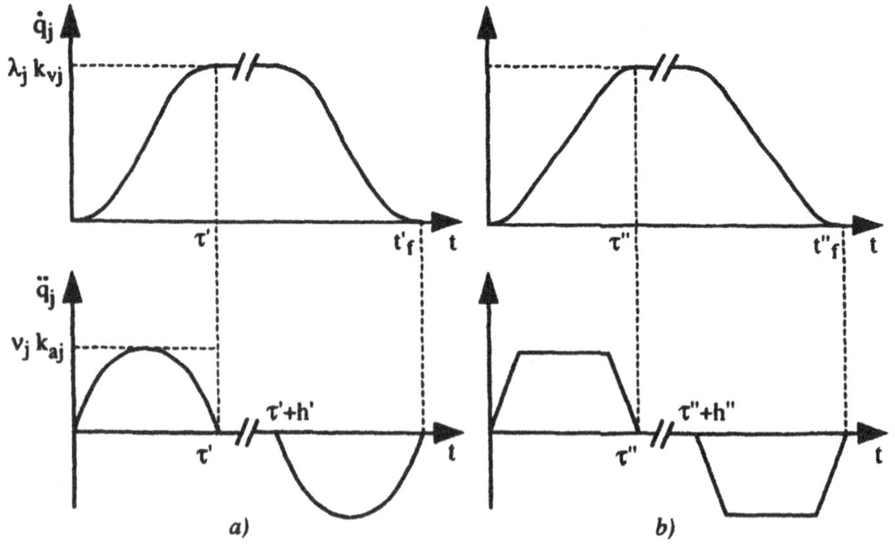

Figure 13.9. *Modification of the acceleration of the trapeze profile to ensure a continuous acceleration*

$$\begin{cases} q_j(0) = q_j^i \\ \dot{q}_j(0) = 0 \\ \dot{q}_j(\tau') = \lambda_j\, k_{vj}\, \mathrm{sign}(D_j) \\ \ddot{q}_j(0) = 0 \\ \ddot{q}_j(\tau') = 0 \end{cases} \qquad [13.36]$$

From these constraints, we derive the equations of position, velocity and acceleration of joint j for $1 \le j \le n$ as $0 \le t \le \tau'$ as follows:

$$q_j(t) = q_j^i - \frac{1}{\tau'^3} \lambda_j\, k_{vj}\, \mathrm{sign}(D_j) \left(\frac{1}{2}t - \tau'\right) t^3 \qquad [13.37]$$

$$\dot{q}_j(t) = -\frac{1}{\tau'^3} \lambda_j\, k_{vj}\, \mathrm{sign}(D_j)\, (2t - 3\tau')\, t^2 \qquad [13.38]$$

$$\ddot{q}_j(t) = -\frac{6}{\tau'^3} \lambda_j\, k_{vj}\, \mathrm{sign}(D_j)\, (t - \tau')\, t \qquad [13.39]$$

The acceleration is maximum at $t = \tau'/2$ and its magnitude is:

$$|\ddot{q}_{jmax}| = \frac{3}{2} \frac{\lambda_j\, k_{vj}}{\tau'} \qquad [13.40]$$

If we take for $|\ddot{q}_{jmax}|$ the value $\upsilon_j\, k_{aj}$ of the velocity trapeze profile, all the joints have the same synchronized duration of acceleration such that:

$$\tau' = \frac{3}{2} \frac{\lambda_j\, k_{vj}}{\upsilon_j\, k_{aj}} \qquad [13.41]$$

Hence, the duration of the acceleration phase is 1.5 times that with a constant acceleration. The joint position equation corresponding to the constant velocity phase, given a duration h', is as follows:

$$q_j(t) = q_j(\tau') + (t - \tau')\, \lambda_j\, k_{vj}\, \mathrm{sign}(D_j) \quad \text{for } \tau' \le t \le \tau' + h' \qquad [13.42]$$

Assuming that the acceleration and deceleration phases are symmetrical ($t'_f = 2\tau' + h'$), the trajectory corresponding to the deceleration phase is defined in the interval $\tau' + h' \leq t \leq t'_f$ as:

$$\begin{cases} q_j(t) = q_j^f + \frac{1}{2}\left[\frac{1}{\tau'^3}(t - 3\tau' - h')(t - \tau' - h')^3 + (2t - 3\tau' - 2h')\right] \lambda_j k_{vj} \operatorname{sign}(D_j) \\ \dot{q}_j(t) = \left[\frac{1}{\tau'^3}(2t - 5\tau' - 2h')(t - \tau' - h')^2 + 1\right] \lambda_j k_{vj} \operatorname{sign}(D_j) \\ \ddot{q}_j(t) = \frac{6}{\tau'^3}(t - 2\tau' - h')(t - \tau' - h')\, \lambda_j k_{vj} \operatorname{sign}(D_j) \end{cases} \qquad [13.43]$$

According to equations [13.37] and [13.41], it should be noted that the distance traveled during the acceleration phase is equal to:

$$|q_j^i - q_j(\tau')| = \frac{3}{4}\frac{(\lambda_j k_{vj})^2}{\upsilon_j k_{aj}} \qquad [13.44]$$

By computing the area under the velocity curve, we verify that:

$$t'_f = \tau' + \frac{|D_j|}{\lambda_j k_{vj}} \qquad [13.45]$$

This expression is similar to equation [13.22] giving the traveling time for the trapeze profile, which suggests that the computation of λ_j and υ_j can be achieved with equations [13.32]. We note as well that to saturate the velocity and the acceleration of a joint trajectory, the distance to travel must be such that:

$$|D_j| > \frac{3}{2}\frac{k_{vj}^2}{k_{aj}} \qquad [13.46]$$

If this condition is not verified, we replace k_{vj} in equations [13.34] and [13.36] by the maximum achievable velocity:

$$k'_{vj} = \sqrt{\frac{2}{3}|D_j|\, k_{aj}} \qquad [13.47]$$

13.4. Point-to-point trajectory in the task space

Let ${}^0\mathbf{T}_E^i$ and ${}^0\mathbf{T}_E^f$ be the homogeneous transformations describing the initial and final desired locations respectively. For convenience, let us note:

$$ {}^0\mathbf{T}_E^i = \begin{bmatrix} \mathbf{A}^i & \mathbf{P}^i \\ 0\ 0\ 0 & 1 \end{bmatrix} \text{ and } {}^0\mathbf{T}_E^f = \begin{bmatrix} \mathbf{A}^f & \mathbf{P}^f \\ 0\ 0\ 0 & 1 \end{bmatrix} $$

The most common way to move from one location to the other is to split the motion into a linear translation between the origins of frames ${}^0\mathbf{T}_E^i$ and ${}^0\mathbf{T}_E^f$, and a rotation α around an axis of the end-effector ${}^E\mathbf{u}$ to align $\mathbf{A}^i$ and $\mathbf{A}^f$. The translation and the rotation should be synchronized.

The distance to travel for the translation is obtained as:

$$ D = \|\mathbf{P}^f - \mathbf{P}^i\| = \sqrt{(P_x^f - P_x^i)^2 + (P_y^f - P_y^i)^2 + (P_z^f - P_z^i)^2} \qquad [13.48] $$

The terms $\mathbf{u}$ and α are computed from the equation:

$$ \mathbf{A}^i\,\mathbf{rot}(\mathbf{u}, \alpha) = \mathbf{A}^f \qquad [13.49] $$

where we recall that $\mathbf{rot}(\mathbf{u}, \alpha)$ is a (3x3) rotation matrix corresponding to a rotation of an angle α about a vector $\mathbf{u}$. Hence, we get:

$$ \mathbf{rot}(\mathbf{u}, \alpha) = [\mathbf{A}^i]^T \mathbf{A}^f = \begin{bmatrix} \mathbf{s}^{iT} \\ \mathbf{n}^{iT} \\ \mathbf{a}^{iT} \end{bmatrix} [\ \mathbf{s}^f \quad \mathbf{n}^f \quad \mathbf{a}^f\ ] = \begin{bmatrix} \mathbf{s}^i.\mathbf{s}^f & \mathbf{s}^i.\mathbf{n}^f & \mathbf{s}^i.\mathbf{a}^f \\ \mathbf{n}^i.\mathbf{s}^f & \mathbf{n}^i.\mathbf{n}^f & \mathbf{n}^i.\mathbf{a}^f \\ \mathbf{a}^i.\mathbf{s}^f & \mathbf{a}^i.\mathbf{n}^f & \mathbf{a}^i.\mathbf{a}^f \end{bmatrix} \qquad [13.50] $$

the symbol "." designating the dot product. Using equations [2.34] through [2.37] yields:

$$ \begin{cases} C\alpha = \frac{1}{2}[\mathbf{s}^i.\mathbf{s}^f + \mathbf{n}^i.\mathbf{n}^f + \mathbf{a}^i.\mathbf{a}^f - 1] \\ S\alpha = \frac{1}{2}\sqrt{(\mathbf{a}^i.\mathbf{n}^f - \mathbf{n}^i.\mathbf{a}^f)^2 + (\mathbf{s}^i.\mathbf{a}^f - \mathbf{a}^i.\mathbf{s}^f)^2 + (\mathbf{n}^i.\mathbf{s}^f - \mathbf{s}^i.\mathbf{n}^f)^2} \\ \alpha = \operatorname{atan2}(S\alpha, C\alpha) \\ \mathbf{u} = \frac{1}{2S\alpha}\begin{bmatrix} \mathbf{a}^i.\mathbf{n}^f - \mathbf{n}^i.\mathbf{a}^f \\ \mathbf{s}^i.\mathbf{a}^f - \mathbf{a}^i.\mathbf{s}^f \\ \mathbf{n}^i.\mathbf{s}^f - \mathbf{s}^i.\mathbf{n}^f \end{bmatrix} \end{cases} \qquad [13.51] $$

When $S\alpha$ is small, the vector $\mathbf{u}$ is computed as indicated in § 2.3.8.

Let k_{v1} and k_{a1} be the maximum velocity and acceleration for the translation motion, and let k_{v2} and k_{a2} be the maximum velocity and acceleration for the rotation motion. The methods described in § 13.3 can be used to generate a synchronized trajectory for the two variables D and α, resulting in the minimum time t_f. The trajectory of the end-effector frame is given by:

$$^{0}\mathbf{T}_E(t) = \begin{bmatrix} \mathbf{A}(t) & & & \mathbf{P}(t) \\ 0 & 0 & 0 & 1 \end{bmatrix} \qquad [13.52]$$

with:

$$\mathbf{P}(t) = \mathbf{P}^i + \frac{s(t)}{D}(\mathbf{P}^f - \mathbf{P}^i) = \mathbf{P}^i + r(t)\,(\mathbf{P}^f - \mathbf{P}^i) \qquad [13.53]$$

$$\mathbf{A}(t) = \mathbf{A}^i\,\mathbf{rot}(\mathbf{u}, r(t)\,\alpha) \qquad [13.54]$$

where $s(t) = D\,r(t)$ is the curvilinear distance traveled at time t and r(t) is the interpolation function.

NOTES.–

– we can specify the rotation from $\mathbf{A}^i$ to $\mathbf{A}^f$ with the three Euler angles φ, θ and ψ. Let $(\phi^i, \theta^i, \psi^i)$ and $(\phi^f, \theta^f, \psi^f)$ designate the Euler angles corresponding to $\mathbf{A}^i$ and $\mathbf{A}^f$ respectively. Thus, equation [13.54] is replaced by:

$$\mathbf{A}(t) = \mathbf{A}^i\,\mathbf{rot}(\mathbf{z}, \phi^i + r(t)\,\phi)\,\mathbf{rot}(\mathbf{x}, \theta^i + r(t)\,\theta)\,\mathbf{rot}(\mathbf{z}, \psi^i + r(t)\,\psi) \qquad [13.55]$$

with $\phi = \phi^f - \phi^i$, $\theta = \theta^f - \theta^i$, $\psi = \psi^f - \psi^i$. The computation of φ, θ and ψ can be carried out as described in § 3.6.1. Thus, we have to deal with four variables: D, φ, θ and ψ;

– we can also choose to specify the rotation around an axis that is fixed with respect to frame R_0. In this case, **u** and α are calculated by solving:

$$\mathbf{rot}(\mathbf{u}, \alpha)\,\mathbf{A}^i = \mathbf{A}^f \qquad [13.56]$$

– the angular velocity ω of the end-effector, with respect to the frame where **u** is defined, is such that:

$$\boldsymbol{\omega} = \mathbf{u}\,\dot{r}(t)\,\alpha = w\,\mathbf{u} \qquad [13.57]$$

13.5. Trajectory generation with via points

We now consider the problem of generating a trajectory when the path includes via points. These via points are inserted between the initial and final points in order to avoid collisions between the robot and its environment. Passing through the via points without stopping reduces the traveling time.

For each variable (joint or Cartesian), we can calculate a single polynomial passing through these points and satisfying the boundary conditions. However, the use of such a polynomial is difficult to exploit with increasing the number of points. Splitting the trajectory in low degree polynomials between the path points provides an elegant way of overcoming this problem and reduces the computational burden of trajectory generation.

In this section, we present three methods based on this principle. The first method consists of specifying linear interpolations with continuous acceleration blends; in the second method, the trajectory between two consecutive points is interpolated by a cubic spline providing continuity of velocity and acceleration; in the third method, the path generation is totally decoupled from the specification of the time history along the path, which gives the possibility of modifying at run-time the velocity of the robot while tracking the desired path.

13.5.1. *Linear interpolations with continuous acceleration blends*

This method can be used for both trajectory generation schemes in the joint space and in the task space. The via points are connected by straight line segments at constant velocity, and the segments are connected around each via point by continuous acceleration motions. This approach was initially described in [Taylor 79], [Paul 81]. The trajectory can be computed on-line, by only looking ahead at a single point. Experimental methods to identify this type of trajectories on a Puma robot have been proposed by [Blanchon 87], [Tondu 94], [Douss 96].

13.5.1.1. *Joint space scheme*

Let the path be represented by the configurations $\mathbf{q}^1$, $\mathbf{q}^2$, ..., $\mathbf{q}^{m-1}$, $\mathbf{q}^m$. First, according to the method presented in § 13.3.3, we compute the terms λ_j^k, υ_j^k and τ_k' for the segment k between points $\mathbf{q}^k$ and $\mathbf{q}^{k+1}$, for k = 1, ..., m – 1, assuming zero velocity at the points.

The constant velocity on the segment k, denoted by $\dot{q}_j^k$, is such that:

$$\dot{q}_j^k = \lambda_j^k k_{vj} \operatorname{sign}(D_j^k) \quad \text{for } j = 1, ..., n \qquad [13.58]$$

with $D_j^k = q_j^{k+1} - q_j^k$.

This velocity allows us, if necessary, to stop at point k + 1 without overshooting. If we assume a constant velocity along the segment k (Figure 13.10), the common traveling time h_k on this segment is given by:

$$h_k = \frac{D_j^k}{\dot{q}_j^k} \quad \text{for } j = 1, \ldots, n \qquad [13.59]$$

To generate a smooth trajectory without stopping at the via points, we connect the trajectories at segments k – 1 and k by a blend (Figure 13.10). The duration of the blend region at point k is equal to $2T_k$. If a velocity continuity is only satisfactory, we can specify a constant acceleration along the blend. Otherwise, it is necessary to use a second degree function providing acceleration continuity. We describe here such a solution, which is a generalization to the one used for a point-to-point trajectory (equations [13.36] through [13.44]). The blend region is traveled at maximum acceleration for each joint in order that the obtained path is as close as possible to the via point. As will be verified further, the blend time is given by:

$$T_{k,j} = \frac{3}{4}\frac{|\dot{q}_j^k - \dot{q}_j^{k-1}|}{k_{aj}} \quad \text{for } k = 2, \ldots, m-1 \text{ and } j = 1, \ldots, n \qquad [13.60]$$

Thus, the blend times are not identical for all joints. The joints are only synchronized at the blends around the initial and final points where we can use equation [13.41] giving $T_{k,j} = \tau_k'/2$.

Let $q_{j,a}^k$ and $q_{j,b}^k$ be the positions of joint j at the beginning and at the end of the blend region around point k respectively (Figure 13.11):

$$\begin{cases} q_{j,a}^k = q_j^k - T_{k,j}\,\dot{q}_j^{k-1} \\ q_{j,b}^k = q_j^k + T_{k,j}\,\dot{q}_j^k \end{cases} \quad \text{for } k = 1, \ldots, m \qquad [13.61]$$

For convenience, $T_{k,j}$ will be denoted by T_k. The equation of the linear motion of joint j at segment k (Figure 13.10) is given by:

$$q_j(t) = (t - t_k - T_k)\,\dot{q}_j^k + q_j^k \quad \text{for } t_k + 2T_k \le t \le t_{k+1} \qquad [13.62]$$

with:

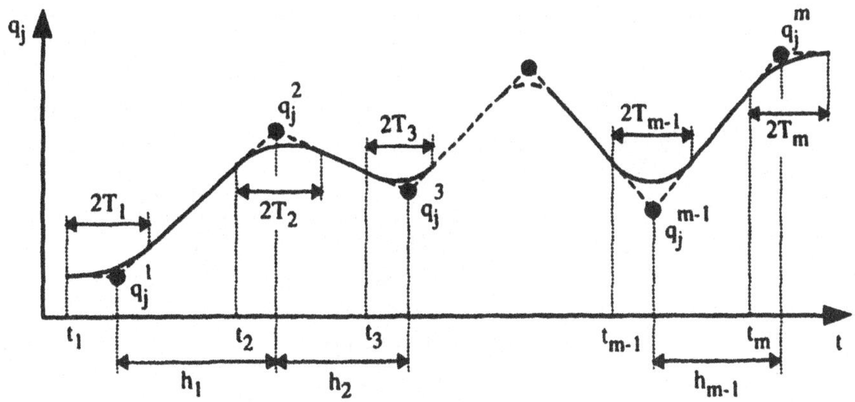

Figure 13.10. *Linear interpolations with continuous acceleration blends (joint space)*

$$\begin{cases} t_k = T_1 - T_k + \sum_{i=1}^{k-1} (h_i) \\ t_1 = 0 \end{cases} \quad \text{for } k = 2, \ldots, m \qquad [13.63]$$

We now consider the blend region around point k. Writing the boundary conditions gives:

$$\begin{cases} q_j(t_k) = q_{j,a}^k \\ q_j(t_k + 2T_k) = q_{j,b}^k \\ \dot{q}_j(t_k) = \dot{q}_j^{k-1} \\ \dot{q}_j(t_k + 2T_k) = \dot{q}_j^k \\ \ddot{q}_j(t_k) = 0 \\ \ddot{q}_j(t_k + 2T_k) = 0 \end{cases} \quad \text{for } k = 1, \ldots, m \qquad [13.64]$$

with $q_j(t_1) = q_j^1 = q_{j,a}^1$ and $q_j(t_m + 2T_m) = q_j^m = q_{j,b}^m$.

Due to the symmetry, the trajectory of joint j along the blend segment k, when it exists ($T_k \neq 0$), is given by the following fourth degree polynomial:

$$q_j(t) = q_j^k - \frac{1}{16(T_k)^3}(t - t_k)^3 (t - t_k - 4T_k)(\dot{q}_j^k - \dot{q}_j^{k-1}) + (t - t_k - T_k)\,\dot{q}_j^{k-1} \qquad [13.65]$$

with $t_k \leq t \leq t_k + 2T_k$ and $1 \leq k \leq m$.

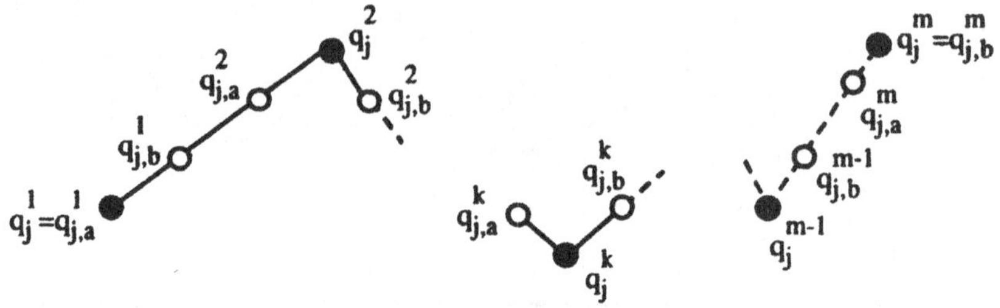

Figure 13.11. *Notations*

The corresponding velocity and acceleration equations are as follows:

$$\dot{q}_j(t) = \dot{q}_j^{k-1} - \frac{1}{4(T_k)^3}(t - t_k)^2 (t - t_k - 3T_k)(\dot{q}_j^k - \dot{q}_j^{k-1}) \quad [13.66]$$

$$\ddot{q}_j(t) = -\frac{3}{4(T_k)^3}(t - t_k)(t - t_k - 2T_k)(\dot{q}_j^k - \dot{q}_j^{k-1}) \quad [13.67]$$

The acceleration is maximum at $t = t_k + T_k$ and has the magnitude:

$$|\ddot{q}_j(t)_{max}| = \frac{3}{4T_k}|\dot{q}_j^k - \dot{q}_j^{k-1}| \quad [13.68]$$

This expression has been used to calculate T_k in terms of the maximum acceleration k_{aj} (equation [13.60]).

NOTES.–

- it is mandatory that $h_k \geq T_k + T_{k+1}$. If this condition does not hold, the velocity at segment $k + 1$ should be scaled down, or even set to zero at point $k + 1$;
- the maximum error around the via point k for joint j is given by:

$$E_j = |q_j^k(t = t_k + T_k) - q_j^k| = \frac{9}{64}\frac{(\dot{q}_j^k - \dot{q}_j^{k-1})^2}{k_{aj}} \quad [13.69]$$

  which justifies high acceleration value to minimize E_j;
- if it is possible to look ahead at several via points, the value of the constant velocity at each segment may be scaled up.

13.5.1.2. *Task space scheme*

The path in the task space is defined by a sequence of end-effector locations ${}^0\mathbf{T}_E^1$, ..., ${}^0\mathbf{T}_E^m$. The results obtained in the joint space may be extended in the task space after splitting the trajectory into:

- a translation between ${}^0\mathbf{P}_E^k$ and ${}^0\mathbf{P}_E^{k+1}$, represented by the distance to travel $D^k = \|{}^0\mathbf{P}_E^{k+1} - {}^0\mathbf{P}_E^k\|$;
- and a rotation represented by the three Euler angles to move from $\mathbf{\Theta}^k$ to $\mathbf{\Theta}^{k+1}$ (where $\mathbf{\Theta}^k = [\,\phi^k \;\; \theta^k \;\; \psi^k\,]^T$ represents the Euler angles corresponding to ${}^0\mathbf{A}_E^k$).

As in the joint space, we first calculate the trajectory parameters at each segment, assuming zero velocity at the via points according to the method presented in § 13.3.3. We thus obtain λ_j^k, υ_j^k and τ_k' for each segment, $j = 1$ designating the translation variable, $j = 2, 3, 4$ indicating the rotation variables. Then, the transition between the constant velocity segments is carried out by a second degree acceleration whose duration is $2T_k$ (Figure 13.12).

a) Translation motion

Let k_{v1} designate the maximum velocity of translation. The magnitude of the constant velocity of translation at segment k, denoted by v^k, is defined as:

$$v^k = \lambda_1^k \, k_{v1} \qquad [13.70]$$

The constant velocity at segment k is given by:

$$\mathbf{V}^k = v^k \frac{\mathbf{P}^{k+1} - \mathbf{P}^k}{\|\mathbf{P}^{k+1} - \mathbf{P}^k\|} \quad \text{for } k = 1, \ldots, m-1 \qquad [13.71]$$

and the traveling time h_k at segment k is equal to $\frac{|D^k|}{v^k}$.

The blend time around point k is such that:

$$T_{k,1} = \frac{3}{4}\frac{\|\mathbf{V}^k - \mathbf{V}^{k-1}\|}{k_{a1}} \quad \text{for } k = 2, \ldots, m-1 \text{ and } \mathbf{V}^0 = \mathbf{V}^m = \mathbf{0} \qquad [13.72]$$

where k_{a1} is the maximum acceleration for the translation motion. For the initial and final points, the blend time is equal to $T_{k,1} = \tau_k'/2$ where τ_k' is obtained by equation [13.41].

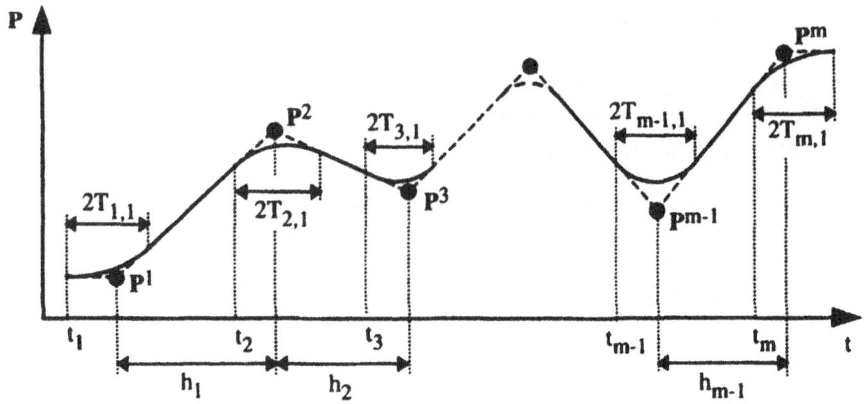

Figure 13.12. *Linear interpolations with continuous acceleration blends (translation* **P** *in the task space)*

The linear trajectory at segments k = 1, ..., m – 1 is given by:

$$\mathbf{P}(t) = (t - t_k - T_{k,1})\, \mathbf{V}^k + \mathbf{P}^k \quad \text{for } t_k + 2T_{k,1} \le t \le t_{k+1} \qquad [13.73]$$

where t_k is defined by equation [13.63].

The blend trajectory around point k for $t_k \le t \le t_k + 2T_{k,1}$ $(T_{k,1} \neq 0)$ and $1 \le k \le m$ is given by:

$$\mathbf{P}(t) = \mathbf{P}^k - \frac{1}{16(T_{k,1})^3}(t - t_k)^3 (t - t_k - 4T_{k,1})(\mathbf{V}^k - \mathbf{V}^{k-1}) + (t - t_k - T_{k,1})\, \mathbf{V}^{k-1}$$

[13.74]

b) Rotation motion

Let β represent one of the Euler variables, and $\dot{\beta}^k$ be the velocity along segment k. The trajectory at constant velocity for $t_k + 2T_{k,j} \le t \le t_{k+1}$ and k = 1, ..., m – 1 is given by:

$$\beta(t) = (t - t_k - T_{k,j})\, \dot{\beta}^k + \beta^k \qquad [13.75]$$

where j = 2, 3, 4 designate φ, θ, ψ respectively, and the trajectory along the blend region for $t_k \le t \le t_k + 2T_{k,j}$ and $1 \le k \le m$ is given by:

$$\beta(t) = \beta^k - \frac{1}{16(T_{k,j})^3}(t-t_k)^3(t-t_k-4T_{k,j})(\dot{\beta}^k-\dot{\beta}^{k-1}) + (t-t_k-T_{k,j})\,\dot{\beta}^{k-1} \quad [13.76]$$

The blend time $T_{k,j}$ is deduced from equation [13.60] by replacing $\dot{q}_j$ with $\dot{\beta}$.

13.5.2. *Trajectory generation with cubic spline functions*

13.5.2.1. *Principle of the method*

As in § 13.5.1, we consider the path defined by a sequence of joint configurations $\mathbf{q}^1$, $\mathbf{q}^2$, ..., $\mathbf{q}^m$ such that $m \geq 4$. We assume that the corresponding traveling times t_1, t_2, ..., t_m are known. On each segment k, i.e. between points k and k + 1, the trajectory is represented by a cubic function (Figure 13.13). This method is also termed *cubic spline function* [Edwall 82], [Lin 83].

The principle is to globally calculate the joint accelerations at the via points to satisfy the velocity and acceleration continuity. The acceleration of the cubic function for joint j (for convenience, we will omit the subscript j) is written as a linear function of time for $t_k \leq t \leq t_{k+1}$ and $k = 1, ..., m-1$:

$$\ddot{F}_k(t) = \frac{(t_{k+1}-t)}{h_k}\ddot{F}_k(t_k) + \frac{(t-t_k)}{h_k}\ddot{F}_k(t_{k+1}) \quad \text{with } h_k = t_{k+1} - t_k \qquad [13.77]$$

Integrating equation [13.77] twice yields the velocity and position equations:

$$\dot{F}_k(t) = -\frac{(t_{k+1}-t)^2}{2h_k}\ddot{F}_k(t_k) + \frac{(t-t_k)^2}{2h_k}\ddot{F}_k(t_{k+1}) + \left[\frac{q^{k+1}}{h_k} - \frac{h_k\ddot{F}_k(t_{k+1})}{6}\right] - \left[\frac{q^k}{h_k} - \frac{h_k\ddot{F}_k(t_k)}{6}\right] \qquad [13.78]$$

$$F_k(t) = \frac{(t_{k+1}-t)^3}{6h_k}\ddot{F}_k(t_k) + \frac{(t-t_k)^3}{6h_k}\ddot{F}_k(t_{k+1}) + (t-t_k)\left[\frac{q^{k+1}}{h_k} - \frac{h_k\ddot{F}_k(t_{k+1})}{6}\right] + (t_{k+1}-t)\left[\frac{q^k}{h_k} - \frac{h_k\ddot{F}_k(t_k)}{6}\right] \qquad [13.79]$$

where $F_k(t_k) = q^k$ and $F_k(t_{k+1}) = q^{k+1}$.

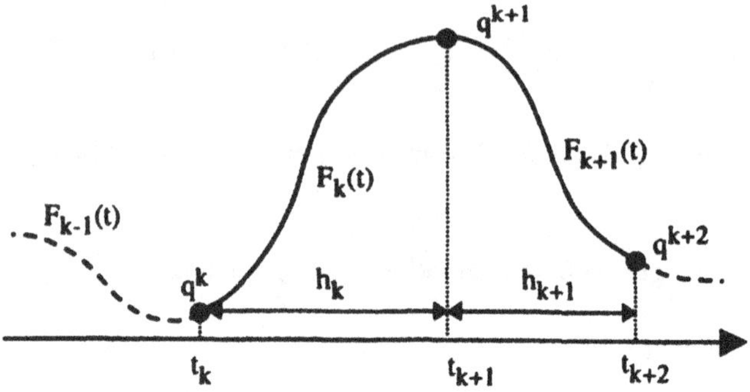

Figure 13.13. *Notations used for the cubic spline functions*

Thus, knowing the accelerations $\ddot{F}_k(t_k)$ allows us to calculate $F_k(t)$ for $k = 1, \ldots, m-1$. Continuous velocity constraint implies that:

$$\dot{F}_k(t_k) = \dot{F}_{k-1}(t_k) \quad \text{for } k = 2, \ldots, m-1 \qquad [13.80]$$

which, after substitution, yields:

$$h_{k-1}\,\ddot{q}^{k-1} + 2\,(h_{k-1} + h_k)\,\ddot{q}^k + h_k\,\ddot{q}^{k+1} = 6\left(\frac{D^k}{h_k} - \frac{D^{k-1}}{h_{k-1}}\right) \qquad [13.81]$$

$$\text{with } \begin{cases} \ddot{F}_k(t_{k+1}) = \ddot{F}_{k+1}(t_{k+1}) = \ddot{q}^{k+1} \\ D^k = q^{k+1} - q^k \end{cases}$$

For convenience, we rewrite equation [13.81] in matrix form:

$$[h_{k-1} \quad 2(h_{k-1}+h_k) \quad h_k] \begin{bmatrix} \ddot{q}^{k-1} \\ \ddot{q}^k \\ \ddot{q}^{k+1} \end{bmatrix} = 6\left(\frac{D^k}{h_k} - \frac{D^{k-1}}{h_{k-1}}\right) \qquad [13.82]$$

Let us assume that $m \geq 4$ and that $\ddot{q}_j^1$ and $\ddot{q}_j^m$ are known[1] (for example, zero). By calculating equation [13.82] for $k = 2, \ldots, m-1$, and combining all the equations together, we obtain for joint j a system of equations that can be written in the following matrix form:

[1] One could assume that $\dot{q}_j^1$ and $\dot{q}_j^m$ are known instead.

$$\mathbf{M}\,\ddot{\mathbf{q}}_j = \mathbf{N}_j \qquad [13.83]$$

with $\ddot{\mathbf{q}}_j = [\ddot{q}_j^2 \;\; \dots \;\; \ddot{q}_j^{m-1}]^T$.

Thus, we can calculate the accelerations at points k = 2, ..., m−1, and consequently the interpolation functions F_k for k = 1, ..., m−1. The matrix **M** is identical for all the joints but the vector $\mathbf{N}_j$ is different. **M** is tridiagonal and regular. Efficient methods to inverse such matrices can be implemented [de Boor 78]. It is worth noting that the initial and final joint velocities are obtained from these equations. To specify desired velocities at the initial and final points, we can either use fourth degree polynomials or two cubic spline functions for the first and the last segments [Edwall 82]. In the following, we develop the second solution. It requires specification of two additional points: one after the initial point and the other before the final point. For convenience, we consider that the total number of points is still denoted by m.

Let us assume that velocities and accelerations on the boundary points are given by $\dot{\mathbf{q}}(t_1)$, $\ddot{\mathbf{q}}(t_1)$, $\dot{\mathbf{q}}(t_m)$, $\ddot{\mathbf{q}}(t_m)$. To satisfy the constraints of continuity, Lin [Lin 83] has shown that the new second point should be defined as:

$$\mathbf{q}^2 = \mathbf{q}^1 + h_1\,\dot{\mathbf{q}}(t_1) + \frac{h_1{}^2}{3}\,\ddot{\mathbf{q}}(t_1) + \frac{h_1{}^2}{6}\,\ddot{\mathbf{q}}(t_2) \qquad [13.84]$$

and the $(m-1)^{th}$ one as:

$$\mathbf{q}^{m-1} = \mathbf{q}^m - h_{m-1}\,\dot{\mathbf{q}}(t_m) + \frac{h_{m-1}{}^2}{3}\,\ddot{\mathbf{q}}(t_m) + \frac{h_{m-1}{}^2}{6}\,\ddot{\mathbf{q}}(t_{m-1}) \qquad [13.85]$$

Thus, the first two and the last two equations of system [13.82] must be modified and the matrix **M** becomes:

$$\begin{bmatrix}
3h_1 + 2h_2 + \frac{h_1^2}{h_2} & h_2 & 0 & \dots & \dots & \dots & 0 \\
h_2 - \frac{h_1^2}{h_2} & 2(h_2+h_3) & h_3 & 0 & \dots & \dots & 0 \\
0 & h_3 & 2(h_3+h_4) & h_4 & 0 & \dots & 0 \\
\dots & \dots & \dots & \dots & \dots & \dots & \dots \\
0 & 0 & \dots & 0 & h_{m-3} & 2(h_{m-3}+h_{m-2}) & h_{m-2} - \frac{h_{m-1}^2}{h_{m-2}} \\
0 & 0 & \dots & 0 & 0 & h_{m-2} & 3h_{m-1} + 2h_{m-2} + \frac{h_{m-1}^2}{h_{m-2}}
\end{bmatrix} \qquad [13.86]$$

while the vector $\mathbf{N}_j$ becomes:

$$\mathbf{N}_j = \begin{bmatrix} 6\left(\frac{q^3}{h_2}+\frac{q^1}{h_1}\right)-6\left(\frac{1}{h_1}+\frac{1}{h_2}\right)\left(q^1+h_1 v_1+\frac{h_1^2}{3}a_1\right)-h_1 a_1 \\ \frac{6}{h_2}\left(q^1+h_1 v_1+\frac{h_1^2}{3}a_1\right)+\frac{6q^4}{h_3}-6\left(\frac{1}{h_2}+\frac{1}{h_3}\right)q^3 \\ 6\left[\frac{(q^5-q^4)}{h_4}-\frac{(q^4-q^3)}{h_3}\right] \\ \cdots \\ \frac{6}{h_{m-2}}\left(q^m-h_{m-1}v_m+\frac{h_{m-1}^2}{3}a_m\right)+\frac{6q^{m-3}}{h_{m-3}}-6\left(\frac{1}{h_{m-2}}+\frac{1}{h_{m-3}}\right)q^{m-2} \\ 6\left(\frac{q^m}{h_{m-1}}+\frac{q^{m-2}}{h_{m-2}}\right)-6\left(\frac{1}{h_{m-1}}+\frac{1}{h_{m-2}}\right)\left(q^m-h_{m-1}v_m+\frac{h_{m-1}^2}{3}a_m\right)-h_{m-1}a_m \end{bmatrix} \qquad [13.87]$$

with $v_1 = \dot{q}_j(t_1)$, $a_1 = \ddot{q}_j(t_1)$, $v_m = \dot{q}_j(t_m)$ and $a_m = \ddot{q}_j(t_m)$.

13.5.2.2. *Calculation of the minimum traveling time on each segment*

If the traveling times ($h_1, \ldots, h_{m-1}$) are not specified, their calculation in order to obtain a minimum global time is not as simple as in the case of a point-to-point trajectory. Optimization techniques must then be implemented [Lin 83]. Nowadays, this problem is facilitated by the existence of efficient optimization softwares.

If $[h_1, h_2, \ldots, h_{m-1}]$ is the vector of variables to be optimized and T the total traveling time, the problem is formulated as follows:

> <u>Minimize the function</u> $T = \sum_{k=1}^{m-1} h_k$ <u>under the constraint that velocities, accelerations and eventually jerks (rate of change of the acceleration) in the joint space remain within their bounds all over the trajectory.</u>

Since cubic spline functions are linear in acceleration, the corresponding inequality constraints are expressed by:

$$|\ddot{q}_j^k| \le k_{aj} \quad \text{for } j = 1, \ldots, n \text{ and } k = 2, \ldots, m-1 \qquad [13.88]$$

The magnitude of the jerk is bounded such that:

$$\frac{|\ddot{q}_j^{k+1} - \ddot{q}_j^k|}{h_k} \le k_{sj} \quad \text{for } j = 1, \ldots, n \text{ and } k = 2, \ldots, m-1 \qquad [13.89]$$

Maximum velocities occur at the time when $\ddot{q}_j(t) = 0$. If for a given function $F_k(t)$, the value of this time is not between t_k and $t_k + h_k$, then the maximum velocity for this function will be $|\dot{q}_j^k|$ or $|\dot{q}_j^{k+1}|$; otherwise it is necessary to calculate the velocity corresponding to this sampling time.

To initialize the optimization procedure, we calculate a lower bound h'_k of the traveling time on each segment k using the equation:

$$h'_k = \max_j \left\{\frac{|D_j^k|}{k_{vj}}\right\} \quad \text{for } j = 1, \ldots, n \text{ and } k = 1, \ldots, m-1 \qquad [13.90]$$

For the first and last two segments, the traveling times are initialized as follows:

$$\begin{cases} h'_1 = h'_2 = \max\left\{\dfrac{|q_j^3 - q_j^1|}{2\,k_{vj}}\right\} \\ h'_{m-2} = h'_{m-1} = \max\left\{\dfrac{|q_j^m - q_j^{m-2}|}{2\,k_{vj}}\right\} \end{cases} \quad \text{for } j = 1, \ldots, n \qquad [13.91]$$

Then, in order to derive an acceptable solution satisfying the constraints, we scale up the time by a factor λ, which modifies the velocity, acceleration and jerk by $1/\lambda$, $1/\lambda^2$ and $1/\lambda^3$ respectively. The time h'_k is thus replaced by h_k:

$$h_k = \lambda h'_k \qquad [13.92]$$

The scale factor λ is selected to saturate the velocity, the acceleration, or the jerk:

$$\lambda = \max\left[\frac{|\dot{q}_{jmax}|}{k_{vj}}, \left(\frac{|\ddot{q}_{jmax}|}{k_{aj}}\right)^{1/2}, \left(\frac{|\dddot{q}_{jmax}|}{k_{sj}}\right)^{1/3}\right] \quad \text{for } j = 1, \ldots, n \qquad [13.93]$$

where $\dot{q}_{jmax}$, $\ddot{q}_{jmax}$ and $\dddot{q}_{jmax}$ denote the maximum velocity, acceleration and jerk of joint j.

NOTES.–

– the minimum traveling time problem is a nonlinear programming problem that can be solved with the "constr" function (Optimization ToolBox) of Matlab (quasi-Newton algorithm);

– instead of calculating the global trajectory for all the points, Chand and Doty [Chand 85] showed that the trajectory could be computed on-line by iteratively considering only a limited number of points at each time.

13.5.3. *Trajectory generation on a continuous path in the task space*

In the previous sections, we showed how to generate a trajectory of the robot endpoint passing through, or close to, a sequence of points. Another procedure, commonly used in continuous processes such as machining or arc welding, consists of first generating a continuous path from the sequence of points at hand, then determining a time history along the path. Processing separately the path generation and the trajectory generation allows the robot to follow the specified spatial path whatever the velocity, which can be modified on-line by an external action (operator or sensor).

Generally, the geometry of the path is described in terms of a parameter u over the interval $0 \leq u \leq 1$. For convenience, let us consider only the position path (the following results are extendable to orientation):

$$\mathbf{P}(u) = [\ P_x(u)\ \ P_y(u)\ \ P_z(u)\ ]^T \tag{13.94}$$

To specify a continuous trajectory in acceleration, the path **P**(u) should be of class C^2 in u, which means a continuity of curvature. For example, cubic splines, cubic B-splines or Bézier curves can be used to represent **P**(u) as a polynomial function in u, as is done in CAD/CAM systems [Bartels 88], [Léon 91].

Then, we determine the trajectory by choosing for u a suitable function of time. The simplest function is $u = \lambda t$, but it is more interesting to specify the velocity of the curvilinear abscissa of the tool along the path. This requires computation of a one-to-one mapping between the curvilinear abscissa, denoted by s, and the parameter u, using the fact that:

$$\frac{ds(u)}{du} = \sqrt{\left[\frac{dPx(u)}{du}\right]^2 + \left[\frac{dPy(u)}{du}\right]^2 + \left[\frac{dPz(u)}{du}\right]^2} = \left\|\frac{d\mathbf{P}(u)}{du}\right\| \tag{13.95}$$

The curvilinear abscissa s is given as a function of u by integration:

$$s = \int_{u_1}^{u_2} \frac{ds(u)}{du}\, du = \int_{u_1}^{u_2} \left\|\frac{d\mathbf{P}(u)}{du}\right\| du \tag{13.96}$$

Thus, the trajectory generation consists of specifying the time history of the curvilinear abscissa s. Cartesian velocities and accelerations are given by:

$$\dot{\mathbf{P}}(s) = \frac{d\mathbf{P}(s)}{dt} = \frac{d\mathbf{P}(s)}{ds}\frac{ds}{dt} = \dot{s}\,\frac{d\mathbf{P}(s)}{ds} \tag{13.97}$$

$$\ddot{\mathbf{P}}(s) = \ddot{s}\,\frac{d\mathbf{P}(s)}{ds} + \dot{s}^2\,\frac{d^2\mathbf{P}(s)}{ds^2} \qquad [13.98]$$

Considering equation [13.95], we obtain:

$$\frac{d\mathbf{P}(s)}{ds} = \frac{d\mathbf{P}(u)}{du}\frac{du}{ds} = \frac{d\mathbf{P}(u)}{du}\frac{1}{\left\|\frac{d\mathbf{P}(u)}{du}\right\|} \qquad [13.99]$$

A closed-form solution for $\mathbf{P}(s)$ exists in the case of straight line and circular paths. For a straight line path for instance, let $\mathbf{P}^i$ and $\mathbf{P}^f$ be the initial and final points, and D be the Cartesian distance to travel. We can write that:

$$\mathbf{P}(s) = \mathbf{P}^i + \frac{s}{D}(\mathbf{P}^f - \mathbf{P}^i) \qquad [13.100]$$

For a circular path in the $(\mathbf{x}_c, \mathbf{y}_c)$ plane of the circle, the equation is given by:

$$\mathbf{P}(s) = \mathbf{P}^c + \begin{bmatrix} R\cos(\frac{s}{R} + \phi_0) \\ R\sin(\frac{s}{R} + \phi_0) \end{bmatrix} \qquad [13.101]$$

where $\mathbf{P}^c$ is the vector of the x and y coordinates of the circle center, R is the circle radius, ϕ_0 is the angle between the vector $\mathbf{P}^c\mathbf{P}^i$ and the axis $\mathbf{x}_c$, and $\mathbf{P}^i$ is the initial point such that:

$$\begin{bmatrix} \cos(\phi_0) \\ \sin(\phi_0) \end{bmatrix} = \frac{1}{R}(\mathbf{P}^c - \mathbf{P}^i) \qquad [13.102]$$

In the general case, we numerically determine a polynomial giving s as a function of u. Indeed, integrating the equation ds/du yields curvilinear abscissa s(u) at regular intervals of u. Then, to evaluate s(u), it is sufficient to interpolate the resulting points with a polynomial function in u, whose coefficients c_i are estimated by a least square procedure. A fourth degree polynomial should provide sufficient accuracy [Froissart 91]:

$$s(u) = \sum_{i=0}^{4} c_i u^i \qquad [13.103]$$

Then, the algorithm can be as follows:

- compute a path **P**(u);
- compute the curvilinear abscissa s(u) along the path;
- compute the time history s(t);
- compute u(t);
- compute the trajectory **P**(t).

The time history can be computed as indicated previously for the curvilinear abscissa s. Another method has been proposed in [Sgarbi 92]: it consists of accelerating until the desired velocity is reached (or velocity and acceleration bounds are attained), maintaining this value, and finally decelerating to finish at zero velocity at the end of the path. Thus, the velocity tracks a trapeze profile. Let $\dot{s}$ be the current curvilinear velocity, $\dot{s}^d$ be the desired one, T_e be the sampling period, and L(i) be the distance between the current point and the final point. The algorithm is as follows:

if $\{\text{abs}\frac{1}{T_e}[\dot{s}(i-1) - \dot{s}^d(i)]\} > \ddot{s}_{max}$, then:

$$\dot{s}(i) = \dot{s}(i-1) + \ddot{s}_{max}\, T_e\, \text{sign}[\dot{s}^d(i) - \dot{s}(i-1)]$$

else $\dot{s}(i) = \dot{s}^d(i)$

if $L(i) < \frac{[\dot{s}(i)]^2}{2\ddot{s}_{max}}$, then begin deceleration phase.

An immediate extension of this algorithm would be to generate a continuous curvilinear acceleration by implementing a trapeze acceleration profile.

13.6. Conclusion

In this chapter, we have presented several methods of trajectory generation that are commonly used in robotics. We have first dealt with point-to-point trajectories: different interpolation functions have been studied, namely the trapeze velocity profile, which is implemented in most of the industrial controllers. For each function, we computed the minimum traveling time, from which it is possible to synchronize the joints so that they reach the final point simultaneously. We have also presented three methods of trajectory generation with via points. In the first method, straight line segments joining the via points are blended together by continuous acceleration phases. In the second method, the path passes through the via points, the trajectory being described by a sequence of cubic spline functions. In the third method, the trajectory is computed on a predefined continuous path.

These methods apply for both joint space and task space. The choice of a space depends on the trajectory specification and on the task description.

The interested reader will find in [Shin 85], [Fourquet 90], [Shiller 94], other techniques using the dynamic model which allows replacement of the constraints of acceleration by those more realistic of actuator torques. Likewise, in [Pledel 96], an approach using an exact model of actuators is considered. However, instead of implementing these techniques, an *a posteriori* verification of the constraint validity and scaling the traveling time may be satisfactory [Hollerbach 84a].

Chapter 14

Motion control

14.1. Introduction

The problem of controlling robots has been extensively addressed in the literature. A great variety of control approaches have been proposed. The most common in use with present industrial robots is a decentralized "proportional, integral, derivative" (PID) control for each degree of freedom. More sophisticated nonlinear control schemes have been developed, such as so-called *computed torque control*, termed *inverse dynamic control*, which linearizes and decouples the equation of motion of the robot. Owing to the modeling uncertainties, nonlinear adaptive techniques have been considered in order to identify on-line the dynamic parameters. More recently, properties of the dynamic model have led Lyapunov-based and passivity-based controls to be proposed.

In this chapter, we first study the classical PID control, then the nonlinear linearizing and decoupling control, which is considered to be the best theoretical solution for the control of robots. Finally, we present some advanced methods related to passivity-based and adaptive controls. Detailed surveys on robot control can be found in [Spong 89], [Samson 91], [Lewis 93], [Zodiac 96].

For simplicity, we will only consider serial robots. The methods presented can easily be generalized to robots with complex structures by employing the results of Chapter 10.

14.2. Equations of motion

In order to understand the basic problem of robot control, it is useful to recall the dynamic model (Chapter 9) whose general form for a robot with n degrees of freedom is the following:

$$\Gamma = \mathbf{A}(\mathbf{q})\,\ddot{\mathbf{q}} + \mathbf{C}(\mathbf{q}, \dot{\mathbf{q}})\,\dot{\mathbf{q}} + \mathbf{Q}(\mathbf{q}) + \mathbf{diag}(\dot{\mathbf{q}})\,\mathbf{F}_v + \mathbf{diag}(\mathrm{sign}(\dot{\mathbf{q}}))\,\mathbf{F}_c \qquad [14.1]$$

or, in a more compact form:

$$\Gamma = \mathbf{A}(\mathbf{q})\,\ddot{\mathbf{q}} + \mathbf{H}(\mathbf{q}, \dot{\mathbf{q}}) \qquad [14.2]$$

and, since the model is linear in the dynamic parameters (equation [12.18]), we can write:

$$\Gamma = \Phi(\mathbf{q}, \dot{\mathbf{q}}, \ddot{\mathbf{q}})\,\chi \qquad [14.3]$$

where Γ is the (nx1) vector of joint torques; $\mathbf{A}(\mathbf{q})$ is the (nxn) inertia matrix of the robot; $\mathbf{C}(\mathbf{q}, \dot{\mathbf{q}})\,\dot{\mathbf{q}}$ is the (nx1) vector of Coriolis and centrifugal torques; $\mathbf{Q}(\mathbf{q})$ is the vector of gravity torques; $\mathbf{F}_v$ and $\mathbf{F}_c$ are the vectors of the viscous friction and Coulomb friction parameters respectively; χ is the vector of the dynamic parameters (inertial parameters and friction parameters).

The torque transmitted to joint j by a current-driven electrical actuator (continuous or synchronous), assuming that the transmissions introduce neither backlash nor flexibility, is expressed by (equation [12.17]):

$$\Gamma_j = N_j\, K_{aj}\, K_{Tj}\, u_j \qquad [14.4]$$

where N_j is the gear transmission ratio, K_{aj} is the current amplifier gain, K_{Tj} is the torque constant of actuator j, and u_j is the control input of the amplifier.

The design of the control consists of computing the joint actuator torques (Γ_j, then u_j) in order to track a desired trajectory or to reach a given position.

14.3. PID control

14.3.1. *PID control in the joint space*

The dynamic model is described by a system of n coupled nonlinear second order differential equations, n being the number of joints. However, for most of today's industrial robots, a local decentralized "proportional, integral, derivative" (PID) control with constant gains is implemented for each joint. The advantages of such a technique are the simplicity of implementation and the low computational cost. The drawbacks are that the dynamic performance of the robot varies according to its configuration, and poor dynamic accuracy when tracking a high velocity trajectory. In many applications, these drawbacks are not of much significance.

Practically, the block diagrams of such a control scheme in the joint space is shown in Figure 14.1. The control law is given by:

$$\Gamma = K_p (q^d - q) + K_d (\dot{q}^d - \dot{q}) + K_I \int_{t0}^{t} (q^d - q) d\tau \qquad [14.5]$$

where $q^d(t)$ and $\dot{q}^d(t)$ denote the desired joint positions and velocities, and where K_p, K_d and K_I are (nxn) positive definite diagonal matrices whose generic elements are the proportional K_{pj}, derivative K_{dj} and integral K_{Ij} gains respectively.

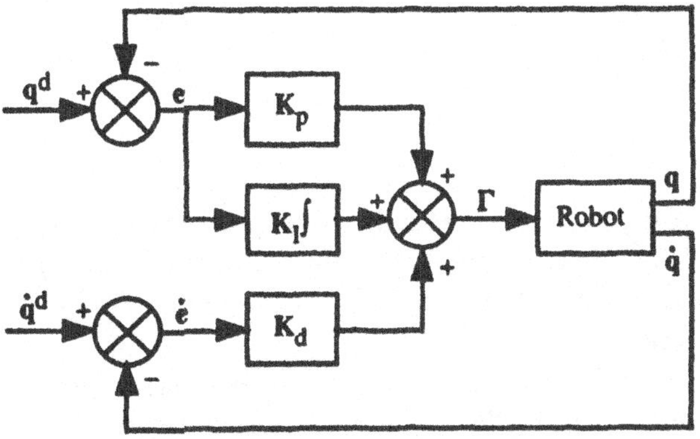

Figure 14.1. *Block diagram of a PID control scheme in the joint space*

The computation of K_{pj}, K_{dj} and K_{Ij} is carried out by considering that joint j is modeled by a linear second order differential equation such that:

$$\Gamma_j = a_j \ddot{q}_j + F_{vj} \dot{q}_j + \gamma_j \qquad [14.6]$$

where $a_j = A_{jj_{max}}$ is the maximum magnitude of the A_{jj} element of the inertia matrix of the robot and γ_j represents a disturbance torque.

Hence, assuming $\gamma_j = 0$, the closed-loop transfer function is given by:

$$\frac{q_j(s)}{q_j^d(s)} = \frac{K_{dj} s^2 + K_{pj} s + K_{Ij}}{a_j s^3 + (K_{dj} + F_{vj}) s^2 + K_{pj} s + K_{Ij}} \qquad [14.7]$$

and the characteristic equation is:

$$\Delta(s) = a_j s^3 + (K_{dj} + F_{vj}) s^2 + K_{pj} s + K_{Ij} \quad [14.8]$$

The most common solution in robotics consists of adjusting the gains in order to obtain a negative real triple pole. This yields the fastest possible response without overshoot. Thus, the characteristic equation is written as:

$$\Delta(s) = a_j (s + \omega_j)^3 \quad [14.9]$$

with $\omega_j > 0$, and after solution, we obtain:

$$\begin{cases} K_{pj} = 3\, a_j\, \omega_j^2 \\ K_{dj} + F_{vj} = 3\, a_j\, \omega_j \\ K_{Ij} = a_j\, \omega_j^3 \end{cases} \quad [14.10]$$

NOTES.–

- high gains K_p and K_d decrease the tracking error but bring the system to the neighborhood of the instability domain. Thus, the frequency ω_j should not be greater than the structural resonance frequency ω_{rj}. A reasonable trade-off is that $\omega_j = \omega_{rj} / 2$;
- in the absence of integral action, a static error due to gravity may affect the final position. Practically, the integral action can be deactivated when the position error is very large, since the proportional action is sufficient. It should also be deactivated if the position error becomes too small in order to avoid oscillations that could be caused by Coulomb frictions;
- the predictive action $K_d\, \dot{q}^d$ of equation [14.5] reduces significantly the tracking errors. In classical control engineering, this action is not often used;
- the gain K_d is generally integrated within the servo amplifier, whereas gain K_p is numerically implemented;
- the performance of a robot controlled in this way is acceptable if high-gear transmission ratios are used (scaling down the time-varying inertias and the coupling torques), if the robot is moving at low velocity, and if high position gains are assigned [Samson 83].

14.3.2. *Stability analysis*

If gravity effects are compensated by an appropriate mechanical design as for the SCARA robot, or by the control software, it can be shown that a PD control law is asymptotically stable for the regulation control problem [Arimoto 84]. The demonstration is based on the definition of the following Lyapunov function candidate (Appendix 9):

$$V = \frac{1}{2}\dot{\mathbf{q}}^T \mathbf{A}(\mathbf{q})\, \dot{\mathbf{q}} + \frac{1}{2}\mathbf{e}^T \mathbf{K}_p\, \mathbf{e} \tag{14.11}$$

where $\mathbf{e} = \mathbf{q}^d - \mathbf{q}$ is the position error, and where $\mathbf{q}^d$ is the desired joint position.

Since $\mathbf{q}^d$ is constant, the PD control law is given by:

$$\Gamma = \mathbf{K}_p\, \mathbf{e} - \mathbf{K}_d\, \dot{\mathbf{q}} + \mathbf{Q}(\mathbf{q}) \tag{14.12}$$

From equations [14.1] and [14.12], and in the absence of friction, we obtain the following closed-loop equation:

$$\mathbf{K}_p\, \mathbf{e} - \mathbf{K}_d\, \dot{\mathbf{q}} = \mathbf{A}\, \ddot{\mathbf{q}} + \mathbf{C}\, \dot{\mathbf{q}} \tag{14.13}$$

Differentiating the Lyapunov function [14.11] with respect to time yields:

$$\dot{V} = \frac{1}{2}\dot{\mathbf{q}}^T \dot{\mathbf{A}}\, \dot{\mathbf{q}} + \dot{\mathbf{q}}^T \mathbf{A}\, \ddot{\mathbf{q}} - \mathbf{e}^T \mathbf{K}_p\, \dot{\mathbf{q}} \tag{14.14}$$

and substituting $\mathbf{A}\, \ddot{\mathbf{q}}$ from equation [14.13] yields:

$$\dot{V} = \frac{1}{2}\dot{\mathbf{q}}^T (\dot{\mathbf{A}} - 2\,\mathbf{C})\, \dot{\mathbf{q}} - \dot{\mathbf{q}}^T \mathbf{K}_d\, \dot{\mathbf{q}} \tag{14.15}$$

Since the matrix $[\dot{\mathbf{A}} - 2\,\mathbf{C}]$ is skew-symmetric [Koditschek 84], [Arimoto 84] (§ 9.3.3.3), the term $\dot{\mathbf{q}}^T [\dot{\mathbf{A}} - 2\,\mathbf{C}]\, \dot{\mathbf{q}}$ is zero, giving:

$$\dot{V} = -\dot{\mathbf{q}}^T \mathbf{K}_d\, \dot{\mathbf{q}} \leq 0 \tag{14.16}$$

This result shows that $\dot{V}$ is negative semi-definite, which is not sufficient to demonstrate that the equilibrium point ($\mathbf{e} = \mathbf{0}$, $\dot{\mathbf{q}} = \mathbf{0}$) is asymptotically stable (Appendix 9). We have now to prove that as $\dot{\mathbf{q}} = \mathbf{0}$, the robot does not reach a configuration $\mathbf{q} \neq \mathbf{q}^d$. This can be done, thanks to the La Salle invariant set theorem [Hahn 67] (Appendix 9). The set $\mathcal{R}$ of points in the neighborhood of the equilibrium that satisfies $\dot{V} = 0$ is such that $\dot{\mathbf{q}} = \mathbf{0}$ and thus $\ddot{\mathbf{q}} = \mathbf{0}$. From equation [14.13], we conclude that necessarily $\mathbf{e} = \mathbf{0}$. Consequently, the equilibrium point ($\mathbf{e} = \mathbf{0}$, $\dot{\mathbf{q}} = \mathbf{0}$) is the only possible equilibrium for the system and is also the largest invariant set in $\mathcal{R}$. Therefore, the equilibrium point is asymptotically stable.

Furthermore, it has been demonstrated that the system is asymptotically stable if in equation [14.12] we replace $\mathbf{Q(q)}$ by the constant term $\mathbf{Q(q^d)}$, corresponding to gravity torque at the desired position $\mathbf{q^d}$. The stability is also proven if one takes $K_{pj} > \|\partial \mathbf{Q(q)}/\partial \mathbf{q}\|$, which represents the 2-norm of the Jacobian matrix of gravity torques with respect to the joint variables [Korrami 88], [Tomei 91]. For more details on the computation of the gains when considering the robot dynamics, interested readers should refer to [Qu 91], [Kelly 95], [Rocco 96], [Freidovich 97].

14.3.3. *PID control in the task space*

When the motion is specified in the task space, one of the following schemes can be used to control the system:

- the control law is designed in the task space;
- the specified trajectory in the task space is transformed into a trajectory in the joint space, then a control in the joint space is performed.

For PID control in the task space, the control law is obtained by replacing $\mathbf{q}$ by $\mathbf{X}$ in equation [14.5] and by transforming the task space error signal into the joint space by multiplying it by $\mathbf{J}^T$(Figure 14.2):

$$\Gamma = \mathbf{J}^T [\mathbf{K}_p (\mathbf{X}^d - \mathbf{X}) + \mathbf{K}_d (\dot{\mathbf{X}}^d - \dot{\mathbf{X}}) + \mathbf{K}_I \int_{t0}^{t} (\mathbf{X}^d - \mathbf{X})\, d\tau] \qquad [14.17]$$

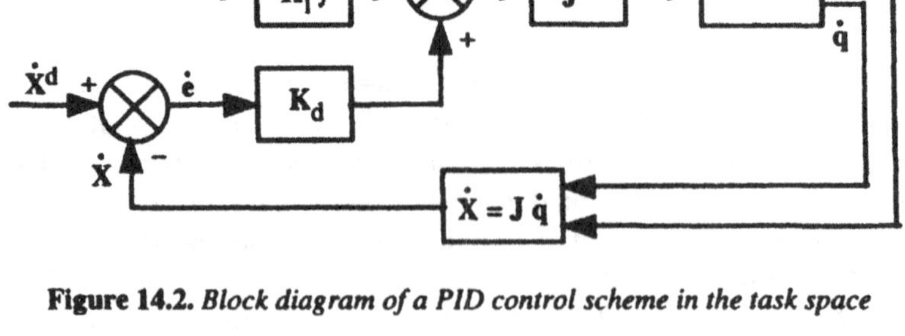

Figure 14.2. *Block diagram of a PID control scheme in the task space*

Two solutions are possible to transform the desired task space trajectory into the joint space: either we use the IGM to compute the joint positions then we compute the velocities and accelerations by differentiating the positions; or we compute the joint positions, velocities and accelerations as indicated below:

i) using the IGM (Chapter 4) to compute the joint positions:

$$\mathbf{q}^d = \mathbf{g}(\mathbf{X}^d) \qquad [14.18]$$

ii) using the IKM (Chapter 6) to compute the joint velocities. In the regular positions:

$$\dot{\mathbf{q}}^d = \mathbf{J}(\mathbf{q}^d)^{-1}\, \dot{\mathbf{X}}^d \qquad [14.19]$$

In singular positions or for redundant robots, the matrix $\mathbf{J}^{-1}$ should be replaced by a generalized inverse as described in Chapter 6;

iii) using the second order IKM (§ 5.10) to compute the joint accelerations (if desired):

$$\ddot{\mathbf{q}}^d = \mathbf{J}^{-1}\,(\ddot{\mathbf{X}}^d - \dot{\mathbf{J}}\,\dot{\mathbf{q}}^d) \qquad [14.20]$$

with:

$$\dot{\mathbf{J}}(\mathbf{q}^d, \dot{\mathbf{q}}^d) = \frac{d}{dt}\mathbf{J}(\mathbf{q}^d) \qquad [14.21]$$

14.4. Linearizing and decoupling control

14.4.1. *Introduction*

When the task requires fast motion of the robot and high dynamic accuracy, it is necessary to improve the performance of the control by taking into account, partially or totally, the dynamic interaction torques. Linearizing and decoupling control is based on canceling the nonlinearities in the robot dynamics [Khalil 78], [Zabala 78], [Raibert 78], [Khatib 80], [Luh 80a], [Freund 82], [Bejczy 85]... Such a control is known as *computed torque control* or *inverse dynamic control* since it is based on the utilization of the dynamic model. Theoretically, it ensures the linearization and the decoupling of the equations of the model, providing a uniform behavior whatever the configuration of the robot.

Implementing this method requires on-line computation of the inverse dynamic model, as well as knowledge of the numerical values of the inertial parameters and friction parameters. Efficient modeling approaches to minimize the computational burden have been presented in § 9.6. With current computer performance, the computation can be handled on-line at a sufficiently high rate and is not anymore a limiting problem. The inertial parameters can be determined off-line with good accuracy by identification techniques as described in Chapter 12.

The linearizing and decoupling techniques consist of transforming a nonlinear control problem into a linear one by using an appropriate feedback law. In the case of rigid robot manipulators, the design of a linearizing and decoupling control law is facilitated by the fact that the number of actuators is equal to the number of joint variables, and that the inverse dynamic model giving the control input Γ of the system as a function of the state vector $(\mathbf{q}, \dot{\mathbf{q}})$ and of $\ddot{\mathbf{q}}$ is naturally obtained. These features ensure that the equations of the robot define a so-called *flat system* whose *flat outputs* are the joint variables $\mathbf{q}$ [Fliess 95]. Since the control law only involves the state variables $\mathbf{q}$ and $\dot{\mathbf{q}}$, it is termed a *static decoupling control law*. In the following, we describe this method both in the joint space and in the task space.

14.4.2. *Computed torque control in the joint space*

14.4.2.1. *Principle of the control*

Let us assume that joint positions and velocities are measurable and that measurements are noiseless. Let $\hat{\mathbf{A}}$ and $\hat{\mathbf{H}}$ be the estimates of $\mathbf{A}$ and $\mathbf{H}$ respectively. Hence, from equation [14.2], if we define a control law Γ such that [Khalil 79]:

$$\Gamma = \hat{\mathbf{A}}(\mathbf{q})\mathbf{w}(t) + \hat{\mathbf{H}}(\mathbf{q}, \dot{\mathbf{q}}) \qquad [14.22]$$

then, after substituting [14.22] into [14.2], we deduce that in the ideal case of perfect modeling and in the absence of disturbances, the problem reduces to that of the linear control of n decoupled double-integrators:

$$\ddot{\mathbf{q}} = \mathbf{w}(t) \qquad [14.23]$$

$\mathbf{w}(t)$ is the new input control vector. In order to define $\mathbf{w}(t)$, we study in the following sections two schemes: the first one is suited for *tracking control* when the trajectory is fully specified, the second one is suited for *position (regulation) control* when just the final point is specified.

14.4.2.2. *Tracking control scheme*

Let $\ddot{\mathbf{q}}^d(t)$, $\dot{\mathbf{q}}^d(t)$ and $\mathbf{q}^d(t)$ be the desired acceleration, velocity and position in the joint space. If we define $\mathbf{w}(t)$ according to the following equation[1]:

$$\mathbf{w}(t) = \ddot{\mathbf{q}}^d + \mathbf{K}_d (\dot{\mathbf{q}}^d - \dot{\mathbf{q}}) + \mathbf{K}_p (\mathbf{q}^d - \mathbf{q}) \qquad [14.24]$$

where $\mathbf{K}_p$ and $\mathbf{K}_d$ are (nxn) positive definite diagonal matrices; hence, referring to equation [14.23], the closed-loop system response is determined by the following decoupled linear error equation:

$$\ddot{\mathbf{e}} + \mathbf{K}_d \dot{\mathbf{e}} + \mathbf{K}_p \mathbf{e} = \mathbf{0} \qquad [14.25]$$

where $\mathbf{e} = \mathbf{q}^d - \mathbf{q}$.

The solution $\mathbf{e}(t)$ of the error equation is globally exponentially stable. The gains K_{pj} and K_{dj} are adjusted to provide the axis j, over the whole set of configurations of the robot, the desired dynamics with a given damping coefficient ξ_j, and a given control bandwidth fixed by a frequency ω_j:

$$\begin{cases} K_{pj} = \omega_j^2 \\ K_{dj} = 2\,\xi_j\,\omega_j \end{cases} \qquad [14.26]$$

Generally, one seeks a critically damped system ($\xi_j = 1$) to obtain the fastest response without overshoot. The block diagram of this control scheme is represented in Figure 14.3. The control input torque to the actuators includes three components: the first compensates for Coriolis, centrifugal, gravity, and friction effects; the second is a proportional and derivative control with variable gains $\hat{\mathbf{A}}\,\mathbf{K}_p$ and $\hat{\mathbf{A}}\,\mathbf{K}_d$ respectively; and the third provides a predictive action of the desired acceleration torques $\hat{\mathbf{A}}\,\ddot{\mathbf{q}}^d$.

In the presence of modeling errors, the closed loop equation corresponding to Figure 14.3 is obtained by combining equations [14.22] and [14.2]:

$$\hat{\mathbf{A}}\,(\ddot{\mathbf{q}}^d + \mathbf{K}_d \dot{\mathbf{e}} + \mathbf{K}_p \mathbf{e}) + \hat{\mathbf{H}} = \mathbf{A}\,\ddot{\mathbf{q}} + \mathbf{H} \qquad [14.27]$$

yielding:

$$\ddot{\mathbf{e}} + \mathbf{K}_d \dot{\mathbf{e}} + \mathbf{K}_p \mathbf{e} = \hat{\mathbf{A}}^{-1}\,[(\mathbf{A} - \hat{\mathbf{A}})\,\ddot{\mathbf{q}} + \mathbf{H} - \hat{\mathbf{H}}] \qquad [14.28]$$

[1] An integral action on $\mathbf{w}(t)$ can also be added.

In this equation, the modeling errors constitute an excitation for the error equation. When these errors are too large, it is necessary to increase the proportional and derivative gains, but their magnitudes are limited by the stability of the system. The robustness and the stability of this control are addressed by Samson et al. [Samson 87]. It is shown namely that the matrix $\hat{\mathbf{A}}$ must be positive definite. It is shown as well that the errors $\mathbf{e}$ and $\dot{\mathbf{e}}$ decrease while the gains increase.

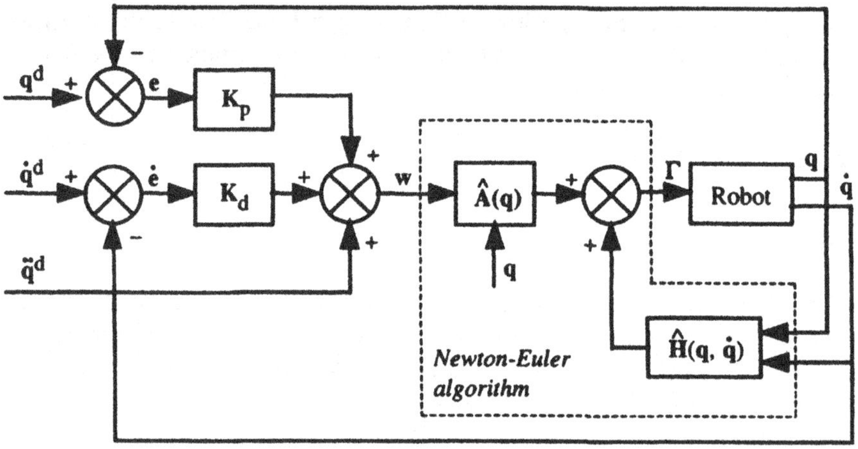

Figure 14.3. *Computed torque: block diagram of the tracking control scheme in the joint space*

14.4.2.3. *Position control scheme*

Let $\mathbf{q}^d$ be the desired position. A possible choice for $\mathbf{w}(t)$ is as follows (Figure 14.4):

$$\mathbf{w}(t) = \mathbf{K}_p (\mathbf{q}^d - \mathbf{q}) - \mathbf{K}_d \dot{\mathbf{q}} \qquad [14.29]$$

From equations [14.23] and [14.29], we obtain the closed-loop equation of the system:

$$\ddot{\mathbf{q}} + \mathbf{K}_d \dot{\mathbf{q}} + \mathbf{K}_p \mathbf{q} = \mathbf{K}_p \mathbf{q}^d \qquad [14.30]$$

which describes a decoupled linear system of second order differential equations. The solution $\mathbf{q}(t)$ is globally exponentially stable by properly choosing $\mathbf{K}_p$ and $\mathbf{K}_d$.

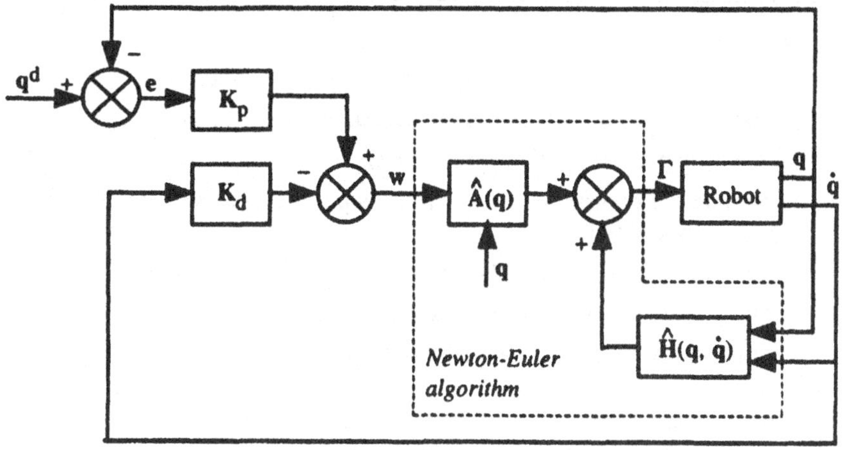

Figure 14.4. *Computed torque: block diagram of the position control scheme in the joint space*

14.4.2.4. *Predictive dynamic control*

Another scheme has been proposed by [Khalil 78] based on a predictive dynamic control: the estimates $\hat{\mathbf{A}}$ and $\hat{\mathbf{H}}$ are no longer computed with the current values of $\mathbf{q}$ and $\dot{\mathbf{q}}$, but rather with the desired values $\mathbf{q}^d$ and $\dot{\mathbf{q}}^d$. Thus, the control law is written as:

$$\Gamma = \hat{\mathbf{A}}\,(\mathbf{q}^d)\,\mathbf{w}(t) + \hat{\mathbf{H}}\,(\mathbf{q}^d, \dot{\mathbf{q}}^d) \qquad [14.31]$$

where $\mathbf{w}(t)$ is given by equation [14.24] or [14.29] according to the desired scheme.

In the case of exact modeling, we can assume that $\hat{\mathbf{A}}(\mathbf{q}) = \hat{\mathbf{A}}(\mathbf{q}^d)$ and $\hat{\mathbf{H}}(\mathbf{q}, \dot{\mathbf{q}}) = \hat{\mathbf{H}}(\mathbf{q}^d, \dot{\mathbf{q}}^d)$. The control law [14.31] linearizes and decouples the equations of the system as in the previous case. The main advantage of this scheme is that the computation of $\hat{\mathbf{A}}(\mathbf{q}^d)$ and $\hat{\mathbf{H}}(\mathbf{q}^d, \dot{\mathbf{q}}^d)$ is not corrupted by noisy variables.

14.4.2.5. *Practical computation of the computed torque control laws*

The control laws [14.22] and [14.31] can be computed by the inverse dynamic Newton-Euler algorithm (§ 9.5) without requiring explicit knowledge of $\mathbf{A}$ and $\mathbf{H}$. The algorithm provides the joint torques as a function of three arguments, namely the vectors of joint positions, velocities and accelerations. By comparing equations [14.2] and [14.22], we can conclude that:

– to compute the control law [14.22] (Figures 14.3 and 14.4), the input arguments of the Newton-Euler algorithm should be:
  - the joint position is equal to the current joint position $\mathbf{q}$;
  - the joint velocity is equal to the current joint velocity $\dot{\mathbf{q}}$;
  - the joint acceleration is equal to $\mathbf{w}(t)$;
– to compute the control law [14.31], the input arguments of the Newton-Euler algorithm should be:
  - the joint position is equal to the desired joint position $\mathbf{q}^d$;
  - the joint velocity is equal to the desired joint velocity $\dot{\mathbf{q}}^d$;
  - the joint acceleration is equal to $\mathbf{w}(t)$.

The computational cost of the computed torque control in the joint space is therefore more or less equal to the number of operations of the inverse dynamic model. As we stated in Chapter 9, the problem of on-line computation of this model at a sufficient rate is now considered solved (Chapter 9). Some industrial robot controllers offer a partial implementation of the computed torque control algorithm.

14.4.3. *Computed torque control in the task space*

The dynamic control in the task space is also known as *resolved acceleration control* [Luh 80a]. The dynamic behavior of the robot in the task space is described by the following equation, obtained after substituting $\ddot{\mathbf{q}}$ from equation [5.44] into equation [14.2]:

$$\mathbf{\Gamma} = \mathbf{A}\,\mathbf{J}^{-1}(\ddot{\mathbf{X}} - \dot{\mathbf{J}}\,\dot{\mathbf{q}}) + \mathbf{H} \qquad [14.32]$$

As in the case of the joint space decoupling control, a control law that linearizes and decouples the equations in the task space is formulated as:

$$\mathbf{\Gamma} = \hat{\mathbf{A}}\,\mathbf{J}^{-1}(\mathbf{w}(t) - \dot{\mathbf{J}}\,\dot{\mathbf{q}}) + \hat{\mathbf{H}} \qquad [14.33]$$

Assuming an exact model, the system is governed by the following equation of a double integrator in the task space:

$$\ddot{\mathbf{X}} = \mathbf{w}(t) \qquad [14.34]$$

Several schemes may be considered for defining $\mathbf{w}$ [Chevallereau 88]. For a tracking control scheme with a PD controller, the control law has the form:

$$\mathbf{w}(t) = \ddot{\mathbf{X}}^d + \mathbf{K}_d (\dot{\mathbf{X}}^d - \dot{\mathbf{X}}) + \mathbf{K}_p (\mathbf{X}^d - \mathbf{X}) \qquad [14.35]$$

The closed-loop behavior of the robot is described by the following error equation:

$$\ddot{\mathbf{e}}_x + \mathbf{K}_d\,\dot{\mathbf{e}}_x + \mathbf{K}_p\,\mathbf{e}_x = \mathbf{0} \qquad [14.36]$$

with:

$$\mathbf{e}_x = \mathbf{X}^d - \mathbf{X} \qquad [14.37]$$

The corresponding block diagram is represented in Figure 14.5. The control input $\mathbf{\Gamma}$ can be computed by the inverse dynamic algorithm of Newton-Euler with the following arguments:

- the joint position is equal to the current joint position $\mathbf{q}$;
- the joint velocity is equal to the current joint velocity $\dot{\mathbf{q}}$;
- the joint acceleration is equal to $\mathbf{J}^{-1}(\mathbf{w}(t) - \dot{\mathbf{J}}\,\dot{\mathbf{q}})$.

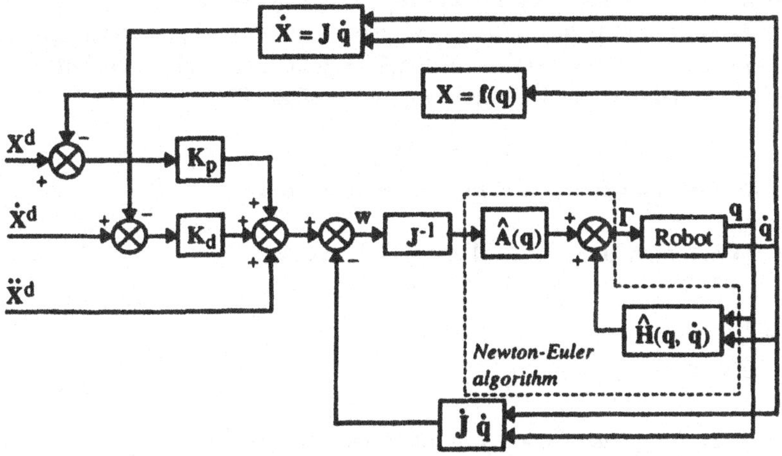

Figure 14.5. *Computed torque control in the task space*

In Appendix 10, we present an efficient algorithm to implement the computed torque control in the task space [Khalil 87a], [Chevallereau 88]. The proposed inverse dynamic algorithm uses many variables that must also be computed for the kinematic models. The computation of $\dot{\mathbf{J}}\,\dot{\mathbf{q}}$ is achieved with a recursive algorithm without differentiating $\mathbf{J}$. In the case of the Stäubli RX-90 robot, the computational cost of such an algorithm is 316 multiplications and 237 additions if we use the simplified base inertial parameters of Table 9.4.

NOTE.– If the robot is redundant, we replace the matrix $\mathbf{J}^{-1}$ in equation [14.33] by a generalized inverse. It can be shown that the robot is also governed by equation [14.36] in non-singular configurations. The homogeneous term of the generalized inverse must be chosen appropriately in order to avoid self joint motions in the null space of **J** [Hsu 88], [de Luca 91a], [Ait Mohamed 95].

14.5. Passivity-based control

14.5.1. *Introduction*

In the previous section, it is shown that the computed torque control exploits the inverse dynamic model to cancel the nonlinearities in the robot dynamics. In this section, we investigate another approach that uses the property of passivity of the robot (system that dissipates energy). Such control laws modify the natural energy of the robot so that it satisfies the desired objectives (position control or tracking control). In what follows, we first describe the robot dynamics with the Hamiltonian formalism, then we introduce the concept of passivity in the case of regulation control (fixed desired point). Finally, we show how to design a tracking controller by using properties of the passive feedback systems (Appendix 11). This section is largely based on the work of [Berguis 93] and [Landau 88].

14.5.2. *Hamiltonian formulation of the robot dynamics*

The Hamiltonian gives the total energy of the robot:

$$H = E + U \qquad [14.38]$$

where:

- $E(\mathbf{q}, \dot{\mathbf{q}})$ is the kinetic energy of the robot equal to $\frac{1}{2}\dot{\mathbf{q}}^T \mathbf{A}(\mathbf{q})\dot{\mathbf{q}}$;
- $U(\mathbf{q})$ is the potential energy of the robot;
- $\mathbf{A}(\mathbf{q})$ is the inertia matrix of the robot.

The (nx1) vector of *generalized momenta* is defined as:

$$\mathbf{p} = \mathbf{A}(\mathbf{q})\dot{\mathbf{q}} \qquad [14.39]$$

Therefore, equation [14.38] becomes:

$$H = \mathbf{p}^T \dot{\mathbf{q}} - L(\mathbf{q}, \dot{\mathbf{q}}) \qquad [14.40]$$

where $L = E - U$ is the Lagrangian of the robot. From equation [14.39], it also follows that:

$$E(\mathbf{p}, \mathbf{q}) = \frac{1}{2} \mathbf{p}^T \mathbf{A}^{-1} \mathbf{p} \qquad [14.41]$$

Defining the state variables by the vectors **q** and **p**, we obtain the Hamiltonian equations of motion in state space form as:

$$\dot{\mathbf{q}} = \frac{\partial H(\mathbf{p}, \mathbf{q})}{\partial \mathbf{p}} = \frac{\partial E(\mathbf{p}, \mathbf{q})}{\partial \mathbf{p}} \qquad [14.42]$$

$$\dot{\mathbf{p}} = -\frac{\partial H(\mathbf{p}, \mathbf{q})}{\partial \mathbf{q}} + \boldsymbol{\Gamma} = -\frac{\partial E(\mathbf{p}, \mathbf{q})}{\partial \mathbf{q}} - \frac{\partial U(\mathbf{q})}{\partial \mathbf{q}} + \boldsymbol{\Gamma} \qquad [14.43]$$

Equation [14.43] is obtained from the Lagrangian equation [9.4], noting that:

$$\frac{\partial E(\mathbf{p}, \mathbf{q})}{\partial \mathbf{q}} = -\frac{\partial E(\mathbf{q}, \dot{\mathbf{q}})}{\partial \mathbf{q}}$$

The time derivative of H is such that:

$$\dot{H} = \frac{dH(\mathbf{p}, \mathbf{q})}{dt} = [\frac{\partial H(\mathbf{p}, \mathbf{q})}{\partial \mathbf{q}}]^T \dot{\mathbf{q}} + [\frac{\partial H(\mathbf{p}, \mathbf{q})}{\partial \mathbf{p}}]^T \dot{\mathbf{p}} = \dot{\mathbf{q}}^T \boldsymbol{\Gamma} \qquad [14.44]$$

which yields:

$$\int_0^{t1} \dot{\mathbf{q}}^T(t)\, \boldsymbol{\Gamma}(t)\, dt = H[\mathbf{p}(t1), \mathbf{q}(t1)] - H[\mathbf{p}(0), \mathbf{q}(0)] \qquad [14.45]$$

A rigid robot is defined as *passive* from the input $\boldsymbol{\Gamma}$ to the output $\dot{\mathbf{q}}$ when there exists a constant $0 < \beta < \infty$ such that:

$$\int_0^{t1} \dot{\mathbf{q}}^T(t)\, \boldsymbol{\Gamma}(t)\, dt \geq -\beta$$

which is true, from equation [14.45], when $\beta = H[\mathbf{p}(0), \mathbf{q}(0)]$. This means that the total energy has a bounded minimum.

14.5.3. *Passivity-based position control*

Let us assume that we want to drive the robot to a desired position $\mathbf{q}^d$. Intuitively, this can be achieved by shifting the open-loop energy minimum from ($\dot{\mathbf{q}} = \mathbf{0}$, $\mathbf{q} = \mathbf{0}$) towards ($\dot{\mathbf{q}} = \mathbf{0}$, $\mathbf{e} = \mathbf{0}$) for the closed-loop system, where $\mathbf{e} = \mathbf{q}^d - \mathbf{q}$ is the position error. This shifting can be obtained by reshaping the potential energy of the system such that it attains the desired minimum at $\mathbf{e} = \mathbf{0}$. To this end, let $U^*(\mathbf{q})$ be an arbitrary function of the desired potential energy for the closed-loop system. Let us define the following control law:

$$\mathbf{\Gamma} = -\frac{\partial U^*(\mathbf{q})}{\partial \mathbf{q}} + \frac{\partial U(\mathbf{q})}{\partial \mathbf{q}} + \mathbf{v} \qquad [14.46]$$

where $\mathbf{v}$ is the (nx1) new input control vector [Takegaki 81b]. The Hamiltonian equations become:

$$\dot{\mathbf{q}} = \frac{\partial E(\mathbf{p}, \mathbf{q})}{\partial \mathbf{p}} \qquad [14.47]$$

$$\dot{\mathbf{p}} = -\frac{\partial E(\mathbf{p}, \mathbf{q})}{\partial \mathbf{q}} - \frac{\partial U^*(\mathbf{q})}{\partial \mathbf{q}} + \mathbf{v} \qquad [14.48]$$

Hence, by using the control law [14.46], the initial Hamiltonian $H(\mathbf{p}, \mathbf{q})$ is modified into the desired one $H^*(\mathbf{p}, \mathbf{q})$ such that:

$$H^* = E + U^* \qquad [14.49]$$

and we can verify that:

$$\dot{H}^* = \dot{\mathbf{q}}^T \mathbf{v} \qquad [14.50]$$

This implies that the robot is passive from the new input $\mathbf{v}$ to the output $\dot{\mathbf{q}}$. To asymptotically stabilize the system, we add a damping in the loop such that:

$$\mathbf{v} = -\mathbf{K}_d \dot{\mathbf{q}} \qquad [14.51]$$

where $\mathbf{K}_d > 0$ is a diagonal matrix. Equation [14.50] becomes:

$$\dot{H}^* = -\dot{q}^T K_d \dot{q} \qquad [14.52]$$

This expression is negative semi-definite. However, it can be verified that the equilibrium point ($e = 0$, $\dot{q} = 0$) is the largest invariant set within the set $\dot{H}^*(q, p) = 0$. Hence, using the La Salle invariance theorem [Hahn 67], the asymptotic stability of the equilibrium can be proven.

Various choices are possible for the desired potential energy function $U^*(q, p)$ [Wen 88]. An obvious one that satisfies the constraint of a strict minimum at $e = 0$ is:

$$U^* = \frac{1}{2} e^T K_p e \qquad [14.53]$$

For this choice, the control law [14.46] becomes:

$$\Gamma = K_p e - K_d \dot{q} + Q(q) \qquad [14.54]$$

which represents gravity compensation and a linear state-feedback loop [Takegaki 81b], as the one presented in § 14.3.

The following choice for $U^*(q, p)$, under the condition that K_p is large enough, is also minimum when $e = 0$ [Takegaki 81b]:

$$U^* = \frac{1}{2} e^T K_p e + U(q) - U(q^d) + e^T Q(q^d) \qquad [14.55]$$

Hence, the control law is:

$$\Gamma = K_p e - K_d \dot{q} + Q(q^d) \qquad [14.56]$$

14.5.4. *Passivity-based tracking control*

For a tracking task, it is necessary to modify the control law so that the strict energy minimum ($q = 0$, $\dot{q} = 0$) of the open-loop system is shifted towards ($e = 0$, $\dot{e} = 0$) for the closed-loop. This can be achieved by modifying both the kinetic energy and the potential energy.

In this section, we analyze the passivity-based control laws of [Paden 88] and [Slotine 87] using the passive system feedback approach proposed by [Landau 88] (Appendix 11). Consider the following control law [Paden 88]:

$$\Gamma = A(q)\,\ddot{q}^d + C(q, \dot{q})\,\dot{q}^d + Q(q) + K_p\,e + K_d\,\dot{e} \qquad [14.57]$$

In the absence of friction, equations [14.57] and [14.1] lead to the closed-loop equation:

$$K_p\,e + K_d\,\dot{e} = \tau \qquad [14.58]$$

with :

$$\tau = -A(q)\,\ddot{e} - C(q, \dot{q})\,\dot{e} \qquad [14.59]$$

Equation [14.58] represents a system of two interconnected feedback blocks (Figure 14.6):

- a linear block B1 in the feedforward chain whose input and output are $\dot{e}$ and τ respectively;
- a nonlinear block B2 in the feedback chain whose input and output are τ and $-\dot{e}$ respectively.

In order to prove that the nonlinear block is passive, let us consider the integral of the input-output dot product:

$$\int_0^{t1} -\dot{e}^T(t)\,\tau(t)\,dt = \int_0^{t1} [\dot{e}^T A(q)\,\ddot{e} + \dot{e}^T\,C(q, \dot{q})\,\dot{e}]\,dt \qquad [14.60]$$

Since:

$$\dot{e}^T A(q)\,\ddot{e} = \frac{1}{2}\frac{d}{dt}[\dot{e}^T A(q)\,\dot{e}] - \frac{1}{2}\dot{e}^T\,\dot{A}(q)\,\dot{e} \qquad [14.61]$$

then:

$$\int_0^{t1} -\dot{e}^T(t)\,\tau(t)\,dt = \int_0^{t1} [\frac{1}{2}\frac{d}{dt}[\dot{e}^T A(q)\dot{e}] - \frac{1}{2}\dot{e}^T\dot{A}(q)\dot{e} + \dot{e}^T C(q, \dot{q})\dot{e}]\,dt \qquad [14.62]$$

Since $[\dot{A}(q) - 2\,C(q, \dot{q})]$ is skew-symmetric, then $\dot{e}^T\,[\dot{A}(q) - 2\,C(q, \dot{q})]\,\dot{e}$ is zero, which reduces equation [14.62] to:

$$\int_0^{t1} -\dot{e}^T(t)\,\tau(t)\,dt = \int_0^{t1} \frac{1}{2}\frac{d}{dt}[\dot{e}^T A(q)\,\dot{e}]\,dt$$

$$= \frac{1}{2}[\dot{e}^T(t1)\,A(q(t1))\,\dot{e}(t1) - \dot{e}^T(0)\,A(q(0))\,\dot{e}(0)] \qquad [14.63]$$

and finally:

$$\int_0^{t1} -\dot{e}^T(t)\,\tau(t)\,dt \geq -\gamma o^2 = -\frac{1}{2}\dot{e}^T(0)\,A(q(0))\,\dot{e}(0) \qquad [14.64]$$

and, since $\gamma o^2 < \infty$, it follows that the block B2 is passive.

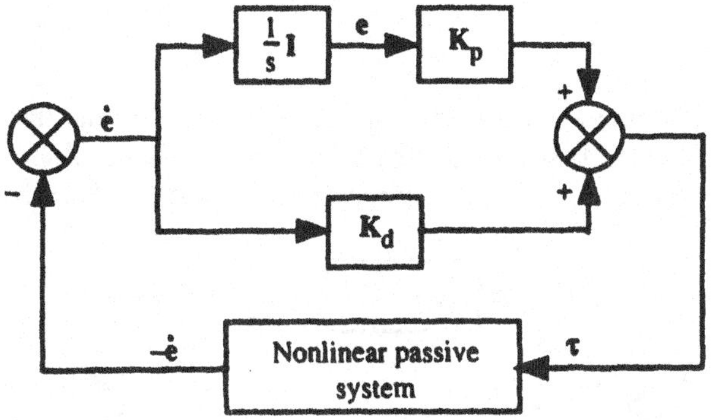

Figure 14.6. *Equivalent feedback representation of the closed-loop equation [14.58] (from [Landau 88])*

The linear block of the feedforward chain is characterized by a positive real transfer matrix:

$$H(s) = K_d + \frac{1}{s}K_p \qquad [14.65]$$

which proves (Appendix 11) that the system represented by equation [14.58] is stable, and more precisely that the error e of the linear system is bounded.

In order to ensure that $\mathbf{e} \rightarrow \mathbf{0}$ as $t \rightarrow \infty$, the control law should be modified so that the transfer function of the feedforward chain be *strictly positive real*[2]. This can be done by removing the pole from the origin, choosing for example $\mathbf{H}(s)$ such that:

$$\mathbf{H}(s) = \mathbf{K}_d + \mathbf{K}_p [s\mathbf{I} + \mathbf{\Lambda}]^{-1} \qquad [14.66]$$

where $\mathbf{\Lambda}$ is a positive definite matrix. Modifying the control law accordingly yields:

$$\Gamma = \mathbf{A}(\mathbf{q})\,\ddot{\mathbf{q}}^d + \mathbf{C}(\mathbf{q}, \dot{\mathbf{q}})\,\dot{\mathbf{q}}^d + \mathbf{Q}(\mathbf{q}) + \mathbf{K}_p\,\tilde{\mathbf{e}} + \mathbf{K}_d\,\dot{\mathbf{e}} \qquad [14.67]$$

with $\frac{d}{dt}\tilde{\mathbf{e}} = -\mathbf{\Lambda}\,\tilde{\mathbf{e}} + \dot{\mathbf{e}}$.

The closed-loop equation becomes:

$$\mathbf{K}_p\,\tilde{\mathbf{e}} + \mathbf{K}_d\,\dot{\mathbf{e}} = \tau \qquad [14.68]$$

The corresponding system is shown in Figure 14.7. As the transfer function of the feedforward chain is strictly positive definite, we can conclude that $\tilde{\mathbf{e}}(t) \rightarrow \mathbf{0}$ as $t \rightarrow \infty$ and $\dot{\mathbf{e}} \rightarrow \mathbf{0}$ as $t \rightarrow \infty$, but unfortunately we cannot conclude that $\mathbf{e} \rightarrow \mathbf{0}$.

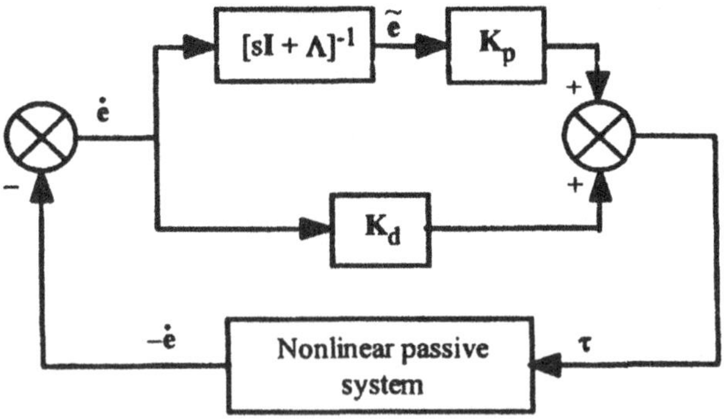

Figure 14.7. *Equivalent feedback representation of the closed-loop equation [14.68] (from [Landau 88])*

In order to ensure that $\mathbf{e} \rightarrow \mathbf{0}$ as $t \rightarrow \infty$, $\mathbf{e}$ should be a state of the feedforward chain. This is achieved with the following control law:

---

2 We can find in [Paden 88] another demonstration proving that this law is asymptotically stable, i.e. that ($\mathbf{e} = \mathbf{0}$, $\dot{\mathbf{e}} = \mathbf{0}$) for arbitrary values $\mathbf{K}_p = \mathbf{K}_p^T > 0$ and $\mathbf{K}_d = \mathbf{K}_d^T > 0$.

$$\Gamma = A(q)\,\ddot{q}^r + C(q, \dot{q})\,\dot{q}^r + Q(q) + K_p\,e + K_d\,\dot{e}_r \qquad [14.69]$$

with:

- $e = q^d - q$
- $\dot{e}_r = \dot{e} + \Lambda\,e = \dot{q}^r - \dot{q}$
- $\dot{q}^r = \dot{q}^d + \Lambda\,e$
- $\Lambda = \Lambda^T > 0$

which implies that e and $\dot{e}_r$ are related through the transfer function $[sI + \Lambda]^{-1}$. The $\dot{q}^r$ vector is called the *reference velocity*.

Combining equations [14.69] and [14.1], and assuming for convenience that friction torques are either compensated or neglected, leads to the following closed-loop equation:

$$K_p\,e + K_d\,\dot{e}_r = -A(q)\,\ddot{e}_r - C(q, \dot{q})\,\dot{e}_r = \tau \qquad [14.70]$$

The corresponding system is shown in Figure 14.8. Note that, in this case, $e(t)$ is the state of the feedforward chain that is strictly positive real. Thus, the system is globally asymptotically stable.

The tracking control law [14.69] presents several interesting analogies with the position control law [14.54]. First, the terms $Q(q)$ and K_p e modify the potential energy as in the control law [14.54]. Then, the compensations for $A(q)$ and $C(q, \dot{q})$ modify the kinetic energy in the desired sense. Finally, the term $K_d\,\dot{e}_r$ introduces a damping that contributes to satisfy the tracking objective.

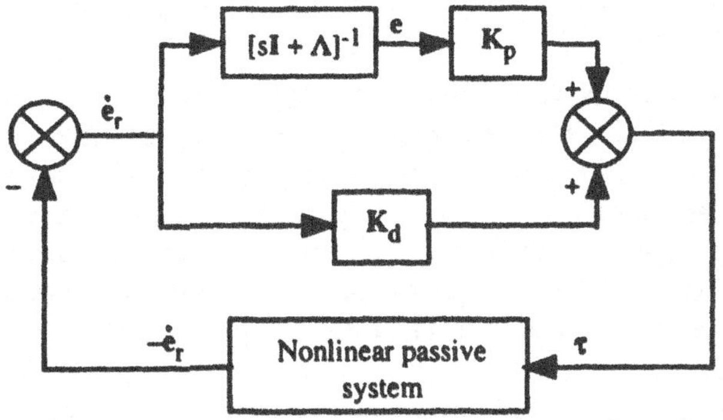

Figure 14.8. *Equivalent feedback representation of the closed-loop equation [14.70] (from [Landau 88])*

NOTE.– The computation of the passivity-based control law [14.69] cannot be achieved by the Newton-Euler inverse dynamic model due to the presence of both $\dot{\mathbf{q}}$ and $\dot{\mathbf{q}}^r$ in the expressions of the Coriolis and centrifugal forces. [Kawasaki 96] proposes an efficient algorithm for its computation.

14.5.5. *Lyapunov-based method*

In [Wen 88], we can find the demonstration of the stability of all the control laws presented in the previous sections with the definition of a suitable Lyapunov function. The exponential stability demonstrations of the following laws are also given:

$$\Gamma = \mathbf{A}(\mathbf{q})\,\ddot{\mathbf{q}}^d + \mathbf{C}(\mathbf{q}, \dot{\mathbf{q}}^d)\,\dot{\mathbf{q}}^d - \frac{\partial U^*(\mathbf{q})}{\partial \mathbf{q}} + \mathbf{Q}(\mathbf{q}) + \mathbf{K}_d\,\dot{\mathbf{e}} \qquad [14.71]$$

$$\Gamma = \mathbf{A}(\mathbf{q})\,\ddot{\mathbf{q}}^d + \mathbf{C}(\mathbf{q}, \dot{\mathbf{q}})\,\dot{\mathbf{q}} - \frac{\partial U^*(\mathbf{q})}{\partial \mathbf{q}} + \mathbf{Q}(\mathbf{q}) + \mathbf{K}_d\,\dot{\mathbf{e}} \qquad [14.72]$$

$$\Gamma = \mathbf{A}(\mathbf{q}^d)\,\ddot{\mathbf{q}}^d + \mathbf{C}(\mathbf{q}^d, \dot{\mathbf{q}}^d)\,\dot{\mathbf{q}}^d - \frac{\partial U^*(\mathbf{q})}{\partial \mathbf{q}} + \mathbf{Q}(\mathbf{q}^d) + \mathbf{K}_d\,\dot{\mathbf{e}} \qquad [14.73]$$

Choosing $U^* = \frac{1}{2}\,\mathbf{e}^T\,\mathbf{K}_p\,\mathbf{e}$ results in $\frac{\partial U^*(\mathbf{q})}{\partial \mathbf{q}} = -\mathbf{K}_p\,\mathbf{e}$, but other choices are also possible. Note that equations [14.71] and [14.72] can be computed by the inverse dynamic algorithm of Newton Euler.

14.6. Adaptive control

14.6.1. *Introduction*

Since the dynamic model is often not exactly known (inaccuracies in the dynamic parameters of the robot, of the payload, high-frequency unmodeled dynamics...), the adaptive control theory has been investigated extensively as an interesting approach to estimate or adjust on-line the dynamic parameter values used in the control. The nonlinear adaptive control of rigid robot manipulators can be considered today to be mature, as is indicated by the large number of methods published over the last two decades [Bayard 88], [Ortega 89]. The different approaches of adaptive control can be classified as:

i) simplification of the dynamic model [Dubowsky 79], [Takegaki 81a];

ii) application of the adaptive techniques, which were developed for linear systems [Horowitz 80], [Nicosia 84], [Hsia 86];

iii) formulation of a nonlinear decoupling and linearizing adaptive control [Craig 86b];

iv) formulation of a nonlinear adaptive control based on the passivity property of the robot [Slotine 87], [Sadegh 87], [Landau 88], [Kelly 88];

v) formulation of parameter adaptation mechanisms that avoid joint acceleration computation as the filtered dynamic model [Middleton 88], [Li 89] or the energy-based model [El Serafi 91a].

The control laws proposed in the first two strategies are only valid for slow motion and do not take into account the full dynamics of the robot. The nonlinear adaptive control law of Craig requires joint accelerations and assumes that the estimated inertia matrix is invertible. The fourth and fifth schemes avoid the joint acceleration estimation and are, at least from a theoretical viewpoint, the most interesting.

In the next sections, we present the principles of the nonlinear linearizing adaptive control and of the passivity-based adaptive control.

14.6.2. *Adaptive feedback linearizing control*

The first version of an adaptive dynamic control has been formulated by Craig et al. [Craig 86b]. The control law has the same structure as the computed torque control law of equation [14.22], and can be written in the following form (Figure 14.9):

$$\mathbf{\Gamma} = \mathbf{A}(\mathbf{q}, \hat{\boldsymbol{\chi}})\, \mathbf{w}(t) + \mathbf{H}(\mathbf{q}, \dot{\mathbf{q}}, \hat{\boldsymbol{\chi}}) \qquad [14.74]$$

where $\hat{\boldsymbol{\chi}}$ is the vector of the estimated base dynamic parameters, and:

$$\mathbf{w}(t) = \ddot{\mathbf{q}}^{d} + \mathbf{K}_{d}\, \dot{\mathbf{e}} + \mathbf{K}_{p}\, \mathbf{e} \qquad [14.75]$$

The control law [14.74] is associated with an on-line identification law, which provides $\hat{\boldsymbol{\chi}}(t)$. For brevity, the control law will be noted:

$$\mathbf{\Gamma} = \hat{\mathbf{A}}(\mathbf{q})\, \mathbf{w}(t) + \hat{\mathbf{H}}(\mathbf{q}, \dot{\mathbf{q}}) \qquad [14.76]$$

Combining equations [14.2], [14.3] and [14.76], leads to the closed-loop error equation (see equation [14.28]):

$$\ddot{\mathbf{e}} + \mathbf{K}_d \dot{\mathbf{e}} + \mathbf{K}_p \mathbf{e} = \hat{\mathbf{A}}^{-1}(\mathbf{q}) [\mathbf{\Phi}(\mathbf{q}, \dot{\mathbf{q}}, \ddot{\mathbf{q}}) \boldsymbol{\chi} - \mathbf{\Phi}(\mathbf{q}, \dot{\mathbf{q}}, \ddot{\mathbf{q}}) \hat{\boldsymbol{\chi}}] = \hat{\mathbf{A}}^{-1}(\mathbf{q}) \mathbf{\Phi}(\mathbf{q}, \dot{\mathbf{q}}, \ddot{\mathbf{q}}) \tilde{\boldsymbol{\chi}} \quad [14.77]$$

with:

$$\tilde{\boldsymbol{\chi}} = \boldsymbol{\chi} - \hat{\boldsymbol{\chi}} \quad [14.78]$$

Let us rewrite equation [14.77] under the state space form:

$$\dot{\mathbf{x}} = \mathbf{a}\,\mathbf{x} + \mathbf{b}\,\hat{\mathbf{A}}^{-1}(\mathbf{q})\,\mathbf{\Phi}(\mathbf{q}, \dot{\mathbf{q}}, \ddot{\mathbf{q}})\,\tilde{\boldsymbol{\chi}} \quad [14.79]$$

$$\mathbf{x} = \begin{bmatrix} \mathbf{e} \\ \dot{\mathbf{e}} \end{bmatrix}, \mathbf{a} = \begin{bmatrix} \mathbf{0}_n & \mathbf{I}_n \\ -\mathbf{K}_p & -\mathbf{K}_d \end{bmatrix}, \mathbf{b} = \begin{bmatrix} \mathbf{0}_n \\ \mathbf{I}_n \end{bmatrix} \quad [14.80]$$

where $\mathbf{0}_n$ and $\mathbf{I}_n$ are the (nxn) null matrix and identity matrix respectively.

Let us consider the following Lyapunov function candidate:

$$V = \mathbf{x}^T \mathbf{P}\,\mathbf{x} + \tilde{\boldsymbol{\chi}}^T \mathbf{\Lambda}\,\tilde{\boldsymbol{\chi}} \quad [14.81]$$

where $\mathbf{\Lambda} = \mathbf{diag}(\lambda_1, \lambda_2, \ldots, \lambda_b)$ is a positive definite adaptation gain matrix.

$\mathbf{P}$ is the unique positive definite matrix, which is the solution of the Lyapunov equation such that:

$$\mathbf{a}^T \mathbf{P} + \mathbf{P}\,\mathbf{a} = -\mathbf{F} \quad [14.82]$$

Differentiating V with respect to time leads to:

$$\dot{V} = -\mathbf{x}^T \mathbf{F}\,\mathbf{x} + 2\,\tilde{\boldsymbol{\chi}}^T [\mathbf{\Phi}^T(\mathbf{q}, \dot{\mathbf{q}}, \ddot{\mathbf{q}})\,\hat{\mathbf{A}}^{-1}(\mathbf{q})\,\mathbf{b}^T \mathbf{P}\,\mathbf{x} + \mathbf{\Lambda}\,\dot{\tilde{\boldsymbol{\chi}}}] \quad [14.83]$$

Assuming the following adaptation law:

$$\dot{\tilde{\boldsymbol{\chi}}} = -\mathbf{\Lambda}^{-1}\,\mathbf{\Phi}^T(\mathbf{q}, \dot{\mathbf{q}}, \ddot{\mathbf{q}})\,\hat{\mathbf{A}}^{-1}(\mathbf{q})\,\mathbf{b}^T \mathbf{P}\,\mathbf{x} = -\dot{\hat{\boldsymbol{\chi}}} \quad [14.84]$$

the expression of $\dot{V}$ becomes:

$$\dot{V} = -\mathbf{x}^T \mathbf{F}\,\mathbf{x} \leq 0 \quad [14.85]$$

Therefore, the vector $\mathbf{x}$ is bounded and $\mathbf{x} \to \mathbf{0}$ as $t \to \infty$. Since $\mathbf{x}$ is composed of $\mathbf{e}$ and $\dot{\mathbf{e}}$, then $\mathbf{e} \to \mathbf{0}$ and $\dot{\mathbf{e}} \to \mathbf{0}$. The adaptive dynamic control algorithm given by equations [14.76] and [14.84] is thus globally asymptotically stable.

This method has two major limitations: the first is that the joint accelerations are required for implementation; the second is that the inverse of the estimated inertia matrix has to be bounded. Craig et al. [Craig 86b] suggested projection of the estimated parameters in a bounded region of the parameter space. However, this projection does not guarantee that the inverse of the inertia matrix exists.

Spong and Ortega [Spong 90] proposed a new version of this algorithm in which the condition of the invertibility of the matrix $\hat{\mathbf{A}}$ is relaxed, but the joint accelerations are still required.

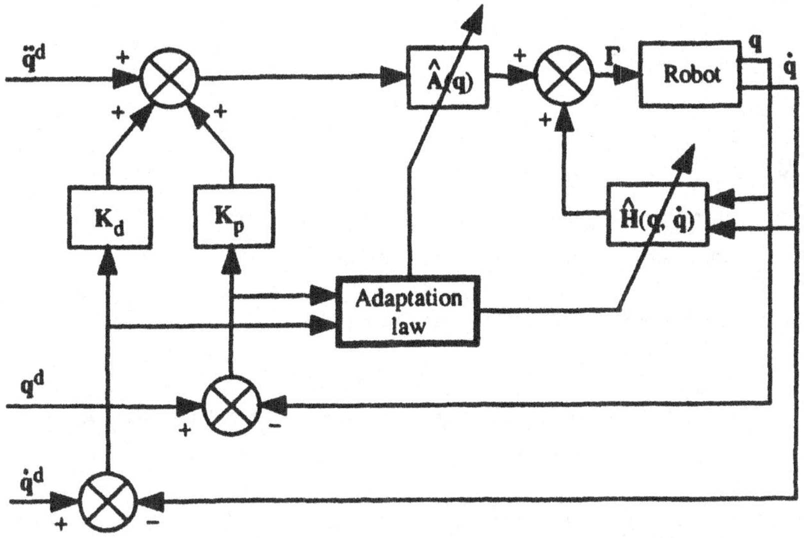

Figure 14.9. *Nonlinear adaptive control (from [Craig 86b])*

14.6.3. *Adaptive passivity-based control*

In order to develop an adaptive algorithm based on the full dynamic model, Slotine and Li [Slotine 87] exploited the property of skew-symmetry of the matrix $[\dot{\mathbf{A}} - 2\,\mathbf{C}]$. This property is a consequence of the passivity of the robot.

The control law is derived from equation [14.69] with $\mathbf{K}_p = \mathbf{0}$:

$$\mathbf{\Gamma} = \mathbf{A}(\mathbf{q}, \hat{\boldsymbol{\chi}})\,\ddot{\mathbf{q}}^r + \mathbf{C}(\mathbf{q}, \dot{\mathbf{q}}, \hat{\boldsymbol{\chi}})\,\dot{\mathbf{q}}^r + \mathbf{Q}(\mathbf{q}, \hat{\boldsymbol{\chi}}) + \mathbf{K}_d\,\dot{\mathbf{e}}_r \qquad [14.86]$$

rewritten as:

$$\mathbf{\Gamma} = \hat{\mathbf{A}}(\mathbf{q})\,\ddot{\mathbf{q}}^r + \hat{\mathbf{C}}(\mathbf{q}, \dot{\mathbf{q}})\,\dot{\mathbf{q}}^r + \hat{\mathbf{Q}}(\mathbf{q}) + \mathbf{K}_d\,\dot{\mathbf{e}}_r \qquad [14.87]$$

with:

- $\mathbf{e} = \mathbf{q}^d - \mathbf{q}$
- $\dot{\mathbf{e}}_r = \dot{\mathbf{e}} + \Lambda\, \mathbf{e}$
- $\dot{\mathbf{q}}^r = \dot{\mathbf{q}} + \dot{\mathbf{e}}_r = \dot{\mathbf{q}}^d + \Lambda\, \mathbf{e}$

$\dot{\mathbf{e}}_r$ may be regarded as a sliding surface in the state space plane defined by $\mathbf{e}$ and $\dot{\mathbf{e}}$. To design the adaptation law, let us consider the following Lyapunov function candidate:

$$V = \frac{1}{2}[\dot{\mathbf{e}}_r^T \mathbf{A}\, \dot{\mathbf{e}}_r + \tilde{\boldsymbol{\chi}}^T \mathbf{F}\, \tilde{\boldsymbol{\chi}}] \qquad [14.88]$$

with:

- $\tilde{\boldsymbol{\chi}} = \boldsymbol{\chi} - \hat{\boldsymbol{\chi}}$;
- $\hat{\boldsymbol{\chi}}$: vector of the estimated base dynamic parameters;
- $\mathbf{F}$: positive definite gain adaptation matrix.

The differentiation of V with respect to time leads to:

$$\dot{V} = \frac{1}{2}\dot{\mathbf{e}}_r^T \dot{\mathbf{A}}\, \dot{\mathbf{e}}_r + \dot{\mathbf{e}}_r^T \mathbf{A}\, \ddot{\mathbf{e}}_r + \tilde{\boldsymbol{\chi}}^T \mathbf{F}\, \dot{\tilde{\boldsymbol{\chi}}} = \dot{\mathbf{e}}_r^T [\frac{1}{2}\dot{\mathbf{A}}\, \dot{\mathbf{e}}_r + \mathbf{A}\,(\ddot{\mathbf{q}}^r - \ddot{\mathbf{q}})] + \tilde{\boldsymbol{\chi}}^T \mathbf{F}\, \dot{\tilde{\boldsymbol{\chi}}}$$

[14.89]

and after substitution of $\mathbf{A}\,\ddot{\mathbf{q}}$, using equation [14.1] while assuming no friction for sake of brevity, it becomes:

$$\dot{V} = \dot{\mathbf{e}}_r^T [\frac{1}{2}\dot{\mathbf{A}}\, \dot{\mathbf{e}}_r + \mathbf{A}\, \ddot{\mathbf{q}}^r - \boldsymbol{\Gamma} + \mathbf{C}(\mathbf{q}, \dot{\mathbf{q}})\,(\dot{\mathbf{q}}^r - \dot{\mathbf{e}}_r) + \mathbf{Q}(\mathbf{q})] + \tilde{\boldsymbol{\chi}}^T \mathbf{F}\, \dot{\tilde{\boldsymbol{\chi}}} \qquad [14.90]$$

Since $[\dot{\mathbf{A}} - 2\mathbf{C}]$ is skew-symmetric [Koditschek 84] (§ 9.3.3.3), we obtain:

$$\dot{V} = \dot{\mathbf{e}}_r^T [\mathbf{A}\, \ddot{\mathbf{q}}^r - \boldsymbol{\Gamma} + \mathbf{C}(\mathbf{q}, \dot{\mathbf{q}})\, \dot{\mathbf{q}}^r + \mathbf{Q}(\mathbf{q})] + \tilde{\boldsymbol{\chi}}^T \mathbf{F}\, \dot{\tilde{\boldsymbol{\chi}}} \qquad [14.91]$$

Substituting the control law [14.87] in equation [14.91] yields:

$$\dot{V} = \dot{\mathbf{e}}_r^T [\tilde{\mathbf{A}}\, \ddot{\mathbf{q}}^r + \tilde{\mathbf{C}}(\mathbf{q}, \dot{\mathbf{q}})\, \dot{\mathbf{q}}^r + \tilde{\mathbf{Q}}(\mathbf{q}) - \mathbf{K}_d\, \dot{\mathbf{e}}_r] + \tilde{\boldsymbol{\chi}}^T \mathbf{F}\, \dot{\tilde{\boldsymbol{\chi}}} \qquad [14.92]$$

where:

$$\tilde{\mathbf{A}} = \mathbf{A} - \hat{\mathbf{A}}, \tilde{\mathbf{C}} = \mathbf{C} - \hat{\mathbf{C}}, \tilde{\mathbf{Q}} = \mathbf{Q} - \hat{\mathbf{Q}} \qquad [14.93]$$

Since **A**, **C** and **Q** are linear in the dynamic parameters, we can write that:

$$\tilde{\mathbf{A}}\,\ddot{\mathbf{q}}^r + \tilde{\mathbf{C}}(\mathbf{q}, \dot{\mathbf{q}})\,\dot{\mathbf{q}}^r + \tilde{\mathbf{Q}}(\mathbf{q}) = \boldsymbol{\Phi}(\mathbf{q}, \dot{\mathbf{q}}, \dot{\mathbf{q}}^r, \ddot{\mathbf{q}}^r)\,\tilde{\boldsymbol{\chi}} \qquad [14.94]$$

By combining equations [14.94] and [14.92], it follows that:

$$\dot{\mathrm{V}} = -\dot{\mathbf{e}}_r^T\,\mathbf{K}_d\,\dot{\mathbf{e}}_r + \tilde{\boldsymbol{\chi}}^T\,[\mathbf{F}\,\dot{\tilde{\boldsymbol{\chi}}} + \boldsymbol{\Phi}^T(\mathbf{q}, \dot{\mathbf{q}}, \dot{\mathbf{q}}^r, \ddot{\mathbf{q}}^r)\,\dot{\mathbf{e}}_r] \qquad [14.95]$$

Let us choose the adaptation law:

$$\dot{\tilde{\boldsymbol{\chi}}} = -\mathbf{F}^{-1}\,\boldsymbol{\Phi}^T(\mathbf{q}, \dot{\mathbf{q}}, \dot{\mathbf{q}}^r, \ddot{\mathbf{q}}^r)\,\dot{\mathbf{e}}_r = -\dot{\hat{\boldsymbol{\chi}}} \qquad [14.96]$$

where the matrix $\mathbf{F}^{-1}$ is the adaptation gain. Equation [14.95] becomes:

$$\dot{\mathrm{V}} = -\dot{\mathbf{e}}_r^T\,\mathbf{K}_d\,\dot{\mathbf{e}}_r \leq 0 \qquad [14.97]$$

From equation [14.97], we conclude that the control law [14.87] associated with the adaptation law [14.96] is stable.

Since $\dot{\mathrm{V}}$ is only negative semi-definite, we cannot conclude on the asymptotic stability of the closed-loop system. Unfortunately, the La Salle invariance theorem cannot be applied to non-autonomous systems, which is the case in tracking tasks. To complete the proof of asymptotic stability, the Barbalat lemma can be used (Appendix 9).

It is worth noting that adding a proportional gain $\mathbf{K}_p\,\mathbf{e}$ to the control law [14.87] makes it possible to use the results on passivity of § 14.5.4. and to prove asymptotic stability with a Lyapunov function. Let us consider the following law [Landau 88]:

$$\boldsymbol{\Gamma} = \hat{\mathbf{A}}(\mathbf{q})\,\ddot{\mathbf{q}}^r + \hat{\mathbf{C}}(\mathbf{q}, \dot{\mathbf{q}})\,\dot{\mathbf{q}}^r + \hat{\mathbf{Q}}(\mathbf{q}) + \mathbf{K}_p\,\mathbf{e} + \mathbf{K}_d\,\dot{\mathbf{e}}_r \qquad [14.98]$$

which can be rewritten as:

$$\Gamma = \mathbf{A}(\mathbf{q}, \chi)\,\ddot{\mathbf{q}}^r + \mathbf{C}(\mathbf{q}, \dot{\mathbf{q}}, \chi)\,\dot{\mathbf{q}}^r + \mathbf{Q}(\mathbf{q}, \chi) + \mathbf{K}_p\,\mathbf{e} + \mathbf{K}_d\,\dot{\mathbf{e}}_r - \mathbf{A}(\mathbf{q}, \tilde{\chi})\,\ddot{\mathbf{q}}^r - \mathbf{C}(\mathbf{q}, \dot{\mathbf{q}}, \tilde{\chi})\,\dot{\mathbf{q}}^r - \mathbf{Q}(\mathbf{q}, \tilde{\chi}) \qquad [14.99]$$

the adaptation law being given by equation [14.96].

From equations [14.1] and [14.99], in the absence of friction, we can represent the system by the three interconnected blocks of Figure 14.10. Blocks B1 and B2 represent the system of Figure 14.8 whose passivity has been demonstrated in § 14.5. To demonstrate the passivity of the block B3, we must verify that:

$$\int_0^{t1} -\dot{\mathbf{e}}_r^T(t)\,\boldsymbol{\Phi}\,\tilde{\chi}\,dt \geq -\gamma_0^2 \quad \text{with } \gamma_0^2 < \infty \qquad [14.100]$$

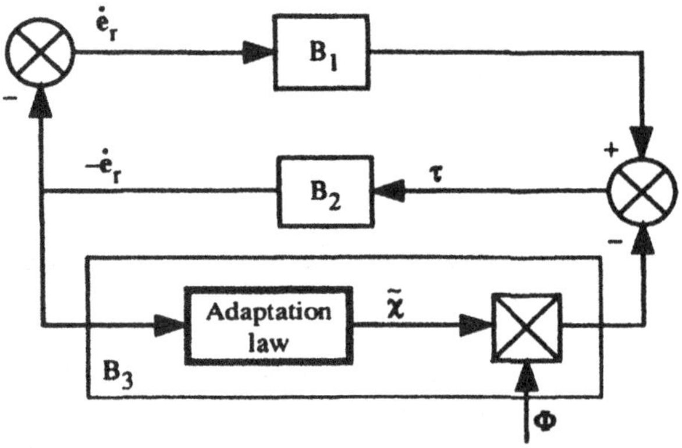

Figure 14.10. *Equivalent feedback representation of the closed-loop equation for the passivity-based adaptive control (from [Landau 88])*

From equation [14.96], and since **F** is symmetric, it follows that:

$$-\dot{\mathbf{e}}_r^T(t)\,\boldsymbol{\Phi} = \dot{\tilde{\chi}}^T\,\mathbf{F} \qquad [14.101]$$

and using equation [14.100], we obtain:

$$\int_0^{t1} -\dot{\mathbf{e}}_r^T(t)\,\boldsymbol{\Phi}\,\tilde{\chi}\,dt = \int_0^{t1} \dot{\tilde{\chi}}^T\,\mathbf{F}\,\tilde{\chi}\,dt = \int_0^{t1} \frac{1}{2}\frac{d}{dt}[\tilde{\chi}^T\,\mathbf{F}\,\tilde{\chi}]\,dt \geq -\frac{1}{2}\tilde{\chi}^T(0)\,\mathbf{F}\,\tilde{\chi}(0) \qquad [14.102]$$

Since $\mathbf{F}$ is positive definite, equation [14.100] is verified and the block B3 is passive, which demonstrates the asymptotic stability since the block B1 is strictly positive real.

To demonstrate the stability of the control law [14.98] with the adaptation law [14.96] using a Lyapunov analysis, let us choose the following Lyapunov function candidate [Sadegh 90]:

$$V = \frac{1}{2}\dot{\mathbf{e}}_r^T \mathbf{A}\, \dot{\mathbf{e}}_r + \frac{1}{2}\mathbf{e}^T \mathbf{K}_p \mathbf{e} + \frac{1}{2}\tilde{\chi}^T \mathbf{F}\, \tilde{\chi} \qquad [14.103]$$

Unlike the function [14.88], equation [14.103] is a function of the transformed state vector $[\mathbf{e}^T \quad \dot{\mathbf{e}}_r^T]^T$. It can be verified that:

$$\dot{V} = -\dot{\mathbf{e}}_r^T \mathbf{K}_d \dot{\mathbf{e}}_r - \mathbf{e}^T \Lambda \mathbf{K}_p \mathbf{e} \leq 0 \qquad [14.104]$$

Since $\dot{V}$ is a function of $\dot{\mathbf{e}}_r$ and $\mathbf{e}$, we can conclude that the closed-loop system is globally asymptotically stable. Note that the two control laws [14.87] and [14.98] are similar, but the second one is more practical for tuning since there is an additional gain $\mathbf{K}_p$.

The passivity-based adaptive control law does not require joint acceleration estimations. However, its drawback is that the inverse dynamics cannot be directly computed by the Newton-Euler algorithm, due to the presence of both $\dot{\mathbf{q}}$ and $\dot{\mathbf{q}}^r$ in the expressions of the Coriolis and centrifugal forces. Kawasaki [Kawasaki 96] proposes an efficient algorithm for its computation.

To avoid this computational problem, Sadegh and Horowitz [Sadegh 90] proposed to calculate both the control and adaptation laws in terms of the desired position, velocity and acceleration:

$$\Gamma = \hat{\mathbf{A}}(\mathbf{q}^d)\, \ddot{\mathbf{q}}^d + \hat{\mathbf{H}}(\mathbf{q}^d, \dot{\mathbf{q}}^d) + \mathbf{K}_d \dot{\mathbf{e}}_r + \mathbf{K}_p \mathbf{e} + \mathbf{K}_n \|\mathbf{e}\|\, \dot{\mathbf{e}}_r \qquad [14.105]$$

$$\dot{\hat{\chi}} = \mathbf{F}^{-1}\, \Phi^T(\mathbf{q}^d, \dot{\mathbf{q}}^d, \ddot{\mathbf{q}}^d)\, \dot{\mathbf{e}}_r \qquad [14.106]$$

where $\mathbf{K}_n \|\mathbf{e}\|\, \dot{\mathbf{e}}_r$ is an additional nonlinear feedback component to compensate for the errors introduced by the modification of the original adaptive control law.

The control law [14.105] can be computed by the Newton-Euler algorithm. The adaptation law [14.106] requires the computation of the elements of Φ related to the dynamic parameters that have to be adapted. The corresponding computational cost of both laws for a six degree-of-freedom robot such as the Stäubli RX-90 is about 700 additions and 950 multiplications. These figures can considerably be reduced if only the parameters of the payload are adapted [El Serafi 91b].

14.7. Conclusion

Although the controllers of most present day industrial robots are merely designed from linear control theory, more advanced methods must be developed to cope with the nonlinear nature of the articulated structures, namely for applications requiring high dynamic performances (cycle time, dynamic accuracy...).

We presented in this chapter three methods achieving this objective: computed-torque or dynamic control, passivity-based control and adaptive control. The implementation of such controls requires on-line computation of the inverse dynamic model, which can be carried out according to the algorithms proposed in Chapter 9. In order to estimate the dynamic parameters, we make use of the techniques described in Chapter 12.

We assumed that the system and the controller are continuous. In practice, the control is achieved by a computer, which introduces time delays due to data acquisition and control law computation. The effect of these delays on the process performance is an issue of the sampling control theory and is out of the scope of this book. However, from an implementation viewpoint, the sampling period should be small enough with respect to the bandwidth of the mechanical system. Typically, a frequency close to 1000 Hz has been used for the controller of the Acma SR400 robot [Restrepo 96]. Note that the use of a high frequency allows us to increase the value of the feedback gains and results in a more robust control [Samson 87].

All the control laws presented in this chapter rely on the availability of joint positions and velocities. All the robots are equipped with high precision sensors for joint position measurements. On the other hand, the tachometers used for joint velocity measurements provide noisy signals. Therefore, it is better to generate the velocity signal by numerical differentiation of the position measurements. Other sophisticated techniques consist of designing a velocity observer from the input torque and the joint position data [Nicosia 90], [Canudas de Wit 92], [Berguis 93], [Khelfi 95], [Cherki 96].

In this chapter, we only considered rigid robots. For further reading about the control of robots with flexible joints, refer for example to [Benallegue 91], [Brogliato 91], [Zodiac 96].

Chapter 15

Compliant motion control

15.1. Introduction

Many industrial applications require the contact of the robot end-effector with an uncertain environment. A long list of such applications could be given, including contour following, pushing, polishing, twisting, deburring, grinding, assembling, etc. Implementation of all these tasks intrinsically necessitates that the robot follows the desired path while providing the force necessary either to overcome the resistance from the environment or to comply with it. In order to control force with purely position-based systems, a precise model of the mechanism and knowledge of the exact location and stiffness of the environment are required. High precision robots can be manufactured only at the expense of size, weight and cost. The ability to control the contact forces generated on the end-effector offers an alternative for extending effective precision. A classification of robot force control algorithms includes:

- methods involving the relation between position and applied force: passive stiffness control, active stiffness control;
- methods using the relation between velocity and applied force: impedance control or accommodation control;
- methods using position and force feedback: parallel hybrid position/force control and external hybrid control;
- methods using force feedback: explicit force control;
- methods based on passivity.

In this chapter we will develop the first three methods that constitute the most commonly used. For more details, the reader can refer to [Siciliano 00].

15.2. Description of a compliant motion

In pure position control, the user has to completely specify the end-effector position and orientation. This implies that the robot moves in free space. The absence of any contact prevents the exertion of forces. On the other hand, in pure force control the manipulator end-effector is constrained by the environment in all directions; hence, there is no motion at all.

Between the extremes of free space and totally constrained space is the workspace with constraint surfaces, termed *C-surfaces* [Mason 82]. In this case, motion is possible along the C-surface tangents, while force can be exerted along the C-surface normals. Thus, position control and force control exclude themselves mutually: we cannot control a force and a position along the same direction simultaneously. Consequently, compliant tasks require control of the end-effector forces along some directions and its motion along others.

Practically, a compliant task is defined in a frame, called a *compliance frame*, providing six degrees of freedom along and around the frame axes. For every degree of freedom, we specify either a position or a force. According to the task, this frame can be attached to the end-effector, to the environment or to the manipulated object (Figure 15.1).

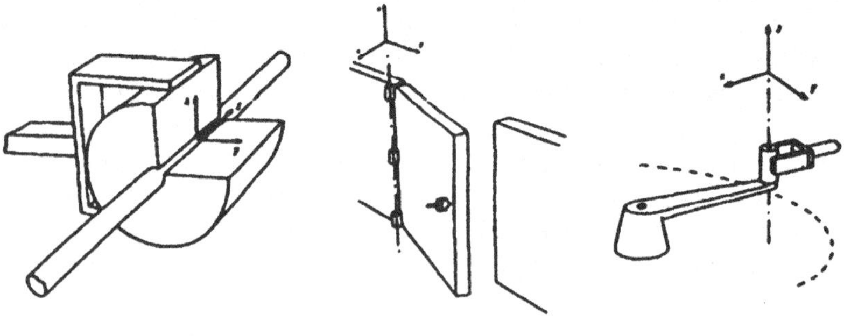

a) *b)* *c)*

Figure 15.1. *Choice of compliance frame according to the task (from [Mason 82])*

15.3. Passive stiffness control

Passive stiffness control or passive compliance is a simple solution to reduce the contact forces between the robot and its environment. It consists of interposing between the manipulated part and the robot a mechanical device able to change its configuration under the effect of contact forces, thus adding to the structure an elastic behavior that compensates for positioning errors [Drake 77], [Whitney 79]. Figure 15.2 shows the principle of such a device, the so-called *RCC (Remote*

Compliance Center) [Nevins 77], that is typically used to handle peg-in-hole assembly problems. The basic compliance formulation follows from a generalization of the linear spring equation and is given by:

$$\mathbf{dX} = C\mathbb{f} \qquad [15.1]$$

where C is the (6x6) compliance matrix, $\mathbb{f} = [\ \mathbf{f}^T \quad \mathbf{m}^T\ ]^T$ represents the wrench that is composed of a force $\mathbf{f}$ and a moment $\mathbf{m}$ (§ 2.6). The differential displacement vector $\mathbf{dX} = [\ \mathbf{dP}^T \quad \boldsymbol{\delta}^T\ ]^T$ is composed of the differential translation vector $\mathbf{dP}$ and the differential orientation vector $\boldsymbol{\delta}$ (§ 2.5).

The compliance matrix C is diagonal with respect to the compliance frame whose origin O_c is called the *compliance center*: the application of a force at O_c along a given direction causes a pure translation in this direction; the application of a moment causes a pure rotation around an axis passing through O_c.

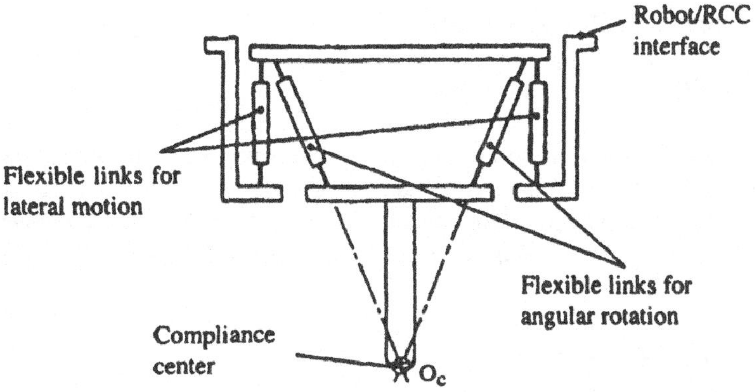

Figure 15.2. *Principle of the RCC device (from [Nevins 77])*

Passive compliance offers some advantages such as fast and accurate insertions of parts without requiring complex strategy (typically, less than 0.2 sec. for tolerance of the order of 1/100 mm). It has achieved success in specific assembly tasks, for example inserting a peg in a hole. The limitation is that each compliant device is devoted to a given task and to a given workpiece.

15.4. Active stiffness control

This method actively controls the apparent stiffness of the robot end-effector and allows simultaneous position and force control. The user can specify the three translational and three rotational stiffnesses of a desired compliance frame. Stiffness may be changed under program control to match varying task requirements [Salisbury 80]. High gain is assigned to the directions that have to be position controlled, while low gain is assigned to the force controlled directions. The basic stiffness formulation is given by:

$$ {}^{c}\mathbf{f}_{c} = \mathbf{K}_{c}\, {}^{c}\mathbf{dX}_{c} \qquad [15.2] $$

where $\mathbf{K}_c$ is the desired (6x6) stiffness matrix, which is diagonal in frame R_c. The wrench ${}^{c}\mathbf{f}_{c}$ and the differential displacement ${}^{c}\mathbf{dX}_{c}$ are expressed in frame R_c, and will be simply denoted by $\mathbf{f}$ and $\mathbf{dX}$ respectively. Assuming that the friction and dynamic forces are compensated or are small enough to be neglected, equation [5.43] gives the joint torque $\mathbf{\Gamma}$ necessary to apply a wrench $\mathbf{f}$:

$$ \mathbf{\Gamma} = \mathbf{J}^{T}\, \mathbf{f} \qquad [15.3] $$

Let us recall the differential model [5.2]:

$$ \mathbf{dX} = \mathbf{J}\, \mathbf{dq} \qquad [15.4] $$

where $\mathbf{J}$ is the Jacobian matrix of the robot describing the differential translational and rotational vectors of the compliance frame as a function of the differential variations of joint positions $\mathbf{dq}$. It should be noted that the Jacobian may be computed for any point fixed in the end-effector frame. Combining equations [15.2], [15.3] and [15.4], we obtain:

$$ \mathbf{\Gamma} = \mathbf{J}^{T}\mathbf{K}_{c}\,\mathbf{J}\,\mathbf{dq} = \mathbf{K}_{q}\,\mathbf{dq} \qquad [15.5] $$

The matrix $\mathbf{K}_q$ is called the *joint stiffness matrix* and is not diagonal but symmetric. It determines the proportional gains of the servo loops in the joint space. It presents the same singular positions as the Jacobian matrix of the kinematic model, which means that, for these configurations, we cannot get the desired stiffness along or about all the degrees of freedom of the compliance frame. The principle of this control scheme is shown in Figure 15.3. The joint torque vector is given by:

$$ \mathbf{\Gamma} = \mathbf{K}_{q}(\mathbf{q}^{d} - \mathbf{q}) + \mathbf{K}_{d}(\dot{\mathbf{q}}^{d} - \dot{\mathbf{q}}) + \mathbf{Q} \qquad [15.6] $$

where $\mathbf{Q}$ represents gravity torque compensation, and $\mathbf{K}_d$ can be interpreted as a damping matrix. A feedforward force term can be added if pure force control is desired in some direction. It is computed using equation [15.3].

The advantage of such an active stiffness control scheme is that it is relatively simple to implement. The stiffness matrix can be changed on-line to adapt the robot behavior to various task constraints.

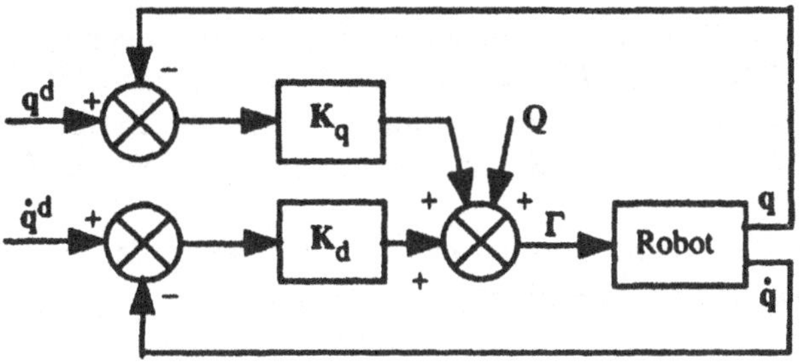

Figure 15.3. *Principle of the active stiffness control scheme*

15.5. Impedance control

According to Hogan [Hogan 85], [Hogan 87], the basic idea of impedance control is to assign a prescribed dynamic behaviour for the robot while its effector is interacting with the environment. The desired performance is specified by a generalized dynamic impedance representing a mass-spring-damper system.

The end-effector velocity $\dot{\mathbf{X}}$ and the applied force are related by a mechanical impedance $\mathbf{Z}$. In the frequency domain, this is represented by:

$$\mathbf{F}(s) = \mathbf{Z}(s)\,\dot{\mathbf{X}}(s) \qquad [15.7a]$$

In terms of position $\mathbf{X}(s)$, we can write:

$$\mathbf{F}(s) = s\,\mathbf{Z}(s)\,\mathbf{X}(s) \qquad [15.7b]$$

The robot should behave like a mechanical system whose impedance $\mathbf{Z}$ is variable according to the different phases of the task. In general, we suppose that the robot is equivalent to a mass-spring-damper second order system, whose transfer function is:

$$s\,\mathbf{Z}(s) = \mathbf{\Lambda}\,s^2 + \mathbf{B}\,s + \mathbf{K} \qquad [15.8]$$

where $\mathbf{\Lambda}$, $\mathbf{B}$ and $\mathbf{K}$ represent the desired inertia, damping and stiffness matrices respectively. The values of these matrices are chosen to obtain the desired performance:

- high values are given to Λ in the directions where a contact is expected in order to limit the dynamics of the robot;
- high values are given to **B** in the directions where it is necessary to dissipate the kinetic energy and therefore to damp the response;
- the stiffness **K** affects the accuracy of the position control: along the force controlled directions, the stiffness should be small enough to limit the contact forces; conversely, along the position controlled directions, the user should set a high stiffness to obtain an accurate positioning of the end-effector.

Two families of control schemes can be implemented depending on whether or not a force sensor is available (Figures 15.4 and 15.5).

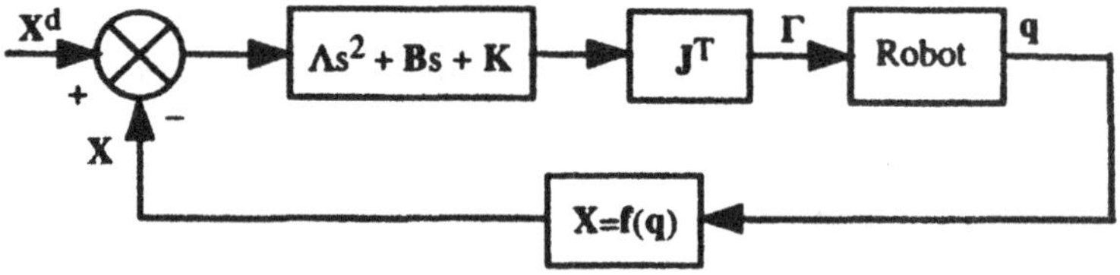

Figure 15.4. *Impedance control scheme without force feedback*

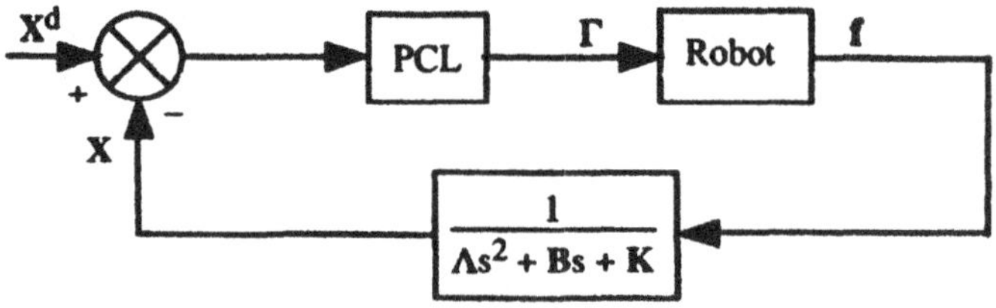

PCL: Position Control Law

Figure 15.5. *Impedance control scheme with force feedback*

Figure 15.6 shows an implementation of the impedance control scheme of the first family. The dynamics of the robot is neglected. The control law is given by:

$$\mathbf{\Gamma} = \mathbf{J}^{T} [\mathbf{B} (\dot{\mathbf{X}}^{d} - \dot{\mathbf{X}}) + \mathbf{K} (\mathbf{X}^{d} - \mathbf{X})] + \mathbf{Q} \qquad [15.9]$$

The **K** and **B** matrices contain the proportional and derivative gains in the task space, which can be interpreted as the stiffness matrix and the damping matrix of the robot respectively. As previously, the vector **Q** represents gravity torque compensation. This control scheme places the compliance center at the desired point $\mathbf{X}^{d}$. It is equivalent to the PD control in the task space (§ 14.3.3).

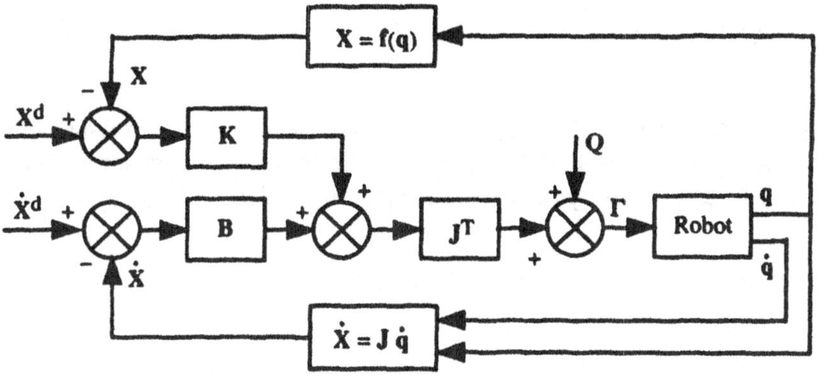

Figure 15.6. *Impedance control scheme without force sensor feedback and with a PD control law in the task space*

In the following, we present two forms of impedance control representative of the second family, using the dynamic model of the robot in the joint space, then in the task space.

Let us note that the dynamic model of a robot exerting a wrench $\mathbf{f}$ on its environment is written as (Chapter 9):

$$\Gamma = \mathbf{A}(\mathbf{q})\,\ddot{\mathbf{q}} + \mathbf{C}(\mathbf{q},\dot{\mathbf{q}})\,\dot{\mathbf{q}} + \mathbf{Q}(\mathbf{q}) + \mathbf{J}^T\mathbf{f} \qquad [15.10]$$

The desired behavior is deduced from equation [15.7b] as:

$$\mathbf{f} = \Lambda\,(\ddot{\mathbf{X}}^d - \ddot{\mathbf{X}}) + \mathbf{B}\,(\dot{\mathbf{X}}^d - \dot{\mathbf{X}}) + \mathbf{K}\,(\mathbf{X}^d - \mathbf{X}) \qquad [15.11]$$

which leads to:

$$\ddot{\mathbf{X}}(t) = \ddot{\mathbf{X}}^d + \Lambda^{-1}\,[\mathbf{B}\,(\dot{\mathbf{X}}^d - \dot{\mathbf{X}}) + \mathbf{K}\,(\mathbf{X}^d - \mathbf{X}) - \mathbf{f}] \qquad [15.12]$$

where $\mathbf{X}^d(t)$ is the desired trajectory.

To achieve this impedance control scheme, let us consider the decoupling nonlinear control law in the task space of equation [14.33] (resolved acceleration control law) in which $\mathbf{w}(t)$ is replaced by equation [15.12], and the external wrench exerted by the robot on the environment $\mathbf{J}^T\mathbf{f}$ is taken into account:

$$\Gamma = \hat{A}J^{-1}\{\ddot{X}^d + \Lambda^{-1}[B(\dot{X}^d - \dot{X}) + K(X^d - X) - f] - \dot{J}\,\dot{q}\} + \hat{H}(q, \dot{q}) + J^T f \tag{15.13}$$

Another formulation of this control law can be derived from the dynamic model in the task space [Zodiac 96]. Combining the first and second order kinematic models (§ 5.10) with the dynamic equation [15.10], yields:

$$J^{-T}\Gamma = A_x(q)\,\ddot{X} + C_x(q, \dot{q})\,\dot{X} + Q_x(q) + f \tag{15.14}$$

where:

- J^{-T} is the inverse of J^T;
- $A_x(q)$ is the inertia matrix in the task space[1] equal to $J^{-T} A(q) J^{-1}$;
- $C_x(q, \dot{q})$ is the vector of Coriolis and centrifugal torques in the task space. It is equal to $[J^{-T} C(q, \dot{q}) J^{-1} - A_x(q)\,\dot{J}\,J^{-1}]$;
- $Q_x(q) = J^{-T} Q(q)$ is the vector of gravity torques in the task space.

We obtain the decoupled control law as indicated in § 14.4:

$$\Gamma = J^T[\hat{A}_x(q)\,w(t) + \hat{C}_x(q, \dot{q})\,\dot{X} + \hat{Q}_x(q) + f] \tag{15.15a}$$

Replacing $w(t)$ by $\ddot{X}(t)$, as given by equation [15.12], leads to:

$$\Gamma = J^T \hat{A}_x(q)\,\{\ddot{X}^d + \Lambda^{-1}[B(\dot{X}^d - \dot{X}) + K(X^d - X)]\} + J^T[\hat{C}_x(q, \dot{q})\,\dot{X} + \hat{Q}_x(q) + (I - \hat{A}_x(q)\,\Lambda^{-1})\,f] \tag{15.15b}$$

This control scheme is represented in Figure 15.7. It is equivalent to the control scheme [15.13], which is easier to implement when the complete control law must be computed. The algorithm is similar to that presented in Appendix 10. The control scheme [15.15] is preferred in quasi-static cases, where the inertia matrix and the Jacobian matrix are roughly constant [Kazerooni 86]. Thus, $\dot{J}$ and $\hat{C}_x(q, \dot{q})\,\dot{X}$ are considered to be equal to zero, and equation [15.15b] becomes:

$$\Gamma = J^T \hat{A}_x \Lambda^{-1}[\Lambda\,\ddot{X}^d + B(\dot{X}^d - \dot{X}) + K(X^d - X) - f] + Q(q) + J^T f \tag{15.16}$$

[1] The reader can find in [Lilly 90] an efficient algorithm for computing A_x without computing the inertia matrix.

Note that the wrench $\mathbf{f}$ appears twice in this equation: once with the term $\mathbf{J}^T \mathbf{f}$ that compensates for the external wrench exerted by the robot and once with the term $-\mathbf{J}^T \hat{\mathbf{A}}_x \mathbf{\Lambda}^{-1} \mathbf{f}$, which represents a force feedback whose gain is $-\mathbf{J}^T \hat{\mathbf{A}}_x \mathbf{\Lambda}^{-1}$. Besides, note that if $\mathbf{\Lambda} = \mathbf{A}_x(\mathbf{q})$, the terms containing $\mathbf{f}$ vanish, which yields to the decoupled control law shown in Figure 15.7:

$$\mathbf{\Gamma} = \hat{\mathbf{A}}(\mathbf{q})\,\mathbf{J}^{-1}[\ddot{\mathbf{X}}^d + \mathbf{\Lambda}^{-1}[\mathbf{B}(\dot{\mathbf{X}}^d - \dot{\mathbf{X}}) + \mathbf{K}(\mathbf{X}^d - \mathbf{X})] - \dot{\mathbf{J}}\,\dot{\mathbf{q}}] + \hat{\mathbf{H}}(\mathbf{q}, \dot{\mathbf{q}}) \qquad [15.17]$$

NOTES.–

- a long time before the formulation of the impedance control by Hogan, particular cases of this control had been proposed in the literature such as those based on a stiffness matrix or on a damping matrix [Whitney 85]. In the former, $\mathbf{\Lambda} = \mathbf{0}$ and $\mathbf{B} = \mathbf{0}$; in the latter, $\mathbf{\Lambda} = \mathbf{0}$ and $\mathbf{K} = \mathbf{0}$;
- the active stiffness control proposed by Salisbury (§ 15.2.2) is also a particular case of impedance control where $\mathbf{\Lambda} = \mathbf{0}$ and $\mathbf{B} = \mathbf{0}$;
- impedance control is similar to resolved acceleration control with the only difference being the inclusion of the desired inertia in the force gain.

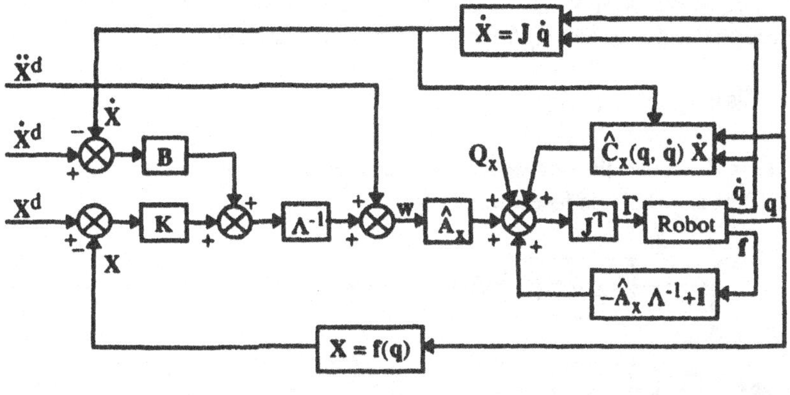

Figure 15.7. *Nonlinear decoupling impedance control without force feedback*

15.6. Hybrid position/force control

Using the previous methods, we can specify a desired dynamic behavior of the robot but we cannot prescribe a desired wrench. In the following, we address some methods where both position and force can be specified. Much work has been carried out on this topic such as that of [Nevins 73], [Reboulet 85], [Merlet 86], [Robert 86], [Perdereau 91], [Dégoulange 93], [Morel 94], etc. Two families of control schemes with force control loops are introduced: parallel hybrid position/force control and external hybrid control.

15.6.1. *Parallel hybrid position/force control*

The parallel hybrid position/force control finds its roots in the work of Raibert and Craig [Raibert 81]. It satisfies simultaneously the desired position and force constraints of the task. Positions and forces are specified according to the Mason formulation: directions that are constrained in position are force controlled, while those that are constrained in force (zero force) are position or velocity controlled. Duffy [Duffy 90] has shown that it is not correct to consider the velocity subspace and the force subspace as orthogonal as suggested in [Raibert 81]. Rather, it is the position or velocity controlled directions and the force controlled directions that have to be orthogonal in the compliance frame.

In the parallel hybrid control method, the robot is controlled by two complementary feedback loops, one for the position, the other for the force. Each has its own sensory system and control law. The control laws of both loops are added before being sent to the actuator as a global control signal **G** (Figure 15.8). Each degree of freedom of the compliant frame is controlled by the position or force loop through the use of a *compliance selection matrix* **S**, which is diagonal such that:

$$\mathbf{S} = \mathbf{diag}(s_1, s_2, \ldots, s_6) \qquad [15.18]$$

where $s_j = 1$ if the jth degree of freedom of the compliance frame is position controlled or $s_j = 0$ if it is force controlled.

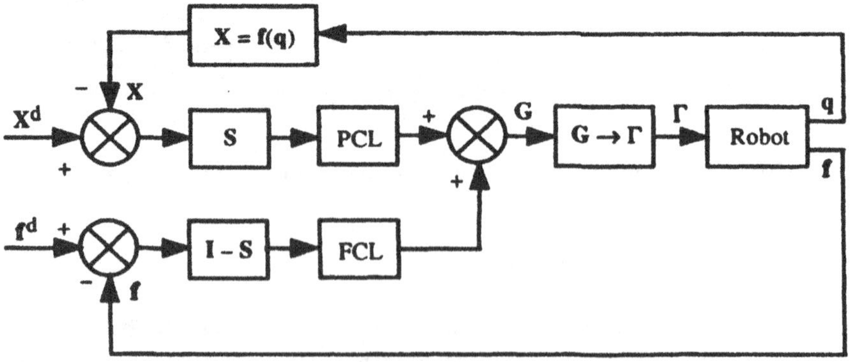

PCL: Position Control Law; FCL: Force Control Law

Figure 15.8. *Principle of the hybrid position/force control*

Since both loops act cooperatively, each joint contributes to the realization of both the position control and the force control.

Three forms of hybrid control schemes can be distinguished according to the type of the global control signal G:

- G is equivalent to joint torques (Figure 15.9);
- G is equivalent to displacements or velocities in the task space and has to be multiplied by the robot inverse Jacobian to obtain joint positions (Figure 15.10);
- G is equivalent to forces in the task space and has to be multiplied by the transpose of the Jacobian matrix (Figure 15.11).

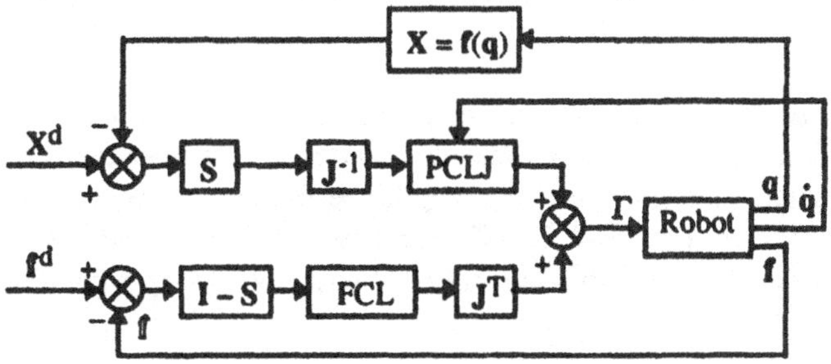

PCLJ: Position Control Law in the Joint space; FCL: Force Control Law

Figure 15.9. *Hybrid force-position control scheme with addition of joint torques (from [Raibert 81])*

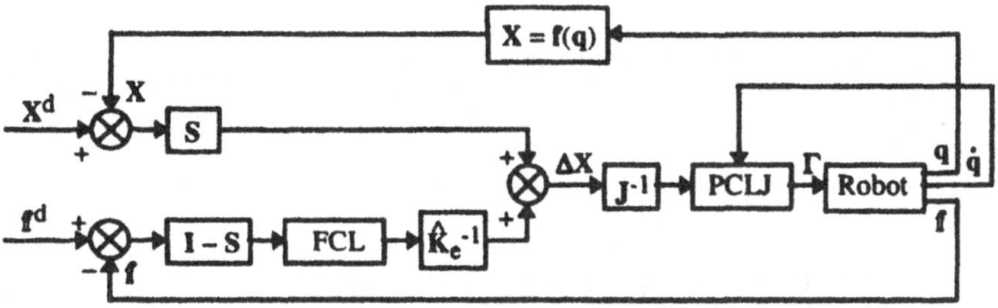

PCLJ: Position Control Law in the Joint space; FCL: Force Control Law

$\hat{K}_e^{-1}$: *estimate of the stiffness matrix of the environment*

Figure 15.10. *Hybrid force-position control scheme with addition of velocities*

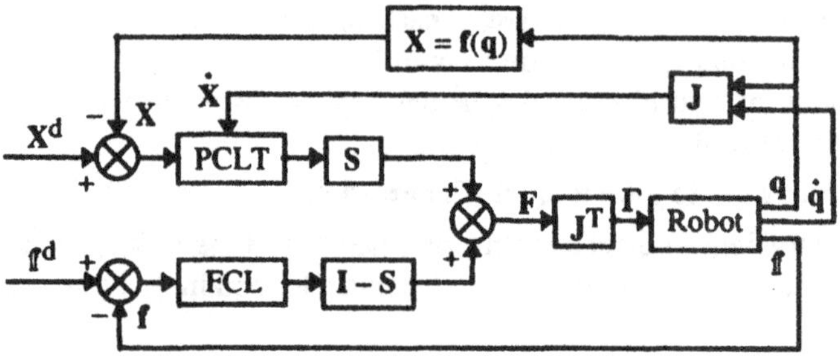

PCLT: Position Control Law in the Task space; FCL: Force Control Law

Figure 15.11. *Hybrid force-position control scheme with addition of task forces*

In these figures, the frame transformation computations for velocities, forces and for the Jacobian matrix are not indicated. Practically, the matrices **S** and (**I** – **S**) are applied to signals expressed in the compliance frame. For position control in the joint space (Figures 15.9 and 15.10), we can use one of the laws presented in Chapter 14, for example the PID controller of equation [14.5], which is:

$$\boldsymbol{\Gamma} = \mathbf{K}_p (\mathbf{q}^d - \mathbf{q}) + \mathbf{K}_d (\dot{\mathbf{q}}^d - \dot{\mathbf{q}}) + \mathbf{K}_I \int_{t0}^{t} (\mathbf{q}^d - \mathbf{q})\, d\tau \qquad [15.19]$$

whereas for a PID control in the task space (Figure 15.11), we have (equation [14.17]):

$$\boldsymbol{\Gamma} = \mathbf{J}^T [\mathbf{K}_p (\mathbf{X}^d - \mathbf{X}) + \mathbf{K}_d (\dot{\mathbf{X}}^d - \dot{\mathbf{X}}) + \mathbf{K}_I \int_{t0}^{t} (\mathbf{X}^d - \mathbf{X})\, d\tau] \qquad [15.20]$$

Normally, the force control law is chosen as:

$$\boldsymbol{\Gamma} = \mathbf{J}^T [\mathbf{f}^d + \mathbf{K}_f (\mathbf{f}^d - \mathbf{f}) - \mathbf{K}_{fd}\, \dot{\mathbf{X}} + \mathbf{K}_{fI} \int_{t0}^{t} (\mathbf{f}^d - \mathbf{f})\, d\tau] \qquad [15.21]$$

Note that, due to the noise of force sensors, the velocity in the task space is used with the derivative gain rather than the derivative of the force.

In these schemes, we can also include feedforward compensation for the nonlinear dynamics of the robot. For example, the position loop of the hybrid control of Figure 15.11 may be realized by the nonlinear decoupling control law in the task space described in § 14.4.3. The corresponding block diagram is given in

Figure 15.12 [Khatib 87]. The computation of the control vector Γ can be achieved with the Newton-Euler algorithm in a similar way to that described in Appendix 10, with the following arguments:

- the joint position is equal to the current joint position $\mathbf{q}$;
- the joint velocity is equal to the current joint velocity $\dot{\mathbf{q}}$;
- the joint acceleration is equal to:

$$\ddot{\mathbf{q}} = \mathbf{J}^{-1}\,\mathbf{S}\,[\ddot{\mathbf{X}}^d + \mathbf{K}_d\,(\dot{\mathbf{X}}^d - \dot{\mathbf{X}}) + \mathbf{K}_p\,(\mathbf{X}^d - \mathbf{X}) - \dot{\mathbf{J}}\,\dot{\mathbf{q}}] \qquad [15.22]$$

- the force exerted by the terminal link on the environment can be taken to be equal to:

$$\mathbf{f}_{en} = (\mathbf{I} - \mathbf{S})\,[\mathbf{f}^d + \mathbf{K}_f\,(\mathbf{f}^d - \mathbf{f}) - \mathbf{K}_{fd}\,\dot{\mathbf{X}} + \mathbf{K}_{fI}\int_{t0}^{t}(\mathbf{f}^d - \mathbf{f})\,d\tau] \qquad [15.23]$$

where all terms are computed in the compliance frame R_c.

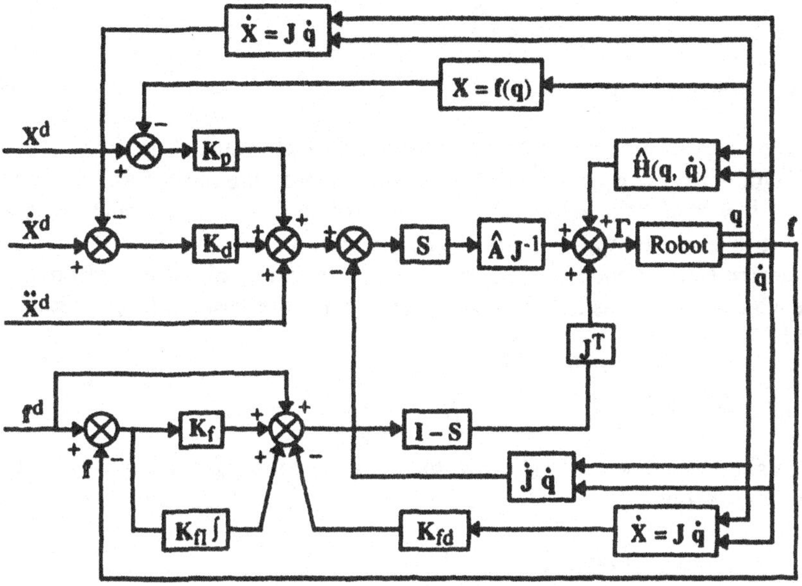

Figure 15.12. *Implementation of the dynamic hybrid position-force control scheme*

An and Hollerbach [An 87] showed that the control schemes of Figures 15.9 and 15.10, which require the inverse of the Jacobian matrix, have an unstable behavior when implemented on a robot with revolute joints, even in non-singular configurations. They assigned this instability to an interaction between the inertia matrix and the inverse kinematic model $\mathbf{J}^{-1}$, whereas the scheme using $\mathbf{J}^T$ (Figure 15.11) always produces stable results. Fisher and Mutjaba [Fisher 92] showed that this instability comes from the formulation of the inverse kinematic model in the position loop of the hybrid scheme. Using the selection matrix **S** to separate position and force requirements in the task space is conceptually straightforward. Geometrically, the selection matrix is a projection that reduces the task space to a desired subspace of interest. Problems may arise when this selected task subspace is mapped onto the joint space using the robot Jacobian matrix. From the classical scheme of Figure 15.9 and equation [5.2], we can write that:

$$\mathbf{S}\,\mathbf{dX} = (\mathbf{SJ})\,\mathbf{dq} \qquad [15.24]$$

From this equation, it can be seen that the selection matrix **S** reduces the task space of the robot, which becomes redundant with respect to the displacement task. Thus, instabilities of the hybrid control scheme of Craig and Raibert are the consequence of an erroneous formulation of the projection of the task error vector into the joint space. In fact, knowing that $(\mathbf{SJ})^+\,\mathbf{S} = (\mathbf{SJ})^+$, the general solution of [15.24] is:

$$\mathbf{dq} = (\mathbf{SJ})^+\,\mathbf{dX} + [\mathbf{I} - (\mathbf{SJ})^+\,(\mathbf{SJ})]\,\mathbf{Z} \qquad [15.25]$$

Fisher and Mutjaba [Fisher 92] showed that choosing $\mathbf{Z} = \mathbf{J}^{-1}\,\mathbf{S}\,\mathbf{dX}$ as the optimization term in equation [15.25] is equivalent to the inverse kinematic relation $\mathbf{dq} = \mathbf{J}^{-1}\,\mathbf{S}\,\mathbf{dX}$. This choice of **Z** does not ensure stability and explains the instabilities that can appear with the hybrid control scheme. Indeed, they showed that the first term of equation [15.25], which is the minimal norm solution, is always stable. Consequently, as indicated in Figure 15.13, the position loop reference input should be:

$$\mathbf{dq} = (\mathbf{SJ})^+\,\mathbf{dX} \qquad [15.26a]$$

In a similar manner, they showed that the force loop reference input could remain as the original one:

$$\mathbf{d}\boldsymbol{\gamma} = [(\mathbf{I} - \mathbf{S})\,\mathbf{J}]^T\,(\mathbf{f}^d - \mathbf{f}) \qquad [15.26b]$$

More general solutions with optimization terms are:

$$\mathbf{dq} = (\mathbf{SJ})^+\,\mathbf{dX} + [\mathbf{I} - \mathbf{J}^+\,\mathbf{J}]\,\mathbf{Z}_q \qquad [15.27a]$$

$$\mathbf{d}\boldsymbol{\gamma} = [(\mathbf{I} - \mathbf{S})\,\mathbf{J}]^T\,(\mathbf{f}^d - \mathbf{f}) + [\mathbf{I} - \mathbf{J}^+\,\mathbf{J}]\,\mathbf{Z}_f \qquad [15.27b]$$

where Z_q and Z_f are arbitrary position and force vectors in the joint space respectively.

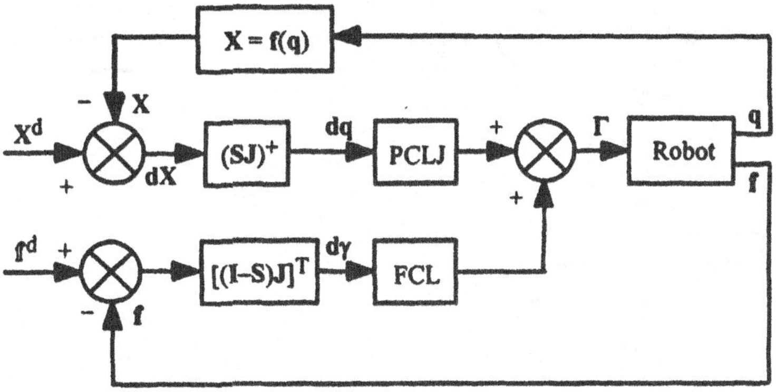

PCLJ: Position Control Law in the Joint space; FCL: Force Control Law

Figure 15.13. *Hybrid force-position control scheme (from [Fisher 92])*

15.6.2. *External hybrid control scheme*

The external hybrid control scheme is composed of two embedded control loops [De Schutter 88], [Perdereau 91]: the outer loop controls force while the inner one controls position (Figure 15.14). The output of the outer loop is transformed into a desired position input for the inner loop. The resulting displacement of the robot permits exertion of the desired contact force on the environment. The external hybrid control scheme is relatively easy to implement and requires a rather small amount of computation. It can be implemented in industrial robots while keeping their conventional controllers [Thérond 96].

The position control loop can be achieved either in the task space or in the joint space by implementing one of the methods presented in Chapter 14.

The additional displacement reference signal is given by:

$$dX_f = \hat{K}_e^{-1} \left[f^d + K_f (f^d - f) + K_{fI} \int_{t0}^{t} (f^d - f)\, d\tau\right] \qquad [15.28]$$

where $\hat{K}_e$ is an estimate of the stiffness of the environment.

Thanks to the integral force action, the wrench error $(f^d - f)$ is allowed to prevail over the position error $(X^d - X)$ at steady state [Pujas 95].

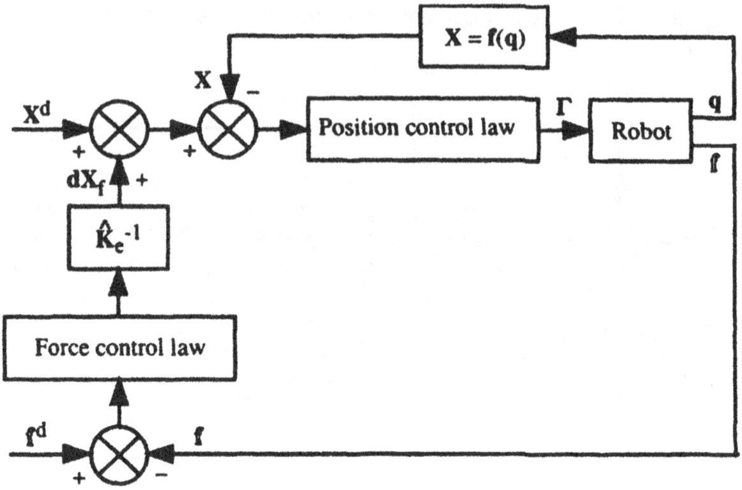

Figure 15.14. *Principle of the external hybrid control scheme*

This control can be applied with a nonlinear decoupling impedance control in the task space [Chiaverini 93] by setting w(t) as the sum of two terms, $w_X(t)$ and $w_F(t)$, which are the contributions of the position loop and the force loop respectively:

$$w(t) = w_X(t) + w_F(t) \qquad [15.29]$$

$$w_X(t) = \ddot{X}^d + \Lambda^{-1}[K_d(\dot{X}^d - \dot{X}) + K_p(X^d - X)] \qquad [15.30]$$

$$w_F(t) = \Lambda^{-1}[K_f(f^d - f) + K_{fI}\int_{t0}^{t}(f^d - f)\,d\tau] \qquad [15.31]$$

Note that $w_F(t)$ is obtained by multiplying the force signal by Λ^{-1} because it is equivalent to an acceleration. The decoupled control law is obtained from equation [15.15] as:

$$\Gamma = J^T[\hat{A}_x(q)(w_X + w_F) + \hat{C}_x(q, \dot{q})\dot{X} + \hat{Q}_x(q) + f] \qquad [15.32]$$

If the robot dynamic model is perfectly known, combining equations [15.30], [15.31] and [15.32] yields:

$$\Lambda(\ddot{X}^d - \ddot{X}) + K_d(\dot{X}^d - \dot{X}) + K_p(X^d - X) + K_f(f^d - f) + K_{fI}\int_{t0}^{t}(f^d - f)\,d\tau = 0$$

[15.33]

When $\mathbf{K_f} = \mathbf{0}$ and $\mathbf{K_{fl}} = \mathbf{0}$, the control law [15.32] becomes equivalent to the impedance control (Figure 15.7). Besides, if $\Lambda = \mathbf{I}$, it reduces to the decoupling nonlinear control in the task space such as that shown in Figure 14.5.

15.7. Conclusion

In this chapter, we have presented the most popular position/force control approaches. For other methods, the reader should refer to [Brogliato 91], [Siciliano 96a], for the passive force control, to [Siciliano 93], [Colbaugh 93], [Arimoto 93], [Siciliano 96b] for the adaptive force control or to [Volpe 93], [Volpe 95] for explicit force control.

We did not address the stability problem of force control and the interested reader should refer to [Wen 91], [Yabuta 92], [Wang 93], [Zodiac 96], [Siciliano 00].

The problem of exerting a force on a moving target, thus of controlling simultaneously force and velocity along the same direction, is addressed in the work of [de Luca 91b].

Among the yet open problems, we have to mention the control of impact when the robot and the environment enter in contact. Another class of problem concerns the programming of compliant tasks: the choice of the axes of the compliance frame and their roles requires from the user a lot of experience and is much more difficult, in terms of abstraction capabilities, than programming displacements. Besides, some physical parameters, like the stiffness of the robot and the environment, are not easy to quantify, which results in instability problems.

Appendix 1

Solution of the inverse geometric model equations (Table 4.1)

A1.1. Type 2

The equation to be solved is:

$$X\,S\theta_i + Y\,C\theta_i = Z \qquad [A1.1]$$

Four cases are possible:

i) if $X = 0$ and $Y \neq 0$, we can write that:

$$C\theta_i = \frac{Z}{Y} \qquad [A1.2]$$

yielding:

$$\theta_i = \mathrm{atan2}(\pm\sqrt{1-(C\theta_i)^2},\ C\theta_i) \qquad [A1.3]$$

ii) if $Y = 0$ and $X \neq 0$, we obtain:

$$S\theta_i = \frac{Z}{X} \qquad [A1.4]$$

yielding:

$$\theta_i = \mathrm{atan2}(S\theta_i,\ \pm\sqrt{1-(S\theta_i)^2}) \qquad [A1.5]$$

iii) if X and Y are not zero, and Z = 0:

$$\begin{cases} \theta_i = \text{atan2}(-Y, X) \\ \theta_i' = \theta_i + \pi \end{cases} \quad [A1.6]$$

(if X = Y = 0, the robot is in a singular configuration);

iv) if X, Y and Z are not zero, we can write that [Gorla 84]:

$$Y\, C\theta_i = Z - X\, S\theta_i \quad [A1.7]$$

Squaring the equation leads to:

$$Y^2 C^2\theta_i = Y^2 (1 - S^2\theta_i) = Z^2 - 2Z\, X\, S\theta_i + X^2 S^2\theta_i \quad [A1.8]$$

Therefore, we have to solve a second degree equation in $S\theta_i$. Likewise, we can write an equation in $C\theta_i$. Finally, we obtain:

$$\begin{cases} S\theta_i = \dfrac{XZ + \varepsilon Y \sqrt{X^2 + Y^2 - Z^2}}{X^2 + Y^2} \\ C\theta_i = \dfrac{YZ - \varepsilon X \sqrt{X^2 + Y^2 - Z^2}}{X^2 + Y^2} \end{cases} \quad [A1.9]$$

with $\varepsilon = \pm 1$ (it is straightforward to verify that two combinations of $S\theta_i$ and $C\theta_i$ can only satisfy the original equation). If $X^2 + Y^2 \leq Z^2$, there is no solution. Otherwise, the solution is given by:

$$\theta_i = \text{atan2}(S\theta_i, C\theta_i) \quad [A1.10]$$

A1.2. Type 3

The system of equations to be solved is the following:

$$\begin{cases} X1\, S\theta_i + Y1\, C\theta_i = Z1 \\ X2\, S\theta_i + Y2\, C\theta_i = Z2 \end{cases} \quad [A1.11]$$

Multiplying the first equation by Y2 and the second by Y1, under the condition that $X1Y2 - X2Y1 \neq 0$, yields:

$$S\theta_i = \frac{Z1\, Y2 - Z2\, Y1}{X1\, Y2 - X2\, Y1} \quad [A1.12]$$

then, multiplying the first equation by X2 and the second by X1, yields:

$$C\theta_i = \frac{Z2\ X1 - Z1\ X2}{X1\ Y2 - X2\ Y1} \qquad [A1.13]$$

Thus:

$$\theta_i = \text{atan2}(S\theta_i, C\theta_i) \qquad [A1.14]$$

The condition $X1Y2 - X2Y1 \neq 0$ means that the two equations of [A1.11] are independent. If it is not the case, we solve one of these equations as a type-2 equation.

In the frequent case where Y1 and X2 are zero, the system [A1.11] reduces to:

$$\begin{cases} X1\ S\theta_i = Z1 \\ Y2\ C\theta_i = Z2 \end{cases} \qquad [A1.15]$$

whose solution is straightforward:

$$\theta_i = \text{atan2}\left(\frac{Z1}{X1}, \frac{Z2}{Y2}\right) \qquad [A1.16]$$

A1.3. Type 4

The system of equations to be solved is given by:

$$\begin{cases} X1\ r_j\ S\theta_i = Y1 \\ X2\ r_j\ C\theta_i = Y2 \end{cases} \qquad [A1.17]$$

We first compute r_j by squaring both equations and adding them; then, we obtain θ_i by solving a type-3 system of equations:

$$\begin{cases} r_j = \pm\sqrt{(Y1/X1)^2 + (Y2/X2)^2} \\ \theta_i = \text{atan2}\left(\frac{Y1}{X1\ r_j}, \frac{Y2}{X2\ r_j}\right) \end{cases} \qquad [A1.18]$$

A1.4. Type 5

The system of equations to be solved is as follows:

$$\begin{cases} X1\, S\theta_i = Y1 + Z1\, r_j \\ X2\, C\theta_i = Y2 + Z2\, r_j \end{cases} \qquad [A1.19]$$

Let us normalize the equations such that:

$$\begin{cases} S\theta_i = V1 + W1\, r_j \\ C\theta i = V2 + W2\, r_j \end{cases} \qquad [A1.20]$$

After squaring both equations and adding them, we obtain a second degree equation in r_j, which can be solved if:

$$[W1^2 + W2^2 - (V1\, W2 - V2\, W1)^2] > 0 \qquad [A1.21]$$

Then, we obtain θ_i by solving a type-3 system of equation.

A1.5. Type 6

The system of equations is given by:

$$\begin{cases} W\, S\theta_j = X\, C\theta_i + Y\, S\theta_i + Z1 \\ W\, C\theta_j = X\, S\theta_i - Y\, C\theta_i + Z2 \end{cases} \qquad [A1.22]$$

with $Z1 \neq 0$ and/or $Z2 \neq 0$. By squaring both equations and adding them, we obtain a type-2 equation in θ_j:

$$B1\, S\theta_i + B2\, C\theta_i = B3 \qquad [A1.23]$$

with:

$B1 = 2\,(Z1\, Y + Z2\, X)$
$B2 = 2\,(Z1\, X - Z2 Y)$
$B3 = W^2 - X^2 - Y^2 - Z1^2 - Z2^2$

Knowing θ_i, we obtain θ_j by solving a type-3 system of equation.

A1.6. Type 7

The system of equations is the following:

$$\begin{cases} W1\, C\theta_j + W2\, S\theta_j = X\, C\theta_i + Y\, S\theta_i + Z1 \\ W1\, S\theta_j - W2\, C\theta_j = X\, S\theta_i - Y\, C\theta_i + Z2 \end{cases} \qquad [A1.24]$$

It is a generalized form of a type-6 system. Squaring both equations and adding them gives a type-2 equation in θ_i:

$$B1\, S\theta_i + B2\, C\theta_i = B3 \qquad [A1.25]$$

where $B3 = W1^2 + W2^2 - X^2 - Y^2 - Z1^2 - Z2^2$. The terms B1 and B2 are identical to those of equation [A1.23].

After solving for θ_i, we compute θ_j as a solution of a type-3 system of equation.

A1.7. Type 8

The system of equations is the following:

$$\begin{cases} X\, C\theta_i + Y\, C(\theta_i + \theta_j) = Z1 \\ X\, S\theta_i + Y\, S(\theta_i + \theta_j) = Z2 \end{cases} \qquad [A1.26]$$

By squaring both equations and adding them, θ_i vanishes, yielding:

$$C\theta_j = \frac{Z1^2 + Z2^2 - X^2 - Y^2}{2XY} \qquad [A1.27]$$

hence:

$$\theta_j = \mathrm{atan2}(\pm\sqrt{1 - (C\theta j)^2}, C\theta j) \qquad [A1.28]$$

Then, [A1.26] reduces to a system of two equations in θ_i such that:

$$\begin{cases} S\theta_i = \dfrac{B1Z2 - B2\, Z1}{B1^2 + B2^2} \\ C\theta_i = \dfrac{B1Z1 + B2Z2}{B1^2 + B2^2} \end{cases} \qquad [A1.29]$$

with $B1 = X + Y\, C\theta_j$ and $B2 = Y\, S\theta_j$. Finally:

$$\theta_i = \mathrm{atan2}(S\theta_i, C\theta_i) \qquad [A1.30]$$

Appendix 2

The inverse robot

The n degree-of-freedom robot whose set of geometric parameters are (σ_j', α_j', d_j', θ_j', r_j') is defined as the inverse of the robot (σ_j, α_j, d_j, θ_j, r_j) if the transformation matrix ${}^0T_n(\sigma_j', \alpha_j', d_j', \theta_j', r_j')$ is equal to ${}^0T_n^{-1}(\sigma_j, \alpha_j, d_j, \theta_j, r_j)$.

Table A2.1 gives the geometric parameters of a general six degree-of-freedom robot. Table A2.2 gives those of the corresponding inverse robot. Indeed, let us write the transformation matrix ${}^0T_6$ under the following form:

$$\begin{aligned}{}^0T_6 = &\mathbf{Rot}(z,\theta_1)\ \mathbf{Trans}(z,r_1)\ \mathbf{Rot}(x,\alpha_2)\ \mathbf{Trans}(x,d_2)\ \mathbf{Trans}(z,r_2)\ \mathbf{Rot}(z,\theta_2)\ \ldots\ \mathbf{Rot}(x,\alpha_6)\\ &\mathbf{Trans}(x,d_6)\ \mathbf{Trans}(z,r_6)\ \mathbf{Rot}(z,\theta_6)\end{aligned} \quad [A2.1]$$

Table A2.1. *Geometric parameters of a general six degree-of-freedom robot*

j	σ_j	α_j	d_j	θ_j	r_j
1	σ_1	0	0	θ_1	r_1
2	σ_2	α_2	d_2	θ_2	r_2
3	σ_3	α_3	d_3	θ_3	r_3
4	σ_4	α_4	d_4	θ_4	r_4
5	σ_5	α_5	d_5	θ_5	r_5
6	σ_6	α_6	d_6	θ_6	r_6

The inverse transformation matrix ${}^6T_0$ can be written as:

$$\begin{aligned}{}^6T_0 = &\mathbf{Rot}(z,-\theta_6)\ \mathbf{Trans}(z,-r_6)\ \mathbf{Trans}(x,-d_6)\ \mathbf{Rot}(x,-\alpha_6)\ \mathbf{Rot}(z,-\theta_5)\\ &\mathbf{Trans}(z,-r_5)\ \ldots\ \mathbf{Trans}(x,-d_2)\ \mathbf{Rot}(x,-\alpha_2)\ \mathbf{Rot}(z,-\theta_1)\ \mathbf{Trans}(z,-r_1)\end{aligned} \quad [A2.2]$$

The parameters of Table A2.2 result from comparing equations [A2.1] and [A2.2]. The corresponding elementary transformation matrices are denoted by ${}^{j-1}T_j{}'$ such that:

$$ {}^{0}T_6{}' = {}^{0}T_1{}' \, {}^{1}T_2{}' \ldots {}^{5}T_6{}' = {}^{0}T_6^{-1} \qquad [A2.3] $$

Table A2.2. *Geometric parameters of the six degree-of-freedom inverse robot*

j	$\sigma_j{}'$	$\alpha_j{}'$	$d_j{}'$	$\theta_j{}'$	$r_j{}'$
1	σ_6	0	0	$-\theta_6$	$-r_6$
2	σ_5	$-\alpha_6$	$-d_6$	$-\theta_5$	$-r_5$
3	σ_4	$-\alpha_5$	$-d_5$	$-\theta_4$	$-r_4$
4	σ_3	$-\alpha_4$	$-d_4$	$-\theta_3$	$-r_3$
5	σ_2	$-\alpha_3$	$-d_3$	$-\theta_2$	$-r_2$
6	σ_1	$-\alpha_2$	$-d_2$	$-\theta_1$	$-r_1$

Appendix 3

Dyalitic elimination

Let us consider the following system of equations in the two unknowns x, y:

$$\begin{cases} a\,x^2\,y^2 + b\,xy = c\,y + d \\ e\,x^2\,y^2 + f\,xy + g = 0 \end{cases} \qquad [A3.1]$$

where the coefficients a, b, ..., g are constants with arbitrary values. The so-called *dyalitic elimination technique* [Salmon 1885] consists of:

i) transforming the system [A3.1] as a linear system such that:

$$\begin{bmatrix} ax^2 & bx-c & -d \\ ex^2 & fx & g \end{bmatrix} \begin{bmatrix} y^2 \\ y \\ 1 \end{bmatrix} = \mathbf{0} \qquad [A3.2]$$

where y^2, y and 1 are termed *power products*;

ii) increasing the number of equations: by multiplying both equations by y, we obtain two new equations that form, together with those of [A3.2], a homogeneous system consisting of four equations in four unknowns (power products):

$$\mathbf{M}\,\mathbf{Y} = \mathbf{0} \qquad [A3.3]$$

where **M** is a function of x:

$$\mathbf{M} = \begin{bmatrix} 0 & ax^2 & bx-c & -d \\ 0 & ex^2 & fx & g \\ ax^2 & bx-c & -d & 0 \\ ex^2 & fx & g & 0 \end{bmatrix} \text{ and } \mathbf{Y} = [\, y^3 \;\; y^2 \;\; y \;\; 1 \,]^T$$

Since one of the elements of **Y** is 1, the system [A3.3] is compatible if, and only if, it is singular, which implies that the determinant of **M** is zero. Applying this condition to the example leads to a fourth degree equation in x. For each of the four roots, we obtain a different matrix **M**. By choosing three equations out of the system [A3.3], we obtain a system of three linear equations of type $\mathbf{A}\,\mathbf{Y}' = \mathbf{B}$ where $\mathbf{Y}' = [\, y^3 \;\; y^2 \;\; y \,]^T$. Doing that, each value of x provides a single value of y.

To summarize, the method requires four steps:

- construct the power product equation in order to minimize the number of unknowns:
- add equations to obtain a homogeneous system;
- from this system, compute a polynomial in a single unknown using the fact that the system is necessarily singular;
- compute the other variables by solving a system of linear equations.

Appendix 4

Solution of systems of linear equations

A4.1. Problem statement

Let us consider the following system of m linear equations in n unknowns:

$$\mathbf{Y} = \mathbf{W}\,\boldsymbol{\zeta} \qquad \text{[A4.1]}$$

where $\mathbf{W}$ is an (mxn) known matrix, $\mathbf{Y}$ is an (mx1) known vector and $\boldsymbol{\zeta}$ is the unknown (nx1) vector.

Let $\mathbf{W}_a$ be the augmented matrix defined by:

$$\mathbf{W}_a = [\mathbf{W} \;\vdots\; \mathbf{Y}]$$

Let r and r_a denote the ranks of $\mathbf{W}$ and $\mathbf{W}_a$ respectively. The relation between r and r_a can be used to analyze the existence of solutions:

a) if $r = r_a$, the system has at least one solution:
   - if $r = r_a = n$, there is a unique solution;
   - if $r = r_a < n$, the number of solutions is infinite; the system is redundant. For example, this case is encountered with the inverse kinematic model (Chapter 6).

b) if $r \neq r_a$, the system [A4.1] is not compatible, meaning that it has no exact solution; it will be written as:

$$\mathbf{Y} = \mathbf{W}\,\boldsymbol{\zeta} + \boldsymbol{\rho} \qquad \text{[A4.2]}$$

where ρ is the residual vector or error vector. This case occurs when identifying the geometric and dynamic parameters (Chapters 11 and 12 respectively) or when solving the inverse kinematic model in the vicinity of singular configurations.

A4.2. Resolution based on the generalized inverse

A4.2.1. *Definitions*

The matrix $\mathbf{W}^{(-1)}$ is a *generalized inverse* of $\mathbf{W}$ if:

$$\mathbf{W}\,\mathbf{W}^{(-1)}\,\mathbf{W} = \mathbf{W} \qquad [A4.3]$$

If $\mathbf{W}$ is square and regular, then $\mathbf{W}^{(-1)} = \mathbf{W}^{-1}$. In addition, $\mathbf{W}^{(-1)}$ is said to be a left inverse or a right inverse respectively if:

$$\mathbf{W}^{(-1)}\,\mathbf{W} = \mathbf{I} \text{ or } \mathbf{W}\,\mathbf{W}^{(-1)} = \mathbf{I} \qquad [A4.4]$$

It can be shown that $\mathbf{W}$ has an infinite number of generalized inverses unless it is of dimension (nxn) and of rank n. A solution of the system [A4.1], when it is compatible, is given by:

$$\zeta = \mathbf{W}^{(-1)}\,\mathbf{Y} \qquad [A4.5]$$

All the solutions are given by the general equation:

$$\zeta = \mathbf{W}^{(-1)}\,\mathbf{Y} + (\mathbf{I} - \mathbf{W}^{(-1)}\,\mathbf{W})\,\mathbf{Z} \qquad [A4.6]$$

where $\mathbf{Z}$ is an arbitrary (nx1) vector. Note that:

$$\mathbf{W}\,(\mathbf{I} - \mathbf{W}^{(-1)}\,\mathbf{W})\,\mathbf{Z} = \mathbf{0} \qquad [A4.7]$$

Therefore, the term $(\mathbf{I} - \mathbf{W}^{(-1)}\,\mathbf{W})\,\mathbf{Z}$ is a projection of $\mathbf{Z}$ on the null space of $\mathbf{W}$.

A4.2.2. *Computation of a generalized inverse*

The matrix $\mathbf{W}$ is partitioned in the following manner:

$$\mathbf{W} = \begin{bmatrix} \mathbf{W}_{11} & \mathbf{W}_{12} \\ \mathbf{W}_{21} & \mathbf{W}_{22} \end{bmatrix} \qquad [A4.8]$$

where $\mathbf{W}_{11}$ is a regular (rxr) matrix, and r is the rank of $\mathbf{W}$. Then, it can be verified that:

$$\mathbf{W}^{(-1)} = \begin{bmatrix} \mathbf{W}_{11}^{-1} & \mathbf{0} \\ \mathbf{0} & \mathbf{0} \end{bmatrix} \quad \text{[A4.9]}$$

This method gives the solution as a function of r components of $\mathbf{Y}$. Thus, the accuracy of the result may depend on the isolated minor. We will see in the next section that the pseudoinverse method allows us to avoid this limitation.

NOTE.– If the (r,r) matrix $\mathbf{W}_{11}$ built up with the first r rows and the first r columns is not regular, it is always possible to define a matrix $\mathbf{W}'$ such that:

$$\mathbf{W}' = \mathbf{R}\,\mathbf{W}\,\mathbf{C} = \begin{bmatrix} \mathbf{W}'_{11} & \mathbf{W}'_{12} \\ \mathbf{W}'_{21} & \mathbf{W}'_{22} \end{bmatrix} \quad \text{[A4.10]}$$

where $\mathbf{W}'_{11}$ is a regular (rxr) matrix. The orthogonal matrices $\mathbf{R}$ and $\mathbf{C}$ permute the rows and columns of $\mathbf{W}$ respectively. The generalized inverse of $\mathbf{W}$ is derived from that of $\mathbf{W}'$ as:

$$\mathbf{W}^{(-1)} = \mathbf{C}\,(\mathbf{W}')^{(-1)}\,\mathbf{R} \quad \text{[A4.11]}$$

A4.3. Resolution based on the pseudoinverse

A4.3.1. *Definition*

The *pseudoinverse* of the matrix $\mathbf{W}$ is the generalized inverse $\mathbf{W}^+$ that satisfies [Penrose 55]:

$$\begin{cases} \mathbf{W}\,\mathbf{W}^+\,\mathbf{W} = \mathbf{W} \\ \mathbf{W}^+\,\mathbf{W}\,\mathbf{W}^+ = \mathbf{W}^+ \\ (\mathbf{W}^+\,\mathbf{W})^T = \mathbf{W}^+\,\mathbf{W} \\ (\mathbf{W}\,\mathbf{W}^+)^T = \mathbf{W}\,\mathbf{W}^+ \end{cases} \quad \text{[A4.12]}$$

It can be shown that the pseudoinverse always exists and is unique. All the solutions of the system [A4.1] are given by:

$$\boldsymbol{\zeta} = \mathbf{W}^+\,\mathbf{Y} + (\mathbf{I} - \mathbf{W}^+\,\mathbf{W})\,\mathbf{Z} \quad \text{[A4.13]}$$

The first term $\mathbf{W}^+\,\mathbf{Y}$ is the solution minimizing the Euclidean norm $\|\boldsymbol{\zeta}\|$. The second term $(\mathbf{I} - \mathbf{W}^+\,\mathbf{W})\,\mathbf{Z}$, also called *optimization term* or *homogeneous solution*, is the projection of an arbitrary vector $\mathbf{Z}$ of $\Re^n$ on $\mathcal{N}(\mathbf{W})$, the null space of $\mathbf{W}$, and therefore, does not change the value of **Y**. It can be shown that $(\mathbf{I} - \mathbf{W}^+ \mathbf{W})$ is of rank $(n - r)$. Consequently, when the robot is redundant, this term may be used to optimize additional criteria satisfying the primary task. This property is illustrated by examples in Chapter 6.

When the system [A4.1] is not compatible, it can be shown that the solution $\mathbf{W}^+ \mathbf{Y}$ gives the least-squares solution minimizing the error $\|\mathbf{W}\,\zeta - \mathbf{Y}\|^2 = \|\rho\|^2$.

A4.3.2. *Pseudoinverse computation methods*

A4.3.2.1. *Method requiring explicit computation of the rank [Gorla 84]*

Let the matrix **W** be partitioned as indicated in equation [A4.8] such that $\mathbf{W}_{11}$ is of full rank r. Using the following notations:

$$\mathbf{W}_1 = \begin{bmatrix} \mathbf{W}_{11} \\ \mathbf{W}_{21} \end{bmatrix} \text{ and } \mathbf{W}_2 = [\, \mathbf{W}_{11} \quad \mathbf{W}_{12} \,]$$

it can be shown that:

$$\mathbf{W}^+ = \mathbf{W}_2^T (\mathbf{W}_1^T \mathbf{W} \mathbf{W}_2^T)^{-1} \mathbf{W}_1^T \qquad \text{[A4.14]}$$

When **W** is of full rank, this equation may be simplified as follows:

- if $m > n$: $\mathbf{W} = \mathbf{W}_1 \rightarrow \mathbf{W}^+ = (\mathbf{W}^T \mathbf{W})^{-1} \mathbf{W}^T$, ($\mathbf{W}^+$ is then the left inverse of **W**);
- if $m < n$: $\mathbf{W} = \mathbf{W}_2 \rightarrow \mathbf{W}^+ = \mathbf{W}^T (\mathbf{W} \mathbf{W}^T)^{-1}$, ($\mathbf{W}^+$ is then the right inverse of **W**);
- if $m = n$: $\mathbf{W} = \mathbf{W}_1 = \mathbf{W}_2 \rightarrow \mathbf{W}^+ = \mathbf{W}^{-1}$.

If $\mathbf{W}_{11}$ is not of rank r, the orthogonal permutation matrices **R** and **C** of equation [A4.10] should be used, yielding:

$$\mathbf{W}^+ = \mathbf{C}\,(\mathbf{W}')^+\,\mathbf{R} \qquad \text{[A4.15]}$$

A4.3.2.2. *Greville method [Greville 60], [Fournier 80]*

This recursive algorithm is based on the pseudoinverse properties of a partitioned matrix. It does not require the explicit computation of the rank of **W**. Let **W** be a partitioned matrix such that:

$$\mathbf{W} = [\, \mathbf{U} \; : \; \mathbf{V} \,] \qquad \text{[A4.16]}$$

Its pseudoinverse $\mathbf{W}^+$ can be written as [Boullion 71]:

$$\mathbf{W}^+ = \begin{bmatrix} \mathbf{U}^+ - \mathbf{U}^+\mathbf{V}\mathbf{C}^+ - \mathbf{U}^+\mathbf{V}(\mathbf{I} - \mathbf{C}^+\mathbf{C})\,\mathbf{M}\,\mathbf{V}^T(\mathbf{U}^+)^T\mathbf{U}^+(\mathbf{I} - \mathbf{V}\mathbf{C}^+) \\ \mathbf{C}^+ + (\mathbf{I} - \mathbf{C}^+\mathbf{C})\,\mathbf{M}\,\mathbf{V}^T(\mathbf{U}^+)^T\mathbf{U}^+(\mathbf{I} - \mathbf{V}\mathbf{C}^T) \end{bmatrix} \quad [A4.17]$$

with:

$$\mathbf{C} = (\mathbf{I} - \mathbf{U}\mathbf{U}^+)\,\mathbf{V}$$
$$\mathbf{M} = [\mathbf{I} + (\mathbf{I} - \mathbf{C}^+\mathbf{C})\,\mathbf{V}^T(\mathbf{U}^+)^T\mathbf{U}^+\mathbf{V}\,(\mathbf{I} - \mathbf{C}^+\mathbf{C})]^{-1}$$

If the matrix **V** reduces to a single column, a recursive algorithm that does not require any matrix inversion may be employed.

Let $\mathbf{W}_k$ contain the first k columns of **W**. If $\mathbf{W}_k$ is partitioned such that the first (k – 1) columns are denoted by $\mathbf{W}_{k-1}$ and the kth column is $\mathbf{w}_k$, then:

$$\mathbf{W}_k = [\,\mathbf{W}_{k-1} \;\vdots\; \mathbf{w}_k\,] \quad [A4.18]$$

The pseudoinverse $\mathbf{W}_k^+$ is derived from $\mathbf{W}_{k-1}^+$ and from the kth column of **W**:

$$\mathbf{W}_k^+ = \begin{bmatrix} \mathbf{W}_{k-1}^+ - \mathbf{d}_k\mathbf{b}_k \\ \mathbf{b}_k \end{bmatrix} \quad [A4.19]$$

where:

$$\mathbf{d}_k = \mathbf{W}_{k-1}^+\,\mathbf{w}_k \quad [A4.20]$$

In order to evaluate $\mathbf{b}_k$, we define:

$$\mathbf{c}_k = \mathbf{w}_k - \mathbf{W}_{k-1}\,\mathbf{d}_k \quad [A4.21]$$

then, we compute:

$$\begin{aligned} \mathbf{b}_k = \mathbf{c}_k^+ &= (\mathbf{c}_k^T\,\mathbf{c}_k)^{-1}\,\mathbf{c}_k^T && \text{if } \mathbf{c}_k \neq \mathbf{0} \\ &= (1 + \mathbf{d}_k^T\,\mathbf{d}_k)^{-1}\,\mathbf{d}_k^T\,\mathbf{W}_{k-1}^+ && \text{if } \mathbf{c}_k = \mathbf{0} \end{aligned} \quad [A4.22]$$

This recursive algorithm is initialized by calculating $\mathbf{W}_1^+$ using equation [A4.14]:

$$\mathbf{W}_1^+ = \mathbf{w}_1^+ = (\mathbf{w}_1^T\,\mathbf{w}_1)^{-1}\,\mathbf{w}_1^T \qquad (\text{if } \mathbf{w}_1 = \mathbf{0}, \text{ then } \mathbf{W}_1^+ = \mathbf{0}^T). \quad [A4.23]$$

The pseudoinverse of **W** can also be calculated by handling recursively rows instead of columns: physically, it comes to consider the equations sequentially. .

• **Example A4.1.** Computation of the pseudoinverse using the Greville method. Let us consider the following matrix:

$$\mathbf{W} = \begin{bmatrix} 1 & 2 & 3 \\ 2 & 3 & 4 \end{bmatrix}$$

i) first iteration (initialization):

$$\mathbf{W}_1^+ = [\ 1/5 \quad 2/5\ ]$$

ii) second iteration:

$$\mathbf{d}_2 = 8/5,\ \mathbf{c}_2 = \begin{bmatrix} 2/5 \\ -1/5 \end{bmatrix},\ \mathbf{b}_2 = [\ 2 \ \ -1\ ],\ \mathbf{W}_2^+ = \begin{bmatrix} -3 & 2 \\ 2 & -1 \end{bmatrix}$$

iii) third iteration:

$$\mathbf{d}_3 = \begin{bmatrix} -1 \\ 2 \end{bmatrix},\ \mathbf{c}_3 = \begin{bmatrix} 0 \\ 0 \end{bmatrix},\ \mathbf{b}_3 = [\ 7/6 \ \ -2/3\ ]$$

Finally, the pseudoinverse is:

$$\mathbf{W}^+ = \begin{bmatrix} -11/6 & 4/3 \\ -1/3 & 1/3 \\ 7/6 & -2/3 \end{bmatrix}$$

A4.3.2.3. *Method based on the singular value decomposition of* **W**

The singular value decomposition theory [Lawson 74], [Dongarra 79], [Klema 80] states that for an (mxn) matrix **W** of rank r, there exist orthogonal matrices **U** and **V** of dimensions (mxm) and (nxn) respectively, such that:

$$\mathbf{W} = \mathbf{U}\,\boldsymbol{\Sigma}\,\mathbf{V}^T \qquad \text{[A4.24]}$$

The (mxn) matrix $\boldsymbol{\Sigma}$ is diagonal and contains the singular values σ_i of **W**. They are arranged in a decreasing order such that $\sigma_1 \geq \sigma_2 \geq \dots \geq \sigma_r$. $\boldsymbol{\Sigma}$ has the following form:

$$\boldsymbol{\Sigma} = \begin{bmatrix} \mathbf{S}_{rxr} & \mathbf{0}_{rx(n-r)} \\ \mathbf{0}_{(m-r)xr} & \mathbf{0}_{(m-r)x(n-r)} \end{bmatrix} \qquad \text{[A4.25]}$$

where **S** is a diagonal (rxr) matrix of rank r, formed by the non-zero singular values σ_i of **W**.

The singular values of **W** are the square roots of the eigenvalues of the matrices $\mathbf{W}^T \mathbf{W}$ or $\mathbf{W}\,\mathbf{W}^T$ depending on whether n < m or n > m respectively.

The columns of **V** are the eigenvectors of $\mathbf{W}^T \mathbf{W}$ and are called *right singular vectors* or *input singular vectors*. The columns of **U** are the eigenvectors of $\mathbf{W}\,\mathbf{W}^T$ and are called *left singular vectors* or *output singular vectors*.

The pseudoinverse is then written as:

$$\mathbf{W}^+ = \mathbf{V}\,\mathbf{\Sigma}^+\mathbf{U}^T \qquad [A4.26]$$

with:

$$\mathbf{\Sigma}^+ = \begin{bmatrix} \mathbf{S}^{-1} & \mathbf{0} \\ \mathbf{0} & \mathbf{0} \end{bmatrix}$$

This method, known as *Singular Value Decomposition* (SVD) [Maciejewski 89], is often implemented for rank determination and pseudoinverse computation in scientific software packages.

The SVD decomposition of **W** makes it possible to evaluate the 2-norm condition number, which can be used to investigate the sensitivity of the linear system to data variations on **Y** and **W**. Indeed, if **W** is a square matrix, and assuming uncertainties $\boldsymbol{\zeta} + \mathbf{d}\boldsymbol{\zeta}$, the system [A4.1] may be written as:

$$\mathbf{Y} + \mathbf{dY} = [\mathbf{W} + \mathbf{dW}]\,[\boldsymbol{\zeta} + \mathbf{d}\boldsymbol{\zeta}] \qquad [A4.27]$$

The relative error of the solution may be bounded such that:

$$\frac{\|\mathbf{d}\boldsymbol{\zeta}\|_p}{\|\boldsymbol{\zeta}\|_p} \le \mathrm{cond}_p(\mathbf{W})\,\frac{\|\mathbf{dY}\|_p}{\|\mathbf{Y}\|_p} \qquad [A4.28a]$$

$$\frac{\|\mathbf{d}\boldsymbol{\zeta}\|_p}{\|\boldsymbol{\zeta} + \mathbf{d}\boldsymbol{\zeta}\|_p} \le \mathrm{cond}_p(\mathbf{W})\,\frac{\|\mathbf{dW}\|_p}{\|\mathbf{W}\|_p} \qquad [A4.28b]$$

$\mathrm{cond}_p(\mathbf{W})$ is the condition number of **W** with respect to the p-norm such that:

$$\mathrm{cond}_p(\mathbf{W}) = \|\mathbf{W}\|_p\,\|\mathbf{W}^+\|_p \qquad [A4.29]$$

where $\|*\|_p$ denotes a vector p-norm or a matrix p-norm.

The 2-norm condition number of a matrix **W** is given by:

$$\mathrm{cond}_2(\mathbf{W}) = \frac{\sigma_{max}}{\sigma_{min}} \qquad [A4.30]$$

Notice that the condition number is such that:

$$\mathrm{cond}_2(\mathbf{W}) \geq 1 \qquad [A4.31]$$

NOTES.–

– the p-norm of a vector $\boldsymbol{\zeta}$ is defined by:

$$\|\boldsymbol{\zeta}\|_p = \left(\sum_{i=1}^{n} |\zeta_i|^p\right)^{1/p} \quad \text{for } p \geq 1 \qquad [A4.32]$$

– the p-norm of a matrix $\mathbf{W}$ is defined by:

$$\|\mathbf{W}\|_p = \max\left\{\frac{\|\mathbf{W}\,\boldsymbol{\zeta}\|_p}{\|\boldsymbol{\zeta}\|_p} : \boldsymbol{\zeta} \neq \mathbf{0}_{n,1}\right\} = \max\left\{\|\mathbf{W}\,\boldsymbol{\zeta}\|_p : \|\boldsymbol{\zeta}\|_p = 1\right\} \qquad [A4.33]$$

– the 2-norm of a matrix is the largest singular value of $\mathbf{W}$. It is given by:

$$\|\mathbf{W}\|_2 = \sigma_{max}$$

– equations similar to [A4.28] can be derived for over determined linear systems.

• **Example A4.2.** Computation of the pseudoinverse with the SVD method. Consider the same matrix as in Example A4.1:

$$\mathbf{W} = \begin{bmatrix} 1 & 2 & 3 \\ 2 & 3 & 4 \end{bmatrix}$$

It can be shown that:

$$\mathbf{V} = \begin{bmatrix} 0.338 & 0.848 & -0.408 \\ 0.551 & 0.174 & 0.816 \\ 0.763 & -0.501 & -0.408 \end{bmatrix}, \boldsymbol{\Sigma} = \begin{bmatrix} 6.55 & 0 & 0 \\ 0 & 0.374 & 0 \end{bmatrix}, \mathbf{U} = \begin{bmatrix} 0.57 & -0.822 \\ 0.822 & 0.57 \end{bmatrix}.$$

The pseudoinverse is obtained by applying equation [A4.26]:

$$W^+ = \begin{bmatrix} -1.83 & 1.33 \\ -0.333 & 0.333 \\ 1.17 & -0.667 \end{bmatrix}$$

A4.4. Resolution based on the QR decomposition

Given the system of equations [A4.1], two cases are to be considered depending on whether **W** is of full rank or not.

A4.4.1. *Full rank system*

Let us assume that **W** is of full rank. The QR decomposition of **W** consists of writing that [Golub 83]:

$$Q^T W = \begin{bmatrix} R \\ 0_{(m-r),n} \end{bmatrix} \quad \text{for } m > n,\ r = n \qquad [A4.34]$$

$$Q^T W = \begin{bmatrix} R & 0_{m,n-r} \end{bmatrix} \quad \text{for } n > m,\ r = m \qquad [A4.35]$$

where **R** is a regular and upper-triangular (rxr) matrix and where **Q** is an orthogonal (mxm) matrix.

For sake of brevity, let us only consider the case m > n, which typically occurs when identifying the geometric and dynamic parameters (Chapters 11 and 12 respectively). The case n > m can be similarly handled. The matrix **Q** is partitioned as follows:

$$Q = \begin{bmatrix} Q1 & Q2 \end{bmatrix} \qquad [A4.36]$$

where the dimensions of **Q1** and **Q2** are (mxr) and mx(m−r) respectively.

Let us define:

$$G = Q^T Y = \begin{bmatrix} Q1^T Y \\ Q2^T Y \end{bmatrix} = \begin{bmatrix} G1 \\ G2 \end{bmatrix} \qquad [A4.37]$$

Since the matrix **Q** is orthogonal, it follows that [Golub 83]:

$$\|Y - W\zeta\|^2 = \|Q^T Y - Q^T W \zeta\|^2 = \|G1 - R\zeta\|^2 + \|G2\|^2 = \|\rho\|^2 \qquad [A4.38]$$

From equation [A4.38], ζ is the unique solution of the system:

$$\mathbf{R}\,\zeta = \mathbf{G1} \qquad \text{[A4.39]}$$

Since **R** is a regular and upper-triangular (rxr) matrix, the system [A4.39] can be easily solved with a backward recursion technique (compute sequentially ζ_n, ζ_{n-1}, ...). The norm of the residual for the optimal solution is derived as:

$$\|\rho\|_{min} = \|\mathbf{G2}\| = \|\mathbf{Q2}^T\,\mathbf{Y}\| \qquad \text{[A4.40]}$$

This solution (when m > n and r = n) is identical to that obtained by the pseudoinverse. In order to speed up the computations for systems of high dimensions (for example, this is the case for the identification of the dynamic parameters), we can partition the system [A4.1] into k sub-systems such that:

$$\mathbf{Y}(i) = \mathbf{W}(i)\,\zeta \quad \text{for } i = 1, \ldots, k \qquad \text{[A4.41]}$$

Let $\mathbf{Q}(i) = [\mathbf{Q1}(i) \quad \mathbf{Q2}(i)]$ and $\mathbf{R}(i)$ be the matrices obtained after a QR decomposition of the matrix $\mathbf{W}(i)$. The global system reduces to the following system of (nxk) equations in n unknowns:

$$\begin{bmatrix} \mathbf{Q1}^T(1)\mathbf{Y}(1) \\ \cdots \\ \mathbf{Q1}^T(k)\mathbf{Y}(k) \end{bmatrix} = \begin{bmatrix} \mathbf{Q1}^T(1)\mathbf{W}(1) \\ \cdots \\ \mathbf{Q1}^T(k)\mathbf{W}(k) \end{bmatrix} \zeta$$

$$\begin{bmatrix} \mathbf{Q1}^T(1)\mathbf{Y}(1) \\ \cdots \\ \mathbf{Q1}^T(k)\mathbf{Y}(k) \end{bmatrix} = \begin{bmatrix} \mathbf{R}(1) \\ \cdots \\ \mathbf{R}(k) \end{bmatrix} \zeta \qquad \text{[A4.42]}$$

A4.4.2. *Rank deficient system*

Again, let us assume that m > n but in this case r < n. We permute the columns of **W** in such a way that the first columns are independent (the independent columns correspond to the diagonal non-zero elements of the matrix **R** obtained after QR decomposition of **W**). We proceed by a QR decomposition of the permutation matrix and we obtain :

$$\mathbf{Q}^T\,\mathbf{W}\,\mathbf{P} = \begin{bmatrix} \mathbf{R1} & \mathbf{R2} \\ \mathbf{0}_{(m-r),r} & \mathbf{0}_{(m-r),(n-r)} \end{bmatrix} \qquad \text{[A4.43]}$$

where **P** is a permutation matrix obtained by permuting the columns of an identity matrix, **Q** is an orthogonal (mxm) matrix, and **R1** is a regular and upper-triangular (rxr) matrix.

Let:

$$\mathbf{P}^T \zeta = \begin{bmatrix} \zeta 1 \\ \zeta 2 \end{bmatrix}$$

From equation [A4.37], we obtain:

$$\|\rho\|^2 = \|\mathbf{Y} - \mathbf{W}\,\zeta\|^2 = \|\mathbf{Q}^T\,\mathbf{Y} - \mathbf{Q}^T\,\mathbf{W}\,\mathbf{P}\,\mathbf{P}^T\,\zeta\|^2$$

$$= \left\| \begin{bmatrix} \mathbf{G1} \\ \mathbf{G2} \end{bmatrix} - \begin{bmatrix} \mathbf{R1}\,\zeta 1 + \mathbf{R2}\,\zeta 2 \\ \mathbf{0}_{(n-r),1} \end{bmatrix} \right\|^2$$

$$= \|\mathbf{G1} - [\mathbf{R1}\,\zeta 1 + \mathbf{R2}\,\zeta 2]\|^2 + \|\mathbf{G2}\|^2 \qquad [A4.44]$$

$\zeta 1$ is the unique solution of the system:

$$\mathbf{R1}\,\zeta 1 = \mathbf{G1} - \mathbf{R2}\,\zeta 2 \qquad [A4.45]$$

Then, we obtain a family of optimal solutions parameterized by the matrices **P** and $\zeta 2$:

$$\zeta = \mathbf{P} \begin{bmatrix} \zeta 1 \\ \zeta 2 \end{bmatrix} \qquad [A4.46]$$

All solutions provide the minimum norm residual given by equation [A4.40]. We obtain a base solution for $\zeta 2 = \mathbf{0}_{(n-r),1}$. Recall that the pseudoinverse solution provides the minimum norm residual together with the minimum norm $\|\,\zeta\,\|^2$.

Appendix 5

Numerical computation of the base parameters

A5.1. Introduction

The base parameters constitute the minimum set of parameters that characterize completely a given system. They also represent the identifiable parameters of the system. They are obtained from the standard parameters by eliminating those that have no effect on the model and by grouping some others in linear combinations. The determination of the base inertial parameters has been carried out in Chapters 9 and 10 by the use of straightforward symbolic methods for serial and tree structured robots. However, the symbolic approach cannot give all the base parameters for robots containing closed loops. This problem can be solved by the use of the numerical method presented in this appendix. In addition, the numerical method can also be applied to determine the base parameters for the geometric calibration of robots.

The symbolic approach of computing the base parameters is based on determining the independent elements of the energy functions represented by the row vector $\mathbf{h}$ (equation [9.41]), or by determining the independent columns of the $\mathbf{D}$ matrix of the dynamic model (equation [9.36]). Numerically this problem is equivalent to the study of the space span by the columns of a matrix $\mathbf{W}$ formed from $\mathbf{h}$ (or $\mathbf{D}$) using r random values of $\mathbf{q}$, $\dot{\mathbf{q}}$ (or $\mathbf{q}$, $\dot{\mathbf{q}}$, $\ddot{\mathbf{q}}$). This study can be carried out using the singular value decomposition (SVD) or the QR decomposition of $\mathbf{W}$ [Gautier 91]. In this appendix, we develop the numerical method that is based on the QR decomposition of a matrix $\mathbf{W}$, which is derived from the energy functions. Both cases of tree structured robots and closed loop robots are treated.

A5.2. Base inertial parameters of serial and tree structured robots

The total energy of the system H is linear in terms of the standard inertial parameters. It is given by the following equation:

$$\mathbf{H} = \mathbf{h}\,\mathbf{K} \qquad [A5.1]$$

where:

- $\mathbf{K}$ represents the (11nx1) vector of the standard inertial parameters of the links and of the rotors of actuators;
- $\mathbf{h}(\mathbf{q}, \dot{\mathbf{q}})$ is the (1x11n) row vector composed of the energy functions;
- $\mathbf{q}$ and $\dot{\mathbf{q}}$ are the (nx1) joint position and joint velocity vectors respectively.

To determine the base parameters, we construct a matrix $\mathbf{W}$ by calculating the energy row $\mathbf{h}$ for r random values of joint positions and velocities such that:

$$\mathbf{W} = \begin{bmatrix} \mathbf{h}(1) \\ \mathbf{h}(2) \\ \cdots \\ \mathbf{h}(r) \end{bmatrix} \qquad [A5.2]$$

with $\mathbf{h}(i) = \mathbf{h}[\mathbf{q}(i), \dot{\mathbf{q}}(i)]$, i = 1, ..., r, and r >> 11n.

An inertial parameter has no effect on the dynamic model if the elements of its corresponding column in $\mathbf{W}$ have the same value, i.e. its function in $\mathbf{h}$ is constant and independent of $\mathbf{q}(i)$, $\dot{\mathbf{q}}(i)$. By eliminating such parameters and the corresponding columns, the matrix $\mathbf{W}$ is reduced to c columns and r rows.

We note that:

- the number of the base inertial parameters b is equal to the rank of $\mathbf{W}$;
- the base parameters are those corresponding to b independent columns of $\mathbf{W}$;
- the grouping equations are obtained by calculating the relationship between the independent columns and the dependent columns of $\mathbf{W}$.

The application of the foregoing statements can be achieved by the use of the QR decomposition of $\mathbf{W}$ (§ A4.4), which is given by:

$$\mathbf{Q}^T\,\mathbf{W} = \begin{bmatrix} \mathbf{R} \\ \mathbf{0}_{(r-c)xc} \end{bmatrix} \qquad [A5.3]$$

where **Q** is an (rxr) orthogonal matrix, **R** is a (cxc) upper-triangular matrix, and $\mathbf{0}_{ixj}$ is the (ixj) matrix of zeros.

Theoretically, the non-identifiable parameters are those whose corresponding elements on the diagonal of the matrix **R** are zero [Forsythe 77], [Golub 83]. Let τ be the numerical zero:

$$\tau = r\,\varepsilon \max_i(|R_{ii}|) \qquad [A5.4]$$

where $|R_{ii}|$ is the absolute value of R_{ii}, and ε is the computer precision.

Thus, if $|R_{ii}| < \tau$, then the ith parameter is not identifiable. On the contrary, if $|R_{ii}| > \tau$, then the corresponding column in **W** is independent and constitutes a base of the space span by **W**. Let the b independent columns be collected in the matrix **W1**, and the corresponding parameters be collected in the vector **K1**. The other columns and parameters are represented by **W2** and **K2** respectively, such that:

$$\mathbf{W\,K} = [\,\mathbf{W1} \;\; \mathbf{W2}\,]\begin{bmatrix}\mathbf{K1}\\ \mathbf{K2}\end{bmatrix} \qquad [A5.5]$$

The matrix **W2** can be written in terms of **W1** as follows:

$$\mathbf{W2} = \mathbf{W1}\,\boldsymbol{\beta} \qquad [A5.6]$$

Consequently:

$$\mathbf{W\,K} = [\,\mathbf{W1} \;\; \mathbf{W2}\,]\begin{bmatrix}\mathbf{K_B}\\ \mathbf{0}\end{bmatrix} = \mathbf{W1\,K_B} \qquad [A5.7]$$

where the base parameter vector $\mathbf{K_B}$ is given by:

$$\mathbf{K_B} = \mathbf{K1} + \boldsymbol{\beta}\,\mathbf{K2} \qquad [A5.8]$$

Thus, the matrix β allows us to obtain the grouping equations of the parameters **K2** with **K1**. In order to determine β, we compute the QR decomposition of the matrix $[\,\mathbf{W1} \;\; \mathbf{W2}\,]$, which is written as:

$$[\,\mathbf{W1} \;\; \mathbf{W2}\,] = [\,\mathbf{Q1} \;\; \mathbf{Q2}\,]\begin{bmatrix}\mathbf{R1} & \mathbf{R2}\\ \mathbf{0}_{(r-b)xb} & \mathbf{0}_{(r-b)x(c-b)}\end{bmatrix} = [\,\mathbf{Q1R1} \;\; \mathbf{Q1R2}\,] \qquad [A5.9]$$

where **R1** is an (rxr) regular upper-triangular matrix, and **R2** is a (bx(c−b)) matrix.

From equation [A5.9], we obtain:

$$\mathbf{Q1} = \mathbf{W1}\,\mathbf{R1}^{-1} \qquad [A5.10]$$

$$\mathbf{W2} = \mathbf{Q1}\,\mathbf{R2} = \mathbf{W1}\,\mathbf{R1}^{-1}\,\mathbf{R2} \qquad [A5.11]$$

and finally, using equation [A5.6]:

$$\beta = \mathbf{R1}^{-1}\,\mathbf{R2} \qquad [A5.12]$$

A5.3. Base inertial parameters of closed loop robots

The geometric description of closed loop robots is given in Chapter 7. The joint position vector of the equivalent tree structure is given by:

$$\mathbf{q}_{ar} = \begin{bmatrix} \mathbf{q}_a \\ \mathbf{q}_p \end{bmatrix} \qquad [A5.13]$$

where $\mathbf{q}_a$ is the (Nx1) vector of the active joint variables, and $\mathbf{q}_p$ is the ((n – N)x1) vector of the passive joint variables.

The energy functions of the inertial parameters of the closed loop structure are the same as those of the equivalent tree structure. This means that we can apply the algorithm of the tree structured robots to the closed loop structures with the difference that the matrix **W** is calculated using random values for the independent active variables $\mathbf{q}_a(i)$ and $\dot{\mathbf{q}}_a(i)$ for $i = 1, ..., r$. The corresponding passive variables $\mathbf{q}_p$ and $\dot{\mathbf{q}}_p$ are evaluated from the constraint equations of the loops.

A5.4. Generality of the numerical method

The numerical method can be used for the determination of the minimum parameters of other applications such as:

- determination of the identifiable parameters for the geometric calibration of the parameters (Chapter 11) [Khalil 91a];
- calculation of the minimum parameters of flexible structures [Pham 91a].

This numerical method is easy to implement, thanks to a software package such as SYMORO+ [Khalil 97] for the automatic computation of the symbolic expressions of the energy functions (to determine the elements of **h**) and thanks to scientific software packages of matrix computation such as Matlab and Mathematica.

Appendix 6

Recursive equations between the energy functions

In this appendix, we establish the recursive equation between the energy functions of the inertial parameters of two consecutive links in an open loop structure (serial or tree structured robots).

A6.1. Recursive equation between the kinetic energy functions of serial robots

The kinetic energy of link j can be written using equation [9.19] as:

$$E_j = \frac{1}{2} {}^j\mathbb{V}_j^T \, {}^j\mathbb{J}_j \, {}^j\mathbb{V}_j \qquad [A6.1]$$

with:

$${}^j\mathbb{V}_j = \begin{bmatrix} {}^j\mathbf{V}_j \\ {}^j\omega_j \end{bmatrix} \qquad [A6.2]$$

and:

$${}^j\mathbb{J}_j = \begin{bmatrix} M_j \mathbf{I}_3 & -{}^j\hat{\mathbf{MS}}_j \\ {}^j\hat{\mathbf{MS}}_j & {}^j\mathbf{J}_j \end{bmatrix} \qquad [A6.3]$$

The recursive equation of the kinematic screw is written using equation [9.22] as:

$${}^j\mathbb{V}_j = {}^j\mathbb{T}_{j-1} \, {}^{j-1}\mathbb{V}_{j-1} + \dot{q}_j \, {}^j\mathbf{a}_j \qquad [A6.4]$$

where ${}^{j}\mathbf{a}_j$ is defined by equation [9.23a].

The kinetic energy of link j is linear in the inertial parameters of link j. Consequently, it can be written as:

$$E_j = \mathbf{e}_j\, \mathbf{K}_j \qquad [A6.5]$$

where $\mathbf{e}_j$ is the (1x10) row matrix containing the energy functions of the inertial parameters of link j. The parameters of link j are given by:

$$\mathbf{K}_j = [\ XX_j \quad XY_j \quad XZ_j \quad YY_j \quad YZ_j \quad ZZ_j \quad MX_j \quad MY_j \quad MZ_j \quad M_j\ ]^T \qquad [A6.6]$$

By substituting for ${}^{j}\mathbb{V}_j$ from equation [A6.4] into equation [A6.1], we obtain:

$$E_j = \frac{1}{2}({}^{j}\mathbb{T}_{j-1}\,{}^{j-1}\mathbb{V}_{j-1} + \dot{q}_j\,{}^{j}\mathbf{a}_j)^T\,{}^{j}\mathbb{J}_j\,({}^{j}\mathbb{T}_{j-1}\,{}^{j-1}\mathbb{V}_{j-1} + \dot{q}_j\,{}^{j}\mathbf{a}_j) \qquad [A6.7]$$

Developing equation [A6.7] gives:

$$E_j = \frac{1}{2}\,{}^{j-1}\mathbb{V}_{j-1}^T\,({}^{j}\mathbb{T}_{j-1}^T\,{}^{j}\mathbb{J}_j\,{}^{j}\mathbb{T}_{j-1})\,{}^{j-1}\mathbb{V}_{j-1} + \dot{q}_j\,{}^{j}\mathbf{a}_j^T\,{}^{j}\mathbb{J}_j\,{}^{j}\mathbb{V}_j - \frac{1}{2}\dot{q}_j^2\,{}^{j}\mathbf{a}_j^T\,{}^{j}\mathbb{J}_j\,{}^{j}\mathbf{a}_j \qquad [A6.8]$$

Let us set:

$$^{j-1}\mathbb{J}_j = {}^{j}\mathbb{T}_{j-1}^T\,{}^{j}\mathbb{J}_j\,{}^{j}\mathbb{T}_{j-1} \qquad [A6.9]$$

$$\dot{q}_j\,\eta_j\,{}^{j}\mathbf{K}_j = {}^{j}\mathbf{a}_j^T\,{}^{j}\mathbb{J}_j\,{}^{j}\mathbb{V}_j - \frac{1}{2}\dot{q}_j\,{}^{j}\mathbf{a}_j^T\,{}^{j}\mathbb{J}_j\,{}^{j}\mathbf{a}_j \qquad [A6.10]$$

where the row vector η_j is given by:

$$\eta_j = \bar{\sigma}_j\,[0 \quad 0 \quad \omega_{1,j} \quad 0 \quad \omega_{2,j} \quad (\omega_{3,j} - \frac{1}{2}\dot{q}_j) \quad V_{2,j} \quad -V_{1,j} \quad 0 \quad 0] +$$
$$\sigma_j\,[0 \quad 0 \quad 0 \quad 0 \quad 0 \quad 0 \quad -\omega_{2,j} \quad \omega_{1,j} \quad 0 \quad (V_{3,j} - \frac{1}{2}\dot{q}_j)] \qquad [A6.11]$$

with:

$${}^{j}\mathbf{V}_j = [V_{1,j} \quad V_{2,j} \quad V_{3,j}]^T$$
$${}^{j}\boldsymbol{\omega}_j = [\omega_{1,j} \quad \omega_{2,j} \quad \omega_{3,j}]^T$$

Equation [A6.9] transforms the inertial parameters of link j from frame R_j into frame R_{j-1}. It can be written as:

$${}^{j-1}\mathbf{K}_j = {}^{j-1}\boldsymbol{\lambda}_j \, {}^j\mathbf{K}_j \qquad [A6.12]$$

The expression of ${}^{j-1}\boldsymbol{\lambda}_j$ is obtained by comparing equations [A6.9] and [A6.12]. It is given in Table 9.1 for serial robots.

Using equations [A6.5], [A6.9], [A6.10] and [A6.12], we rewrite equation [A6.8] as follows:

$$E_j = (\mathbf{e}_{j-1} \, {}^{j-1}\boldsymbol{\lambda}_j + \dot{q}_j \, \boldsymbol{\eta}_j) \, \mathbf{K}_j \qquad [A6.13]$$

Finally, from equations [A6.5] and [A6.13], we deduce that:

$$\mathbf{e}_j = \mathbf{e}_{j-1} \, {}^{j-1}\boldsymbol{\lambda}_j + \dot{q}_j \, \boldsymbol{\eta}_j \qquad [A6.14]$$

with $\mathbf{e}_0 = \mathbf{0}_{1x10}$.

A6.2. Recursive equation between the potential energy functions of serial robots

The potential energy of link j is written as (equation [9.25b]):

$$U_j = -{}^0\mathbf{g}^T [{}^0\mathbf{P}_j \, M_j + {}^0\mathbf{A}_j \, {}^j\mathbf{MS}_j] \qquad [A6.15]$$

where ${}^0\mathbf{g} = [\, g_1 \quad g_2 \quad g_3 \,]$ indicates the acceleration due to gravity.

This expression is linear in the inertial parameters. It can be written as:

$$U_j = \mathbf{u}_j \, \mathbf{K}_j \qquad [A6.16]$$

Using equations [A6.15] and [A6.12], we can write that:

$$U_j = \mathbf{g}_u \, {}^0\boldsymbol{\lambda}_j \, \mathbf{K}_j \qquad [A6.17]$$

where:

$$\mathbf{g}_u = [\, \mathbf{0}_{1x6} \quad -{}^0\mathbf{g}^T \quad 0 \,] \qquad [A6.18]$$

From equation [A6.17], we deduce that:

$$\mathbf{u}_j = \mathbf{g}_u \, {}^0\boldsymbol{\lambda}_j \qquad [A6.19]$$

Since ${}^0\boldsymbol{\lambda}_j = {}^0\boldsymbol{\lambda}_{j-1} \, {}^{j-1}\boldsymbol{\lambda}_j$, we obtain the following recursive equation for the potential energy functions:

$$\mathbf{u}_j = \mathbf{u}_{j-1}\,{}^{j-1}\boldsymbol{\lambda}_j \qquad [A6.20]$$

with $\mathbf{u}_0 = \mathbf{g}_u$.

A6.3. Recursive equation between the total energy functions of serial robots

The total energy of link j is written as:

$$H_j = E_j + U_j = (\mathbf{e}_j + \mathbf{u}_j)\,\mathbf{K}_j = \mathbf{h}_j\,\mathbf{K}_j \qquad [A6.21]$$

with:

$$\mathbf{h}_j = \mathbf{e}_j + \mathbf{u}_j \qquad [A6.22]$$

From equations [A6.14] and [A6.20], we obtain the following recursive equation:

$$\mathbf{h}_j = \mathbf{h}_{j-1}\,{}^{j-1}\boldsymbol{\lambda}_j + \dot{q}_j\,\boldsymbol{\eta}_j \qquad [A6.23]$$

with $\mathbf{h}_0 = \mathbf{g}_u$.

A6.4. Expression of ${}^{a(j)}\boldsymbol{\lambda}_j$ in the case of the tree structured robot

In the case of the tree structured robot, equation [A6.23] is valid after replacing j – 1 by i = a(j). The (10x10) matrix ${}^i\boldsymbol{\lambda}_j$ represents the matrix transforming the inertial parameters $\mathbf{K}_j$ from frame R_j to frame R_i and can be obtained by developing the following equation:

$${}^i\mathbb{J}_j = {}^j\mathbb{T}_i^T\,{}^j\mathbb{J}_j\,{}^j\mathbb{T}_i \qquad [A6.24]$$

which is equivalent to:

$${}^i\mathbf{K}_j = {}^i\boldsymbol{\lambda}_j\,{}^j\mathbf{K}_j \qquad [A6.25]$$

By comparing equations [A6.24] and [A6.25], we obtain the expressions of the elements of ${}^i\boldsymbol{\lambda}_j$ in terms of the elements of the matrix ${}^i\mathbf{T}_j$, which are functions of the geometric parameters (γ_j, b_j, α_j, d_j, θ_j, r_j) defining frame R_j relative to frame R_i (Chapter 7), as follows:

$${}^i\boldsymbol{\lambda}_j = \begin{bmatrix} \lambda 11 & \lambda 12 & \lambda 13 \\ 0_{3x6} & {}^i\mathbf{A}_j & {}^i\mathbf{P}_j \\ 0_{1x6} & 0_{1x3} & 1 \end{bmatrix} \qquad [A6.26]$$

The dimensions of the matrices **λ11**, **λ12** and **λ13** are (6x6), (6x3) and (6x1) respectively. To simplify the writing, let:

$$ {}^{i}A_j = [\ s \ \ n \ \ a\ ] \qquad [A6.27] $$

$$ {}^{i}P_j = [\ P_x \ \ P_y \ \ P_z\ ]^T \qquad [A6.28] $$

Thus:

$$ \lambda 11 = \begin{bmatrix} s_x s_x & 2s_x n_x & 2s_x a_x & n_x n_x & 2n_x a_x & a_x a_x \\ s_x s_y & s_y n_x + s_x n_y & s_y a_x + s_x a_y & n_x n_y & n_y a_x + n_x a_y & a_x a_y \\ s_z s_x & s_z n_x + s_x n_z & s_z a_x + s_x a_z & n_x n_z & n_z a_x + n_x a_z & a_z a_x \\ s_y s_y & 2n_y s_y & 2s_y a_y & n_y n_y & 2n_y a_y & a_y a_y \\ s_y s_z & s_z n_y + s_y n_z & a_y s_z + s_y a_z & n_y n_z & n_z a_y + n_y a_z & a_y a_z \\ s_z s_z & 2n_z s_z & 2s_z a_z & n_z n_z & 2n_z a_z & a_z a_z \end{bmatrix} \qquad [A6.29] $$

$$ \lambda 12 = \begin{bmatrix} 2(s_z P_z + s_y P_y) & 2(n_z P_z + n_y P_y) & 2(a_z P_z + a_y P_y) \\ -s_y P_x - s_x P_y & -n_y P_x - n_x P_y & -a_y P_x - a_x P_y \\ -s_z P_x - s_x P_z & -n_z P_x - n_x P_z & -a_z P_x - a_x P_z \\ 2(s_z P_z + s_x P_x) & 2(n_z P_z + n_x P_x) & 2(a_z P_z + a_x P_x) \\ -s_z P_y - s_y P_z & -n_z P_y - n_y P_z & -a_z P_y - a_y P_z \\ 2(s_y P_y + s_x P_x) & 2(n_y P_y + n_x P_x) & 2(a_y P_y + a_x P_x) \end{bmatrix} \qquad [A6.30] $$

$$ \lambda 13 = [\ P_z P_z + P_y P_y \ \ -P_x P_y \ \ -P_x P_z \ \ P_z P_z + P_x P_x \ \ -P_z P_y \ \ P_x P_x + P_y P_y\ ]^T \qquad [A6.31] $$

Appendix 7

Dynamic model of the Stäubli RX-90 robot

In this appendix, we present the simplified Newton-Euler inverse dynamic model of the Stäubli RX-90 robot. This model is obtained automatically using the software package SYMORO+ [Khalil 97]. The inertial parameters correspond to the case of symmetric links, which are given in Table 9.4. The components of the force and moments exerted by the end-effector on the environment are denoted by FX6, FY6, FZ6, CX6, CY6, and CZ6. The joint friction forces are neglected. The velocity, acceleration and torque of joint j are denoted by QPj, QDPj and GAMj respectively. The acceleration of gravity is denoted by G3. As already mentioned, Sj and Cj denote $\sin(\theta_j)$ and $\cos(\theta_j)$ respectively.

Noting that the equations with an asterisk (*) on their left are constants and can be evaluated off-line, the computational cost of this model is 160 multiplications and 113 additions.

```
No31=QDP1*ZZ1R
WI12=QP1*S2
WI22=C2*QP1
WP12=QDP1*S2 + QP2*WI22
WP22=C2*QDP1 - QP2*WI12
DV222=-WI22**2
DV332=-QP2**2
DV122=WI12*WI22
DV132=QP2*WI12
DV232=QP2*WI22
U112=DV222 + DV332
U212=DV122 + QDP2
U312=DV132 - WP22
VP12=- G3*S2
VP22=- C2*G3
PIS22=XXR2 - ZZR2
```

```
No12=WP12*XXR2 + DV232*ZZR2
No22=DV132*PIS22
No32=-DV122*XXR2 + QDP2*ZZR2
WI13=C3*WI12 + S3*WI22
WI23=-S3*WI12 + C3*WI22
W33=QP2 + QP3
WP13=QP3*WI23 + C3*WP12 + S3*WP22
WP23=-QP3*WI13 - S3*WP12 + C3*WP22
WP33=QDP2 + QDP3
DV113=-WI13**2
DV333=-W33**2
DV123=WI13*WI23
DV133=W33*WI13
DV233=W33*WI23
U123=DV123 - WP33
U223=DV113 + DV333
U323=DV233 + WP13
VSP13=d3*U112 + VP12
VSP23=d3*U212 + VP22
VSP33=d3*U312
VP13=C3*VSP13 + S3*VSP23
VP23=-S3*VSP13 + C3*VSP23
F13=MYR3*U123
F23=MYR3*U223
F33=MYR3*U323
*PIS23=XXR3 - ZZR3
No13=WP13*XXR3 + DV233*ZZR3
No23=DV133*PIS23
No33=-DV123*XXR3 + WP33*ZZR3
WI14=-S4*W33 + C4*WI13
WI24=-C4*W33 - S4*WI13
W34=QP4 + WI23
WP14=QP4*WI24 + C4*WP13 - S4*WP33
WP24=-QP4*WI14 - S4*WP13 - C4*WP33
WP34=QDP4 + WP23
DV124=WI14*WI24
DV134=W34*WI14
DV234=W34*WI24
VSP14=RL4*U123 + VP13
VSP24=RL4*U223 + VP23
VSP34=RL4*U323 + VSP33
VP14=C4*VSP14 - S4*VSP34
VP24=-S4*VSP14 - C4*VSP34
```

```
*PIS24=XXR4 – ZZR4
No14=WP14*XXR4 + DV234*ZZR4
No24=DV134*PIS24
No34=–DV124*XXR4 + WP34*ZZR4
WI15=S5*W34 + C5*WI14
WI25=C5*W34 – S5*WI14
W35=QP5 – WI24
WP15=QP5*WI25 + C5*WP14 + S5*WP34
WP25=–QP5*WI15 – S5*WP14 + C5*WP34
WP35=QDP5 – WP24
DV115=–WI15**2
DV335=–W35**2
DV125=WI15*WI25
DV135=W35*WI15
DV235=W35*WI25
U125=DV125 – WP35
U225=DV115 + DV335
U325=DV235 + WP15
VP15=C5*VP14 + S5*VSP24
F15=MYR5*U125
F25=MYR5*U225
F35=MYR5*U325
*PIS25=XXR5 – ZZR5
No15=WP15*XXR5 + DV235*ZZR5
No25=DV135*PIS25
No35=–DV125*XXR5 + WP35*ZZR5
WI16=–S6*W35 + C6*WI15
WI26=–C6*W35 – S6*WI15
W36=QP6 + WI25
WP16=QP6*WI26 + C6*WP15 – S6*WP35
WP36=QDP6 + WP25
DV126=WI16*WI26
DV136=W36*WI16
DV236=W36*WI26
*PIS26=XXR6 – ZZ6
No16=WP16*XXR6 + DV236*ZZ6
No26=DV136*PIS26
No36=–DV126*XXR6 + WP36*ZZ6
N16=CX6 + No16
N26=CY6 + No26
N36=CZ6 + No36
FDI16=C6*FX6 – FY6*S6
FDI36=–C6*FY6 – FX6*S6
```

```
E15=F15 + FDI16
E25=F25 + FZ6
E35=F35 + FDI36
N15=C6*N16 + No15 – N26*S6 – MYR5*VP24
N25=N36 + No25
N35=-(C6*N26) + No35 – N16*S6 – MYR5*VP15
FDI15=C5*E15 – E25*S5
FDI35=C5*E25 + E15*S5
N14=C5*N15 + No14 – N25*S5
N24=-N35 + No24
N34=C5*N25 + No34 + N15*S5
FDI14=C4*FDI15 + E35*S4
FDI34=C4*E35 – FDI15*S4
E13=F13 + FDI14
E23=F23 + FDI35
E33=F33 + FDI34
N13=C4*N14 + No13 + FDI34*RL4 – N24*S4 + MYR3*VSP33
N23=N34 + No23
N33=-(C4*N24) + No33 – FDI14*RL4 – N14*S4 – MYR3*VP13
FDI23=C3*E23 + E13*S3
N12=C3*N13 + No12 – N23*S3
N22=-(d3*E33) + C3*N23 + No22 + N13*S3
N32=d3*FDI23 + N33 + No32 – MY2*VP12 + MXR2*VP22
N31=C2*N22 + No31 + N12*S2
GAM1=N31
GAM2=N32
GAM3=N33 + IA3*QDP3
GAM4=N34 + IA4*QDP4
GAM5=N35 + IA5*QDP5
GAM6=N36 + IA6*QDP6
```

Appendix 8

Computation of the inertia matrix of tree structured robots

In this appendix, we develop a method to compute efficiently the inertia matrix of tree structured robots. Note that a serial robot is a special case of the tree structured robot. This method is based on the utilization of a simplified special case of Newton-Euler algorithm and on the concept of composite links [Khalil 90b].

A8.1. Inertial parameters of a composite link

The composite link j^+ is composed of link j and of the links supported by link j (Figure A8.1). The inertial parameters of the composite link j^+ can be calculated in terms of the standard parameters (or base parameters) of its links using the following recursive algorithm:

i) initialization. For j = 1, ..., n:

$${}^{j}\mathbf{J}_j^+ = {}^{j}\mathbf{J}_j,\ {}^{j}\mathbf{MS}_j^+ = {}^{j}\mathbf{MS}_j,\ M_j^+ = M_j$$

We recall that a(j) indicates the link that is antecedent to link j;

ii) for j = n, ..., 2 and a(j) ≠ 0:

$${}^{a(j)}\mathbf{J}_{a(j)}^+ = {}^{a(j)}\mathbf{J}_{a(j)}^+ + {}^{a(j)}\mathbf{A}_j\,{}^{j}\mathbf{J}_j^+\,{}^{j}\mathbf{A}_{a(j)} - [{}^{a(j)}\hat{\mathbf{P}}_j\,{}^{a(j)}\hat{\mathbf{MS}}_j^+ + ({}^{a(j)}\hat{\mathbf{P}}_j\,{}^{a(j)}\hat{\mathbf{MS}}_j^+)^T] + {}^{a(j)}\hat{\mathbf{P}}_j\,{}^{a(j)}\hat{\mathbf{P}}_j^T\,M_j^+ \qquad [A8.1a]$$

$${}^{a(j)}\mathbf{MS}_{a(j)}^+ = {}^{a(j)}\mathbf{MS}_{a(j)}^+ + {}^{a(j)}\mathbf{MS}_j^+ + {}^{a(j)}\mathbf{P}_j\,M_j^+ \qquad [A8.1b]$$

$$M_{a(j)}^+ = M_{a(j)}^+ + M_j^+ \qquad [A8.1c]$$

with:

- ${}^{a(j)}\mathbf{MS}_j^+ = {}^{a(j)}\mathbf{A}_j\,{}^j\mathbf{MS}_j^+$;
- $\hat{\mathbf{v}}$: (3x3) skew-symmetric matrix of the components of the vector **v**;
- ${}^{a(j)}\mathbf{T}_j = \begin{bmatrix} {}^{a(j)}\mathbf{A}_j & {}^{a(j)}\mathbf{P}_j \\ \mathbf{0}_{3,1} & 1 \end{bmatrix}$;
- ${}^j\mathbf{J}_j^+$: inertia tensor of the composite link j^+ referred to frame R_j;
- ${}^j\mathbf{MS}_j^+$: first moments of the composite link j^+ referred to frame R_j;
- M_j^+: mass of the composite link j^+.

We note that equations [A8.1] are equivalent to the following:

$$ {}^{a(j)}\mathbf{K}_{a(j)}^+ = {}^{a(j)}\mathbf{K}_{a(j)}^+ + {}^{a(j)}\boldsymbol{\lambda}_j\,{}^j\mathbf{K}_j^+ \qquad [A8.2] $$

where ${}^{a(j)}\boldsymbol{\lambda}_j$ and ${}^j\mathbf{K}_j$ are defined in Chapters 9 and 10 and Appendix 7.

NOTE.– The relationship between the concept of composite link parameters and base inertial parameters is considered in [Khalil 90a].

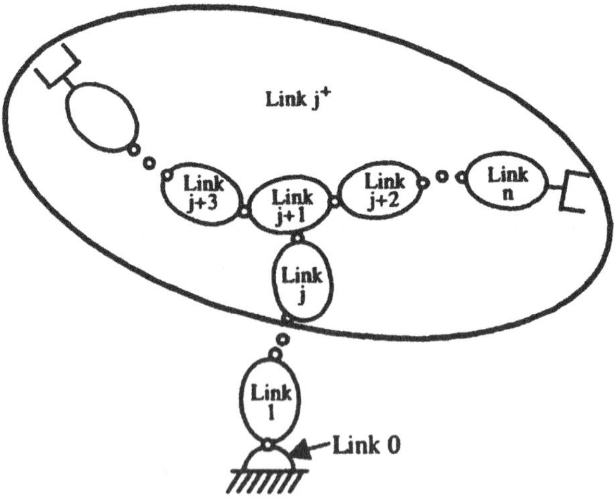

Figure A8.1. *Composite link j^+*

A8.2. Computation of the inertia matrix

We have seen in § 9.7.1 that the jth column of the inertia matrix $\mathbf{A}$ can be computed by the Newton-Euler inverse dynamic algorithm by setting:

$$\ddot{\mathbf{q}} = \mathbf{u}_j,\ \dot{\mathbf{q}} = \mathbf{0},\ \mathbf{g} = \mathbf{0},\ \mathbf{F}_c = \mathbf{0}\ (\mathbf{f}_{ei} = \mathbf{0},\ \mathbf{m}_{ei} = \mathbf{0} \quad \text{for } i = 1,\ldots, n)$$

where $\mathbf{u}_j$ is an (nx1) vector with a 1 in the jth row and zeros elsewhere.

Under these conditions, the forward recursive equations of the Newton-Euler inverse dynamic (Chapter 9) are only applied to link j^+:

$$^k\omega_k = 0,\ ^k\dot{\omega}_k = 0,\ ^k\dot{\mathbf{V}}_k = 0,\ ^k\mathbf{F}_k = 0,\ ^k\mathbf{M}_k = 0 \quad \text{for } k < j \qquad \text{[A8.3]}$$

$$^j\omega_j = 0 \qquad \text{[A8.4]}$$

$$^j\dot{\omega}_j = \bar{\sigma}_j\, {}^j\mathbf{a}_j \qquad \text{[A8.5]}$$

$$^j\dot{\mathbf{V}}_j = \sigma_j\, {}^j\mathbf{a}_j \qquad \text{[A8.6]}$$

$$^j\mathbf{F}_j = M_j^+\, {}^j\dot{\mathbf{V}}_j + {}^j\dot{\omega}_j \times {}^j\mathbf{MS}_j^+ \qquad \text{[A8.7]}$$

$$^j\mathbf{M}_j = {}^j\mathbf{J}_j^+\, {}^j\dot{\omega}_j + {}^j\mathbf{MS}_j^+ \times {}^j\dot{\mathbf{V}}_j \qquad \text{[A8.8]}$$

We deduce that:

- if joint j is prismatic ($^j\dot{\omega}_j = 0$, $^j\mathbf{M}_j = 0$ and $^j\dot{\mathbf{V}}_j = [\,0 \quad 0 \quad 1\,]^T$), then:

$$^j\mathbf{F}_j = [0 \quad 0 \quad M_j^+]^T \qquad \text{[A8.9]}$$

$$^j\mathbf{M}_j = [MY_j^+ \quad -MX_j^+ \quad 0]^T \qquad \text{[A8.10]}$$

- if joint j is revolute ($^j\dot{\mathbf{V}}_j = 0$ and $^j\dot{\omega}_j = [\,0 \quad 0 \quad 1\,]^T$), then:

$$^j\mathbf{F}_j = [-MY_j^+ \quad MX_j^+ \quad 0]^T \qquad \text{[A8.11]}$$

$$^j\mathbf{M}_j = [XZ_j^+ \quad YZ_j^+ \quad ZZ_j^+]^T \qquad \text{[A8.12]}$$

The recursive backward computation starts by link j and ends with link s, where a(s) = 0. The algorithm is given by the following equations:

- if joint j is prismatic, then:

$$^j\mathbf{f}_j = {}^j\mathbf{F}_j = [0 \quad 0 \quad M_j^+]^T \qquad \text{[A8.13]}$$

$$^{j}\mathbf{m}_j = {}^{j}\mathbf{M}_j = [MY_j^{+} \quad -MX_j^{+} \quad 0]^T \qquad [A8.14]$$

$$A_{j,j} = M_j^{+} + Ia_j \qquad [A8.15]$$

– if joint j is revolute, then:

$$^{j}\mathbf{f}_j = {}^{j}\mathbf{F}_j = [-MY_j^{+} \quad MX_j^{+} \quad 0]^T \qquad [A8.16]$$

$$^{j}\mathbf{m}_j = {}^{j}\mathbf{M}_j = [XZ_j^{+} \quad YZ_j^{+} \quad ZZ_j^{+}]^T \qquad [A8.17]$$

$$A_{j,j} = ZZ_j^{+} + Ia_j \qquad [A8.18]$$

Then, the following equations are computed for $k = j$, $a(j)$, $a(a(j))$, ..., s, where $a(s) = 0$:

$$^{a(k)}\mathbf{f}_{a(k)} = {}^{a(k)}\mathbf{A}_k \, {}^{k}\mathbf{f}_k \qquad [A8.19]$$

$$^{a(k)}\mathbf{m}_{a(k)} = {}^{a(k)}\mathbf{A}_k \, {}^{k}\mathbf{m}_k + ({}^{a(k)}\mathbf{P}_k \times {}^{a(k)}\mathbf{f}_{a(k)}) \qquad [A8.20]$$

$$\Gamma_{a(k)} = A_{a(k),j} = (\sigma_{a(k)} \mathbf{f}_{a(k)} + \bar{\sigma}_{a(k)} \mathbf{m}_{a(k)})^T \, {}^{a(k)}\mathbf{a}_{a(k)} \qquad [A8.21]$$

NOTES.–

– the element $A_{i,j}$ of the inertia matrix is set to zero if link i does not belong to the path between the base and link j;

– this algorithm provides the elements of the lower part of the inertia matrix. The other elements are deduced using the fact that the inertia matrix **A** is symmetric.

Appendix 9

Stability analysis using Lyapunov theory

In this appendix, we present some results about the stability analysis of nonlinear systems using Lyapunov theory. It is largely based on [Slotine 91] and [Zodiac 96].

A9.1. Autonomous systems

Let us consider the autonomous system (i.e. time-invariant) represented by the following state equation:

$$\dot{\mathbf{x}} = \mathbf{f}(\mathbf{x}) \qquad [A9.1]$$

A9.1.1. *Definition of stability*

An equilibrium point $\mathbf{x} = \mathbf{0}$ such that $\mathbf{f}(\mathbf{x}) = \mathbf{0}$ is said to be:

a) stable if for any $\varepsilon > 0$, there exists $R > 0$ such that if $\|\mathbf{x}(0)\| < \varepsilon$, then $\|\mathbf{x}(t)\| < R$;

b) asymptotically stable if for any $\varepsilon > 0$ and if $\|\mathbf{x}(0)\| < \varepsilon$, then $\|\mathbf{x}(t)\| \to 0$ as $t \to \infty$;

c) exponentially stable if there exist two strictly positive numbers α and λ such that:
$\|\mathbf{x}(t)\| \leq \alpha \exp(-\lambda\, t)\, \|\mathbf{x}(0)\|$

d) an equilibrium point is globally asymptotically (exponentially) stable if it is asymptotically (exponentially) stable for any initial value $\mathbf{x}(0)$. A linear system is always globally exponentially stable or unstable.

Some of these definitions are illustrated in Figure A9.1.

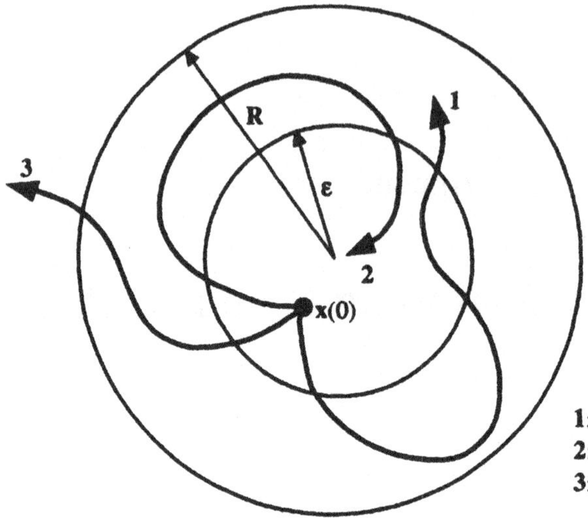

Figure A9.1. *Stability definition*

A9.1.2. *Positive definite and positive semi-definite functions*

The real function V(**x**) is positive definite (PD) in a ball B at the equilibrium point **x** = **0** if V(**x**) > 0 and V(**0**) = 0. The function V(**x**) should have continuous partial derivatives. Moreover, for some ε > 0, V(**x**) should be less than ε in a finite region at the origin.

If V(**x**) ≥ 0, then the function is positive semi-definite (PSD).

A9.1.3. *Lyapunov direct theorem (sufficient conditions)*

If there exists V(**x**) PD in a ball B around the equilibrium point **x** = **0** and if:

- $\dot{V}(\mathbf{x})$ is negative semi-definite (NSD), then **0** is a stable equilibrium point;
- $\dot{V}(\mathbf{x})$ is negative definite (ND), then **0** is asymptotically stable;
- $\dot{V}(\mathbf{x})$ is NSD and ≠ 0 along all the trajectory, then **0** is asymptotically stable.

Moreover, if V(**x**) is PD all over the state space $\forall \mathbf{x} \neq \mathbf{0}$, $V(\mathbf{x}) \rightarrow 0$ as $\mathbf{x} \rightarrow \mathbf{0}$, $\lim V(\mathbf{x}) \rightarrow \infty$ as $\|\mathbf{x}\| \rightarrow \infty$ and if $\dot{V}(\mathbf{x})$ is ND, then **0** is globally asymptotically stable.

A Lyapunov function can be interpreted as an energy function.

A9.1.4. *La Salle theorem and invariant set principle*

If $\dot{V}(\mathbf{x})$ is only NSD, it is yet possible to prove that the system is asymptotically stable, thanks to La Salle theorem [Hahn 67].

Definition. The set G is invariant for a dynamic system if every trajectory starting in G remains in G $\forall t$.

Theorem. Let R be the set of all points where $\dot{V} = 0$ and M be the largest invariant set of R; then every solution originating from R tends to M as $t \to \infty$.

A9.2. Non-autonomous systems

Let us consider the non-autonomous system (i.e. time-varying) represented by the following state equation:

$$\dot{\mathbf{x}} = \mathbf{f}(\mathbf{x}, t) \qquad [A9.2]$$

A9.2.1. *Definition of stability*

The equilibrium point $\mathbf{x} = \mathbf{0}$ such that $\mathbf{f}(\mathbf{x}, 0) = \mathbf{0}$ $\forall t \geq t_0$ is said to be:

a) stable at $t = t_0$, if for any $\varepsilon > 0$ there exists $R(\varepsilon, t_0) > 0$ such that if $\|\mathbf{x}(t_0)\| < \varepsilon$ then $\|\mathbf{x}(t)\| < R$ $\forall t \geq t_0$;

b) asymptotically stable at $t = t_0$, if it is stable and if there exists $R(t_0) > 0$ such that $\|\mathbf{x}(t_0)\| < R(t_0) \Rightarrow \mathbf{x}(t) \to \mathbf{0}$ as $t \to \infty$;

c) exponentially stable, if there exist two positive numbers α and λ such that:

$\|\mathbf{x}(t)\| \leq \alpha \exp(-\lambda (t-t_0)) \|\mathbf{x}(t_0)\|$, $\forall t \geq t_0$ for $\mathbf{x}(t_0)$ sufficiently small;

d) globally asymptotically stable, if it is stable and if $\mathbf{x}(t) \to \mathbf{0}$ as $t \to \infty$, $\forall \mathbf{x}(t_0)$;

e) uniformly stable, if $R = R(\varepsilon)$ can be chosen independently of t_0.

A9.2.2. *Lyapunov direct method*

Definition 1 (Function of class K)

A continuous function $\alpha : \Re^+ \to \Re^+$ is of class K if $\alpha(0) = 0$, $\alpha(\sigma) > 0$ $\forall \sigma > 0$, and α is non-decreasing.

Definition 2 (PD function)

A function $V(\mathbf{x}, t)$ is locally (globally) PD if and only if there exists a function α of class K such that $V(\mathbf{0}, t) = 0$, $V(\mathbf{x}, t) \geq \alpha(\|\mathbf{x}\|)$, $\forall t \geq 0$ and $\forall \mathbf{x}$ in a ball B.

Definition 3 (Decreasing function)

A function $V(\mathbf{x}, t)$ is locally (globally) decreasing if there exists a function α of class K such that $V(\mathbf{0}, t) = 0$ and $V(\mathbf{x}, t) \leq \alpha(\|\mathbf{x}\|)$, $\forall t > 0$ and $\forall \mathbf{x}$ in a ball B.

Lyapunov theorem

Let us assume that in a ball B around the equilibrium point $\mathbf{x} = \mathbf{0}$:

- there exists a Lyapunov function $V(\mathbf{x}, t)$ whose first derivatives are continuous;
- there exist functions α, β, γ of class K;

then, the equilibrium point is:

a) stable if $V(\mathbf{x}, t) \geq \alpha(\|\mathbf{x}\|)$, $\dot{V}(\mathbf{x}, t) \leq 0$;

b) uniformly stable if $\alpha(\|\mathbf{x}\|) \leq V(\mathbf{x}, t) \leq \beta(\|\mathbf{x}\|)$, $\dot{V}(\mathbf{x}, t) \leq 0$;

c) uniformly asymptotically stable if:

$\alpha(\|\mathbf{x}\|) \leq V(\mathbf{x}, t) \leq \beta(\|\mathbf{x}\|)$, $\dot{V}(\mathbf{x}, t) \leq -\gamma(\|\mathbf{x}\|) < 0$;

d) globally uniformly asymptotically stable if:

$\alpha(\|\mathbf{x}\|) \leq V(\mathbf{x}, t) \leq \beta(\|\mathbf{x}\|)$, $\dot{V}(\mathbf{x}, t) \leq -\gamma(\|\mathbf{x}\|) < 0$, $\alpha(\|\mathbf{x}\|) \to \infty$ as $\mathbf{x} \to \infty$.

Barbalat lemma. If $\dot{\mathbf{f}}(t)$ is a uniformly continuous function such that lim $\mathbf{f}(t)$ is bounded as $t \to \infty$, then $\dot{\mathbf{f}}(t) \to 0$ as $t \to \infty$.

Barbalat theorem. If $V(\mathbf{x}, t)$ has a lower bound such that $V(\mathbf{x}, t) \geq \alpha(\|\mathbf{x}\|)$ and if $\dot{V}(\mathbf{x}, t) \leq 0$, then $\dot{V}(\mathbf{x}, t) \to 0$ as $t \to \infty$ if $\dot{V}(\mathbf{x}, t)$ is uniformly continuous with respect to time.

Appendix 10

Computation of the dynamic control law in the task space

In this appendix, we present the computation of the decoupling nonlinear control in the task space [Khalil 87a]. The dynamic model is computed by a specialized Newton-Euler algorithm, which takes into account many variables that are evaluated for the kinematic models.

The number of operations of this control law for the Stäubli RX-90 robot, assuming symmetrical links whose inertial parameters are given in Table 9.4, is 316 multiplications and 237 additions.

A10.1. Calculation of the location error $\mathbf{e_x}$

The current location of the terminal link is given by the homogeneous transformation matrix ${}^0\mathbf{T}_n$, which can be obtained using a customized symbolic algorithm (Chapter 3):

$$
{}^0\mathbf{T}_n = {}^0\mathbf{T}_1 \, {}^1\mathbf{T}_2 \ldots {}^{n-1}\mathbf{T}_n = \begin{bmatrix} {}^0\mathbf{s}_n & {}^0\mathbf{n}_n & {}^0\mathbf{a}_n & {}^0\mathbf{P}_n \\ 0 & 0 & 0 & 1 \end{bmatrix} \qquad \text{[A10.1]}
$$

Let the desired position and orientation be given by ${}^0\mathbf{P}_n^d$ and $[{}^0\mathbf{s}_n^d \;\; {}^0\mathbf{n}_n^d \;\; {}^0\mathbf{a}_n^d]$ respectively. The location error, denoted by $\mathbf{e_x}$, is given by:

$$
\mathbf{e_x} = [\; \mathbf{dX}_p^T \;\; \mathbf{dX}_r^T \;]^T \qquad \text{[A10.2]}
$$

where $\mathbf{dX}_p$ and $\mathbf{dX}_r$ indicate the position error and orientation error respectively.

The position error is obtained by:

$$dX_p = dP = {}^0P^d_n - {}^0P_n \qquad [A10.3]$$

The orientation error is given by [Luh 80a]:

$$dX_r = {}^0u\,\alpha \qquad [A10.4]$$

where ${}^0u$ and α are obtained by solving $rot(u,\alpha)\,{}^0A_n = {}^0A^d_n$.

Let us assume that:

$${}^0A^d_n\,{}^0A^T_n = [\,s \;\; n \;\; a\,] \qquad [A10.5]$$

If α is small, the orientation error dX_r can be considered to be equal to $u\sin(\alpha)$ or equal to ${}^0\delta_n$ (§ 2.5). Using equation [2.35], we obtain:

$$dX_r = u\sin(\alpha) = \frac{1}{2}\begin{bmatrix} n_z - a_y \\ a_x - s_z \\ s_y - n_x \end{bmatrix} \qquad [A10.6]$$

Using equation [5.59], we obtain:

$$dX_r = {}^0\delta_n = \Omega^+_{CD}\begin{bmatrix} {}^0s^d_n - {}^0s_n \\ {}^0n^d_n - {}^0n_n \\ {}^0a^d_n - {}^0a_n \end{bmatrix} \qquad [A10.7]$$

$$dX_r = \frac{1}{2}\,[{}^0s_n \times {}^0s^d_n + {}^0n_n \times {}^0n^d_n + {}^0a_n \times {}^0a^d_n] \qquad [A10.8]$$

A10.2. Calculation of the velocity of the terminal link $\dot{X}$

The terminal link velocity is composed of the linear velocity ${}^0V_n$ and of the angular velocity ${}^0\omega_n$:

$$\dot{X} = \begin{bmatrix} {}^0V_n \\ {}^0\omega_n \end{bmatrix} \qquad [A10.9]$$

${}^0\mathbf{V}_n$ and ${}^0\omega_n$ are calculated using the following recursive equations, which are developed in Chapter 9:

For j = 1, ..., n:

$${}^j\mathbf{V}_j = {}^j\mathbf{A}_{j-1}({}^{j-1}\mathbf{V}_{j-1} + {}^{j-1}\omega_{j-1} \times {}^{j-1}\mathbf{P}_j) + \sigma_j \dot{q}_j \, {}^j\mathbf{a}_j \quad [A10.10]$$

$${}^j\omega_{j-1} = {}^j\mathbf{A}_{j-1} \, {}^{j-1}\omega_{j-1} \quad [A10.11]$$

$${}^j\omega_j = {}^j\omega_{j-1} + \bar{\sigma}_j \dot{q}_j \, {}^j\mathbf{a}_j \quad [A10.12]$$

The vectors ${}^0\mathbf{V}_n$ and ${}^0\omega_n$ are obtained by ${}^0\mathbf{V}_n = {}^0\mathbf{A}_n \, {}^n\mathbf{V}_n$ and ${}^0\omega_n = {}^0\mathbf{A}_n \, {}^n\omega_n$. We note that ${}^1\omega_1, \ldots, {}^n\omega_n$ are also required for the inverse dynamic algorithm.

A10.3. Calculation of $\dot{\mathbf{J}}\,\dot{\mathbf{q}}$

The calculation of this vector by differentiating the Jacobian matrix with respect to time and the multiplication of the result by $\dot{\mathbf{q}}$ would need a prohibitive number of operations. We propose to use an efficient recursive algorithm derived from the second order kinematic model. Many intermediate variables of this algorithm are used for the computation of the inverse dynamic model. The second order kinematic model is given by:

$$\ddot{\mathbf{X}}_n = \begin{bmatrix} \dot{\mathbf{V}}_n \\ \dot{\omega}_n \end{bmatrix} = \mathbf{J}(\mathbf{q})\,\ddot{\mathbf{q}} + \dot{\mathbf{J}}(\mathbf{q}, \dot{\mathbf{q}})\,\dot{\mathbf{q}} \quad [A10.13]$$

From equation [A10.13], we deduce that $\dot{\mathbf{J}}(\mathbf{q}, \dot{\mathbf{q}})\,\dot{\mathbf{q}}$ is equal to $\ddot{\mathbf{X}}_n$ when setting $\ddot{\mathbf{q}} = \mathbf{0}$.

$$\dot{\mathbf{J}}(\mathbf{q}, \dot{\mathbf{q}})\,\dot{\mathbf{q}} = \begin{bmatrix} \dot{\mathbf{V}}_n\,(\ddot{\mathbf{q}} = \mathbf{0}) \\ \dot{\omega}_n\,(\ddot{\mathbf{q}} = \mathbf{0}) \end{bmatrix} = \begin{bmatrix} \phi_n \\ \Psi_n \end{bmatrix} \quad [A10.14]$$

Consequently, $\dot{\mathbf{J}}\,\dot{\mathbf{q}}$ can be computed using equations [9.86], [9.87] and [9.90] after setting $\ddot{\mathbf{q}} = \mathbf{0}$. The algorithm is given as follows:

For j = 1, ..., n:

$${}^j\Psi_j = {}^j\mathbf{A}_{j-1} \, {}^{j-1}\Psi_{j-1} + \bar{\sigma}_j \, ({}^j\omega_{j-1} \times \dot{q}_j \, {}^j\mathbf{a}_j) \quad [A10.15]$$

$${}^j\mathbf{U}_j^* = {}^j\hat{\Psi}_j + {}^j\hat{\omega}_j \, {}^j\hat{\omega}_j \quad [A10.16]$$

$${}^j\phi_j = {}^j\mathbf{A}_{j-1} \, ({}^{j-1}\phi_{j-1} + {}^{j-1}\mathbf{U}_{j-1}^* \, {}^{j-1}\mathbf{P}_j) + 2\,\sigma_j \, {}^j\omega_{j-1} \times (\dot{q}_j \, {}^j\mathbf{a}_j) \quad [A10.17]$$

The initial values are: ${}^0\Psi_0 = \mathbf{0}$ and ${}^0\phi_0 = \mathbf{0}$.

A10.4. Calculation of $\mathbf{J}(\mathbf{q})^{-1}\,\mathbf{y}$

The vector $\mathbf{y}$ indicates the term $(\mathbf{w}(t) - \dot{\mathbf{J}}\,\dot{\mathbf{q}})$ as given in equation [14.33]. This problem is treated in Chapter 6. The solution for the regular case of the Stäubli RX-90 robot is developed in Example 6.1.

A10.5. Modified dynamic model

We modify the inverse dynamic model developed in Chapter 9 to take into account the availability of $\mathbf{\Phi}_j$ and $\mathbf{\Psi}_j$. Equations [9.86] and [9.87], giving $\dot{\omega}_j$ and $\dot{\mathbf{V}}_j$, are replaced by:

$$^j\dot{\omega}_j = {}^j\mathbf{\Psi}_j + {}^j\boldsymbol{\varepsilon}_j \qquad [A10.18]$$

$$^j\dot{\mathbf{V}}_j = {}^j\boldsymbol{\phi}_j + {}^j\mathbf{b}_j \qquad [A10.19]$$

where ${}^j\boldsymbol{\varepsilon}_j$ and ${}^j\mathbf{b}_j$ represent ${}^j\dot{\omega}_j$ and ${}^j\dot{\mathbf{V}}_j$ respectively, when $\dot{\mathbf{q}} = \mathbf{0}$:

$$^j\boldsymbol{\varepsilon}_j = {}^j\mathbf{A}_{j-1}\,{}^{j-1}\boldsymbol{\varepsilon}_{j-1} + \bar{\sigma}_j\,\ddot{q}_j\,{}^j\mathbf{a}_j \qquad [A10.20]$$

$$^j\mathbf{b}_j = {}^j\mathbf{A}_{j-1}({}^{j-1}\mathbf{b}_{j-1} + {}^{j-1}\hat{\boldsymbol{\varepsilon}}_{j-1}\,{}^{j-1}\mathbf{P}_j) + \sigma_j\,\ddot{q}_j\,{}^j\mathbf{a}_j \qquad [A10.21]$$

with the initial values ${}^0\boldsymbol{\varepsilon}_0 = \mathbf{0}$ and ${}^0\mathbf{b}_0 = -\mathbf{g}$.

The matrix ${}^j\mathbf{U}_j$, defined in equation [9.90], is computed using ${}^j\mathbf{U}_j^*$ (equation [A10.16]) and ${}^j\boldsymbol{\varepsilon}_j$ (equation [A10.20]) such that:

$$^j\mathbf{U}_j = {}^j\mathbf{U}_j^* + {}^j\hat{\boldsymbol{\varepsilon}}_j \qquad [A10.22]$$

Taking into account that the angular velocities have been evaluated while computing $\dot{\mathbf{X}}$, the modified dynamic model needs 110 multiplications and 82 additions instead of 160 multiplications and 113 additions.

Appendix 11

Stability of passive systems

In this appendix, we present some useful results for the analysis and the design of passive and adaptive control laws [Landau 88]. For more details, the reader is referred to [Popov 73], [Desoer 75], [Landau 79].

A11.1. Definitions

Definition 1: positive real function

A rational function $h(s)$ of the complex variable $s = \sigma + j\omega$ is positive real if:

a) $h(s)$ is real when s is real;

b) $h(s)$ has no poles in the Laplace right half plane $Re(s) > 0$ (Re denotes the real part of s);

c) the possible poles of $h(s)$ along the axis $Re(s) = 0$ (when $s = j\omega$) are separate and the corresponding residuals are positive real or zero;

d) for any ω for which $s = j\omega$ is not a pole of $h(s)$, then $Re(h(s)) \geq 0$.

Definition 2: strictly positive real function

A rational function $h(s)$ of the complex variable $s = \sigma + j\omega$ is strictly positive real if:

a) $h(s)$ is real when s is real;

b) $h(s)$ has no poles in the Laplace right half plane $Re(s) \geq 0$;

c) $Re[h(j\,\omega)] > 0$ for any real value of ω, $-\infty < \omega < \infty$.

Definition 3: Hermitian matrix

A matrix **H**(s) of rational real functions in the complex variable $s = \sigma + j\omega$ is Hermitian if:

$$\mathbf{H}(s) = \mathbf{H}^T(s^*) \qquad [A11.1]$$

where s^* is the conjugate of s.

Definition 4: positive real matrix of functions

An (mxm) transfer matrix **H**(s) of rational real functions is positive real if:

a) no poles of the elements of **H**(s) are in the Laplace right half plane Re(s) > 0;

b) the possible poles of **H**(s) along the axis Re(s) = 0 are distinct and the corresponding matrix of residuals is Hermitian positive semi-definite;

c) the matrix $\mathbf{H}(j\omega) + \mathbf{H}^T(-j\omega)$ is Hermitian positive semi-definite for any real value ω that is not a pole of any element of **H**(s).

Definition 5: strictly positive real matrix of functions

An (mxm) transfer matrix **H**(s) of rational real functions is strictly positive real if:

a) no poles of the elements of **H**(s) are in the Laplace right half plane Re(s) ≥ 0;

b) the matrix $\mathbf{H}(j\omega) + \mathbf{H}^T(-j\omega)$ is Hermitian positive definite for any real value of ω.

A11.2. Stability analysis of closed-loop positive feedback

Let us consider the closed-loop system of Figure A11.1, where the linear and time-invariant feed-forward block is described by the following state equations [Landau 88]:

$$\begin{cases} \dot{\mathbf{x}} = \mathbf{A}\mathbf{x} + \mathbf{B}\mathbf{u}_1 = \mathbf{A}\mathbf{x} - \mathbf{B}\mathbf{y}_2 \\ \mathbf{y}_1 = \mathbf{C}\mathbf{x} + \mathbf{D}\mathbf{u}_1 = \mathbf{C}\mathbf{x} - \mathbf{D}\mathbf{y}_2 \end{cases} \qquad [A11.2]$$

in which (**A**, **B**) and (**A**, **C**) are controllable and observable respectively. The system is characterized by the transfer matrix **H**(s) defined by:

$$\mathbf{H}(s) = \mathbf{D} + \mathbf{C}[s\,\mathbf{I} - \mathbf{A}]^{-1}\,\mathbf{B} \qquad [A11.3]$$

The nonlinear time-varying feedback block is such that:

$$\mathbf{y}_2 = \mathbf{f}(\mathbf{u}_2, t, \tau) \quad \text{with } \tau \leq t \qquad [A11.4]$$

and satisfies the Popov inequality (proving the block passivity):

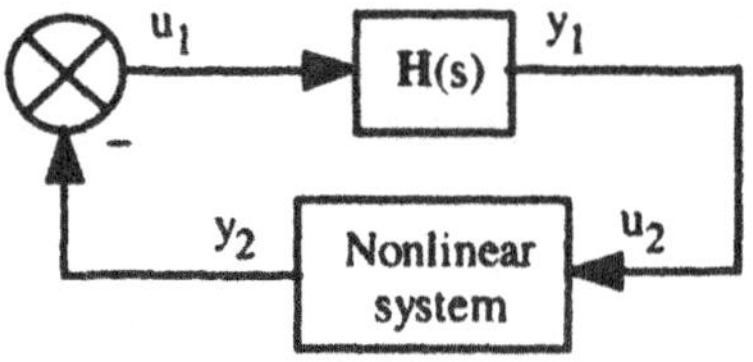

Figure A11.1. *Closed-loop positive feedback system*

$$\int_{t0}^{t1} y_2^T(t)\ u_2(t)\,dt \geq -\gamma_0^2 \quad \text{with } \gamma_0^2 < \infty, \text{ for } t_1 \geq t_0 \qquad [A11.5]$$

Theorem 1 (hyperstability)

For the closed-loop system of Figure A11.1 described by equations [A11.2], [A11.3] and [A11.4], and for any feedback block satisfying the inequality [A11.5], all solutions $\mathbf{x}(\mathbf{x}(0), t)$ satisfy the inequality:

$$\|\mathbf{x}(t)\| < \delta[\|\mathbf{x}(0)\| + \alpha_0] \quad \text{for } \delta > 0, \alpha_0 > 0, t \geq 0 \qquad [A11.6]$$

if, and only if, $\mathbf{H}(s)$ is a positive real transfer matrix.

Theorem 2 (asymptotic hyperstability)

For the closed-loop system of Figure A11.1 described by equations [A11.2], [A11.3] and [A11.4], and for any feedback block satisfying the inequality [A11.6], all solutions $\mathbf{x}(\mathbf{x}(0), t)$ satisfy both the inequality [A11.6] and $\lim \mathbf{x}(t) \to \mathbf{0}$ as $t \to \infty$ for any bounded input $\mathbf{u}_1(t)$ if, and only if, $\mathbf{H}(s)$ is a strictly positive real transfer matrix.

NOTE.– Theorems 1 and 2 provide sufficient conditions to prove the stability and asymptotic stability respectively in the case where the Popov inequality is satisfied by the feedback block.

A11.3. Stability properties of passive systems

Lemma 1

A feedback combination of two strictly passive (positive) systems is always asymptotically stable.

Lemma 2

Any system obtained by a parallel combination of two passive (positive) blocks is itself a passive (positive) system.

Lemma 3

Any system obtained by a feedback combination of two passive (positive) blocks is itself a passive (positive) system.

References

[Ait Mohamed 95] Ait Mohamed A., "Commande dynamique de robots redondants dans l'espace opérationnel", Thèse de Doctorat, Université de Nantes et Ecole Centrale de Nantes, France, February 1995.

[Aldon 82] Aldon M.J., "Elaboration automatique de modèles dynamiques de robots en vue de leur conception et de leur commande", Thèse d'Etat, USTL, Montpellier, France, October 1982.

[Aldon 86] Aldon M.J., "Identification des paramètres structuraux des robots manipulateurs", *Proc. Outils Mathématiques pour la Modélisation et la Commande des Robots*, Paris, France, September 1986, p. 243-296.

[An 85] An C.H., Atkeson C.G., Hollerbach J.M., "Estimation of inertial parameters of rigid body links of manipulators", *Proc. 24th IEEE Conf. on Decision and Control*, Fort Lauderdale, USA, December 1985, p. 990-995.

[An 87] An C.H., Hollerbach J.M., "Kinematic stability issues in force control of manipulators, *Proc. IEEE Int. Conf. on Robotics and Automation*, Raleigh, USA, March-April 1987, p. 897-903.

[Angeles 88] Angeles J., Gosselin C., "Détermination du degré de liberté des chaînes cinématiques", *Trans. of the CSME*, Vol. 12, 1977, p. 219-226.

[Arimoto 84] Arimoto S., Miyazaki F., "Stability and robustness of PID feedback control for robots manipulators of sensory capability", *The 1st Int. Symp. of Robotics Research*, MIT Press, Cambridge, USA, 1984.

[Arimoto 93] Arimoto S., Liu Y.H., Naniwa T., "Model-based adaptive hybrid control for geometrically constrained robots", *Proc. IEEE Int. Conf. on Robotics and Automation*, Atlanta, USA, May 1993, p. 618-623.

[Armstrong 79] Armstrong W.W., "Recursive solution to the equation of motion of an N-links manipulator", *Proc. 5th World Congress on Theory of Machines and Mechanisms*, Montréal, Canada, 1979, p. 1343-1346.

[Armstrong 86] Armstrong B., Khatib O., Burdick J., "The explicit dynamic model and inertial parameters of the PUMA 560 Arm", *Proc. IEEE Int. Conf. on Robotics and Automation*, San Francisco, USA, April 1986, p. 510-518.

[Armstrong 88] Armstrong B., "Dynamics for robot control: friction modeling and ensuring excitation during parameter identification", Ph. D Thesis, Dept. of Electrical Engineering, Stanford University, Stanford, USA, May 1988.

[Armstrong 89] Armstong B., "On finding exciting trajectories for identification experiments involving systems with non linear dynamics", *The Int. J. of Robotics Research*, Vol. 8(6), 1989, p. 28-48.

[Armstrong 91] Armstrong B., *Control of Machines with frictions*, Kluwer Academic Publ., Boston, USA, 1991.

[Armstrong 94] Armstrong-Hélouvry B., Dupont P., Canudas de Wit C., "A survey of analysis tools and compensation methods for the control of machines with friction", *Automatica*, Vol. 30(10), 1994, p. 1083-1138.

[Asada 86] Asada H., Slotine J.-J.E., *Robot analysis and control*, John Wiley & Sons, New York, USA, 1986.

[Atkeson 86] Atkeson C.G., An C.H., Hollerbach J.M., "Estimation of inertial parameters of manipulator loads and links", *The Int. J. of Robotics Research*, Vol. 5(3), 1986, p. 101-119.

[Aubin 91] Aubin A., "Modélisation, identification et commande du bras manipulateur TAM", Thèse de Doctorat, INPG, Grenoble, France, 1991.

[Baillieul 84] Baillieul J., Hollerbach J.M., Brockett R., "Programming and control of kinematically redundant manipulators", *Proc. 23rd IEEE Conf. on Decision and Control*, Las Vegas, USA, December 1984, p. 768-774.

[Baillieul 85] Baillieul J., "Kinematic programming alternatives for redundant manipulators", *Proc. IEEE Int. Conf. on Robotics and Automation*, St Louis, USA, March 1985, p. 722-728.

[Baillieul 86] Baillieul J., "Avoiding obstacles and resolving kinematic redundancy", *Proc. IEEE Int. Conf. on Robotics and Automation*, San Francisco, USA, April 1986, p. 1698-1704.

[Baron 94] Baron L., Angeles J., "The decoupling of the direct kinematics of parallel manipulators using redundant sensors", *Proc. IEEE Int. Conf. on Robotics and Automation*, San Diego, USA, May 1994, p. 974-979.

[Bartels 88] Bartels R.H., Beaty J.C., Barsky B.A., *Mathématiques et CAO 6, B-splines*, Hermès, Paris, France, 1988.

[Baumgarte 72] Baumgarte J., "Stabilization of constraints and integral of motion", *Computer Methods in Applied Mech. Eng.*, Vol. 1(1), 1972, p. 1-16.

[Bayard 88] Bayard D., Wen J.T., "New class of control laws for robotic manipulators; Part 2: Adaptive case", *Int. J. Control*, Vol. 47(5), 1988, p. 1387-1406.

[Bejczy 74] Bejczy A.K., "Robot arm dynamics and control", NASA Technical Memorandum 33-669, Jet Propulsion Laboratory, Pasadena, USA, 1974.

[Bejczy 79] Bejczy A.K., "Dynamic models and control equations for manipulators", Tutorial Workshop, *18th IEEE Conf. on Decision and Control*, Fort Lauderdale, USA, December 1979.

[Bejczy 85] Bejczy A.K., Tarn T.J., Yun X., Hans S., "Non linear feedback control of Puma 560 robot arm by computer", *Proc. 24th IEEE Conf. on Decision and Control*, Fort Lauderdale, USA, December 1985, p. 1680-1688.

[Beji 97] Beji L., "Modélisation, identification et commande d'un robot parallèle", Thèse de Doctorat, Université d'Evry-Val d'Essone, France, June 1997.

[Benallegue 91] Bennallegue A., "Contribution à la commande dynamique adaptative des robots manipulateurs rapides", Thèse de Doctorat, Université Pierre et Marie Curie, Paris, France, November 1991.

[Benhlima 93] Benhlima, S., "Identification des paramètres dynamiques des systèmes mécaniques articulés complexes", Thèse de Doctorat, ENSAM, Paris, France, 1993.

[Bennett 03] Bennett G.T., "A new mechanism", *Engineering*, Vol. 76, 1903., p. 777-778.

[Bennis 91a] Bennis F., "Contribution à la modélisation géométrique et dynamique des robots à chaîne simple et complexe", Thèse de doctorat, E.N.S.M, Nantes, France, 1991.

[Bennis 91b] Bennis F., Khalil W., "Minimum inertial parameters of robots with parallelogram closed-loops", *IEEE Trans. on Systems, Man, and Cybernetics*, Vol. SMC-21(2), 1991, p. 318-326.

[Bennis 93] Bennis F., Khalil W., "Modèle géométrique inverse des robots à chaîne découplable : application aux équations de contraintes des boucles fermées", *Trans. of the Canadian Society for Mechanical Engineering*, Vol. 17(4A), 1993, p. 473-491.

[Benoit 75] Benoit M., Briot M., Donnarel H., Liégeois A., Meyer M.A., Renaud M., "Synthèse de la commande dynamique d'un téléopérateur redondant", *Revue RAIRO*, Série J-2, May 1975, p. 89-103.

[Berghuis 93] Berghuis H., "Model-based robot control: from theory to practice", Ph. D. Thesis, University of Twente, Enschede, The Netherlands, 1993.

[Bernhardt 93] Bernhardt R., Albright S.L., *Robot calibration*, Chapman & Hall, London, UK, 1993.

[Besnard 99] Besnard S., Khalil W., "Calibration of parallel robots using two inclinometers", *Proc. IEEE Int. Conf. on Robotics and Automation*, Detroit, USA, May 1999, p. 1758-1763.

[Besnard 00a] Besnard S., "Etalonnage géométrique des robots série et parallèles", Thèse de Doctorat, Université de Nantes, France, 1996.

[Besnard 00b] Besnard S., Khalil W., Garcia G., "Robot calibration using multiple plane constraints", *Proc. 7th Int. Symp. on Advances in Robot Kinematics, ARK 2000*, Slovenia, June 2000, p. 61-70.

[Besnard 01] Besnard S., Khalil W., "Identifiable parameters for parallel robots kinematic calibration", *Proc. IEEE Int. Conf. on Robotics and Automation*, Seoul, Korea, May 2001, p. 2859-2866.

[Bicchi 98] Bicchi A., "Manipulability of cooperating robots with passive joints", *Proc. IEEE Int. Conf. on Robotics and Automation*, Louvain, Belgium, May 1998, p. 1038-1044.

[Binford 77] Binford T.O. et al., "Discussion of trajectory calculation methods", *in* 'Exploratory study of computer integrated assembly system', Stanford Artificial Intelligence Lab., Progress Report, Memo AIM-285.4, Stanford, USA, June 1977.

[Blanchon 87] Blanchon J.-L., "Génération et identification des lois de mouvement en robotique : application à l'identification et à la correction de trajectoires du manipulateur PUMA 560, Thèse de Doctorat, USTL, Montpellier, France, March 1987.

[Borm 91] Borm J.H., Menq C.H., "Determination of optimal measurement configurations for robot calibration based on observability measure", *The Int. J. of Robotics Research*, Vol. 10(1), p. 51-63, 1991.

[Borrel 79] Borrel P., "Modèle de comportement de manipulateurs ; application à l'analyse de leurs performances et à leur commande automatique", Thèse de Troisième Cycle, USTL, Montpellier, France, December 1979.

[Borrel 86] Borrel P., "Contribution à la modélisation géométrique des robots-manipulateurs ; application à la conception assistée par ordinateur", Thèse d'Etat, USTL, Montpellier, France, July 1986.

[Boullion 71] Boullion T.L., Odell P.L., *Generalized inverse matrices*, John Wiley & Sons, New York, USA, 1971.

[Bouzouia 89] Bouzouia B., "Commande des robots manipulateurs : identification des paramètres et étude des stratégies adaptatives", Thèse de Doctorat, UPS, Toulouse, France, May 1989.

[Boyer 94] Boyer F., "Contribution à la modélisation et à la commande dynamique des robots flexible", Thèse de Doctorat, Université Pierre et Marie Curie, Paris, France, 1994.

[Boyer 98] Boyer F., Khalil W., "An efficient calculation of flexible manipulator inverse dynamics", *The Int. J. of Robotics Research*, Vol. 17(3), 1998, p. 282-293.

[Brandl 86] Brandl H., Johanni R., Otter M., "A very efficient algorithm for the simulation of robots and multibody systems without inversion of the mass matrix", *Proc. IFAC Symp. on Theory of Robots*, Vienna, Austria, December1986, p. 365-370.

[Brogliato 91] Brogliato B., "Systèmes passifs et commande adaptative des manipulateurs", Thèse de Doctorat, INPG, Grenoble, France, 1991.

[Bruyninckx 98] Bruyninckx H., "Closed-form forward position kinematics for a $(3\text{-}1\text{-}1\text{-}1)^2$ fully parallel manipulator", *IEEE Trans. on Robotics and Automation*, Vol. RA-14(2), 1998, p. 326-328.

[Burdick 86] Burdick J.W., "An algorithm for generation of efficient manipulator dynamic equations", *Proc. IEEE Int. Conf. on Robotics and Automation*, San Francisco, USA, April 1986, p. 212-218.

[Burdick 88] Burdick J.W., "Kinematic analysis and design of redundant manipulators", Ph. D Thesis, Stanford University, Stanford, USA, 1988.

[Caenen 93] Caenen J.-L., "Contribution à l'identification de paramètres géométriques et non géométriques d'un modèle de robot. Application à l'amélioration de la précision de positionnement statique",. Thèse de Doctorat, Université de Valenciennes et de Hainault-Cambrésis, France, January 1993.

[Cannon 84] Cannon H., Schmitz E., "Initial experiments on the end-point control of a flexible one-link robot", *The Int. J. of Robotics Research*, Vol. 3(3), 1984, p. 62-75.

[Canudas de Wit 89] Canudas de Wit C., Noël P., Aubin A., Brogliato B., Drevet P., "Adaptive Friction compensation in robot manipulators: low-velocities", *Proc. IEEE Int. Conf. on Robotics and Automation*, Scottsdale, USA, May 1989, p. 1352-1357.

[Canudas de Wit 90] Canudas de Wit C., Seront V., "Robust adaptive friction compensation", *Proc. IEEE Int. Conf. on Robotics and Automation*, Cincinnati, USA, May 1990, p. 1383-1389.

[Canudas de Wit 92] Canudas de Wit C., Fixot N., Aström K.J., "Trajectory tracking in robot manipulators via nonlinear estimated feedback", *IEEE Trans. on Robotics and Automation*, Vol. RA-8(1), 1992, p. 138-144.

[Castain 84] Castain R.H., Paul R.P., "An on-line dynamic trajectory generator", *The Int. J. of Robotics Research*, Vol. 3(1), 1984, p. 68-72.

[Castelain 86] Castelain J.M., "Application de la méthode hypercomplexe aux modélisations géométriques et différentielles des robots constitués d'une chaîne cinématique simple", Thèse d'Etat, Université de Valenciennes et du Hainaut-Cambresis, France, December 1986.

[Catia] Dassault Systèmes, 308 Bureaux de la Colline, 92210 Saint Cloud, France.

[Cesareo 84] Cesareo G., Nicolo F., Nicosia S., "DYMIR: a code for generating dynamic model of robots", *Proc. IEEE Int. Conf. on Robotics*, Atlanta, USA, March 1984, p. 115-120.

[Chace 67] Chace M.A., "Analysis of the time dependance of multi-freedoom mechanical system in relative coordinate", *Trans. of ASME, J. of Engineering for Industry*, Vol. 89, February 1967, p. 119-125.

[Chace 71] Chace M.A., Bayazitoglu Y.O., "Development and application of a generalized d'Alembert force for multi-freedom mechanical system", *Trans. ASME, J. of Engineering for Industry*, Vol. 93, February 1971, p. 317-327.

[Chand 85] Chand S., Doty K.L., "On-line polynomial trajectories for robot manipulators", *The Int. J. of Robotics Research*, Vol. 4(2), 1985, p. 38-48.

[Chang 86] Chang P.H., "A closed form solution for the control of manipulators with kinematic redundancy", *Proc. IEEE Int. Conf. on Robotics and Automation*, San Francisco, USA, April 1986, p. 9-14.

[Charentus 90] Charentus S., "Modélisation et commande d'un robot manipulateur redondant composé de plusieurs plates-formes de Stewart", Thèse de Doctorat, UPS, Toulouse, France, April 1990.

[Chedmail 86] Chedmail P., Gautier M., Khalil W., "Automatic modelling of robots including parameters of links and actuators", *Proc. IFAC Symp. on Theory of Robots*, Vienna, Austria, December 1986, p. 295-299.

[Chedmail 90a] Chedmail P.,"Contribution à la conception des robots et à la modélisation et commande de robots souples", Thèse d'Etat, ENSM, Nantes, France, January 1990.

[Chedmail 90b] Chedmail P., Gautier M., "Optimum choice of robot actuators", *Trans. of ASME, J. of Engineering for Industry*, Vol. 112(4), 1990, p. 361-367.

[Chedmail 92] Chedmail P., Dombre E., "CAO et robotique : conception et programmation des cellules robotisées", *Revue d'Automatique et de Productique Appliquées*, Vol. 5(2), 1992, p. 27-38.

[Chedmail 98] Chedmail P., Dombre E., Wenger P., *La CAO en robotique : outils et méthodologies*, Collection 'Etudes en Mécanique des Matériaux et des Structures', Hermès, Paris, France, 1998.

[Cheok 93] Cheok K.C., Overholt J.L., Beck R.R., "Exact methods for determining the kinematics of a Stewart platform using additional displacement sensors", *J. of Robotic Systems*, Vol. 10(5), 1993, p. 974-979.

[Cherki 96] Cherki B, "Commande des robots manipulateurs par retour d'état estimé", Thèse de Doctorat, Université de Nantes et Ecole Centrale de Nantes, France, 1996.

[Chevallereau 87] Chevallereau C., Khalil W., "Efficient method for the calculation of the pseudo inverse kinematic problem", *Proc. IEEE Int. Conf. on Robotics and Automation*, Raleigh, USA, March-April 1987, p. 1842-1848.

[Chevallereau 88] Chevallereau C., "Contribution à la commande des robots-manipulateurs dans l'espace opérationnel", Thèse de Doctorat, ENSM, Nantes, France, May 1988.

[Chevallereau 98] Chevallereau C., "Feasible trajectories in task space from a singularity for a non redundant or redundant robot manipulator", *The Int. J. of Robotic Research*, Vol. 17(1), 1998, p. 56-69.

[Chiaverini 93] Chiaverini S., Sciavicco L., "The parallel approach to force/position control of robotic manipulators", *IEEE Trans. on Robotics and Automation*, Vol. RA-9(4), 1993, p. 361-373.

[Clavel 89] Clavel R., "Une nouvelle structure de manipulateur parallèle pour la robotique légère", *Revue APII-AFCET*, Vol. 23, 1989, p. 501-519.

[Cloutier 95] Cloutier B.P., Pai D.K., Ascher U.M., "The formulation stiffness of forward dynamics algorithms and implications for robot simulation", *Proc. IEEE Int. Conf. on Robotics and Automation*, Nagoya, Japan, May 1995, p. 2816-2822.

[Coiffet 81] Coiffet P., *Les Robots ; Tome 1 : Modélisation et commande*, Hermès, Paris, France, 1981.

[Colbaugh 93] Colbaugh R., Seraji H., Glass K., "Direct adaptive impedance control of robot manipulators", *J. of Robotic Systems*, Vol. 10, 1993, p. 217-248.

[Craig 86a] Craig J.J., *Introduction to robotics: mechanics and control*, Addison Wesley Publishing Company, Reading, USA, 1986.

[Craig 86b] Craig J.J., Hsu P., Sastry S., "Adaptive control of mechanical manipulators", *Proc. IEEE Int. Conf. on Robotics and Automation*, San Francisco, USA, April 1986, p. 190-195.

[Craig 93] Craig J.J., "Calibration of industrial robots", *Proc. 24th Int. Symp. on Industrial Robots*, Tokyo, Japan, November 1993, p. 889-893.

[Dafaoui 94] Dafaoui E., "Modélisation et commande d'un robot parallèle : application au suivi de contour", Thèse de Doctorat, Paris XII-Val de Marne, France, September 1994.

[Dahl 77] Dahl P.R.,"Measurements of solid friction parameters of ball bearings", *Proc. of the 6th Annual Symp. on Incremental Motion Control Systems and Devices*, University of Illinois, USA, 1977.

[Damak 96] Damak M., "Théorie et instrumentation pour l'étalonnage statique des robots : vers une programmation hors-ligne industriellement plus efficace", Thèse de Doctorat, ENSAM, Lille, France, July 1996.

[Daney 00] Daney D., "Etalonnage géométrique des robots parallèles", Thèse de Doctorat, Université de Nice-Sophia Antipolis, France, February 2000.

[de Boor 78] de Boor C., *A practical guide to splines*, Springer-Verlag, New York, USA, 1978.

[de Casteljau 87] de Casteljau P., *Les quaternions*, Hermès, Paris, France, 1987.

[Dégoulange 93] Dégoulange E., "Commande en effort d'un robot manipulateur à deux bras : application au contrôle de la déformation d'une chaîne cinématique fermée", Thèse de Doctorat, Université Montpellier II, France, December 1993.

[de Larminat 77] de Larminat P., Thomas Y., *Automatique des systèmes linéaires ; Tome 2 : Identification*, Editions Flammarion, Paris, France, 1977.

[Delignières 87] Delignières S., "Choix de morphologies de robot", Thèse de Docteur-Ingénieur, ENSM, Nantes, France, November 1987.

[de Luca 91a] de Luca A., Oriolo G., "Issues in acceleration resolution of robot redundancy", *Proc. IFAC Symp. on Robot Control, SYROCO'91*, Vienna, Austria, 1991, p. 665-670.

[de Luca 91b] de Luca A., Manes C., "Hybrid force-position control for robots in contact with dynamic environments", *Proc. IFAC Symp. on Robot Control, SYROCO'91*, Vienna, Austria, 1991, p. 177-182.

[Denavit 55] Denavit J., Hartenberg R.S., "A kinematic notation for lower pair mechanism based on matrices", *Trans. of ASME, J. of Applied Mechanics*, Vol. 22, June 1955, p. 215-221.

[Desbats 90] Desbats P., "Modélisation et commande dynamique des robots rapides", Thèse de Doctorat, Université de Paris Sud, Orsay, France, June 1990.

[de Schutter 88] de Schutter J., van Brussel H., "Compliant robot motion - II - a control approach based on external control loop", *The Int. J. of Robotics Research*, Vol. 7(4), 1988, p. 18-33.

[Desoer 75] Desoer C.A., Vidyaasagar M., *Feedback systems: input-output properties*, Academic Press, New York, USA, 1975.

[Dietmaier 98] Deitmaier P., "The Stewart-Gough platform of general geometry can have 40 real postures", *in Advances in Robot Kinematics: Analysis and Control*, J. Lenarčič, M.L. Husty Eds, Klumer Academic Publishers, 1998, p. 7-16.

[Dillon 73] Dillon S.R., "Computer assisted equation generation in linkage dynamics", Ph. D. Thesis, Ohio State University, USA, August 1973.

[Dombre 81] Dombre E., "Analyse des performances des robots-manipulateurs flexibles et redondants ; contribution à leur modélisation et à leur commande", Thèse d'Etat, USTL, Montpellier, France, June 1981.

[Dombre 85] Dombre E., Borrel P., Liégeois A., "A CAD system for programming and simulating robot's actions", *in Computing Techniques for Robots*, I. Aleksander Ed., Kogan Page, London, UK, 1985, p. 222-247.

[Dombre 88a] Dombre E., Khalil W., *Modélisation et commande des robots*, Hermès, Paris, France, 1988.

[Dombre 88b] Dombre E., "Outils d'aide à la conception de cellules robotisées", *in Techniques de la robotique : perception et planification*, Hermès, Paris, France, 1988, p. 255-291.

[Dombre 94] Dombre E., Fournier A., "Yet another calibration technique to reduce the gap between CAD world and real world", *Proc. 1st WAC'94 on Intelligent Automation and Soft Computing*, Hawaï, USA, August 1994, p. 47-52.

[Dongarra 79] Dongarra J.J., Moler C.B., Bunch J.R., Stewart G.W., "LINPACK User's Guide", Philadelphia, USA, SIAM, 1979.

[Douss 96] Douss M., "Programmation hors ligne par CAO-Robotique : caractérisation de lois de contrôleurs de robots et étalonnage de cellules robotisées en vue d'améliorer la précision", Thèse de Doctorat, Université de Franche-Comté, Besançon, France, November 1996.

[Drake 77] Drake S.H., "Using compliance in lieu of sensory feedback for automatic assembly", Ph. D. Thesis, Dept. of Mechanical Engineering, MIT, USA, September 1977.

[Driels 90] Driels M. R., Pathre U.S., "Significance of observation strategy on the design of robot calibration experiment", *J. of Robotic Systems*, Vol. 7, 1990, p. 197-223.

[Dubowsky 79] Dubowsky S., Des Forges D.T., "The application of model-referenced adaptive control to robotic manipulators", *Trans. of ASME, J. of Dynamic Systems, Measurement, and Control*, Vol. 101, 1979, p. 193-200.

[Duffy 90] Duffy J., "The fallacy of modern hybrid control theory that is based on 'orthogonal complements' of twist and wrench spaces", *J. of Robotic Systems*, Vol. 7, 1990, p. 139-144.

[Edwall 82] Edwall C.W., Pottinger H.J., Ho C.Y., "Trajectory generation and control of a robot arm using spline functions", *Proc. Robot-6*, Detroit, USA, 1982, p. 421-444.

[Egeland 91] Egeland O., Spangelo I., "Manipulator control in singular configurations-Motion in degenerate directions", *in Advanced Robot Control, Lecture Notes in Control and Information Sciences*, Springer-Verlag, New York, USA, 1991, p. 296-306.

[El Omri 96] El Omri J., "Analyse géométrique et cinématique des mécanismes de type manipulateur", Thèse de Doctorat, Université de Nantes et Ecole Centrale de Nantes, France, February 1996.

[El Serafi 91a] El Serafi K., Khalil W., "Energy based indirect adaptive control of robots", *Proc. IEEE Int. Conf. on Robotics and Automation*, Sacramento, USA, April 1991, p. 2142-2147.

[El Serafi 91b] El Serafi K., "Contribution à la commande adaptative des robots manipulateurs", Thèse de Doctorat, ENSM, Nantes, France, May 1991.

[Everett 88] Everett L.J, Suryohadiprojo A.H., "A study of kinematic models for forward calibration of manipulators", *Proc. IEEE Int. Conf. on Robotics and Automation*, Philadelphia, USA, April 1988, p. 798-800.

[Eykhoff 74] Eykhoff P., *System identification: parameter and state estimation*, John Wiley & Sons, London, UK, 1974.

[Fages 98] Fages G., "Statistiques 97", *RobAut*, n° 21, March-April 1998, p. 28-32.

[Featherstone 83a] Featherstone R., "Position and velocity transformations between robot end-effector coordinates and joint angles", *The Int. J. of Robotics Research*, Vol. 2(2), 1983, p. 35-45.

[Featherstone 83b] Featherstone R., "The calculation of robot dynamics using articulated-body inertias", *The Int. J. of Robotics Research*, Vol. 2(3), 1983, p. 87-101.

[Ferreira 84] Ferreira E.P., "Contribution à l'identification de paramètres et à la commande des robots manipulateurs", Thèse de Docteur-Ingénieur, UPS, Toulouse, France, July 1984.

[Fichter 86] Fichter E.F., "A Stewart platform-based manipulator: general theory and practical construction", *The Int. J. of Robotic Research*, Vol. 5(2), 1986, p. 157-181.

[Fisher 92] Fisher W., Mujtaba M.S., "Hybrid position / force control: a correct formulation", *The Int. J. of Robotics Research*, Vol. 11(4), 1992, p. 299-311.

[Fliess 95] Fliess M., Lévine J., Martin P., Rouchon P., "Flatness and defect of nonlinear systems: introductory theory and examples", *Int. J. Control*, Vol. 61, 1995, p. 1327-1361.

[Forsythe 77] Forsythe G.E., Malcom M.A., Moler C.B., *Computer methods for mathematical computations*, Prentice Hall, Englewood Cliffs, USA, 1977.

[Fournier 80] Fournier A., "Génération de mouvements en robotique ; application des inverses généralisées et des pseudo-inverses", Thèse d'Etat, USTL, Montpellier, France, April 1980.

[Fourquet 90] Fourquet J.-Y., "Mouvement en temps minimal pour les robots manipulateurs en tenant compte de leur dynamique", Thèse de Doctorat, UPS, Toulouse, France, 1990.

[Freidovich 97] Freidovich L.B., Pervozvanski A.A., "Some estimates of performance for PID-like control of robotic manipulators", *Proc. IFAC Symp. on Robot Control, SYROCO'97*, Nantes, France, September 1997, p. 85-90.

[Freund 82] Freund E., "Fast nonlinear control with arbitrary pole placement for industrial robots and manipulators", *The Int. J. of Robotics Research*, Vol. 1(1), 1982, p. 65-78.

[Froissart 91] Froissart C., "Génération adaptative de mouvement pour processus continus ; application au suivi de joint", Thèse de Doctorat, Université Pierre et Marie Curie, Paris, France, December 1991.

[Gaudin 92] Gaudin H., "Contribution à l'identification *in situ* des constantes d'inertie et des lois de frottement articulaire d'un robot manipulateur en vue d'une application expérimentale au suivi de trajectoires optimales", Thèse de Doctorat, Université de Poitiers, France, November 1992.

[Gautier 86] Gautier M., "Identification of robots dynamics", *Proc. IFAC Symp. on Theory of Robots*, Vienna, Austria, December 1986, p. 351-356.

[Gautier 88] Gautier M., Khalil W., "On the identification of the inertial parameters of robots", *Proc. 27th IEEE Conf. on Decision and Control*, Austin, USA, December 1988, p. 2264-2269.

[Gautier 90a] Gautier M., "Contribution à la modélisation et à l'identification des robots", Thèse de Doctorat d'Etat, ENSM, Nantes, France, May 1990.

[Gautier 90b] Gautier M., Khalil W., "Direct calculation of minimum set of inertial parameters of serial robots", *IEEE Trans. on Robotics and Automation*, Vol. RA-6(3), 1990, p. 368-373.

[Gautier 91] Gautier M., "Numerical calculation of the base inertial parameters", *J. of Robotic Systems*, Vol. 8(4), August 1991, p. 485-506.

[Gautier 92a] Gautier M., Khalil W., "Exciting trajectories for inertial parameters identification", *The Int. J. of Robotics Research*, Vol. 11(4), 1992, p. 362-375.

[Gautier 92b] Gautier M., "Optimal motion planning for robot's inertial parameters identification", *Proc. 31st IEEE Conf. on Decision and Control*, Tucson, USA, December 1992, Vol. 1, p. 70-73.

[Gautier 93] Gautier M., Janin C., Pressé C., "On the validation of robot dynamic model", *Proc. 2nd European Control Conf., ECC'93*, Groningen, The Netherlands, June-July 1993, p. 2291-2295.

[Gautier 94] Gautier M., Vandanjon P.-O., Pressé C., "Identification of inertial and drive gain parameters of robots", *Proc. IEEE 33th Conf. on Decision and Control*, Orlando, USA, December 1994, p.3764-3769.

[Gautier 95] Gautier M., Khalil W., Restrepo P. P.,"Identification of the dynamic parameters of a closed loop robot", *Proc. IEEE Int. Conf. on Robotics and Automation*, Nagoya, Japan, May 1995, p. 3045-3050.

[Gautier 96] Gautier M., "A comparison of filtered models for dynamic identification of robots", *Proc. IEEE 35th Conf. on Decision and Control*, Kobe, Japan, December 1996, p. 875-880.

[Gautier 97] Gautier M., "Dynamic identification of robots with power model", *Proc. IEEE Int. Conf. on Robotics and Automation*, Albuquerque, USA, April 1997, p. 1922-1927.

[Geffard 00] Geffard F., "Etude et conception de la commande en effort d'un télémanipulateur équipé d'un capteur d'effort à sa base et son extrémité", Thèse de Doctorat, Université de Nantes, France, December 2000.

[Giordano 86] Giordano M., "Dynamic model of robots with a complex kinematic chain", *Proc.16th Int. Symp. on Industrial Robots*, Brussels, Belgium, September-October 1986, p. 377-388.

[Goldenberg 85] Goldenberg A.A., Benhabib B., Fenton R.G., "A complete generalized solution to inverse kinematics of robots", *IEEE J. of Robotics and Automation*, Vol. RA-1(1), 1985, p. 14-20.

[Golub 83] Golub G.H., Van Loan C.F., *Matrix computations*, Johns Hopkins University Press, Baltimore, USA, 1983.

[Gorla 84] Gorla B., Renaud M., *Modèles des robots-manipulateurs ; application à leur commande*, Cepadues Editions, Toulouse, France, 1984.

[Gosselin 88] Gosselin C., "Kinematic analysis, optimization and programming of parallel robotic manipulators", Ph. D. Thesis, McGill University, Montréal, Canada, June 1988.

[Gosselin 90] Gosselin C., Angeles J., "Singularity analysis of closed-loop kinematic chains", *IEEE Trans. on Robotics and Automation*, Vol. RA-6(3), 1990, p. 281-290.

[Goswami 93] Goswami A., Quaid A., Peshkin M., "Identifying robot parameters using partial pose information", *Proc. IEEE Int. Conf. on Systems, Man, and Cybernetics*, Chicago, USA, October 1993, p. 6-14.

[Goudali 96] Goudali A., Lallemand J.-P, Zeghloul S., "Modeling of the 2-delta decoupled parallel robot", *Proc. 6th Int. Symp. on Robotics and Manufacturing, WAC'96*, Vol. 6, Montpellier, France, May 1996, p. 243-248.

[Gough 56] Gough V.E., "Contribution to discussion of papers on research in automobile stability, control and tyre performance", *Proc. Auto. Div. Inst. Mech. Eng.*, 1956-1957.

[Greville 60] Greville T.N., "Some applications of the pseudo-inverse of a matrix", *SIAM Review*, Vol. 2(1), 1960, p. 15-22.

[Grudić 93] Grudić G.Z., Lawrence P.D., "Iteratitive inverse kinematics with manipulator configuration control", *IEEE Trans. on Robotics and Automation*, Vol. RA-9(4), August 1993, p. 476-483.

[Guglielmi 87] Guglielmi M., Jonker E., Piasco J.-M., "Modelisation and identification of a two degrees of freedom SCARA robot using extended Kalman filtering", *Proc. Int. Conf. on Advanced Robotics, ICAR'87*, Versailles, France, October 1987, p. 137-148.

[Guyot 95] Guyot G., "Contribution à l'étalonnage géométrique des robots manipulateurs", Thèse de Doctorat, Université de Nice-Sophia Antipolis, France, January 1995.

[Ha 89] Ha I.J., Ko M.S., Kwon S.K., "An efficient estimation algorithm for the model parameters of robotic manipulators", *IEEE Trans. on Robotics and Automation*, Vol. RA-5(6), 1989, p. 386-394.

[Hahn 67] Hahn W., *Stability of Motion*, Springer-Verlag, New York, USA, 1967.

[Han 95] Han K., Chung W.K., Youm Y., "Local structurization for the forward kinematics of parallel manipulators using extra sensor data", *Proc. IEEE Int. Conf. on Robotics and Automation*, Nayoga, Japan, May 1995 p. 514-520.

[Hayati 83] Hayati S.A.,"Robot arm geometric link parameter estimation", *Proc. 22nd IEEE Conf. Decision and Control*, San Antonio, USA, December 1983, p. 1477-1483.

[Held 88] Held V., Maron C., "Estimation of friction characteristics, inertial and coupling coefficients in robotic joints based on current and speed measurements", *Proc. IFAC Symp. on Robot Control, SYROCO'88*, 1988, p. 86.1-86.6.

[Hervé 78] Hervé J.-M., "Analyse structurelle des mécanismes par groupe de déplacement", *J. of Mechanism and Machine Theory*, Vol. 13(4), 1978, p. 437-450.

[Hervé 91] Hervé J.-M., Sparacino F., "Structural synthesis of parallel robot generating spatial translation", *Proc. Int. Conf. on Advanced Robotics, ICAR'91*, Pise, Italy, 1991, p. 808-813.

[Hogan 85] Hogan N., "Impedance control: an approach to manipulation", *Trans. of ASME, J. of Dynamic Systems, Measurement, and Control*, Vol. 107, March 1985, p. 1-24.

[Hogan 87] Hogan N., "Stable execution of contact tasks using impedance control", *Proc. IEEE Int. Conf. on Robotics and Automation*, Raleigh, USA, March-April 1987, p. 1047-1054.

[Hollerbach 80] Hollerbach J.M., "An iterative lagrangian formulation of manipulators dynamics and a comparative study of dynamics formulation complexity", *IEEE Trans. on Systems, Man, and Cybernetics*, Vol. SMC-10(11), 1980, p. 730-736.

[Hollerbach 84a] Hollerbach J.M., "Dynamic scaling of manipulator trajectories", *Trans. of ASME, J. of Dynamic Systems, Measurement, and Control*, Vol. 106(1), March 1984, p. 102-106.

[Hollerbach 84b] Hollerbach J.M., "Optimum kinematic design for a seven degree of freedom manipulator", *Proc. 2nd Int. Symp. of Robotics Research*, Kyoto, Japan, August 1984, p. 349-356.

[Hollerbach 85] Hollerbach J.M., Suh K.C., "Redundancy resolution of manipulators through torque optimization", *Proc. IEEE Int. Conf. on Robotics and Automation*, St Louis, USA, March 1985, p. 1016-1021.

[Hollerbach 89] Hollerbach J.M., "A survey of kinematic calibration", in *The Robotics Review n°1*, MIT Press, Cambridge, USA, 1989, p. 207-242.

[Hollerbach 95] Hollerbach J.M., Lokhorst D., "Closed-loop kinematic calibration of the RSI 6-dof hand controller", *IEEE Trans. on Robotics and Automation*, Vol. RA-11, 1995, p. 352-359.

[Hollerbach 96] Hollerbach J.M., Wampler C.W., "The calibration index and taxonomy of kinematic calibration methods", *The Int. J. of Robotics Research*, Vol. 14, 1996, p. 573-591.

[Hooker 65] Hooker W.W., Margulies G., "The dynamical attitude equations for a n-body satellite", *The Journal of the Astronautical Sciences*, Vol. 12(4), 1965, p. 123-128.

[Horowitz 80] Horowitz R., Tomizuka M., "An adaptive control scheme for mechanical manipulators; compensation of non linearity and decoupling control", *Presentation at the Winter Meeting of ASME, Dynamic Systems and Control Division*, Chicago, USA, 1980.

[Hsia 86] Hsia T.C., "Adaptive control of robot manipulators; a review", *Proc. IEEE Int. Conf. on Robotics and Automation*, San Francisco, USA, April 1986, p. 183-189.

[Hsu 88] Hsu P., Hauser J., Sastry S., "Dynamic control of redundant manipulators", *Proc. IEEE Int. Conf. on Robotics and Automation*, Philadelphia, USA, April 1988, p. 183-187.

[Hunt 83] Hunt K. H., "Structural kinematics of in-parallel-actuated robot-arms", *Trans of ASME, J. of Mechanisms, Transmissions, and Automation in Design*, Vol. 105, March 1983, p. 705-712.

[Husty 96] Husty M., "An algorithm for solving the direct kinematics of Stewart-Gough-type platforms", *J. of Mechanisms and Machine Theory*, Vol. 31(4), 1996, p. 365-379.

[Ikits 97] Ikits M, Hollerbach J.M., "Kinematic calibration using a plane constraint", *Proc. IEEE Int. Conf. on Robotics and Automation*, Albuquerque, USA, April 1997, p. 3191-3196.

[Innocenti 93] Innocenti C., Parenti-Castelli V., "Direct kinematics in analytical form of a general geometry 5-4 fully parallel manipulator", *in Computational kinematics*, J. Angeles *et al.* Eds., Klumer Academic Publishers, 1993, p. 141-152.

[Innocenti 95] Innocenti C., "Analytical-form direct kinematics for the second scheme of a 5-5 general-geometry fully parallel manipulator", *J. of Robotic Systems*, Vol. 12(10), 1995, p. 661-676.

[Inoue 85] Inoue H., Tsusaka Y., Fukuizumi T., "Parallel manipulator", *Proc. 3rd Int. Symp. of Robotics Research*, Gouvieux, France, October 1985, p. 69-75.

[Izaguirre 86] Izaguirre A., Paul R.C.P., "Automatic generation of the dynamic equations of the robot manipulators using a LISP program", *Proc. IEEE Int. Conf. on Robotics and Automation*, San Francisco, USA, April 1986, p. 220-226.

[Judd 90] Judd R., Knasinski A., "A technique to calibrate industrial robots with experimental verification", *IEEE Trans. on Robotics and Automation*, Vol. RA-6(1), 1990, p. 20-30.

[Kahn 69] Kahn M.E., "The near minimum time control of open loop articulated kinematic chains", Ph. D. Thesis, Stanford University, Stanford, USA, December 1969.

[Kanade 84] Kanade T., Khosla P.,Tanaka N., "Real-time control of the CMU direct-drive arm II using customized inverse dynamics", *Proc. 23rd IEEE Conf. on Decision and Control*, Las Vegas, USA, December 1984, p. 1345-1352.

[Kane 83] Kane T.R., Levinson D., "The use of Kane's dynamical equations in robotics", *The Int. J. of Robotics Research*, Vol. 2(3), 1983, p. 3-21.

[**Kawasaki 88**] Kawasaki H., Nishimura K., "Terminal-link parameter estimation and trajectory control of robotic manipulators", *IEEE J. of Robotics and Automation*, Vol. RA-4(5), p. 485-490, 1988.

[**Kawasaki 96**] Kawasaki H., Takahiro B., Kazuo K., "An efficient algorithm for the model-based adaptive control of robotic manipulators", *IEEE Trans. on Robotics and Automation*, Vol. RA-12(3), 1996, p. 496-501.

[**Kazerooni 86**] Kazerooni H., Sheridan T., Houpt P., "Robust compliant motion for manipulators", Parts I and II, *IEEE J. of Robotics and Automation*, Vol. RA-2(2), 1986, p.83-105.

[**Kazerounian 86**] Kazerounian K., Gupta K.C., "Manipulator dynamics using the extended zero reference position description", *IEEE J. of Robotics and Automation*, Vol. RA-2(4), 1986, p. 221-224.

[**Kelly 88**] Kelly R., Ortega R., "Adaptive control of robot manipulators: an input-output approach", *Proc. IEEE Int. Conf. on Robotics and Automation*, Philadelphia, USA, April 1988, p. 699-703.

[**Kelly 95**] Kelly R., "A tuning procedure for stable PID control of robot manipulators", *Robotica*, Vol. 13, 1995, p. 141-148.

[**Khalil 76**] Khalil W., "Modélisation et commande par calculateur du manipulateur MA-23 ; extension à la conception par ordinateur des manipulateurs", Thèse de Docteur-Ingénieur, USTL, Montpellier, France, September 1976.

[**Khalil 78**] Khalil W., "Contribution à la commande automatique des manipulateurs avec l'aide d'un modèle mathématique des mécanismes", Thèse d'Etat, USTL, Montpellier, France, October 1978.

[**Khalil 79**] Khalil W., Liegeois A., Fournier A., "Commande dynamique des robots", *Revue RAIRO Automatique / Systems Analysis and Control*, Vol. 13(2), 1979, p. 189-201.

[**Khalil 85a**] Khalil W., Kleinfinger J.-F., "Une modélisation performante pour la commande dynamique de robots", *Revue RAIRO, APII*, Vol. 6, 1985, p. 561-574.

[**Khalil 85b**] Khalil W., Gautier M., "Identification of geometric parameters of robots", *Proc. IFAC Symp. on Robot Control, SYROCO'85*, Barcelona, Spain, November 1985, p. 191-194.

[**Khalil 86a**] Khalil W., Kleinfinger J.-F., "A new geometric notation for open and closed-loop robots", *Proc. IEEE Int. Conf. on Robotics and Automation*, San Francisco, USA, April 1986, p. 1174-1180.

[**Khalil 86b**] Khalil W., "On the explicit derivation of the inverse geometric models of robots", *Proc. IMACS-IFAC Symp.*, Villeneuve d'Ascq, France, June 1986, p. 541-546.

[**Khalil 86c**] Khalil W., Kleinfinger J.-F., Gautier M., "Reducing the computational burden of the dynamic model of robots", *Proc. IEEE Int. Conf. on Robotics and Automation*, San Francisco, USA, April 1986, p. 525-531.

[**Khalil 87a**] Khalil W., Chevallereau C., "An efficient algorithm for the dynamic control of robots in the cartesian space", *Proc. 26th IEEE Conf. on Decision and Control*, Los Angeles, USA, December 1987, p. 582-588.

[**Khalil 87b**] Khalil W., Kleinfinger J.-F., "Minimum operations and minimum parameters of the dynamic model of tree structure robots", *IEEE J. of Robotics and Automation*, Vol. RA-3(6), December 1987, p. 517-526.

[Khalil 89a] Khalil W., Bennis F., Chevallereau C., Kleinfinger J.-F., "SYMORO: a software package for the symbolic modelling of robots", *Proc. 20th Int. Symp. on Industrial Robots*, Tokyo, Japan, October 1989, p. 1023-1030.

[Khalil 89b] Khalil W., Bennis F., Gautier M., "Calculation of the minimum inertial parameters of tree structure robots", *Proc. Int. Conf. on Advanced Robotics, ICAR'89*, Columbus, USA, 1989, p. 189-201.

[Khalil 89c] Khalil. W., Caenen. J.-L., Enguehard. C., "Identification and calibration of the geometric parameters of robots", *Proc. 1st Experimental Robot Conference*, Montreal, Canada, 1989, Springer-Verlag, New York, Vol. 139, p. 528-538.

[Khalil 90a] Khalil W., Bennis F., Gautier M., "The use of the generalized links to determine the minimum inertial parameters of robots", *J. of Robotic Systems*, Vol. 7(2), 1990, p. 225-242.

[Khalil 90b] Khalil W., Bennis F., "Calcul de la matrice d'inertie des robots à chaîne ouverte simple ou arborescente", Rapport interne n° 90-09, LAN-ENSM, Nantes, France, April 1990.

[Khalil 90c] Khalil W., Bennis F., "Automatic generation of the inverse geometric model of robots", *J. of Robotics and Autonomous Systems*, Vol. 7, 1991, p. 1-10.

[Khalil 91a] Khalil W., Gautier M., "Calculation of the identifiable parameters for robot calibration", *Proc. IFAC Symp. on Identification and System Parameter Estimation*, Budapest, Hungary, 1991, p. 888-892.

[Khalil 91b] Khalil W., Gautier, M., Enguehard C.,"Identifiable parameters and optimum configurations for robot calibration", *Robotica*, Vol. 9, 1991, p. 63-70.

[Khalil 93] Khalil W., Gautier M., "Computed current control of robots", *Proc. IFAC 12th World Congress*, Sydney, Australia, July 1993, Vol. IV, p. 129-132.

[Khalil 94a] Khalil W., Bennis F., "Comments on Direct Calculation of Minimum Set of Inertial Parameters of Serial Robots", *IEEE Trans. on Robotics and Automation*, Vol. RA-10(1), 1994, p. 78-79.

[Khalil 94b] Khalil W., Murareci D., "On the general solution of the inverse kinematics of six-degrees-of-freedom manipulators", *Proc. Int. Workshop on Advances in Robot Kinematics, ARK 94*, Slovenia, July 1994.

[Khalil 95a] Khalil W., Bennis F., "Symbolic calculation of the base inertial parameters of closed-loop robots", *The Int. J. of Robotics Research*, Vol. 14(2), April 1995, p. 112-128.

[Khalil 95b] Khalil W., Garcia G., Delagarde J.-F. "Calibration of the geometric parameters of robots without external sensors", *Proc. IEEE Int. Conf. on Robotics and Automation*, Nayoga, Japan, May 1995, p. 3039-3044.

[Khalil 96a] Khalil W., Restrepo P.P., "An efficient algorithm for the calculation of the filtred dynamic model of robots", *Proc. IEEE Int. Conf. on Robotics and Automation*, Minneapolis, USA, April 1996, p. 323-329.

[Khalil 96b] Khalil W., Lemoine P., Gautier M., "Autonomous calibration of robots using planar points", *Proc. 6th Int. Symp. on Robotics and Manufacturing, WAC'96*, Vol. 3, Montpellier, France, May 1996, p. 383-388.

[Khalil 96c] Khalil W., Murareci D., "Kinematic analysis and singular configuration of a class of parallel robots", *J. of Mathematic and Computer in Smulation*, n°1245, 1996, p. 1-14.

[Khalil 97] Khalil W., Creusot D., "SYMORO+: a system for the symbolic modelling of robots", *Robotica*, Vol. 15, 1997, p. 153-161.

[Khalil 99a] Khalil W., Lemoine P., "Gecaro: a system for the geometric calibration of robots", *Revue APII-JESA*, Vol. 33(5-6), 1999, p. 717-739.

[Khalil 99b] Khalil W., Besnard S., "Self calibration of Stewart-Gough parallel robots without extra sensors", *IEEE Trans. on Robotics and Automation*, Vol. RA-15(6), p. 1116-1121, December 1999.

[Khalil 00a] Khalil W., Gautier M., "Modeling of mechanical systems with lumped elasticity", *Proc. IEEE Int. Conf. on Robotics and Automation*, San Francisco, April 2000, p. 3965-3970.

[Khalil 00b] Khalil W., Besnard S., Lemoine P., "Comparison study of the geometric parameters calibration methods ", *Int. J. of Robotics and Automation*, ", Vol. 15(2), 2000, p. 56- 67.

[Khatib 80] Khatib O., "Commande dynamique dans l'espace opérationnel des robots-manipulateurs en présence d'obstacles", Thèse de Docteur-Ingénieur, Ecole Nationale Supérieure de l'Aéronautique et de l'Espace, Toulouse, France, December 1980.

[Khatib 87] Khatib O., "A unified approach for motion and force control of robot manipulators: the operational space formulation", *IEEE J. of Robotics and Automation*, Vol. RA-3(1), February 1987, p. 43-53.

[Khelfi 95] Khelfi M.-F, "Observateurs non linéaires : application à la commande des robots manipulateurs", Thèse de Doctorat, Université Poincaré Nancy I, France, 1995.

[Kholi 85] Kholi D., Spanos J., "Workspace analysis of mechanical manipulators using polynomial discriminants", *J. of Mechanisms, Transmissions, and Automation in Design*, Vol. 107, June 1985, p. 209-215.

[Kholi 87] Kholi D., Hsu M.S., "The jacobian analysis of workspaces of mechanical manipulators", *J. of Mechanisms and Machine Theory*, Vol. 22(3), 1987, p. 265-275.

[Khosla 85] Khosla P.K., Kanade T., "Parameter identification of robot dynamics", *Proc. 24th IEEE Conf. on Decision and Control*, Fort-Lauderdale, USA, December 1985, p. 1754-1760.

[Khosla 86] Khosla P.K., "Real-time control and identification of direct drive manipulators", Ph. D. Thesis, Carnegie Mellon University, Pittsburgh, USA, 1986.

[Kircánski 85] Kircánski M., Vukobratovic M., "Computer-aided generation of manipulator kinematic models in symbolic form", *Proc. 15th Int. Symp. on Industrial Robots*, Tokyo, Japan, September 1985, p. 1043-1049.

[Klein 83] Klein C.A., Huang C., "Review of pseudo inverse control for use with kinematically redundant manipulators", *IEEE Trans. on Systems, Man, and Cybernetics*, Vol. SMC-13(2), 1983, p. 245-250.

[Klein 84] Klein C.A., "Use of redundancy in the design of robotic systems", *Proc. 2nd Int. Symp. of Robotic Research*, Kyoto, Japan, August 1984, p. 58-65.

[Klein 95] Klein C.A., Chu-Jenq, Ahmed S., "A new formulation of the extended Jacobian method and its use in mapping algorithmic singularities for kinematically redundant manipulators", *IEEE Trans. on Robotics and Automation*, Vol. RA-11(1), 1995, p. 50-55.

[Kleinfinger 86a] Kleinfinger J.-F, "Modélisation dynamique de robots à chaîne cinématique simple, arborescente ou fermée, en vue de leur commande", Thèse de Doctorat, ENSM, Nantes, France, May 1986.

[Kleinfinger 86b] Kleinfinger J.-F, Khalil W., "Dynamic modelling of closed-chain robots", *Proc. 16th Int. Symp. on Industrial Robots*, Brussels, Belgium, September-October 1986, p. 401-412.

[Klema 80] Klema V.C., Lanio A.J., "The singular value decomposition: its computation and some applications", *IEEE Trans. on Automatic Control*, Vol. AC-25(2), 1980, p. 164-176.

[Koditschek 84] Koditschek D.E., "Natural motion for robot arms", *Proc. 23rd IEEE Conf. on Decision and Control*, Las Vegas, USA, December 1984, p. 737-735.

[Konstantinov 81] Konstantinov M.S., Markov M.D., Nenchev D.N., "Kinematic control of redundant manipulators", *Proc. 11th Int. Symp. on Industrial Robots*, Tokyo, Japan, October 1981, p. 561-568.

[Korrami 88] Korrami F., Özgnüner U., "Decentralized control of robot manipulators via state and proportional-integral feedback", *Proc. IEEE Int. Conf. on Robotics and Automation*, Philadelphia, USA, April 1988, p. 1198-1203.

[Kreuzer 79] Kreuzer E.J., "Dynamical analysis of mechanisms using symbolical equation manipulation", *Proc. 5th World Congress on Theory of Machines and Mechanisms*, Montréal, Canada, 1979, p. 599-602.

[Landau 79] Landau I. D, *Adaptive control; The model reference approach*, Marcel Dekker Inc. New York, USA, 1979.

[Landau 88] Landau I. D., Horowitz R., "Synthesis of adaptive controllers for robot manipulators using a passive feedback system approach". *Proc. IEEE Int. Conf. on Robotics and Automation*, Philadelphia, USA, April 1988, p. 1028-1033.

[Lavallée 92] Lavallée S., "Procédé d'étalonnage d'un robot", Brevet n° FR 2 696 969 du 21/10/92.

[Lawson 74] Lawson C.L., Hanson R.J., *Solving Least Squares Problems*, Prentice Hall, Englewood Cliffs, USA, 1974.

[Lazard 92] Lazard D., "Stewart platforms and Grübner basis", *Proc. 3rd Int. Workshop on Advances in Robot Kinematics, ARK 92*, Ferrare, Italy, 1992, p. 136-142.

[Le Borzec 75] Le Borzec R., Lotterie J., *Principes de la théorie des mécanismes*, Dunod, Paris, 1975.

[Lee 83] Lee C.S.G., Ziegler M., "A geometric approach in solving the inverse kinematics of PUMA robots", *Proc. 13th Int. Symp. on Industrial Robots*, Chicago, USA, April 1983, p. (16-1)-(16-18).

[Lee 88] Lee H.Y, Liang C.G.,"Displacement analysis of the general 7-link 7R mechanism", *J. of Mechanism and Machine Theory*, Vol. 23(3), 1988, p. 219-226.

[Lee 93] Lee H.Y., Roth B., "A closed-form solution of the forward displacement analysis of a class of in-parallel robots", *Proc. IEEE Int. Conf. on Robotics and Automation*, Atlanta, USA, May 1993, p. 720-724.

[Léon 91] Léon J.-C., *Modélisation et construction de surfaces pour la CFAO*, Hermès, Paris, France, 1991.

[Lewis 93] Lewis F.L., Abdallah C.T., Dawson D.M., *Control of robot manipulators*, Macmillan, New York, USA, 1993.

[Li 89] Li W., Slotine J.-J.E., "An indirect adaptive robot controller", *Systems & Control Letters*, Vol. 12, 1989, p. 259-266.

[Liégeois 79] Liégeois A., Dombre E., "Analyse des robots industriels ; relations entre structures, performances et fonctions", Rapport IRIA n° 79102, Projet SURF, LAM, Montpellier, France, 1979.

[Lilly 90] Lilly K.W., Orin D.E., "Efficient O(N) computation of the operational space inertia matrix", *Proc. IEEE Int. Conf. on Robotics and Automation*, Cincinnati, USA, May 1990, p. 1014-1019.

[Lin 83] Lin C.S., Chang P.R., Luh J.Y.S., "Formulation and optimization of cubic polynomial joint trajectories for industrial robots", *IEEE Trans. on Automatic Control*, Vol. AC-28(12), December 1983, p. 1066-1073.

[Lin 92] Lin W., Griffis M., Duffy J., "Forward displacement analysis of the 4-4 Stewart platforms", *Trans. of the ASME, J. of Mechanical Design*, Vol. 114, September 1992, p. 444-450.

[Llibre 83] Llibre M., Mampey R., Chrétien J.P., "Simulation de la dynamique des robots manipulateurs motorisés", *Congrès AFCET : Productique et Robotique Intelligente*, Besançon, France, November 1983, p. 197-207.

[Lloyd 96] Lloyd J.E., "Using Puiseux series to control non-redundant robots at singularities", *Proc. IEEE Int. Conf. on Robotics and Automation*, Minneapolis, USA, April 1996, p. 1877-1882.

[Lu 93] Lu Z., Shimoga K.B., Goldberg A., "Experimental determination of dynamic parameters of robotic arms", *J. of Robotic Systems*, Vol. 10(8), 1993, p. 1009-1029.

[Luh 80a] Luh J.Y.S., Walker M.W., Paul R.C.P., "Resolved-acceleration control of mechanical manipulators", *IEEE Trans. on Automatic Control*, Vol. AC-25(3), June 1980, p. 468-474.

[Luh 80b] Luh J.Y.S., Walker M.W., Paul R.C.P., "On-line computational scheme for mechanical manipulators", *Trans. of ASME, J. of Dynamic Systems, Measurement, and Control*, Vol. 102(2), 1980, p. 69-76.

[Luh 85a] Luh J.Y.S., Gu Y.L., "Industrial robots with seven joints", *Proc. IEEE Int. Conf. on Robotics and Automation*, St Louis, USA, March 1985, p. 1010-1015.

[Luh 85b] Luh J.Y.S., Zheng Y.F., "Computation of input generalized forces for robots with closed kinematic chain mechanisms", *IEEE J. of Robotics and Automation*, Vol. RA-1(2), 1985, p. 95-103.

[Ma 91] Ma O., Angeles J., "Architecture singularities of platform manipulators", *Proc. IEEE Int. Conf. on Robotics and Automation*, Sacramento, USA, April 1991, p. 1542-1547.

[Maciejewski 85] Maciejewski A.A., Klein C.A., "Obstacle avoidance for kinematically redundant manipulators in dynamically varying environments", *The Int. J. of Robotics Research*, Vol. 4(3), Fall 1985, p. 109-117.

[Maciejewski 88] Maciejewski A.A., Klein C.A., "Numerical filtering operation of robotic manipulators throuh kinematically singular configurations", *J. of Robotic Systems*, Vol. 5(6), 1988, p. 527-552.

[Maciejewski 89] Maciejewski A.A., Klein C.A., "The singular value decomposition: computation and applications to robotics", *The Int. J. of Robotics Research*, Vol. 8(6), 1989, p. 63-79.

[Manaoui 85] Manaoui O., "Calcul automatique de transformateurs de coordonnées analytiques de robots à partir de classes de solutions préétablies", Rapport de DEA, USTL, Montpellier, France, July 1985.

[Mason 82] Mason M.T., "Compliant motion", *in Robot motion: planning and control*, M. Brady *et al.* Eds., MIT Press, Cambridge, USA, 1982.

[Maurine 96] Maurine P., "Développement et mise en œuvre de méthodologies d'étalonnage de robots manipulateurs industriels", Thèse de Doctorat, Université Montpellier II, France, December 1996.

[Mavroidis 93] Mavroidis C., "Résolution du problème géométrique inverse pour les manipulateurs série à six degrés de liberté", Thèse de Doctorat, Université Pierre et Marie Curie, Paris, France, May 1993.

[Mayeda 84] Mayeda H., Osuka K., Kangawa A., "A new identification method for serial manipulator arms", *Proc. IFAC 9th World Congress*, Budapest, Hungary, July 1984, p. 74-79.

[Mayeda 90] Mayeda H., Yoshida K., Osuka K., "Base parameters of manipulator dynamic models", *IEEE Trans. on Robotics and Automation*, Vol. RA-6(3), 1990, p. 312-321.

[Megahed 82] Megahed S., Renaud M., "Minimization of the computation time necessary for the dynamic control", *Proc. 12th Int. Symp. on Industrial Robots*, Paris, France, June 1982, p. 469-478.

[Megahed 84] Megahed S., "Contribution à la modélisation géométrique et dynamique des robots manipulateurs ayant une structure de chaîne cinématique simple ou complexe ; application à leur commande", Thèse d'Etat, UPS, Toulouse, France, July 1984.

[Mendel 73] Mendel J.M., *Discrete techniques of parameter estimation: the equation error formulation*, Marcel Dekker Inc. New York, USA, 1973.

[Merlet 86] Merlet J.-P., "Contribution à la formalisation de la commande par retour d'effort en robotique ; application à la commande de robots parallèles", Thèse de Doctorat, Université Pierre et Marie Curie, Paris, France, June 1986.

[Merlet 89] Merlet J.-P., "Singular configurations of parallel manipulators and Grassmann geometry", *The Int. J. of Robotics Research*, Vol. 8(5), October 1989, p. 45-56.

[Merlet 93] Merlet J.-P, "Closed-form resolution of the direct kinematics of parallel manipulators using extra sensor data", *Proc. IEEE Int. Conf. on Robotics and Automation*, Atlanta, USA, May 1993, p. 200-204.

[Merlet 00] Merlet J.-P., *Parallel robots*, Kluwer Academic Publ., Dordrecht, The Netherlands, 2000.

[Middleton 88] Middleton R.H., Goodwin G.C., "Adaptive computed torque control for rigid link manipulators", *Systems & Control Letters*, Vol. 10, 1988, p. 9-16.

[Milenkovic 83] Milenkovic V., Huang B., "Kinematics of major robot linkage", *Proc. 13th Int. Symp. on Industrial Robots*, Chicago, USA, April 1983, p. 16.31-16.47.

[Mooring 91] Mooring B.W., Roth Z.S., Driels M.R., *Fundamentals of manipulator calibration*, John Wiley & Sons, New York, USA, 1991.

[Morel 94] Morel G., "Programmation et adaptation d'impédance de manipulateurs au contact", Thèse de Doctorat, Université Pierre et Marie Curie, Paris, France, June 1994.

[Murareci 97] Murareci D., "Contribution à la modélisation géométrique et à l'étalonnage des robots séries et parallèles", Thèse de Doctorat, Université de Nantes et Ecole Centrale de Nantes, France, March 1997.

[Murphy 93] Murphy S.H., Wen J.T.U, "Analysis of active manipulator elements in space manipulation", *IEEE Trans. on Robotics and Automation*, Vol. RA-9(5), October 1993, p. 544-552.

[Murray 84] Murray J.J., Newman C.P., "ARM: an algebraic robot dynamic modeling program", *Proc. IEEE Int. Conf. on Robotics*, Atlanta, USA, March 1984, p. 103-104.

[Nahvi 94] Nahvi A., Hollerbach J.M., Hayward V., "Calibration of parallel robot using multiple kinematic closed loops", *Proc. IEEE Int. Conf. on Robotics and Automation*, San Diego, USA, May 1994, p. 407-412.

[Nakamura 86] Nakamura Y., Hanafusa Y., "Inverse kinematic solutions with singularity robustness for robot manipulator control", *Trans. of ASME, J. of Dynamic Systems, Measurement, and Control*, Vol. 108, 1986, p. 163-171.

[Nakamura 87] Nakamura Y., Hanafusa Y., Yoshikawa T., "Task-priority based redundancy control of a robot manipulator", *The Int. J. of Robotics Research*, Vol. 6(2), 1987, p. 3-15.

[Nanua 90] Nanua P., Waldron K.J., Murthy V., "Direct kinematic solution of a Stewart platform", *IEEE Trans. on Robotics and Automation*, Vol. RA-6(4), 1990, p. 438-444.

[Nenchev 92] Nenchev D.N., "Restricted jacobian matrices of redundant manipulators in constrained motion tasks", *The Int. J. of Robotics Research*, Vol. 11(6), 1992, p. 584-597.

[Nevins 73] Nevins J.L., Whitney D.E., "The force vector assembler concept", *Proc. 1st CISM-IFToMM Symp. on Theory and Practice of Robots and Manipulators*, Udine, Italy, September 1973, p. 273-288.

[Nevins 77] Nevins J.L. *et al.*, "Exploratory research in industrial modular assembly", Charles Stark Draper Lab., Cambridge, USA, Report R-1111, 1977.

[Newman 79] Newman W.H., Sproull R.F., *Principles of interactive computer graphics*, McGraw Hill, New York, USA, 1979.

[Nicosia 84] Nicosia S., Tomei P., "Model reference adaptive control algorithms for industrial robots", *Automatica*, Vol. 20(5), 1984, p. 635-644.

[Nicosia 90] Nicosia S, Tomei P., "Robot control by using only joint position measurements", *IEEE Trans. on Automatic Control*, Vol. AC-35(5), 1990, p. 1058-1061.

[Nielsen 91] Nielsen L., Canudas de Wit C., Hagander P., "Controllability issues of robots near singular configurations", *in Advanced Robot Control, Lecture Notes in Control and Information Sciences*, Springer-Verlag, New York, USA, 1991, p. 307-314.

[Olsen 86] Olsen H.B., Bekey G.A., "Identification of robot dynamics", *Proc. IEEE Int. Conf. on Robotics and Automation*, San Francisco, USA, April 1986, p. 1004-1010.

[Orin 79] Orin D.E., McGhee R.B., Vukobratovic M., Hartoch G., "Kinematic and kinetic analysis of open-chain linkages utilizing Newton-Euler methods", *Mathematical Biosciences*, Vol. 43, 1979, p. 107-130.

[Ortega 89] Ortega R., Spong M.W., "Adaptive motion control of rigid robots: a tutorial", *Automatica*, Vol. 25(6), 1989, p. 877-888.

[Paden 88] Paden B, Panja R., "Globally asymptotically stable 'PD+' controller for robot manipulator", *Int. J. Control*, Vol. 47, 1988, p. 877-888.

[Paul 72] Paul R.C.P., "Modelling, trajectory calculation, and servoing of a computer controlled arm", Ph. D. Thesis, Stanford Artificial Intelligence Lab., Stanford, USA, 1972.

[Paul 81] Paul R.C.P., *Robot manipulators: mathematics, programming and control*, MIT Press, Cambridge, USA, 1981.

[Payannet 85] Payannet D., "Modélisation et correction des erreurs statiques des robots manipulateurs", Thèse de Doctorat, USTL, Montpellier, France, 1985.

[Penrose 55] Penrose R., "A generalized inverse for matrices", *Proc. Cambridge Philos. Soc.*, Vol. 51, 1955, p. 406-413.

[Perdereau 91] Perdereau V., "Contribution à la commande hybride force-position", Thèse de Doctorat, Université Pierre et Marie Curie, Paris, France, February 1991.

[Pham 91a] Pham C., Chedmail P., Gautier M., "Determination of base parameters of flexible links manipulators", *Proc. IMACS MCTS Symposium, Modelling and Control of Technological Systems*, Lille, France, May 1991, Vol. 1, p. 524-529.

[Pham 91b] Pham C., Gautier M., "Essential parameters of robots", *Proc. 30th IEEE Conf. on Decision and Control*, Brighton, UK, December 1991, p. 2769-2774.

[Pieper 68] Pieper D.L., "The kinematics of manipulators under computer control", Ph. D. Thesis, Stanford University, Stanford, USA, 1968.

[Pierrot 91a] Pierrot F., "Robots pleinement parallèles légers : conception, modélisation et commande", Thèse de Doctorat, Université Montpellier II, France, April 1991.

[Pierrot 91b] Pierrot F., Fournier A., Dauchez P.,"Towards a fully-parallel six dof robot for high-speed applications", *Proc. IEEE Int. Conf. on Robotics and Automation*, Sacramento, USA, April 1991, p. 1288-1293.

[Pledel 96] Pledel P., "Génération de mouvements optimaux pour un robot manipulateur", Thèse de doctorat, Université de Nantes et Ecole Centrale de Nantes, France, September 1996.

[Poignet 00] Poignet P., Gautier M., "Comparison of weighted least squares and extended Kalman filtering methods for dynamic identification of robots", *Proc. IEEE Int. Conf. on Robotics and Automation*, San Francisco, USA, April 2000, p. 3622-3627.

[Popov 73] Popov V.M., *Hyperstability of control systems*, Springer-Verlag, New York, USA, 1973.

[Potkonjak 86] Potkonjak V., "Thermal criterion for the selection of DC drives for industrial robots", *Proc. 16th Int. Symp. on Industrial Robots*, Brussels, Belgium, September-October 1986, p. 129-140.

[Powell 64] Powell, M.J.D., "An efficient method for finding the minimum of a function of several variables without calculating derivatives", *The Computer Journal*, Vol. 7, 1964, p. 155-162.

[Pressé 93] Pressé C., Gautier M., "New criteria of exciting trajectories for robot identification", *Proc. IEEE Int. Conf. on Robotics and Automation*, Atlanta, USA, May 1993, p. 907-912.

[Pressé 94] Pressé C., Identification des paramètres dynamiques des robots", Thèse de Doctorat, Université de Nantes et Ecole Centrale de Nantes, France, October 1994.

[Priel 90] Priel M., *Les robots industriels : caractéristiques, performance et choix*, Collection AFNOR Technique, Paris, France, 1990.

[Prüfer 94] Prüfer M.,Schmidt C., Wahl F., "Identification of robots dynamics with differential and integral models: a comparison", *Proc. IEEE Int. Conf. on Robotics and Automation*, San Diego, USA, May 1994, p. 340-345.

[Pujas 95] Pujas A., "Etude de la robustesse de schémas de commande position / force pour robots à deux bras", Thèse de Doctorat, Université Montpellier II, France, June 1995.

[Qu 91] Qu Z., Dorsey J., "Robust PID control of robots", *Int. J. Robotics and Automation*, Vol. 6(4), 1991, p. 228-235.

[Raghavan 90] Raghavan M., Roth B., "Inverse kinematics of thegeneral 6R manipulator and related linkages", *Trans. of the ASME, J. of Mechanical Design*, Vol. 115, 1990, p. 502-508.

[Raghavan 91] Raghavan M., "The Stewart platform of general geometry has 40 configurations", *in Advances in Design Automation*, ASME Press, New-York, USA, 1991, p. 397-402.

[Raibert 77] Raibert M.H., "Analytic equations vs. table look-up for manipulation: a unifying concept", *Proc. 16th IIEEE Conf. on Decision and Control*, New Orleans, USA, 1977, p. 576-579.

[Raibert 78] Raibert M.H., Horn B.K.P., "Manipulator control using the configuration space method", *The Industrial Robot*, Vol. 5(2), 1978, p. 69-73.

[Raibert 81] Raibert M.H., Craig J.J., "Hybrid position/force control of manipulators", *Trans. of the ASME, J. of Dynamic Systems, Measurement, and Control*, Vol. 103, June 1981, p. 126-133.

[Raucent 90] Raucent B., "Identification des paramètres dynamiques des robots manipulateurs", Thèse de Doctorat, Université de Louvain, Belgium, 1990.

[Raucent 92] Raucent B., Bastin G., Campion G., Samin J.-C., Willems P.Y., "Identification of barycentric parameters of robotic manipulators from external measurements", *Automatica*, Vol. 28(5), 1992, p. 1011-1016.

[Reboulet 85] Reboulet C., Robert A., "Hybrid control of a manipulator equipped with an active compliant wrist", *Proc. 3rd Int. Symp. of Robotics Research*, Gouvieux, France, October 1985, p. 76-80.

[Reboulet 88] Reboulet C., "Modélisation des robots parallèles", *in Techniques de la robotique : architectures et commandes*, Hermès, Paris, France, 1988.

[Renaud 75] Renaud M., "Contribution à l'étude de la modélisation et de la commande des systèmes mécaniques articulés", Thèse de Docteur-Ingénieur, UPS, Toulouse, France, December 1975.

[Renaud 80a] Renaud M., "Contribution à la modélisation et à la commande dynamique des robots manipulateurs", Thèse d'Etat, UPS, Toulouse, France, September 1980.

[Renaud 80b] Renaud M., "Calcul de la matrice jacobienne nécessaire à la commande coordonnée d'un manipulateur", *J. of Mechanism and Machine Theory*, Vol. 15(1), 1980, p. 81-91.

[Renaud 85] Renaud M., "A near minimum iterative analytical procedure for obtaining a robot-manipulator dynamic model", *IUTAM/IFToMM Symp. on Dynamics of Multi-body Systems*, Udine, Italy, 1985.

[Renaud 87] Renaud M., "Quasi-minimal computation of the dynamic model of a robot manipulator utilizing the Newton-Euler formalism and the notion of augmented body", *Proc. IEEE Int. Conf. on Robotics and Automation*, Raleigh, USA, March-April 1987, p. 1677-1682.

[Restrepo 95] Restrepo P.P., Gautier M., "Calibration of drive chain of robot joints", *Proc. 4th IEEE Conf. on Control Applications, CCA'95*, Albany, USA, 1995, p. 526-531.

[Restrepo 96] Restrepo P.P., Contribution à la modélisation, identification et commande des robots à structures fermées : application au robot Acma SR400", Thèse de Doctorat, Université de Nantes et Ecole Centrale de Nantes, France, October 1996.

[Richalet 98] Richalet J., *Pratique de l'identification*, 2ième édition, Hermès, Paris, France, 1998.

[Robert 86] Robert A., "Commande hybride position-force ; mise en œuvre et expérimentation sur un micro-ordinateur parallèle", Thèse de Docteur-Ingénieur, Ecole Nationale Supérieure de l'Aéronautique et de l'Espace, Toulouse, France, December 1986.

[Roberts 65] Roberts L.G., "Homogeneous matrix representation and manipulation of N-dimensional constructs", MIT Lincoln Lab., USA, MS 1405, May 1965.

[Rocco 96] Rocco P., "Stability of PID control for industrial robot arms", *IEEE Trans. on Robotics and Automation*, Vol. RA-12(4), 1996, p. 607-614.

[Roth 76] Roth B., "Performance evaluation of manipulators from a kinematic viewpoint", Cours de Robotique, IRIA, Toulouse, France, 1976, p. 233-263.

[Roth 87] Roth Z.S., Mooring B.W., Ravani B., "An overview of robot calibration", *IEEE J. of Robotics and Automation*, Vol. RA-3(5), October 1987, p. 377-385.

[Sadegh 87] Sadegh N., Horowitz R., "Stability analysis of an adaptive controller for robotic manipulators", *Proc. IEEE Int. Conf. on Robotics and Automation*, Raleigh, USA, March-April 1987, p. 1223-1229.

[Sadegh 90] Sadegh N., Horowitz R., "Stability and robustness analysis of a class of adaptive controllers for robotic manipulators", *The Int. J. of Robotics Research*, Vol. 9(3), 1990, p. 74-92.

[Salisbury 80] Salisbury J.K., "Active stiffness control of a manipulator in cartesian coordinates", *Proc. 19th IEEE Conf. on Decision and Control*, Albuquerque, USA, December 1980, p. 95-100.

[Salmon 1885] Salmon G., Lessons introductory to the modern higher algebra, Chelsea Publishing CO., New York, USA, 1885.

[Samson 83] Samson C., "Problèmes en identification et commande de systèmes dynamiques", Thèse d'Etat, Rennes, France, 1983.

[Samson 87] Samson C., "Robust control of a class of non-linear systems and applications to robotics", *Int. J. of Adaptive Control and Signal Processing*, Vol. 1, 1987, p. 49-68.

[Samson 91] Samson C., Le Borgne M., Espiau B., *Robot Control*, Oxford University Press, Oxford, UK, 1991.

[Schefer 82] Schefer B., "Geometric control and calibration method of an industrial robot", *Proc. 12th Int. Symposium on Industrial Robotics*, Paris, France, 1982, p. 331-339.

[Sciavicco 94] Sciavicco L., Siciliano B., Villani L., "On dynamic modelling of gear-driven rigid robot manipulators", *Proc. 4th IFAC Symp. on Robot Control, SYROCO'94*, Capri, Italy, September 1994, p. 543-549.

[Sgarbi 92] Sgarbi F., Cammoun R., "Real time trajectory generation using filtering techniques", *Proc. 2nd Int. Conf. on Automation, Robotics, and Computer Vision, ICARCV'92*, Singapour, September 1992, p. RO-8.5.1-RO-8.5.6.

[Sheth 71] Sheth P.N., Uicker J.J., "A generalized symbolic notation for mechanism", *Trans. of ASME, J. of Engineering for Industry*, Vol. 93, 1971, p. 102-112.

[Shiller 94] Shiller Z., "On singular time-optimal control along specified paths", *IEEE Trans. on Robotics and Automation*, Vol. RA-10(4), 1994, p. 561-566.

[Shin 85] Shin K.G., McKay N.D., "Minimum time control of robotic manipulators with geometric path constraints", *IEEE Trans. on Automatic Control*, Vol. AC-30(6), 1985, p. 531-541.

[Siciliano 93] Siciliano B., Villani L., "An adaptive force/position regulator for robot manipulators", *Int. J. of Adaptive Control and Signal Processing*, Vol. 7, 1993, p. 389-403.

[Siciliano 96a] Siciliano B., Villani L., "A passivity-based approach to force regulation and motion control of robot manipulators", *Automatica*, Vol. 32, 1996, p. 443-447.

[Siciliano 96b] Siciliano B., Villani L., "Adaptive compliant control of robot manipulators", *Control Engineering Practice*, Vol. 4, 1996, p. 705-712.

[Siciliano 00] Siciliano B., Villani L., *Robot force control*, Kluwer Academic Publ., Boston, USA, 2000.

[Slotine 87] Slotine J.-J.E., Li W., "Adaptive manipulator control: a case study", *Proc. IEEE Int. Conf. on Robotics and Automation*, Raleigh, USA, March-April 1987, p. 1312-1400.

[Slotine 91] Slotine J.-J.E., Li W., *Applied Nonlinear Control*, Prentice Hall, Englewood Cliffs, USA, 1991.

[Spanos 85] Spanos J., Kholi D., "Workspace analysis of regional structures of manipulators", *J. of Mechanisms, Transmissions, and Automation in Design*, Vol. 107, June 1985, p. 219-225.

[Spetch 88] Spetch R., Isermann R., "On-line identification of inertia, friction and gravitational forces applied to an industrial robot", *Proc. IFAC Symp. on Robot Control, SYROCO'88*, 1988, p. 88.1-88.6.

[Spong 87] Spong M.W., "Modeling and control of elastic joint robots", *Trans. of the ASME, J. of Dynamic Systems, Measurement, and Control*, Vol. 109, 1987, p. 310-319.

[Spong 89] Spong M.W., Vidyasagar M., *Robot dynamics and control*, John Wiley & Sons, New York, USA, 1989.

[Spong 90] Spong M.W., Ortega R., "On adaptive inverse dynamics control of rigid robots", *IEEE Trans. on Automatic Control*, Vol. AC-35(1), 1990, p. 92-95.

[Stepanenko 93] Stepanenko Y., Yuan J., "A reduced order regressor and its application to adaptive robotic control", *The Int. J. of Robotics Research*, Vol. 12(2), April 1993, p. 180-187.

[Stewart 65] Stewart D., "A platform with six degrees of freedom", *Proc. of Institution of Mechanical Engineers*, Vol. 180, Part 1, n° 15, 1965-1966, p. 371-385.

[Sugimoto 85] Sugimoto K., Okada T., "Compensation of positionning errors caused by geometric deviations in robot systems", *The 3rd Int. Symp. of Robotics Research*, MIT Press, Cambridge, USA, 1985, p. 231-236.

[Takegaki 81a] Takegaki M., Arimoto S., "An adaptive trajectory control of manipulators", *Int. J. Control*, Vol. 34(2), 1981, p. 219-230.

[Takegaki 81b] Takegaki M., Arimoto S., "A new feedback method for dynamic control of manipulators", *Trans. of ASME, J. of Dynamic Systems, Measurement, and Control*, Vol. 102, 1981, p. 119-125.

[Tancredi 95] Tancredi L., "De la simplification et la résolution du modèle géométrique direct des robots parallèles", Thèse de Doctorat, Ecole Nationale Supérieure des Mines de Paris, France, December 1995.

[Tang 94] Tang G.R., Lieu L.S., "A study of three robot calibration methods based on flat surfaces", *J. of Mechanism and Machine Theory*, Vol. 29(2), 1994, p. 195-206.

[Taylor 79] Taylor R.H., "Planning and execution of straight line manipulator trajectories", *IBM J. of Research and Development*, Vol. 23(4), July 1979, p. 424-436.

[Thérond 96] Thérond X., Dégoulange E., Dombre E., Pierrot F., "Force control of a medical robot for arterial diseases detection", *Proc. 6th Int. Symp. on Robotics and Manufacturing, WAC'96*, Montpellier, France, May 1996, Vol. 6, p. 793-798.

[Tomei 91] Tomei P., "Adaptive PD controller for robot manipulators", *IEEE Trans. on Robotics and Automation*, Vol. RA-7(4), 1991, p. 565-570.

[Tondu 94] Tondu B., El Zorkany H., "Identification of a trajectory model for the PUMA-560 Robot", *J. of Robotic Systems*, Vol. 11(2), 1994, p. 77-90.

[Touron 84] Touron P., "Modélisation de la dynamique des mécanismes polyarticulés ; application à la CAO et à la simulation de robots", Thèse de Docteur-Ingénieur, USTL, Montpellier, France, July 1984.

[Uicker 69] Uicker J.J., "Dynamic behavior of spatial linkages", *Trans. of ASME, J. of Engineering for Industry*, Vol. 91, 1969, p. 251-258.

[Vandanjon 95] Vandanjon P.-O, Gautier M., Desbats P.,"Identification of inertial parameters of robots by means of spectrum analysis", *Proc. IEEE Int. Conf. on Robotics and Automation*, Nayoga, Japan, May 1995, p. 3033-3038.

[Veitschegger 86] Veitschegger W.K., Wu C.H., "Robot accuracy analysis based on kinematics", *IEEE J. of Robotics and Automation*, Vol. RA-2(3), September 1986, p. 171-179.

[Volpe 93] Volpe R., Khosla P., "A theoretical and experimental investigation of explicit force control for manipulators", *IEEE Trans. on Automatic Control*, Vol. AC-38(11), 1993, p. 1634-1650.

[Volpe 95] Volpe R., Khosla P., "The equivalence of second-order impedance control and proportional gain explicit force control", *The Int. J. of Robotics Research*, Vol. 14(6), 1995, p. 574-589.

[Vukobratovic 82] Vukobratovic M., Potkonjak V., *Dynamics of manipulation robots; Vol. 1: Theory and applications*, Springer-Verlag, New York, USA, 1982.

[Vukobratovic 85] Vukobratovic M., Kircanski N., *Real-time dynamics of manipulation robots*, Springer-Verlag, New York, USA, 1985.

[Walker 82] Walker M.W., Orin D.E., "Efficient dynamic computer simulation of robotics mechanism", *Trans. of ASME, J. of Dynamic Systems, Measurement, and Control*, Vol. 104, 1982, p. 205-211.

[Wampler 86] Wampler C.W.,"Manipulator inverse kinematic solutions based on vector formulation and damped least-squares methods", *IEEE Trans. on Systems, Man, and Cybernetics*, Vol. SMC-16, 1986, p. 93-101.

[Wang 83] Wang L.T., Ravani B., "Recursive computations of kinematic and dynamic equations for mechanical manipulators", *IEEE J. of Robotics and Automation*, Vol. RA-1(3), September 1983, p. 124-131.

[Wang 91] Wang L.C.T., Chen C.C., "A combined optimisation method for solving the inverse kinematics problem of mechanical manipulators", *IEEE Trans. on Robotics and Automation*, Vol. RA-7(4), 1991, p. 489-499.

[Wang 93] Wang D., McClamroch N.H., "Position and force control for constrained manipulator motion: Lyapunov's direct method", *IEEE Trans. on Robotics and Automation*, Vol. RA-9(3), 1993, p. 308-313.

[Wen 88] Wen J. T., Bayard D., "New class of control laws for robotic manipulators; Part 1: non-adaptive case", *Int. J. Control*, Vol. 47(5), 1988, p. 1361-1385.

[Wen 91] Wen J., Murphy S., "Stability analysis of position and force control for robot arms", *IEEE Trans. on Automatic Control*, Vol. AC-36, 1991, p. 365-371.

[Wenger 89] Wenger P., "Aptitude d'un robot manipulateur à parcourir son espace de travail en présence d'obstacles", Thèse de Doctorat, ENSM, Nantes, France, September 1989.

[Wenger 92] Wenger P., "A new general formalism for the kinematic analysis of all non redundant manipulators", *Proc. IEEE Int. Conf. on Robotics and Automation*, Nice, France, May 1992, p. 442-447.

[Wenger 93] Wenger P., "A classification of manipulator geometries based on singularity avoidance ability", *Proc. Int. Conf. on Advanced Robotics, ICAR'93*, Tokyo, Japan, November 1993, p. 649-654.

[Wenger 98] Wenger P., "Classification of 3R positioning manipulators", *ASME J. of Mechanical Design*, Vol. 120(2), June 1998, p. 327-332.

[West 89] West H., Papadopoulos E., Dubowsky S., Cheah H., "A method for estimating the mass properties of a manipulator by measuring the reaction moments at its base", *Proc. IEEE Int. Conf. on Robotics and Automation*, Scottsdale, USA, May 1989, p. 1510-1516.

[Whitney 69] Whitney D.E., "Resolved motion rate control of manipulators and human prostheses", *IEEE Trans. on Man Machine Systems*, Vol. MMS-10(2), June 1969, p. 47-53.

[Whitney 79] Whitney D.E., Nevins J.L, "What is the remote center compliance (RCC) and what can it do", *Proc. 9th Int. Symp. on Industrial Robots*, Washington, USA, March 1979, p. 135-147.

[Whitney 85] Whitney D.E., "Historical perspective and state of the art in robot force control", *Proc. IEEE Int. Conf. on Robotics and Automation*, St Louis, USA, March 1985, p. 262-268.

[Whitney 86] Whitney D.E., Lozinski C.A., Rourke J.M., "Industrial robot forward calibration method and results", *J. Dynamic Systems, Measurements, and Control*, Vol. 108, p. 1-8, 1986.

[Wittenburg 77] Wittenburg J., *Dynamics of system of rigid bodies*, B.G. Teubner, Stuttgart, Germany, 1977.

[Wittenburg 85] Wittenburg J., Holtz U., "The program MESA VERDE for robot dynamics simulations", *The 3rd Int. Symp. of Robotics Research*, MIT Press, Cambridge, USA, 1985, p. 197-204.

[Wu 84] Wu C.H., "A kinematic CAD tool for the design and control of robot manipulators", *The Int. J. of Robotics Research*, Vol. 3(1), 1984, p. 74-85.

[Yabuta 92] Yabuta T., "Nonlinear basic stability concept of the hybrid position/force control scheme for robot manipulators", *IEEE Trans. on Robotics and Automation*, Vol. RA-8(5), 1992, p. 663-670.

[Yang 66] Yang A.T., Freudenstein F., "Application of dual number quaternion algebra in the analysis of spatial mechanisms", *Trans. of ASME, J. of Applied Mechanics*, Vol. 33, 1966. p. 300-308.

[Yoshida 00] Yoshida K., Khalil W., "Verification of the positive defiteness of the inertial matrix of manipulators using base inertial parameters", *The Int. J. of Robotics Research*, Vol. 19(5), 2000, p. 498-510.

[Yoshikawa 84a] Yoshikawa T., "Analysis and control of robot manipulators with redundancy", *The 1st Int. Symp. of Robotics Research*, MIT Press, Cambridge, USA, 1984, p. 735-748.

[Yoshikawa 84b] Yoshikawa T., "Manipulability of robotics mechanisms", *Proc. 2nd Int. Symp. of Robotics Research*, Kyoto, Japan, August 1984, p. 91-98.

[Zabala 78] Zabala Iturralde J., "Commande des robots-manipulateurs à partir de la modélisation de leur dynamique", Thèse de Troisième Cycle, UPS, Toulouse, France, July 1978.

[Zeghloul 91] Zeghloul S., "Développement d'un système de CAO Robotique intégrant la planification de tâches et la synthèse de sites robotisés", Thèse d'Etat, Université de Poitiers, France, February 1991.

[Zgaib 92] Zghaib W., "Génération symbolique automatique des équations de la dynamique des systèmes mécaniques complexes avec contraintes cinématiques", Thèse de Doctorat, ENSAM, Paris, France, 1992.

[Zhang 92] Zhang C., Song S.M., "Forward kinematics of a class of parallel (Stewart) platforms with closed-form solutions", *J. of Robotic Systems*, Vol. 9(1), 1992, p. 93-112.

[Zhong 95] Zhong X.L., Lewis J.M., "A new method for autonomous robot calibration", *Proc. IEEE Int. Conf. on Robotics and Automation*, Nayoga, Japan, May 1995, p. 1790-1795.

[Zhuang 95] Zhuang H., Masory O., Yan J., "Kinematic calibration of a Stewart platform using pose measurements obtained by a single theodolite", *Proc. 1995 Int. Conf. on Intelligent Robots and Systems, IROS'95*, Pittsburg, USA, August 1995, p. 329-334.

[Zhuang 97] Zhuang H., "Self-calibration of a class of parallel mechanisms with a case study on Stewart platform", *IEEE Trans. on Robotics and Automation*, Vol. RA-13(3), 1997, p. 387-397.

[Zhuang 99] Zhuang H., Motaghedi S.H., Roth Z.S., "Robot calibration with planar constraints", *Proc. IEEE Int. Conf. on Robotics and Automation*, Detroit, USA, May 1999, p. 805-810.

[Zodiac 96] The Zodiac, *Theory of robot control*, C. Canudas de Wit, B. Siciliano, G. Bastin Eds., Springer-Verlag, Berlin, Germany, 1996.

Index

Printed in the United Kingdom
by Lightning Source UK Ltd.
118378UK00001B/15

Made in United States
North Haven, CT
03 November 2026

21719578R00280